SOBOLEV SPACES IN MATHEMATICS II

APPLICATIONS IN ANALYSIS AND PARTIAL DIFFERENTIAL EQUATIONS

INTERNATIONAL MATHEMATICAL SERIES

Series Editor: **Tamara Rozhkovskaya**
Novosibirsk, Russia

1. **Nonlinear Problems in Mathematical Physics and Related Topics I. In Honor of Professor O.A. Ladyzhenskaya** • M.Sh. Birman, S. Hildebrandt, V.A. Solonnikov, N.N. Uraltseva Eds. • 2002
2. **Nonlinear Problems in Mathematical Physics and Related Topics II. In Honor of Professor O.A. Ladyzhenskaya** • M.Sh. Birman, S. Hildebrandt, V.A. Solonnikov, N.N. Uraltseva Eds. • 2003
3. **Different Faces of Geometry** • S. Donaldson, Ya. Eliashberg, M. Gromov Eds. • 2004
4. **Mathematical Problems from Applied Logic I. Logics for the XXIst Century** • D. Gabbay, S. Goncharov, M. Zakharyaschev Eds. • 2006
5. **Mathematical Problems from Applied Logic II. Logics for the XXIst Century** • D. Gabbay, S. Goncharov, M. Zakharyaschev Eds. • 2007
6. **Instability in Models Connected with Fluid Flows I** • C. Bardos, A. Fursikov Eds. • 2008
7. **Instability in Models Connected with Fluid Flows II** • C. Bardos, A. Fursikov Eds. • 2008
8. **Sobolev Spaces in Mathematics I. Sobolev Type Inequalities** • V. Maz'ya Ed. • 2009
9. **Sobolev Spaces in Mathematics II. Applications in Analysis and Partial Differential Equations** • V. Maz'ya Ed. • 2009
10. **Sobolev Spaces in Mathematics III. Applications in Mathematical Physics** • V. Isakov Ed. • 2009

SOBOLEV SPACES IN MATHEMATICS II

Applications in Analysis and Partial Differential Equations

Editor: Vladimir Maz'ya

Ohio State University, USA
University of Liverpool, UK
Linköping University, SWEDEN

Springer
Tamara Rozhkovskaya Publisher

Editor

Prof. Vladimir Maz'ya
Ohio State University
Department of Mathematics
Columbus, USA

University of Liverpool
Department of Mathematical Sciences
Liverpool, UK

Linköping University
Department of Mathematics
Linköping, Sweden

This series was founded in 2002 and is a joint publication of Springer and "Tamara Rozhkovskaya Publisher." Each volume presents contributions from the Volume Editors and Authors exclusively invited by the Series Editor Tamara Rozhkovskaya who also prepares the Camera Ready Manuscript. This volume is distributed by "Tamara Rozhkovskaya Publisher" (tamara@mathbooks.ru) in Russia and by Springer over all the world.

ISBN 978-0-387-85649-0 e-ISBN 978-0-387-85650-6
ISBN 978-5-901873-26-7 (Tamara Rozhkovskaya Publisher)
ISSN 1571-5485

Library of Congress Control Number: 2008937494

Printed on acid-free paper.

9 8 7 6 5 4 3 2 1

springer.com

To the memory of
Sergey L'vovich Sobolev
on the occasion of his centenary

Main Topics

Sobolev's discoveries of the 1930's have a strong influence on development of the theory of partial differential equations, analysis, mathematical physics, differential geometry, and other fields of mathematics. The three-volume collection *Sobolev Spaces in Mathematics* presents the latest results in the theory of Sobolev spaces and applications from leading experts in these areas.

I. Sobolev Type Inequalities
In 1938, exactly 70 years ago, the original Sobolev inequality (an embedding theorem) was published in the celebrated paper by S.L. Sobolev "On a theorem of functional analysis." By now, the Sobolev inequality and its numerous versions continue to attract attention of researchers because of the central role played by such inequalities in the theory of partial differential equations, mathematical physics, and many various areas of analysis and differential geometry. The volume presents the recent study of different Sobolev type inequalities, in particular, inequalities on manifolds, Carnot–Carathéodory spaces, and metric measure spaces, trace inequalities, inequalities with weights, the sharpness of constants in inequalities, embedding theorems in domains with irregular boundaries, the behavior of maximal functions in Sobolev spaces, etc. Some unfamiliar settings of Sobolev type inequalities (for example, on graphs) are also discussed. The volume opens with the survey article "My Love Affair with the Sobolev Inequality" by David R. Adams.

II. Applications in Analysis and Partial Differential Equations
Sobolev spaces become the established language of the theory of partial differential equations and analysis. Among a huge variety of problems where Sobolev spaces are used, the following important topics are in the focus of this volume: boundary value problems in domains with singularities, higher order partial differential equations, nonlinear evolution equations, local polynomial approximations, regularity for the Poisson equation in cones, harmonic functions, inequalities in Sobolev–Lorentz spaces, properties of function spaces in cellular domains, the spectrum of a Schrödinger operator with negative potential, the spectrum of boundary value problems in domains with cylindrical and quasicylindrical outlets to infinity, criteria for the complete integrability of systems of differential equations with applications to differential geometry, some aspects of differential forms on Riemannian manifolds related to the Sobolev inequality, a Brownian motion on a Cartan–Hadamard manifold, etc. Two short biographical articles with unique archive photos of S.L. Sobolev are also included.

III. Applications in Mathematical Physics

The mathematical works of S.L. Sobolev were strongly motivated by particular problems coming from applications. The approach and ideas of his famous book "Applications of Functional Analysis in Mathematical Physics" of 1950 turned out to be very influential and are widely used in the study of various problems of mathematical physics. The topics of this volume concern mathematical problems, mainly from control theory and inverse problems, describing various processes in physics and mechanics, in particular, the stochastic Ginzburg–Landau model with white noise simulating the phenomenon of superconductivity in materials under low temperatures, spectral asymptotics for the magnetic Schrödinger operator, the theory of boundary controllability for models of Kirchhoff plate and the Euler–Bernoulli plate with various physically meaningful boundary controls, asymptotics for boundary value problems in perforated domains and bodies with different type defects, the Finsler metric in connection with the study of wave propagation, the electric impedance tomography problem, the dynamical Lamé system with residual stress, etc.

Contents

I. Sobolev Type Inequalities

Vladimir Maz'ya Ed.

II. Applications in Analysis and Partial Differential Equations

Vladimir Maz'ya Ed.

III. Applications in Mathematical Physics

Victor Isakov Ed.

Contributors

Editors

Vladimir Maz'ya

Ohio State University
Columbus, OH 43210
USA

University of Liverpool
Liverpool L69 7ZL
UK

Linköping University
Linköping SE-58183
SWEDEN

vlmaz@mai.liu.se

vlmaz@math.ohio-state.edu

Victor Isakov

Wichita State University
Wichita, KS 67206
USA

victor.isakov@wichita.edu

Contributors

Authors

Daniel Aalto
Institute of Mathematics
Helsinki University of Technology
P.O. Box 1100, FI-02015
FINLAND
e-mail: daniel.aalto@tkk.fi

David R. Adams
University of Kentucky
Lexington, KY 40506-0027
USA
e-mail: dave@ms.uky.edu

Hiroaki Aikawa
Hokkaido University
Sapporo 060-0810
JAPAN
e-mail: aik@math.sci.hokudai.ac.jp

Vasili Babich
Steklov Mathematical Institute
Russian Academy of Sciences
27 Fontanka Str., St.-Petersburg 191023
RUSSIA
e-mail: babich@pdmi.ras.ru

Mikhail Belishev
Steklov Mathematical Institute
Russian Academy of Sciences
27 Fontanka Str., St.-Petersburg 191023
RUSSIA
e-mail: belishev@pdmi.ras.ru

Sergey Bobkov
University of Minnesota
Minneapolis, MN 55455
USA
e-mail: bobkov@math.umn.edu

Yuri Brudnyi

Technion – Israel Institute of Technology
Haifa 32000
ISRAEL
e-mail: ybrudnyi@math.technion.ac.il

Victor Burenkov

Università degli Studi di Padova
63 Via Trieste, 35121 Padova
ITALY
e-mail: burenkov@math.unipd.it

Andrea Cianchi

Università di Firenze
Piazza Ghiberti 27, 50122 Firenze
ITALY
e-mail: cianchi@unifi.it

Serban Costea

McMaster University
1280 Main Street West
Hamilton, Ontario L8S 4K1
CANADA
e-mail: secostea@math.mcmaster.ca

Stephan Dahlke

Philipps–Universität Marburg
Fachbereich Mathematik und Informatik
Hans Meerwein Str., Lahnberge 35032 Marburg
GERMANY
e-mail: dahlke@mathematik.uni-marburg.de

Donatella Danielli

Purdue University
150 N. University Str.
West Lafayette, IN 47906
USA
e-mail: danielli@math.purdue.edu

David E. Edmunds

School of Mathematics Cardiff University
Senghennydd Road CARDIFF
Wales CF24 4AG
UK
e-mail: davideedmunds@aol.com

W. Desmond Evans

School of Mathematics Cardiff University
Senghennydd Road CARDIFF
Wales CF24 4AG
UK
e-mail: EvansWD@cf.ac.uk

Andrei Fursikov

Moscow State University
Vorob'evy Gory, Moscow 119992
RUSSIA
e-mail: fursikov@mtu-net.ru

Victor Galaktionov

University of Bath
Bath, BA2 7AY
UK
e-mail: vag@maths.bath.ac.uk

Nicola Garofalo

Purdue University
150 N. University Str.
West Lafayette, IN 47906
USA
e-mail: garofalo@math.purdue.edu

Friedrich Götze

Bielefeld University
Bielefeld 33501
GERMANY
e-mail: goetze@math.uni-bielefeld.de

Vladimir Gol'dshtein

Ben Gurion University of the Negev
P.O.B. 653, Beer Sheva 84105
ISRAEL
e-mail: vladimir@bgu.ac.il

Alexander Grigor'yan

Bielefeld University
Bielefeld 33501
GERMANY
e-mail: grigor@math.uni-bielefeld.de

Max Gunzburger

Florida State University
Tallahassee, FL 32306-4120
USA
e-mail: gunzburg@scs.fsu.edu

Piotr Hajłasz

University of Pittsburgh
301 Thackeray Hall, Pittsburgh, PA 15260
USA
e-mail: hajlasz@pitt.edu

Elton Hsu

Northwestern University
2033 Sheridan Road, Evanston, IL 60208-2730
USA
e-mail: ehsu@math.northwestern.edu

Victor Isakov

Wichita State University
Wichita, KS 67206
USA
e-mail: victor.isakov@wichita.edu

Victor Ivrii

University of Toronto
40 St.George Str., Toronto, Ontario M5S 2E4
CANADA
e-mail: ivrii@math.toronto.edu

Tünde Jakab

University of Virginia
Charlottesville, VA 22904
USA
e-mail: tj8y@virginia.edu

Nanhee Kim

Wichita State University
Wichita, KS 67206
USA
e-mail: kim@math.wichita.edu

Juha Kinnunen

Institute of Mathematics
Helsinki University of Technology
P.O. Box 1100, FI-02015
FINLAND
e-mail: juha.kinnunen@tkk.fi

Pier Domenico Lamberti

Universitá degli Studi di Padova
63 Via Trieste, 35121 Padova
ITALY
e-mail: lamberti@math.unipd.it

Irena Lasiecka

University of Virginia
Charlottesville, VA 22904
USA
e-mail: il2v@virginia.edu

Vladimir Maz'ya

Ohio State University
Columbus, OH 43210
USA

University of Liverpool
Liverpool L69 7ZL
UK
Linköping University
Linköping SE-58183
SWEDEN
e-mail: vlmaz@mai.liu.se
e-mail: vlmaz@math.ohio-state.edu

Enzo Mitidieri

Università di Trieste
Via Valerio 12/1, 34127 Trieste
ITALY
e-mail: mitidier@units.it

Irina Mitrea

University of Virginia
Charlottesville, VA 22904
USA
e-mail: im3p@virginia.edu

Marius Mitrea

University of Missouri
Columbia, MO
USA
e-mail: marius@math.missouri.edu

Alexander Movchan

University of Liverpool
Liverpool L69 3BX
UK
e-mail: abm@liverpool.ac.uk

Sergey Nazarov

Institute of Problems in Mechanical Engineering
Russian Academy of Sciences
61, Bolshoi pr., V.O., St.-Petersburg 199178
RUSSIA
e-mail: serna@snark.ipme.ru

Janet Peterson

Florida State University
Tallahassee FL 32306-4120
USA
e-mail: peterson@scs.fsu.edu

Nguyen Cong Phuc

Purdue University
150 N. University Str.
West Lafayette, IN 47906
USA
e-mail: pcnguyen@math.purdue.edu

Luboš Pick

Charles University
Sokolovská 83, 186 75 Praha 8
CZECH REPUBLIC

e-mail: pick@karlin.mff.cuni.cz

Yehuda Pinchover

Technion – Israel Institute of Technology
Haifa 32000
ISRAEL

e-mail: pincho@techunix.technion.ac.il

Stanislav Pokhozhaev

Steklov Mathematical Institute
Russian Academy of Sciences
8, Gubkina Str., Moscow 119991
RUSSIA

e-mail: pokhozhaev@mi.ras.ru

Yuri Reshetnyak

Sobolev Institute of Mathematics
Siberian Branch
Russian Academy of Sciences
4, Pr. Koptyuga, Novosibirsk 630090
RUSSIA

Novosibirsk State University
2, Pirogova Str., Novosibirsk 630090
RUSSIA

e-mail: Reshetnyak@math.nsc.ru

Grigori Rozenblum

University of Gothenburg
S-412 96, Gothenburg
SWEDEN

e-mail: grigori@math.chalmers.se

Laurent Saloff-Coste

Cornell University
Mallot Hall, Ithaca, NY 14853
USA

e-mail: lsc@math.cornell.edu

Nageswari Shanmugalingam

University of Cincinnati
Cincinnati, OH 45221-0025
USA

e-mail: nages@math.uc.edu

Tatyana Shaposhnikova

Ohio State University
Columbus, OH 43210
USA

Linköping University
Linköping SE-58183
SWEDEN
e-mail: tasha@mai.liu.se

Winfried Sickel

Friedrich-Schiller-Universität Jena
Mathematisches Institut
Ernst–Abbe–Platz 2, D-07740 Jena
GERMANY
e-mail: sickel@minet.uni-jena.de

Michael Solomyak

The Weizmann Institute of Science
Rehovot, 76100
ISRAEL
e-mail: michail.solomyak@weizmann.ac.il

Michael Taylor

University of North Carolina
Chapel Hill, NC 27599
USA
e-mail: met@email.unc.edu

Kyril Tintarev

Uppsala University
P.O. Box 480, SE-751 06 Uppsala
SWEDEN
e-mail: kyril.tintarev@math.uu.se

Hans Triebel

Mathematisches Institut
Friedrich-Schiller-Universität Jena
D-07737 Jena
GERMANY
e-mail: triebel@minet.uni-jena.de

Roberto Triggiani

University of Virginia
Charlottesville, VA 22904
USA
e-mail: rt7u@virginia.edu

Marc Troyanov

Institute of Geometry, Algebra, and Topology
École Polytechnique Fédérale de Lausanne
1015 Lausanne
SWITZERLAND
e-mail: marc.troyanov@epfl.ch

Applications in Analysis and Partial Differential Equations

Vladimir Maz'ya Ed.

Contents

On the Mathematical Works of S.L. Sobolev in the 1930s

Vasilii Babich

Abstract A review of the works of S.L. Sobolev that have played a fundamental role in the development of the theory of partial differential equations and mathematical analysis in the second part of the 20th century. Supplied with a short biographical note.

Birth and Education

S.L. Sobolev, 1939.

Sergey Sobolev was born on 6 October 1908 in Petersburg (Leningrad, Petrograd). His parents, Lev Aleksandrovich and Natalia Georgievna, met in the provincial town of Saratov where both were sent into exile for their revolutionary activities in the 1900's. His father originated from the Siberian Cossacks. Unfortunately, he passed away when Sergey was only 14 years old. His mother graduated with a gold medal from the prestigious Bestuzhev courses of higher education for women and later she graduated from the medical institute. Sergey inherited the shining talents of his parents. In his childhood, he was interested in the French language, music, poetry, chess, and photography, among other things. But his mission was mathematics. At the age of 12, he was already familiar with elementary algebra, geometry, and trigonometry.

Vasili Babich
St.-Petersburg Department of the Steklov Mathematical Institute, 27 Fontanka, St. Petersburg 191023, Russia, e-mail: babich@pdmi.ras.ru

V. Maz'ya (ed.), *Sobolev Spaces in Mathematics II*, *International Mathematical Series*.
doi: 10.1007/978-0-387-85650-6,

In 1925, S. Sobolev entered the Physics and Mathematics Faculty of Leningrad State University.

As a student, his bright talent was spotted by Professors V.I. Smirnov and N.M. Gyunter, prominent specialists in mathematical physics. Sergey published his first scientific paper [1] when he was a final year student. In 1929, four people graduated from the Faculty: S.L. Sobolev, S.A. Khristianovich, S.G. Mikhlin, and V.N. Zamyatina (married name Faddeva). Each of them has left a deep impact on science.

At the Seismological Institute and Steklov Mathematical Institute

After their university education, Solomon Mikhlin and Sergey Sobolev were invited by Professor Smirnov, who always supported talented young mathematicians, to work at the Theoretical Department of the Seismological Institute of the USSR Academy of Sciences (Leningrad).

S. Sobolev (right) and his student at the Leningrad Electrotechnic Institute (1930-31).

The period of working at the Seismological Institute and then at the Steklov Institute of Physics and Mathematics was extremely fruitful for Sobolev. In one decade, the young mathematician not only solved several difficult problems of mathematical physics but also introduced and developed new theories and constructions. In particular, we associate the name of Sobolev with outstanding advantages in the study of the well-posedness of the Cauchy problem and boundary value problems for elliptic operators, construction of functional-invariant solutions for the wave equation (in collaboration with V. Smirnov), theory of distributions (generalized functions), weak solutions of problems of mathematical physics, Sobolev spaces, and embedding theorems or Sobolev inequalities.

Sobolev was elected a corresponding member (1933) and then a full member (1939) of the USSR Academy of Sciences. He was the youngest member of the Academy (in 1933, he was only 25 years old!).

In 1939, the Steklov Mathematical Institute moved from Leningrad to Moscow, and Sergey Sobolev moved with it. In 1941, at the beginning of World War II, Sobolev was appointed as Director of the Institute. Under his leadership, the Institute was evacuated from Moscow to Kazan', where it continued to work despite extreme wartime conditions. In 1943, the Institute came back to Moscow. In the same year, Sobolev was involved (together with other prominent Soviet scientists) in the realization of the nuclear project and was appointed as a deputy director of the Institute for Atomic Energy. A new chapter of his life had begun.

S. Sobolev with his children.

Hyperbolic Equations of Second Order [2, 3, 5, 9, 14, 15, 19, 23, 25, 27, 28, 29]

The first scientific works of Sobolev were devoted to the Cauchy problem for a linear hyperbolic equation with smooth coefficients. The well-posedness of this problem had been established by J. Hadamard by the method based on smooth asymptotics of the so-called elementary solution (the fundamental solution in the modern terminology). Having constructed such asymptotics, it is simple to derive an integral equation similar to the Volterra integral equation which can be solved by successive approximations. This method leads to the well-posedness of the Cauchy problem and also provides information about the analytic structure of the solution. However, this method works only if the number of spatial variables is even. Therefore, in the odd-dimensional case, Hadamard used an elegant, but rather artificial trick, the so-called *descent method.*

Sobolev developed a new method, covering the case of odd dimensions, for reducing the Cauchy problem to an integral equation and proved the well-posedness of the Cauchy problem for the model wave equation with variable velocity [2, 3, 5, 9, 14] and then for general hyperbolic equations [15, 19, 23, 25, 27, 28, 29]. Moreover, his method yields a solution to the classical wave equation in an explicit form. Sobolev's method was based on the analysis of certain relations on characteristic conoids satisfied by the solution to a hyperbolic equation.

The remarkable results of the young mathematician were highly regarded by J. Hadamard and A.N. Krylov.

Three Russian academicians.
Left to right: S.A. Chaplygin, S.L. Sobolev, A.N. Krylov.

The theory of distributions easily explains why Hadamard's method "did not work" in odd dimensions. The issue is that, in the physical language, Hadamard's elementary solution is a solution to the problem about a point source of oscillations of impulse type. Singularities of such a solution are described by distributions

$$\frac{\gamma_+^\lambda}{\Gamma(\lambda+1)}, \quad \gamma = \gamma(t,x), \quad x = (x^1, x^2, \dots, x^m), \quad \operatorname{grad}\gamma\big|_{\gamma=0} \neq 0,$$

where $\gamma = 0$ is the equation of the wavefront ($\gamma < 0$ in front of the wave), Γ is the Γ-function, $\lambda = -(m-1)/2$, and m is the number of spatial variables. If $m \geqslant 3$ is odd, then $\dfrac{\gamma_+^\lambda}{\Gamma(\lambda+1)}$ goes to $\delta^{(\frac{m-3}{2})}(\gamma)$, where δ is the delta-function. In other words, constructing the elementary solution by Hadamard's method, one necessarily arrives at a formula containing δ-functions. But, in the early 1930s, very little was known about distributions! Moreover, while physicists freely manipulated with the δ-function introduced by P. Dirac for the needs of quantum mechanics, mathematicians did not regard the Dirac function as a mathematical object at all. S. Mikhlin told me that once Sobolev had attempted to introduce expressions like $\delta^{(l)}(\gamma)$ at his lecture, but his innovation had not been accepted by mathematical attendance.

Functional-Invariant Solutions [4, 6, 7, 8, 10, 11, 12, 13, 16, 18, 21, 22]

V. Smirnov and S. Sobolev constructed and described an important class of the so-called *functional-invariant* solutions to the wave equation

$$\frac{1}{a^2}\frac{\partial^2 U}{\partial t^2} - \frac{\partial^2 U}{\partial x^2} - \frac{\partial^2 U}{\partial y^2} = 0. \qquad (1)$$

To explain the notion of a functional-invariant solution, let us consider smooth functions $\ell(\tau)$, $m(\tau)$, $n(\tau)$, and $p(\tau)$ of variable τ and define

$$\delta := \ell(\tau)\tau + m(\tau)x + n(\tau)y + p(\tau).$$

Under the conditions

$$\frac{\partial \delta}{\partial \tau} \neq 0 \text{ and } \ell^2 \equiv a^2(m^2 + n^2),$$

the function $\tau = \tau(x, y)$ determined from the equation

$$\delta \equiv \ell + mx + ny + p = 0$$

turns out to be a solution to Eq. (1). Furthermore, any function of the form $\mathfrak{F}(\tau(t, x, y))$, where $\mathfrak{F}(\tau)$ is a smooth function, is also a solution to Eq. (1), which expresses the *functional-invariant property.* If ℓ, m, n, p, $\mathfrak{F}$ are smooth functions of complex variable τ, we obtain complex functional-invariant solutions. In the case $p \equiv 0$, the functional-invariant solutions form the important class of homogeneous solutions of zero degree. Smirnov and Sobolev solved (in many cases, explicitly) a number of important problems of mathematical physics (in particular, a remarkable example of the solution to the problem of diffraction of a plane wave on a semi-infinite plane screen was included by Smirnov in his “A Course of Higher Mathematics,” Vol. III, Part 2[1]).

Weak Solutions [18, 20, 29]

The necessity to generalize the notion of a classical (smooth) solution was dictated by the physical nature of mathematical problems. Of course, attempts to extend too restricted frameworks of smooth solutions were taken far before Sobolev's definition [18, 20, 29] and can be traced in the literature. For example, Leonhard Euler considered the solution to the vibrating string equation $u_{tt} - a^2 u_{xx} = 0$ in the form of a curve arbitrarily drawn by hand

[1] English transl.: Smirnov, V.I.: A Course of HIgher Mathematics. Vol. III. Part two. Complex Variables. Special Functions. Pergamon Press, Oxford-Edinburg et al. (1964).

and moving without deformation with velocity a along the x-axis. A natural question arises: Whether such a curve can serve as a solution if it is not twice differentiable? Similar questions arise in the consideration of the plane waves

$$u = f(-at + \alpha_j x_j), \quad \sum_{j=1}^{m} \alpha_j^2 = 1,$$

where $f \in C^2$, satisfying the wave equation

$$\Box u \equiv \sum_{j=1}^{m} \frac{\partial^2 u}{\partial x_j^2} - \frac{1}{a^2}\frac{\partial^2 u}{\partial t^2} = 0 \tag{2}$$

and the spherically symmetric solution of (2) with $m = 3$:

$$u = \frac{1}{r} f\left(t - \frac{r}{a}\right), \quad r = \sqrt{\sum_{j=1}^{3} x_j^2}.$$

The triumph of Sobolev's notion of a weak solution (which is in common use today) is explained by its ability to be adapted to numerous problems of mathematical physics and various fields of mathematics. Another important reason is that Sobolev's notion essentially weakens the smoothness requirements on solutions, so that a large class of discontinuous functions, including the Dirac function and many other outsiders of the classical approach had been involved and become allowable in mathematics.

To be complete, let us recall Sobolev's definition of a weak solution. Let $D \subset \mathbb{R}^3$ be a domain with boundary admitting integration by parts. A function $u \in L_1(D)$ is called a *weak solution* to Eq. (2) in D if for any function $v \in C^2(D)$ vanishing, together with its normal derivative, on ∂D the following integral equality holds:

$$\int_D u \,\Box v \, dD = 0.$$

It is evident how to extend the definition to unbounded domains, more general operators, etc.etc.

Distributions [17, 19, 23]

In 1935/36, Sobolev introduced [17, 19] the notion of a distribution (Sobolev called it a *solution in functionals*; the term "distribution" appeared much later). A more detailed description of fundamentals of the theory of distributions is contained in his celebrated paper [23] of 1936 (see also [29]).

In particular, Sobolev proposed to understand a (generalized) function as a functional on the space of compactly supported functions of class C^k and introduced differential operators on the space of generalized functions.

The theory of distributions was further developed by efforts of many outstanding mathematicians (I.M. Gel'fand, L. Schwartz, and many others). As is known, the theory of distributions, originated from the theory of partial differential equations, turned out to be closely linked with harmonic analysis, linear group representations, integral geometry, and many other fields of mathematics.

Commenting on the paper [23] by S.L. Sobolev in the review (see [29]) of the theory of distributions, V.P. Palamodov wrote that the theory of distributions become one of the main events in the development of Analysis of the 20th century. We completely joint to his viewpoint.

Sobolev Spaces and Sobolev Inequalities [24, 26, 29]

A feature of the mathematical physics of the 20th century is that there are a lot of researches devoted to the proof of the well-posedness of different boundary value problems and initial-boundary value problems. The proof of the well-posedness is usually based on a priori estimates for solutions in various norms. At present, mathematicians dealing with questions of well-posedness follow the rule to choose spaces that are natural for the problem under consideration.

One of the most important problems of mathematical physics was the problem (coming from the 19th century) of proving the well-posedness of the Dirichlet and Neumann problems for the Laplace equation and more general elliptic equations of second order, so that the proof must be based on the fact that the solutions to these problems are minimizer of the Dirichlet integral. This problem attracted the attention of outstanding scientists of that time: D. Hilbert, K.O. Friedrichs, R. Courant, G. Weyl. S. Sobolev overcame the central difficulty in this problem: he found adequate function spaces, known now as *Sobolev spaces* $W_p^\ell(\Omega)$, where $p > 1$, $\ell = 0, 1, 2, \cdots$, Ω is a domain in $\mathbb{R}^n$. The Sobolev space $W_p^\ell(\Omega)$ is defined as the space of functions in $L_1(\Omega)$ whose distributional derivatives of order up to ℓ exist and belong to $L_p(\Omega)$.

In his celebrated paper [26] of 1938, Sobolev also proved the first embedding theorems (or Sobolev inequalities) which established relations between $W_p^\ell(\Omega)$ and the spaces $L_p(\Omega)$, $C^m(\Omega)$.

Sobolev spaces and Sobolev inequalities have played a fundamental role in the further development of the theory of partial differential equations, mathematical physics, differential geometry, and various fields of mathematical analysis.

Conclusion

Sobolev's advantages in the 1930's are very impressed. He dealt with particular problems of mathematical physics. But general methods and tools he introduced and developed for solving these problems are applicable to many other different situations. Owing to this fact, the discoveries of Sobolev (weak statements of problems, Sobolev spaces, embeddings, etc.) are very attractive for researchers and are generalized in different settings.

Acknowledgement. I thank Professors M.Sh. Birman and G.A. Chechkin (the grandson of S.L. Sobolev) for our helpful discussions on the subject. I used the book *Dinastiya*, Piligrim. Moscow (2002) written by A.D. Soboleva and the bibliography of S.L. Sobolev published by "Nauka," Novosibirsk (1969). Pictures, kindly presented by Grigori Chechkin, were compiled and prepared for printing by Tamara Rozhkovskaya and are reproduced here by permission.

References

1. Sobolev, S.L.: Remarks on the works of N.N. Saltykov "Research in the theory of first order partial differential equations with one unknown function" and "On the development of the theory of first order partial differential equations with one unknown function" (Russian). Dokl. Akad. Nauk SSSR, no. 7, 168-170 (1929)
2. Sobolev, S.L.: The wave equation in an inhomogeneous medium (Russian). Tr. Seismol. Inst. Akad. Nauk SSSR, no. 6 (1930)
3. Sobolev, S.L.: Sur l'équation d'onde pour le cas d'un milieu hétérogène isotrope (French). Dokl. Akad. Nauk SSSR, no. 7, 163-167 (1930)
4. Smirnov, V.I., Sobolev S.L.: On a new method for solving the plane problem of elastic oscillations (Russian). Tr. Seismol. Inst. Akad. Nauk SSSR, no. 16, 14-15 (1931)
5. Sobolev, S.L.: The wave equation in an inhomogeneous medium (Russian). Tr. Seismol. Inst. Akad. Nauk SSSR, no. 16, 15-18 (1931)
6. Smirnov, V.I., Sobolev, S.L.: Sur une méthode nouvelle dans le problème plan des vibrations élastiques (French). Tr. Seismol. Inst. Akad. Nauk SSSR, no. 18 (1932)
7. Smirnov, V.I., Sobolev, S.L.: Sur le problème plan des vibrations élastiques (French). C.R. Acad. Sci. Paris **194**, 1437-1439 (1932)
8. Smirnov, V.I., Sobolev, S.L.: Sur l'application de la méthode nouvelle a l'étude des vibrations élastiques dans l'espace à symmetrie axiale (French). Tr. Seismol. Inst. Akad. Nauk SSSR, no. 29 (1933)
9. Sobolev, S.L.: On a generalization of the Kirchhoff formula (Russian, French). Dokl. Akad. Nauk SSSR **1**, no. 6, 256-258, 258-262 (1933)
10. Sobolev, S.L.: On vibration of a half-plane and of a lamina under arbitrary initial conditions (French). Mat. Sb. **40**, no. 2, 236-266 (1933); English transl.: Russ. Math. Surv. **23**, no. 5, 95-129 (1968)
11. Sobolev, S.L.: On a method for solving the problem of oscillation propagation (Russian). Prikl. Mat. Mekh. **1**, no. 2, 290-309 (1933)

12. Sobolev, S.L.: L'équation d'onde sur la surface logarithmique de Riemann (French). C. R. Acad. Sci. Paris **196**, 49-51 (1933)
13. Sobolev, S.L.: Theory of diffraction of plane waves (Russian). Tr. Seismol. Inst. Akad. Nauk SSSR, no. 41 (1934)
14. Sobolev, S.L.: On integration of the wave equation in an inhomogeneous medium (Russian). Tr. Seismol. Inst. Akad. Nauk SSSR, no. 42 (1934)
15. Sobolev, S.L.: A new method for solving the Cauchy problem for partial differential equations of second order (Russian, French). Dokl. Akad. Nauk SSSR **1**, no. 8, 433-435, 435-438 (1934)
16. Sobolev, S.L.: Functional-invariant solutions to the wave equation (Russian). Tr. Steklov Phys. Mat. Inst. **5**, 259-264 (1934)
17. Sobolev, S.L.: The Cauchy problem in the space of functionals (Russian, French). Dokl. Akad. Nauk SSSR **3**, no. 7, 291-294 (1935)
18. Sobolev, S.L.: General theory of wave diffraction on Riemannian surfaces (Russian). Tr. Steklov Mat. Inst. **9**, 39-105 (1935)
19. Sobolev, S.L.: A new method for solving the Cauchy problem for partial differential equations of hyperbolic type (Russian). In: Proc. the 2nd All-Union Math. Congr. (Leningrad, 24-30 June 1934), Vol. 2, pp. 258-259. Akad. Nauk SSSR, Moscow–Leningrad (1936)
20. Sobolev, S.L.: Generalized solutions to the wave equation (Russian). In: Proc. the 2nd All-Union Math. Congr. (Leningrad, 24-30 June 1934), Vol. 2, p. 259. Akad. Nauk SSSR, Moscow–Leningrad (1936)
21. Sobolev, S.L.: On the diffraction problem on Riemannian surfaces (Russian). In: Proc. the 2nd All-Union Math. Congr. (Leningrad, 24-30 June 1934), Vol. 2, p. 364. Akad. Nauk SSSR, Moscow–Leningrad (1936)
22. Sobolev, S.L.: The Schwarz algorithm in the theory of elasticity (Russian, French). Dokl. Akad. Nauk SSSR **4**, no. 6, 235-238, 243-246 (1936)
23. Sobolev, S.L.: Méthode nouvelle à résoudre le problème de Cauchy pour les équations linéaires hyperboliques normales (French). Mat. Sb. **1**, no. 1, 39-72 (1936); English transl.: In: Sobolev, S.L.: Some Applications of Functinal Analysis in Mathematical Physics. Third edition. Appendix. Am. Math. Soc., Providence, RI (1991)
24. Sobolev, S.L.: On a theorem of functional analysis (Russian, French). Dokl. Akad. Nauk SSSR **20**, no. 1, 5-9 (1938)
25. Sobolev, S.L.: On the Cauchy problem for quasilinear hyperbolic equations (Russian, French). Dokl. Akad. Nauk SSSR **20**, no. 2-3, 79-83 (1938)
26. Sobolev, S.L.: On a theorem of functional analysis (Russian). Mat. Sb. **4**, no. 3, 471-496 (1938); English transl.: In: Eleven Papers in Analysis. Am. Math. Soc. Transl. (2) **34**, 39-68 (1963)
27. Sobolev, S.L.: On the theory of nonlinear hyperbolic partial differential equations (Russian). Mat. Sb. **5**, no. 1, 71-79 (1939)
28. Sobolev, S.L.: On the stability of solutions to boundary value problems for partial differential equations of hyperbolic type (Russian, French). Dokl. Akad. Nauk SSSR **32**, no. 7, 459-462 (1941)
29. Sobolev, S.L.: Some Applications of Functional Analysis in Mathematical Physics (Russian). 1st ed. Leningrad State Univ., Leningrad (1950); 2nd ed. SB Akad. Nauk SSSR, Novosibirsk (1962); 3rd ed. Nauka, Moscow (1988); English transl. of the first edition: Am. Math. Soc., Providence, RI (1963); English transl. of the third edition, with Appendix (English transl. of the paper [23] by S.L. Sobolev) and comments on the appendix by V.P. Palamodov: Am. Math. Soc., Providence, RI (1991)

Sobolev in Siberia

Yuri Reshetnyak

Abstract Reminiscence. Supplied with photos.

Scientists and representatives of the USSR Government choose a place for building Akademgorodok. Left to right: 4) S.L. Sobolev, 7) N.S. Khruschev, 8) M.A. Lavrent'ev (November, 1957)

Yuri Reshetnyak
Sobolev Institute of Mathematics SB RAS, 4 Pr. Koptyga, 630090 Novosibirsk, Russia,
e-mail: Reshetnyak@math.nsc.ru

V. Maz'ya (ed.), *Sobolev Spaces in Mathematics II, International Mathematical Series.*
doi: 10.1007/978-0-387-85650-6,

On my 20th birthday, as a birthday present, my friends gave me a book entitled "Mathematics in the USSR in 30 years." It was in this book that I read a review of particular interest on partial differential equations written by S.L. Sobolev. The paper contained a definition of objects, now called *distributions*, and formulations of theorems, now known as *Sobolev inequalities* or *embedding theorems*. I must confess that I did not realize the significance of this discovery at that time.

From 1948, I attended the seminar on geometry at the Leningrad State University. The seminar was supervised by Professor A.D. Aleksandrov, an outstanding expert in geometry. One of the principles Aleksandrov asserted was that the main notions of geometry must be formulated by purely geometric means, without any intervention of Analysis. Personally, I believed that Analysis may be useful for the greater understanding of geometric ideas, and at the seminar, I initiated to review some studies in order to familiarize myself with the latest ideas taking place in Analysis.

In 1950, in the book-shop of the Leningrad State University, I noted the recently published book "Some Applications of Functional Analysis in Mathematical Physics" by S.L. Sobolev. I looked through the book, and my initial feeling was that Sobolev's theory was exactly what we needed. Then we reviewed Sobolev's book at Aleksandrov's seminar.

S.L. Sobolev at the foundation of the Institute of Mathematics.

The ideas of Sobolev's book strongly influenced my research, in particular, my main result that concerns isothermic coordinates in two-dimensional manifolds of bounded curvature (for which I was later awarded with the Lobachevski Prize of the Russian Academy of Sciences). In Novosibirsk, I obtained results on different aspects of the theory of Sobolev spaces. My last challenge is the definition of the spaces $W^{1,p}$ of mappings with the values in general metric spaces.

But let us return to events of the 1950's. When, in the spring of 1957, I heard that S.L. Sobolev was appointed the Director of the Institute of Mathematics (in a new scientific center at Novosibirsk), my decision to follow him to Siberia was instant. My wife supported my choice, and my whole family including two babies moved from Leningrad to Siberia. We arrived at Novosibirsk at the end of November 1957, and on the 1st December 1957, I was appointed to the Institute of Mathematics where I have now worked for more than 50 years.

In 1957, the Institute consisted of 5 persons (including the director) and occupied one small room of a building in the center of Novosibirsk. The construction of Akademgorodok started in 1958 and continued very rapidly. Soon we moved to Akademgorodok, a very nice place in the forest near the artificial Ob' Sea.

S.L. Sobolev during his lecture.

S.L. Sobolev and I.N. Vekua. 1965.

S.L. Sobolev during his lecture.

There were three founders of the Siberian Branch of the USSR Academy of Sciences: M.A. Lavrent'ev, S.A. Khristianovich, and S.L. Sobolev. The idea was to create a new center for fundamental research in mathematics, physics, chemistry, etc. The center was planned to be in Siberia (which was not rich with scientists the time). So, initially, it was proposed that the scientific staff should consist of specialists from Moscow and Leningrad. Now we can see that this ambitious project was successfully achieved.

Sobolev organized the Institute of Mathematics with great enthusiasm. He personally recruited researchers for the Institute. A.I. Mal'tsev, I.N. Vekua, A.V. Bitsadze, L.V. Kantorovich, and other prominent mathematicians accepted Sobolev's invitation. In 1959, the Novosibirsk State University in Akademgorodok opened the doors to students. The first Rector of the University was I.N. Vekua. The first lecture was given by S.L. Sobolev.

"That was one of the best meetings I ever went to."

Interview with Louis Nirenberg. Notice of the AMS, **49** no. 4, p. 447 (2002)

International Meetings at Akademgorodok

⇐ **Soviet-American Conference in Partial Differential Equations. 1963** Front row (left to right): J. Moser, A. Zygmund, L. Ahlfors, N. Brunswick, S.L. Sobolev, C.B. Morrey, C. Loewner, R. Courant, M.A. Lavrent'ev, I.N. Vekua, S. Bergman, A.N. Tikhonov, G.I. Marchuk, D. Spencer, A. Dynin; Second row: T.I. Zelenyak, M.G. Krein, O.A. Oleinik, H. Weinberger, H. Grad, M. Schechter, J. Douglas, F. Browder, M.H. Protter, A.D. Myshkis, Yu.M. Berezanskii, V.A. Il'in, A.P. Calderón, P.D. Lax, E.B. Dynkin; Third row: A.Ya. Povzner, B.L. Rozhdestvenskii, P.E. Sobolevskii, –, V.G. Maz'ya, B.V. Shabat, L.D. Kudryavtsev, –, T.I. Amanov, P.P. Belinskii, S.G. Krein, I.D. Safronov, R. Richtmeyer, A.A. Lyapunov, S.K. Godunov, L.I. Kamynin; Fourth row: Abalashvilli, M.S. Salakhatdinov, R. Finn, –, –, L. Nirenberg, –, D.K. Fage, B.P. Vainberg, L.V. Ovsyannikov, –, I.I. Danilyuk, M.I. Vishik, S.M. Nikol'skii, V.N. Maslennikova, Yu.V. Egorov; Fifth row: M.S. Agranovich, –, N.D. Vedenskaya, L.R. Volevich, T.D. Ventsel', A.M. Il'in, A.F. Sidorov, Ya.A. Roitberg, T.G. Golenopol'skii, I.A. Shishmarev, Ya.S. Bugrov, –, –, A.V. Sychev, V.S. Ryaben'kii, O.V. Besov, S.V. Uspenskii, V.N. Dulov, –, V.K. Ivanov, S.N. Kruzhkov; Sixth row: V.A. Solonnikov, M.M. Lavrent'ev, A.K. Gerasimov, –, –, A. Dzhuraev, –, P.I. Lizorkin, –, T. Dzhuraev, –, G. Salikhov, V.M. Babich, L.D. Faddeev, A.I. Koshelev, M.Sh. Birman, S.I. Pokhozhaev, –, –, A.I. Prilepko, –, A.M. Molchanov, –, –, –, Yu.I. Gil'derman, –. L.G. Mikhailov, Yu.V. Sidorov

S.L. Sobolev and J. Leray (1978).

S.L. Sobolev (1983).

Left to right: R. Finn, O.A. Oleinik, –, Yu.G. Reshetnyak, N.N. Uraltseva, T.N. Rozhkovskaya (1983).

In 1960, the Institute launched the *Siberian Mathematical Journal.* From 1966 the journal was translated into English and published in the USA by Plenum Publishing Corporation (now known as Springer). The journal made the latest research of Siberian mathematicians available to the international mathematical community.

By 1962, the organizational structure of the Institute of Mathematics was complete: the departments were comprised of algebra and mathematical logic, analysis, the theory of partial differential equations, the theory of functions of a complex variable, geometry and topology, mathematical economics, cybernetics, theoretical physics and computational center.

S.L. Sobolev and A.A. Trofimuk.

S.L. Sobolev was Director of the Institute of Mathematics from the day of foundation until 1981. Under his leadership, the Institute became a world renowned center for mathematics. The international scientific authority of Sergey L'vovich Sobolev played a key role.

There were many remarkable events in the history of the Institute. I recall a Soviet–American Conference in Partial Differential Equations held in August 1963. It was a great event! It seemed as though all the best experts from both countries gathered at this meeting. The American contingent was represented by R. Cournat, A. Zygmund, A.P. Calderón, L. Nirenberg, P.D. Lax and others. Even the number of foreign participants was extraordinary for a meeting in the Soviet Union at that time! It was a unique opportunity to discuss recent studies with foreign colleagues personally. At this meeting, I lectured about aspects of the theory of spatial quasiconformal mappings and then had useful discussions with C.B. Morrey and A. Zigmund.

The life of S.L. Sobolev can be distinctly divided into three periods. The first period covers the youth of Sergey Sobolev. All the major results which established his name in the mathematical community were obtained by S.L. Sobolev in the 1930's. Note that the idea to generalize the notion of derivatives of solutions to problems in mathematical physics was actual in the 1920-1930's. In particular, N.M. Gyunter, the supervisor of S. Sobolev at the Leningrad State University, tried to realize this idea by applying set functions to equations of mathematical physics. Other researchers used the classes ACT_p of functions absolutely continuous in the sense of Tonelli. The definition was based on the behavior of functions on almost all lines parallel to the coordinate axes. However, the definition of a distribution that we use today was first proposed by S.L. Sobolev who also introduced function classes, known as *Sobolev spaces*, and established relationships (embeddings) between these spaces. The Sobolev approach is very elegant and simple. Moreover, Sobolev showed how to use weakly differentiable functions in applications. Owing to S.L. Sobolev, the common viewpoint on a solution had been revised and new possibilities have now been discovered in the theory of PDEs and mathematical physics. The ideas of S.L. Sobolev became very attractive and were developed by many mathematicians in different ways.

The second period could be dated from 1941 to 1957. At the beginning of the World War II, the main task of the Director of the Steklov Mathematical Institute, S.L. Sobolev, was to evacuate the Institute from Moscow to Kazan' and to organize conditions for work. In 1943, S.L. Sobolev was appointed as the deputy director of the Institute for Atomic Energy. The best Soviet mathematicians were gathered under the roof of this institute to work on the nuclear project. S.L. Sobolev worked there for more than 14 years, however not much is known about these years.

The third period of Sobolev's life was devoted to Akademgorodok. S.L. Sobolev was the Director of the Institute of Mathematics for 25 years and then returned to Moscow. Sergey L'vovich Sobolev passed away on 3 January 1989. Now the Institute of Mathematics in Novosibirsk bears his name.

Acknowledgement. Pictures of the press-photographers Rashid Akhmerov and Vladimir Novikov were compiled and prepared for printing by Tamara Rozhkovskaya and are reproduced here by kind permission.

Sergey L'vovich and Ariadna Dmitrievna Sobolev. Autumn in Akademgorodok.

Boundary Harnack Principle and the Quasihyperbolic Boundary Condition

Hiroaki Aikawa

Abstract We discuss the global boundary Harnack principle for domains in $\mathbb{R}^n$ $(n \geqslant 2)$ satisfying some conditions related to the quasihyperbolic metric. For this purpose, we reformulate the global boundary Harnack principle and the global Carleson estimate in terms of the Green function. Based on our other result asserting the equivalence between the global boundary Harnack principle and the global Carleson estimate, we obtain four equivalent conditions. Using the box argument for the estimate of the harmonic measure, we obtain the global Carleson estimate in terms of the Green function (and thus the global boundary Harnack principle) for a domain satisfying a condition on the quasihyperbolic metric and the capacity density condition, as well as for a Hölder domain whose boundary is locally given by the graph of a Hölder continuous function in $\mathbb{R}^{n-1}$. Our argument is purely analytic and elementary, unlike the probabilistic approach of Bass–Burdzy–Bañuelos.

1 Introduction

One of the most important partial differential equations is the Laplace equation: $\Delta u = 0$. A solution to this equation is called a *harmonic function*. When we impose further the boundary condition, say $u = f$, the problem is said to be the Dirichlet problem. The notions of weak derivatives and *Sobolev spaces* play a crucial role in the variational method for solving the Dirichlet problem. Harmonic functions and Sobolev spaces are inextricably related.

In this paper, we deal with positive harmonic functions. Our purpose is to show the global boundary Harnack principle for domains satisfying some conditions related to the quasihyperbolic metric. Throughout the paper, D is a bounded domain in $\mathbb{R}^n$ with $n \geqslant 2$ and $\delta_D(x) = \operatorname{dist}(x, \partial D)$. We write

Hiroaki Aikawa
Hokkaido University, Sapporo 060-0810, Japan, e-mail: aik@math.sci.hokudai.ac.jp

V. Maz'ya (ed.), *Sobolev Spaces in Mathematics II*, *International Mathematical Series*.
doi: 10.1007/978-0-387-85650-6,

$B(x,r)$ and $S(x,r)$ for the open ball and the sphere of center at x and radius r respectively. We denote by the symbol A an absolute positive constant whose value is unimportant and may change from one occurrence to the next. If necessary, we use $A_0, A_1, \dots,$ to specify them. If two positive quantities f and g satisfies $A^{-1} \leqslant f/g \leqslant A$ with some constant $A \geqslant 1$, then we say that f and g are *comparable* and write $f \approx g$.

We consider a pair (V, K) of a bounded open set $V \subset \mathbb{R}^n$ and a compact set $K \subset \mathbb{R}^n$ such that

$$K \subset V, \ K \cap D \neq \varnothing, \text{ and } K \cap \partial D \neq \varnothing. \tag{1.1}$$

Definition 1.1. We say that a domain D enjoys the *global boundary Harnack principle* if for each pair (V, K) with (1.1) there exists a constant A_0 depending only on D, V and K with the following property: If u and v are positive superharmonic functions on D such that

(i) u and v are bounded, positive, and harmonic in $V \cap D$,

(ii) u and v vanish on $V \cap \partial D$ except for a polar set,

then

$$\frac{u(x)/u(y)}{v(x)/v(y)} \leqslant A_0 \quad \text{for } x, y \in K \cap D. \tag{1.2}$$

Remark 1.1. Since $K \cap D$ may be disconnected, the superharmonicity of u and v over the whole D is needed. Usually, the harmonicity and positivity of u and v over the whole D are assumed for the global boundary Harnack principle. Our boundary Harnack principle is slightly stronger.

For a Lipschitz domain the global boundary Harnack principle was proved by Ancona [3], Dahlberg [11] and Wu [16] independently. Caffarelli–Fabes–Mortola–Salsa [10] and Jerison–Kenig [13] gave significant extensions. From the probabilistic point of view, Bass–Burdzy–Bañuelos [8, 7] proved the global boundary Harnack principle for a *Hölder domain*, a domain whose boundary is locally given by the graph of a Hölder continuous function in $\mathbb{R}^{n-1}$.

The term "Hölder domain" was used for a different type domainin [14, 15]. We define the quasihyperbolic metric $k_D(x,y)$ by

$$k_D(x,y) = \inf_\gamma \int_\gamma \frac{ds(z)}{\delta_D(z)},$$

where the infimum is taken over all rectifiable curves γ connecting x to y in D and $ds(z)$ stands for the line element on γ. In [14, 15], a domain D is called a *Hölder domain* if

$$k_D(x, x_0) \leqslant A \log \frac{\delta_D(x_0)}{\delta_D(x)} + A' \quad \text{for all } x \in D \tag{1.3}$$

with some positive constants A and A'. In [6], such a domain is called a *Hölder domain of order* 0. To avoid the confusion, in this paper, we say that D satisfies the *quasihyperbolic boundary condition* (*of order 0*) if (1.3) holds.

Extending (1.3), we consider the following condition:

$$k_D(x, x_0) \leqslant A\Big(\frac{\delta_D(x_0)}{\delta_D(x)}\Big)^{\alpha} + A' \quad \text{for all } x \in D \tag{1.4}$$

with some positive constants A and A'. We say that D satisfies the *quasi-hyperbolic boundary condition of order* α if (1.4) holds. The above condition and (1.3) are *interior* conditions. Let us consider an *exterior* condition.

Definition 1.2. By Cap we denote the logarithmic capacity if $n = 2$ and the Newtonian capacity if $n \geqslant 3$. We say that the *capacity density condition* holds if there exist constants $A > 1$ and $r_0 > 0$ such that

$$\mathrm{Cap}(B(\xi, r) \setminus D) \geqslant \begin{cases} A^{-1}r & \text{if } n = 2, \\ A^{-1}r^{n-2} & \text{if } n \geqslant 3, \end{cases}$$

whenever $\xi \in \partial D$ and $0 < r < r_0$ (see [5, p. 150] for the logarithmic capacity, which illustrates the inhomogeneity between the cases $n = 2$ and $n \geqslant 3$).

In [6], a domain D is called a *uniformly Hölder domain of order* α if (1.4) and the capacity density condition hold. It seems that the capacity density condition is needed for $\alpha > 0$ because of the lack of the exponential integrability of the quasihyperbolic metric which was proved in [15] for a domain with the quasihyperbolic boundary condition.

While the main interest in [6] was the intrinsic ultra-contractivity for a domain satisfying the quasihyperbolic boundary condition of order α and the capacity density condition, Bass–Burdzy–Bañuelos [8, 7] proved the boundary Harnack principle for a Hölder domain. The main tool was the so-called *box argument.* They also claimed that the boundary Harnack principle may hold for a domain satisfying the quasihyperbolic boundary condition of order α, $0 < \alpha < 1$, and the capacity density condition. However, no proof has not been provided so far.

Though the arguments of [8, 7] were very probabilistic, the author [1] managed to understand analytically the box argument and applied it to the proof of the local (or scale-invariant) boundary Harnack principle for a uniform domain. In [2], the author showed a completely different approach to the boundary Harnack principle: the Domar method and the equivalence between the boundary Harnack principle and the Carleson estimate.

Definition 1.3. We say that a domain D enjoys the *global Carleson estimate* if for each pair (V, K) with (1.1) and a point $x_0 \in K \cap D$ there exists a constant A_1 depending only on D, V, K and x_0 with the following property: If u is a positive superharmonic function on D such that

(i) u is bounded, positive and harmonic in $V \cap D$,

(ii) u vanishes on $V \cap \partial D$ except for a polar set,

then

$$u(x) \leqslant A_1 u(x_0) \quad \text{for } x \in K \cap D. \tag{1.5}$$

Theorem 1.1 ([2]). *The global boundary Harnack principle and the global Carleson estimate are equivalent.*

As a consequence, we obtain the following assertion.

Theorem 1.2 ([2]). *The global boundary Harnack principle holds for a domain satisfying the quasihyperbolic boundary condition (of order 0).*

The purpose of this paper is to prove analytically the boundary Harnack principle for a domain satisfying the quasihyperbolic boundary condition of order $0 < \alpha < 1$, and the capacity density condition. The proof will clarify the role of the capacity density condition.

Theorem 1.3. *Let D be a bounded domain satisfying the capacity density condition and* (1.4) *for $0 < \alpha < 1$. Then the global boundary Harnack principle holds for D.*

Let $0 < \beta \leqslant 1$. We say that D is a *β–John domain* if there is a point $x_0 \in D$ and every point $x \in D$ can be connected to x_0 by a rectifiable curve γ with

$$\delta_D(y)^\beta \geqslant A\ell(\gamma(x, y)) \quad \text{for all } y \in \gamma,$$

where $\ell(\gamma(x, y))$ is the arc length of the subcurve $\gamma(x, y)$ connecting x and y along γ (see [4, 9.2]). If $\beta = 1$, then a β-John domain is a classical John domain and the quasihyperbolic boundary condition of order 0 is satisfied. In general, a β-John domain satisfies the quasihyperbolic boundary condition of order $1 - \beta$. Hence we obtain the following assertion.

Corollary 1.1. *Let $0 < \beta \leqslant 1$. Then a β-John domain with the capacity density condition enjoys the global boundary Harnack principle.*

Remark 1.2. In view of Theorem 1.2, the capacity density condition is superfluous in case $\beta = 1$.

A typical example of a β-John domain is a β-Hölder domain, whose boundary is given locally by the graph of a β-Hölder continuous function in $\mathbb{R}^{n-1}$. A 1-Hölder domain is a Lipschitz domain, so that the boundary Harnack principle is classically established. A β-John domain need not to satisfy the capacity density condition if $0 < \beta < 1$. However, Bass–Burdzy–Bañuelos [8] showed that the global boundary Harnack principle holds for a β-Hölder domain without the capacity density condition. Their proof is very probabilistic. We give an elementary analytic proof.

Theorem 1.4. *Let $0 < \beta < 1$. A β-Hölder domain enjoys the global boundary Harnack principle.*

In the next section, we restate the boundary Harnack principle and the Carleson estimate by using the Green function. Several equivalent conditions will be given in Theorem 2.3. We prove that one of them holds for a domain in Theorems 1.3 and 1.4. Our argument will rely on the capacitary width, estimates of harmonic measures, and the box argument. It will be purely analytic.

2 Boundary Harnack Principle and Carleson Estimate in Terms of the Green Function

Let us recall the definition of the global Carleson estimate (Definition 1.3). The Riesz decomposition theorem says that, in $D \cap V$, u can be represented as the Green potential $G\mu$ of a measure μ on $D \cap \partial V$. Therefore, we can easily see that the global Carleson estimate can be restated in terms of the Green function. One more geometrical observation is relevant. Let $\overline{B}$ be a closed ball including D. Then $F = \overline{B} \setminus V$ is a compact set and $D \setminus V = D \cap F$ and $K \cap F = \varnothing$. Thus, instead of a pair of a bounded open set V and a compact set K, we can consider a pair of disjoint compact sets K and F with

$$K \cap \partial D \neq \varnothing, \quad K \cap D \neq \varnothing, \quad F \cap \partial D \neq \varnothing, \text{ and } F \cap D \neq \varnothing. \tag{2.1}$$

Definition 2.1. We say that a domain D enjoys the *global Carleson estimate in terms of the Green function* if for each pair of disjoint compact sets K and F with (2.1) and a point $x_0 \in K \cap D$, there exists a constant A_1 depending only on D, K, F and x_0 such that

$$G(x,y) \leqslant A_1 G(x_0,y) \quad \text{for } x \in K \cap D \text{ and } y \in F \cap D \tag{2.2}$$

(see Fig. 1(a)).

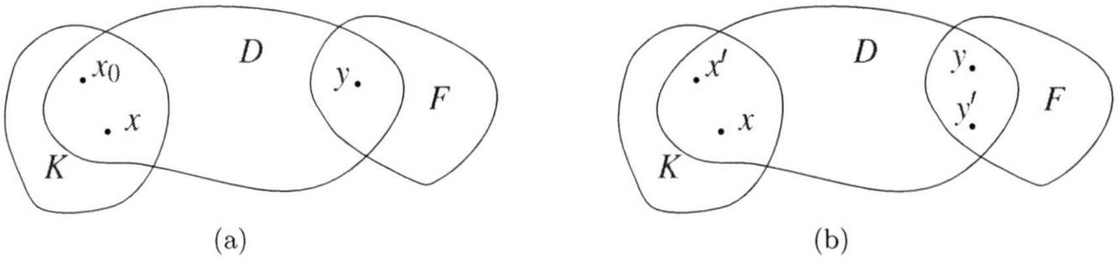

Fig. 1 (a) Carleson estimate in terms of the Green function. (b) Boundary Harnack principle in terms of the Green function.

Theorem 2.1. *The global Carleson estimate and the global Carleson estimate in terms of the Green function are equivalent.*

Using the Riesz decomposition theorem, we similarly obtain a counterpart of the boundary Harnack principle.

Definition 2.2. We say that a domain D enjoys the *global boundary Harnack principle in terms of the Green function* if for each pair of disjoint compact sets K and F with (2.1) there exists a constant A_0 depending only on D, K, and F such that

$$\frac{G(x,y)/G(x',y)}{G(x,y')/G(x',y')} \leqslant A_0 \quad \text{for } x,\ x' \in K \cap D \text{ and } y, y' \in F \cap D \tag{2.3}$$

(see Fig. 1(b)).

Theorem 2.2. *The global boundary Harnack principle and the global boundary Harnack principle in terms of the Green function are equivalent.*

Remark 2.1. This boundary Harnack principle in terms of the Green function resembles the parabolic boundary Harnack principle due to Bass–Burdzy [9, Theorem 1.2]. Nevertheless, there is a significant difference. In the elliptic case, the positions of points are limited whereas they are free for the parabolic case. In other words, positive harmonic functions vanishing on some *portion of the boundary* are treated in the elliptic case, whereas positive solutions to the heat equation vanishing on the *whole lateral boundary* are treated in the parabolic case. It seems that the background of these phenomena is a backward Harnack inequality [12].

Just for the convenience sake, we combine Theorems 1.1, 2.1, and 2.2 to obtain the following assertion.

Theorem 2.3. *Let D be a bounded domain in $\mathbb{R}^n$. Then the following statements are equivalent:*

(i) *D enjoys the global boundary Harnack principle.*

(ii) *D enjoys the global boundary Harnack principle in terms of the Green function.*

(iii) *D enjoys the global Carleson estimate.*

(iv) *D enjoys the global Carleson estimate in terms of the Green function.*

3 Proof of the Main Result

Lemmas

We write $\omega(E,U)$ for the harmonic measure over an open set U of $E \subset \partial U$, i.e., $\omega(E,U)$ is the Dirichlet solution in U of the boundary function χ_E (see, for example, [5, Chapt. 6]). Let $0 < \eta < 1$. For an open set $U \subset \mathbb{R}^n$ we define the *capacitary width* $w_\eta(U)$ by

$$w_\eta(U) = \inf\left\{r > 0 : \frac{\operatorname{Cap}(B(x,r) \setminus U)}{\operatorname{Cap}(B(x,r))} \geqslant \eta \quad \text{for all } x \in U\right\}.$$

Under the capacity density condition, the capacitary width of certain sets can be estimated in a straightforward fashion.

Lemma 3.1. *If the capacity density condition holds, then* $w_\eta(\{x \in D : \delta_D(x) < r\}) \leqslant 2r$ *for some* $\eta > 0$.

The capacitary width is useful for the estimate of harmonic measure (see [1, Lemma 1]).

Lemma 3.2. *There is a positive constant* $A_2 > 0$ *depending only on* η *and* n *with the following property: if* $x \in U$ *and* $R > 0$, *then*

$$\omega^x(U \cap S(x,R); U \cap B(x,R)) \leqslant \exp\left(2 - A_2\frac{R}{w_\eta(U)}\right).$$

We say that $x, y \in D$ are connected by a *Harnack chain* $\{B(x_j, \frac{1}{2}\delta_D(x_j))\}_{j=1}^k$ if

$$x \in B(x_1, \frac{1}{2}\delta_D(x_1)), \quad y \in B(y_k, \frac{1}{2}\delta_D(y_k))$$

and

$$B(x_j, \frac{1}{2}\delta_D(x_j)) \cap B(x_{j+1}, \frac{1}{2}\delta_D(x_{j+1})) \neq \varnothing$$

for $j = 1, \ldots, k-1$. The number k is called the *length* of the Harnack chain. We observe that the shortest length of the Harnack chain connecting x and y in D is comparable to $k_D(x,y)$. Therefore, the Harnack inequality yields the following assertion.

Lemma 3.3. *There is a constant* $A_3 > 1$ *depending only on the dimension* n *(even independent of* D*) such that*

$$\exp(-A_3(k_D(x,y)+1)) \leqslant \frac{h(x)}{h(y)} \leqslant \exp(A_3(k_D(x,y)+1)) \qquad (3.1)$$

for every positive harmonic function h *on* D.

Proof of Theorem 1.3

In view of Theorem 2.3, it suffices to show the global Carleson estimate in terms of the Green function. Let D be as in Theorem 1.3. Let K and F be a pair of disjoint compact sets with (2.1), and let $x_0 \in K \cap D$. For simplicity, we put $U(r) = \{x \in \mathbb{R}^n : \text{dist}(x, F) < r\}$ for $r > 0$. Let $2R = \text{dist}(K, F)$. We put $U = U(R)$. It is obvious that U is an open set including F and yet its closure is apart from K. Let $\omega_0 = \omega(D \cap \partial U, D \cap U)$ be the harmonic measure of $D \cap \partial U$ in $D \cap U$. First, wes compare ω_0 and $G(x_0, \cdot)$.

Lemma 3.4. *Let* ω_0 *be as above. Then*

$$\omega_0(y) \leqslant AG(x_0, y) \quad \textit{for } y \in F \cap D.$$

Proof. We employ the box argument. Let $R_0 = R$, and let

$$R_j = \Big(1 - \frac{3}{\pi^2} \sum_{k=1}^{j} \frac{1}{k^2}\Big) R \quad \text{for } j \geqslant 1.$$

Then it is easy to see that $R_{j-1} - R_j = (3R)/(\pi^2 j^2)$, so that

$$\sum_{j=1}^{\infty} R_j = \frac{R}{2}. \tag{3.2}$$

It is obvious that $G(x_0, \cdot)$ is bounded on $U \cap D$. Hence, by a suitable choice of $A > 1$, we may assume that $u = G(x_0, \cdot)/A$ is bounded by 1 on $U \cap D$. Now let

$$U_j = \{y \in U(R_j) \cap D : 0 < u(y) < \exp(-2^j)\},$$
$$D_j = \{y \in U(R_j) \cap D : \exp(-2^{j+1}) \leqslant u(y) < \exp(-2^j)\}$$

for $j \geqslant 1$. Let $q_j = \sup_{D_j} \omega_0/u$ if $D_j \neq \varnothing$ and $q_j = 0$ if $D_j = \varnothing$. It is clear that q_j is finite. Since $0 < u < 1$ on $U \cap D$, from (3.2) it follows that $F \cap \partial D$ is included in the closure of $\bigcup_{j=1}^{\infty} D_j$. Hence, by the Harnack principle, it suffices to show that q_j is bounded.

By (1.4) and (3.1), we have

$$\exp(-2^j) > u(y) \geqslant \exp(-A_3(k_D(y, x_0)+1)) \geqslant A \exp\Big(-\frac{A}{\delta_D(y)^\alpha}\Big) \quad \text{for } y \in U_j.$$

In other words, $\delta_D(y) \leqslant A2^{-j/\alpha}$ for $y \in U_j$. Hence Lemma 3.1 implies that $w_\eta(U_j) \leqslant A2^{-j/\alpha}$. Observe that $\text{dist}(D \cap \partial U_{j-1}, U_j) \geqslant R_{j-1} - R_j$. Applying the maximum principle on U_{j-1}, we obtain

$$\omega_0 \leqslant \omega(D \cap \partial U_{j-1} \setminus \overline{D_{j-1}}, U_{j-1}) + q_{j-1} u \quad \text{on } U_{j-1}$$

(see Fig. 2).

We divide both sides by u and take the supremum over D_j. Then Lemma 3.2 yields

$$q_j \leqslant A \exp\Big(2^{j+1} - ARj^{-2} 2^{j/\alpha}\Big) + q_{j-1}.$$

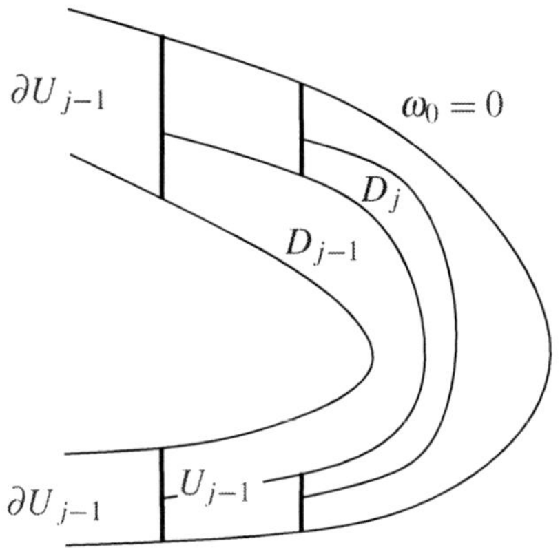

Fig. 2 Box argument

Since $0 < \alpha < 1$, it follows that $\sum_{j=1}^{\infty} \exp(2^{j+1} - ARj^{-2}2^{j/\alpha}) < \infty$, so that q_j is bounded. □

Proof of Theorem 1.3. In view of Theorem 2.3, it suffices to obtain the global Carleson estimate in terms of the Green function. Let K and F be disjoint compact sets with (2.1). Let U and ω_0 be the same as in Lemma 3.4. Since there is a distance between K and U, it is obvious that $G(x,y) \leqslant A$ uniformly for $x \in K \cap D$ and $y \in U \cap D$. For a while we fix $x \in K \cap D$. The maximum principle implies that

$$G(x,y) \leqslant A\omega_0(y) \quad \text{for } y \in U \cap D.$$

Combining with Lemma 3.4, we obtain

$$G(x,y) \leqslant AG(x_0,y) \quad \text{for } y \in F \cap D.$$

Thus, the Carleson estimate in terms of the Green function holds. □

Proof of Theorem 1.4

Let $0 < \beta < 1$. Because of the local nature of the boundary Harnack principle, we may assume that D is above the graph of a β-Hölder continuous function φ in $\mathbb{R}^{n-1}$. We may assume that $\|\varphi\|_\infty \leqslant 1$ and $\varphi(x) = 0$ for $|x| \geqslant 1$. For a point $x \in D$ we define $d(x) = x_n - \varphi(x')$, where $x = (x', x_n)$. Let x_0 be *high above* from the boundary. Suppose x is just below x_0 and $0 < d(x) < 1$. In general, $\delta_D(x) \approx d(x)$ does not hold. We can assert only that $\delta_D(x)^\beta \geqslant Ad(x)$. Considering the line segment connecting x and x_0, we obtain the following estimate of the quasihyperbolic metric:

$$k_D(x, x_0) \leqslant A \int_{d(x)}^{d(x_0)} \frac{dt}{t^{1/\beta}} \leqslant Ad(x)^{1-1/\beta} + A \tag{3.3}$$

(see Fig. 3). The same estimate holds if $x \in D$ is close to 0, say $|x| < 1$. Now

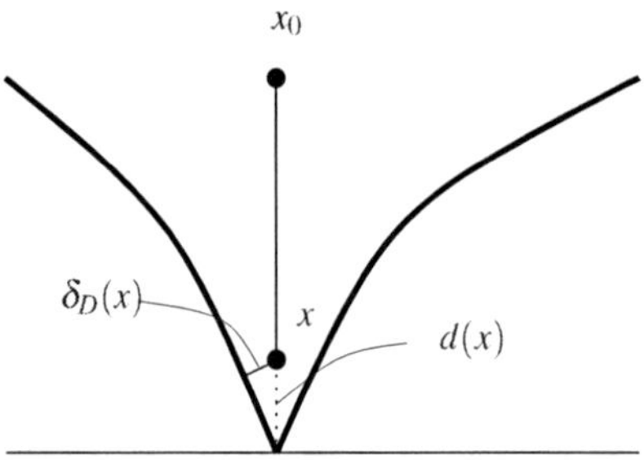

Fig. 3 Quasihyperbolic metric in a Hölder domain

we easily obtain a counterpart of Lemma 3.1.

Lemma 3.5. *Let D be as above. If $0 < r < 1$, then $w_\eta(\{x \in D : d(x) < r\}) \leqslant Ar^\beta$ for some $\eta > 0$.*

In view of Theorem 2.3, it suffices to obtain the global Carleson estimate in terms of the Green function. Let K and F be a pair of disjoint compact sets with (2.1). We may assume that F is included in the ball of radius 1 and center at the origin. As above, we put $U(r) = \{x \in \mathbb{R}^n : \mathrm{dist}(x, F) < r\}$ for $r > 0$. Let $2R = \mathrm{dist}(K, F)$. We put $U = U(R)$. It is obvious that U is an open set including F and yet its closure is apart from K. Let $\omega_0 = \omega(D \cap \partial U, D \cap U)$ be the harmonic measure of $D \cap \partial U$ in $D \cap U$. First, we compare ω_0 and $G(x_0, \cdot)$.

Lemma 3.6. *Let ω_0 be as above. Then*

$$\omega_0(y) \leqslant AG(x_0, y) \quad \textit{for } y \in F \cap D.$$

Proof. We employ the box argument again. Let $R_0 = R$, and let

$$R_j = \Big(1 - \frac{3}{\pi^2} \sum_{k=1}^{j} \frac{1}{k^2}\Big) R \quad \text{for } j \geqslant 1.$$

Then it is easy to see that $R_{j-1} - R_j = (3R)/(\pi^2 j^2)$, so that (3.2) holds. It is obvious that $G(x_0, \cdot)$ is bounded on $U \cap D$. Hence, by a suitable choice of

$A > 1$, we may assume that $u = G(x_0, \cdot)/A$ is bounded by 1 on $U \cap D$. Now let

$$U_j = \{y \in U(R_j) \cap D : 0 < u(y) < \exp(-2^j)\},$$
$$D_j = \{y \in U(R_j) \cap D : \exp(-2^{j+1}) \leqslant u(y) < \exp(-2^j)\}$$

for $j \geqslant 1$. Let $q_j = \sup_{D_j} \omega_0/u$ if $D_j \neq \varnothing$ and $q_j = 0$ if $D_j = \varnothing$. It is clear that q_j is finite. Since $0 < u < 1$ on $U \cap D$, from (3.2) it follows that $F \cap \partial D$ is included in the closure of $\bigcup_{j=1}^{\infty} D_j$. Hence, by the Harnack principle, it suffices to show that q_j is bounded.

By (3.3) and (3.1), we have

$$\exp(-2^j) > u(y) \geqslant \exp(-A_3(k_D(y, x_0) + 1)) \geqslant A \exp\big(-Ad(y)^{1-1/\beta}\big)$$

for $y \in U_j$. In other words, $d(y) \leqslant A2^{-j/(1/\beta - 1)}$ for $y \in U_j$. Hence Lemma 3.5 implies that $w_\eta(U_j) \leqslant A2^{-j/(1-\beta)}$. Applying the maximum principle on U_{j-1}, we obtain

$$\omega_0 \leqslant \omega(D \cap \partial U_{j-1} \setminus \overline{D_{j-1}}, U_{j-1}) + q_{j-1} u \quad \text{on } U_{j-1}.$$

We divide both sides by u and take the supremum over D_j. Then Lemma 3.2 yields

$$q_j \leqslant A \exp\left(2^{j+1} - ARj^{-2} 2^{j/(1-\beta)}\right) + q_{j-1}.$$

Since $0 < 1 - \beta < 1$, it follows that $\sum_{j=1}^{\infty} \exp(2^{j+1} - ARj^{-2}2^{j/(1-\beta)}) < \infty$, so that q_j is bounded. □

Proof of Theorem 1.4. In view of Theorem 2.3 , it suffices to show the global Carleson estimate in terms of the Green function. Let K and F be disjoint compact sets with (2.1). Let U and ω_0 be as in Lemma 3.6. Since there is a distance between K and U, it is obvious that $G(x, y) \leqslant A$ uniformly for $x \in K \cap D$ and $y \in U \cap D$. For a while we fix $x \in K \cap D$. The maximum principle implies that

$$G(x, y) \leqslant A\omega_0(y) \quad \text{for } y \in U \cap D.$$

Combining Lemma 3.6, we obtain

$$G(x, y) \leqslant AG(x_0, y) \quad \text{for } y \in F \cap D.$$

Thus the Carleson estimate in terms of the Green function holds. □

Acknowledgement. This work was partially supported by Grant-in-Aid for Exploratory Research (No. 19654023) Japan Society for the Promotion of Science.

References

1. Aikawa, H.: Boundary Harnack principle and Martin boundary for a uniform domain. J. Math. Soc. Japan **53**, no. 1, 119–145 (2001)
2. Aikawa, H.: Equivalence between the boundary Harnack principle and the Carleson estimate, Math. Scad. [To appear]
3. Ancona, A.: Principe de Harnack à la frontière et théorème de Fatou pour un opérateur elliptique dans un domaine lipschitzien. Ann. Inst. Fourier (Grenoble) **28**, no. 4, 169–213, x (1978)
4. Ancona, A.: First eigenvalues and comparison of Green's functions for elliptic operators on manifolds or domains. J. Anal. Math. **72**, 45–92 (1997)
5. Armitagem, D.H., Gardiner, S.J.: Classical Potential Theory. Springer, London (2001)
6. Bañuelos, R.: Intrinsic ultracontractivity and eigenfunction estimates for Schrödinger operators. J. Funct. Anal. **100**, no. 1, 181–206 (1991)
7. Bañuelos, R., Bass, R.F., Burdzy, K.: Hölder domains and the boundary Harnack principle. Duke Math. J. **64**, no. 1, 195–200 (1991)
8. Bass, R.F., Burdzy, K.: A boundary Harnack principle in twisted Hölder domains. Ann. Math. (2) **134**, no. 2, 253–276 (1991)
9. Bass, R.F., Burdzy, K.: Lifetimes of conditioned diffusions. Probab. Theory Related Fields **91**, no. 3-4, 405–443 (1992)
10. Caffarelli, L., Fabes, E., Mortola, S., Salsa, S.: Boundary behavior of nonnegative solutions of elliptic operators in divergence form. Indiana Univ. Math. J. **30**, no. 4, 621–640 (1981)
11. Dahlberg, B.E.J.: Estimates of harmonic measure. Arch. Ration. Mech. Anal. **65**, no. 3, 275–288 (1977)
12. Fabes, E.B., Garofalo, N., Salsa, S.: A backward Harnack inequality and Fatou theorem for nonnegative solutions of parabolic equations. Illinois J. Math. **30**, no. 4, 536–565 (1986)
13. Jerison, D.S., Kenig, C.E.: Boundary behavior of harmonic functions in nontangentially accessible domains. Adv. Math. **46**, no. 1, 80–147 (1982)
14. Smith, W., Stegenga, D.A.: Hölder domains and Poincaré domains. Trans. Am. Math. Soc. **319**, no. 1, 67–100 (1990)
15. Smith, W., Stegenga, D.A.: Exponential integrability of the quasi-hyperbolic metric on Hölder domains. Ann. Acad. Sci. Fenn. Ser. A I Math. **16**, no. 2, 345–360 (1991)
16. Wu, J.M.G.: Comparisons of kernel functions, boundary Harnack principle and relative Fatou theorem on Lipschitz domains. Ann. Inst. Fourier (Grenoble) **28**, no. 4, 147–167, vi (1978)

Sobolev Spaces and their Relatives: Local Polynomial Approximation Approach

Yuri Brudnyi

Dedicated to the memory of Sergey L'vovich Sobolev

Abstract The paper presents a survey of the theory of local polynomial approximation and its applications to the study of the classical spaces of smooth functions. The study includes such topics as embeddings and extensions, pointwise differentiability and Luzin type theorems, nonlinear approximation by piecewise polynomials and splines, and the real interpolation.

Introduction

The first part of the paper surveys topics in the theory of Local Polynomial Approximation developed in my work during the 1960–1969's. The results were announced on several occasions; the detailed account was then presented in two long paged manuscripts, of which only the second one was published [9]. The first one was, however, never published for reasons irrelevant to its content (see, for example, the letter of the leading American mathematicians [1] on the situation in Soviet Mathematics at that time). Some results and their generalizations of this manuscript have appeared in the subsequent publications of the author and his students and collaborators.

In this survey, I attempt to present the theory systematically restricting myself, however, to topics that can be described without a lot of technical details.

The remaining part of the paper is devoted to applications of the theory to the study of different properties of functions belonging to Sobolev spaces and other commonly used spaces of smooth functions (BV, Besov, Triebel–

Yuri Brudnyi
Technion – Israel Institute of Technology, Haifa 32000, Israel, e-mail: ybrudnyi@math.technion.ac.il

V. Maz'ya (ed.), *Sobolev Spaces in Mathematics II*, *International Mathematical Series*.
doi: 10.1007/978-0-387-85650-6, © Springer Science + Business Media, LLC 2009

Ligorkin, etc). The key point of this approach is the possibility to represent all of these spaces via behavior of local approximation of their members. Using such representations, we rather easily derive from the results of Local Polynomial Approximation theory extensive information on properties of the spaces. In particular, we discuss such topics as embeddings, pointwise differentiability and Luzin type "correction" theorems, extensions and traces, the real interpolation and nonlinear approximation by piecewise polynomials and wavelets. Some of these results are refined versions of those presented in [9], others are rather new and their proofs will appear elsewhere.

The relation of the present paper to Sobolev's scientific heritage is clear to the reader well acquainted with the theory of smooth spaces. But I have a personal motivation for publishing the paper in Sobolev's volume. By a lucky chance, I had a possibility to recognize Sobolev's attitude to an early version of the theory presented here. His support strongly influenced my scientific career.

Throughout the text, we freely use standard facts of smooth space theory. The reader can find required information in one of numerous books on the subject (see, in particular, [33, 47] for comprehensive study of Sobolev and BV space and the encyclopedic two-volume treatise [44, 45] containing, for example, the theory of Besov and Lizorkin–Triebel spaces).

1 Topics in Local Polynomial Approximation Theory

Local polynomial approximation of continuously differentiable functions is one of the cornerstones of classical analysis. Modern analysis deals with functions of much more complicated structure, such as functions of Sobolev and Lipschitz classes, fractal functions, etc. In this case, the main tool of the classical approach, Taylor approximation, cannot be used since Taylor polynomials may not exist or be unstable.

It would be natural to replace Taylor polynomials by those of *local best approximation*. The area investigating approximation of this kind may be conventionally called *local approximation theory*. However, it cannot be seen as a field of Approximation Theory for reasons indicated below. To this end, we introduce several basic notions of local approximation theory.

Definition 1.1. Let $f \in L_p^{\mathrm{loc}}(\mathbb{R}^n)$, $0 < p \leqslant \infty$. *Local approximation* of f of order $k \in \mathbb{Z}_+$ is a set-function given for a measurable set $S \subset \mathbb{R}^n$ by

$$E_k(S\,;\, f\,;\, L_p) := \inf_m \|f - m\|_{L_p(S)}, \tag{1.1}$$

where m runs over the space $\mathcal{P}_{k-1}(\mathbb{R}^n)$ of polynomials of degree $k-1$.

Note that order of approximation differs from degree of approximating polynomials by one. In particular,

$$E_0(S\,;\,f\,;\,L_p) = \|f\|_{L_p(S)}$$

We denote by $P_k(S\,;\,f\,;\,L_p)$ an optimal for (1.1) polynomial called a *polynomial of best approximation.* If $p \notin (1,\infty)$, such a polynomial is not unique and we choose one of them arbitrarily.

Hence, unlike Approximation Theory where approximation is done on a *fixed* domain by polynomials of *increasing* degree, we do it on a *variable* set by polynomial of *fixed* (maybe small) degree and study the behavior of best approximation and relative approximating polynomials as a function of the position and size of the variable set. This simple change of the point leads, however, to the essential transformation of the subject and methods of the theory.

The following example [5], a local version of the Markov–Bernstein inequality, demonstrates the interaction of the basic concepts of Approximation Theory and geometric measure theory.

Let V be a convex body in $\mathbb{R}^n$, and let S be its subset whose Hausdorff ℓ-measure satisfies

$$\mathcal{H}_\ell(S) \geqslant \lambda |V|^{\frac{\ell}{n}} \tag{1.2}$$

for some $\lambda \in (0,1)$. Hereafter, $|\Omega|$ stands for the Lebesgue n-measure of a set $\Omega \subset \mathbb{R}^n$.

Theorem 1.2. *Assume that* (1.2) *holds with* $n-1 < \ell \leqslant n$. *Given* $0 < p, q \leqslant \infty$, $k \in \mathbb{N}$, *and* $\alpha \in \mathbb{Z}^n_+$, *there exists a constant*[1] $c = c(k,n,p,q,\lambda) > 1$ *such that for every polynomial* $m \in \mathcal{P}_k(\mathbb{R}^n)$

$$\left\{\frac{1}{|V|}\int\limits_V |D^\alpha m|^p dx\right\}^{\frac{1}{p}} \leqslant c(\operatorname{diam} V)^{-|\alpha|} \cdot \left\{\frac{1}{\mathcal{H}_\ell(S)}\int\limits_S |m|^q d\mathcal{H}_\ell\right\}^{\frac{1}{q}}. \tag{1.3}$$

Remark 1.3. (a) The result is clearly untrue for $\ell \leqslant n-1$.

(b) The constant c is increasing in $\frac{1}{\lambda}$.

The sharp asymptotic of c as $\lambda \to 0$ and $k, n \to \infty$ is known for $\ell = n$; in particular, $c(\lambda) \approx \lambda^{-k}$ in this case (see [16, 23]).

The basic properties of local approximation regarded as a set-function is described by the following assertions.

Theorem 1.4. (Continuity) *Assume that a sequence* $\{S_j\}$ *of measurable subsets in a fixed n-cube converges in measure. Then for* $0 < p < \infty$

$$\lim E_k(S_j\,;\,f\,;\,L_p) = E_k(\lim S_j\,;\,f\,;\,L_p).$$

---

[1] The notation $c = c(x, y \ldots)$ throughout the paper means that the constant c depends *only* on the parameters in the brackets.

For $p = \infty$ the same is true for every nondecreasing sequence of measurable subsets of the cube.

(Partial subadditivity) *Assume that subsets $S_1, S_2 \subset \mathbb{R}^n$ satisfy for some $\lambda \in (0,1)$*

$$\frac{|S_1 \cap S_2|}{|\mathrm{conv}\,(S_1 \cup S_2)|} \geqslant \lambda.$$

Then there exists a constant $c = c(n,k,p^*)$, where hereafter $p^* := \min(1,p)$, such that

$$E_k\big(S_1 \cup S_2\,;\, f\,;\, L_p\big) \leqslant c\lambda^{-k+1} \sum_{i=1,2} E_k(S_i\,;\, f\,;\, L_p).$$

(Pointwise convergence) *Let $S \subset \mathbb{R}^n$ be measurable and $0 < p < \infty$. Then for almost all $x_0 \in S$*

$$\lim_{Q \to x_0} P_k(Q \cap S\,;\, f\,;\, L_p) = f(x_0).$$

(Linearization) *Let $S \subset \mathbb{R}^n$ be a bounded measurable subset of positive n-measure, and let $1 \leqslant p \leqslant \infty$. Then there exists a projection $T_k(S\,;\, L_p)$ from $L_p(S)$ onto the subspace $\mathcal{P}_{k-1}(\mathbb{R}^n)\big|_S$ with a norm bounded by $\sqrt{\dim \mathcal{P}_{k-1}(\mathbb{R}^n)}$. In particular,*

$$\big\|f - T_k(S\,; f;\, L_p)\big\|_{L_p(S)} \leqslant c(k,n) E_k(S\,;\, f\,;\, L_p). \tag{1.4}$$

Remark 1.5. (a) All but the last assertions are direct consequences of the inequality (1.3) (see, for example, [7]). The last assertion follows from the Kadets–Snobar theorem [31].

(b) Using the inequality (1.3), one can show that the orthogonal projection $\widehat{T}_k(S) : L_1(S) \to \mathcal{P}_{k-1}(\mathbb{R}^n)\big|_S$ satisfies for every $1 \leqslant p \leqslant \infty$ the inequality (1.4) with the constant $c(k,n)\Big(\frac{|\mathrm{conv}\,(S)|}{|S|}\Big)^{k-1}$.

(c) Clearly, there is no such linear projections if $0 < p < 1$. Using the Brown–Lucier theorem [4], one can find a homogeneous nonlinear projection from $L_p(S)$ onto $\mathcal{P}_{k-1}(\mathbb{R}^n)\big|_S$ which is independent of p and satisfies (1.4) with the constant $c(k,n,p)\Big(\frac{|\mathrm{conv}\,(S)|}{|S|}\Big)^{k-1+\frac{1}{p}}$.

For the effective applications of geometric analysis methods and results the set-functions $S \mapsto E_k(S\,;\, f\,;\, L_p)$ should be restricted to a subclass of geometrically better organized measurable subsets (cubes, Ahlfors regular sets, etc.). In the sequel, we confine ourselves to the class $\mathcal{K}(\mathbb{R}^n)$ of cubes homothetic to the unit cube $Q_0 := [0,1]^n$. Then we denote by $Q_r(x)$ a closed cube from $\mathcal{K}(\mathbb{R}^n)$ of center x and side length $2r$ (or of *radius* r if we regard $Q_r(x)$ as an ℓ_∞^n-ball).

In the case of functions over a measurable subset $S \subset \mathbb{R}^n$, we deal with a subclass $\mathcal{K}(S)$ of $\mathcal{K}(\mathbb{R}^n)$ consisting of all cubes centered at S that do not contain S. Hence, $\mathcal{K}(S)$ consists of quasicubes $Q \cap S$ instead of cubes.

The following theorem, a variant of an extension result proved in [8] in a special case and generalized by Shvartsman [39], allows us to work in the sequel only with cubes.

Theorem 1.6. *Assume that a measurable subset $S \subset \mathbb{R}^n$ satisfies the condition*

$$|Q \cap S|/|Q| \geqslant \lambda \tag{1.5}$$

for all $Q \in \mathcal{K}(S)$ and some $\lambda \in (0,1)$. There exists an extension operator $T : L_p^{\mathrm{loc}}(S) \to L_p(\mathbb{R}^n)$, linear if $1 \leqslant p \leqslant \infty$ and homogeneous otherwise, such that for some constants $c = c(n,k,p^*)$ and $\gamma = \gamma(n) \in \left(0, \frac{1}{2}\right)$ and every cube $Q \in \mathcal{K}(S)$

$$E_k(\gamma Q\,;\, Tf\,;\, L_p) \leqslant c(k,n,p^*)\lambda^{-k+1} E_k(S \cap Q\,;\, f\,;\, L_p).$$

Hereafter, γQ is the γ-homothety of Q with respect to its center (so $|\gamma Q| = \gamma^n |Q|$).

This result allows us to restrict our consideration to the case of functions over $\mathbb{R}^n$ and to use $\mathcal{K}(\mathbb{R}^n)$ as a class of supports. To simplify the notation, we fix numbers $p \in (0, +\infty]$ and $k \in \mathbb{N}$ and write for $f \in L_p^{\mathrm{loc}}(\mathbb{R}^n)$ and $Q \in \mathcal{K}(\mathbb{R}^n)$

$$E(Q\,;\, f) := E_k(Q\,;\, f\,;\, L_p), \;\; P(Q\,;\, f) := P_k(Q\,;\, f\,;\, L_p).$$

In the subsequent text, it is also convenient to use a *normed* local approximation defined by

$$\mathcal{E}_k(S\,;\, f\,;\, L_p) := |S|^{-\frac{1}{p}} E_k(S\,;\, f\,;\, L_p).$$

As above, we shorten this notation for $f \in L_p^{\mathrm{loc}}(\mathbb{R}^n)$ and $Q \in \mathcal{K}(\mathbb{R}^n)$ by writing

$$\mathcal{E}(Q\,;\, f) := \mathcal{E}(Q\,;\, f\,;\, L_p).$$

It is the matter of definition to check that for $p \leqslant q$ and $k \geqslant \ell$,

$$\mathcal{E}(Q\,;\, f\,) := \mathcal{E}_k(Q\,;\, f\,;\, L_p) \leqslant \mathcal{E}(Q\,;\, f\,;\, L_q). \tag{1.6}$$

Our following results give a partial conversion of this inequality.

We begin with a conversion in k which may be seen as a local version of the so-called Marchaud inequality (see, for example, [43, Chapt. 2]).

Theorem 1.7. *Given $f \in L_p^{\mathrm{loc}}(\mathbb{R}^n)$, $Q = Q_r(x)$ and an integer $\ell \in [0, k)$, there exists a constant $c = c(k, n, p^*)$ such that*

$$\mathcal{E}_\ell(r, x;\, f) \leqslant cr^\ell \left\{ \int_r^\infty \left(\frac{\mathcal{E}_k(t\,; x\,; f)}{t^\ell} \right)^{p^*} \frac{dt}{t} \right\}^{\frac{1}{p^*}},$$

where we set for brevity

$$\mathcal{E}_k(r\,; x\,; f) := \mathcal{E}_k(Q_r(x)\,; f\,; L_p).$$

Clearly, there is not direct conversions of (1.6) in p, but the results of this kind do exist for some regularizations of normed local approximation. The first of them, *q-average* ($q \geqslant p$) of local approximation, is a function $E^{\langle q \rangle} : \mathbb{R}_+ \times \mathcal{K}(\mathbb{R}^n) \times L_p^{\mathrm{loc}}(\mathbb{R}^n) \to \mathbb{R}_+$ given by

$$E^{\langle q \rangle}(t\,;\, Q\,;\, f) := \sup_\pi \left\{ \sum_{K \in \pi} E(K\,;\, f)^q \right\}^{\frac{1}{q}}, \tag{1.7}$$

where π runs over all disjoint families of congruent subcubes $K \subset Q$ of volume $\min\{(2t)^n, |Q|\}$.

Hence $E^{\langle q \rangle}$ is nondecreasing in the first two arguments and is a norm (quasinorm for $p < 1$) in f on $L_p(Q)/\mathcal{P}_{k-1}(\mathbb{R}^n)$. Using the Besicovich covering lemma (see, for example, [28, Theorem 1.1]) and partial subadditivity of local approximation, one can show that $E^{\langle q \rangle}$ also satisfies the Δ_2-condition [2] in t.

The same argument leads to equivalence of $E^{\langle q \rangle}$ to an *integral q-average* of local approximation, a function $\widehat{E}^{\langle q \rangle}$ of the same arguments as $E^{\langle q \rangle}$ given by

$$\widehat{E}^{\langle q \rangle}(t\,;\, Q\,;\, f) := \left\{ \int_Q \mathcal{E}\big(Q_t(x); f\big)^q dx \right\}^{\frac{1}{q}}.$$

Actually, the following is true:

Proposition 1.8. *There exist constants $c_1, c_2 > 0$ depending only on k, n, p^* such that, for all arguments of $E^{\langle q \rangle}$,*

$$c_1 E^{\langle q \rangle}\Big(\frac{r}{2}\,;\, Q\,;\, f\Big) \leqslant \widehat{E}^{\langle q \rangle}(r\,;\, Q\,;\, f) \leqslant c_2 E^{\langle q \rangle}(r\,;\, Q\,;\, f).$$

The desired conversion of the inequality (1.6) in p can be derived from an estimate for the nonincreased rearrangement of the restriction $(f - P(Q\,;\, f))\big|_Q$. Denoting this rearrangement by f_Q^*, we then have the following result [13, Appendix II].

[2] recall that $\varphi : \mathbb{R}_+ \to \mathbb{R}_+$ satisfies this condition if $\sup\limits_{t>0} \frac{\varphi(2t)}{\varphi(t)} < \infty$.

Theorem 1.9. *There exists a constant $c = c(k, n, p^*)$ such that for $0 < t \leqslant |Q|^{\frac{1}{n}}$*

$$f_Q^*(t^n) \leqslant c \int\limits_t^{|Q|^{\frac{1}{n}}} \frac{E^{\langle p \rangle}(s\,;\,Q\,;\,f)}{s^{\frac{n}{p}}}\,\frac{ds}{s}\,. \tag{1.8}$$

In turn, $E^{\langle q \rangle}$ computed in the L_p and L_q norms with $p < q \leqslant \infty$ satisfies for $0 < t < |Q|^{\frac{1}{n}}$

$$E^{\langle q \rangle}(t\,;\,Q\,;\,f\,;L_q) \leqslant c \left\{ \int\limits_0^t \left[\frac{E^{\langle q \rangle}(s\,;Q\,;f)}{s^{\sigma}} \right]^{p^*} \frac{ds}{s} \right\}^{\frac{1}{p^*}},$$

where $c = c(k, n, p^*, q)$ and

$$\sigma := n\left(\frac{1}{p} - \frac{1}{q}\right).$$

As a consequence, we obtain a partial conversion of the inequality (1.6) given by

Corollary 1.10. *There exists a constant $c = c(k, n, p^*, q)$ such that for $p < q \leqslant \infty$ and for every cube $Q := Q_r(x)$*

$$\mathcal{E}(Q\,;\,f\,;\,L_q) \leqslant c \left\{ \int\limits_0^r \left(\frac{E^{\langle p \rangle}(s\,;\,Q\,;\,f)}{s^{\sigma}} \right)^q \frac{ds}{s} \right\}^{\frac{1}{q}}.$$

Remark 1.11. Apparently, the inequality (1.8) holds for $E^{\langle q \rangle}$ with $q > p$ substituted for $E^{\langle p \rangle}$. Note, however, that for $q = \infty$ this result would imply the John–Nirenberg exponential estimate [30] for BMO-functions defined over L_p with $p < 1$. The latter is true, but requires a much more elaborate argument than that applied in the aforementioned paper, see [42].

Another kind of regularization for local approximation, *weighted q-variation*, is defined as follows.

Let $\omega : \mathbb{R}_+ \to \mathbb{R}_+$ be a *k-majorant*, i.e., a nondecreasing function such that $t \mapsto \omega(t)/t^k$, $t > 0$, is nonincreasing (in particular, ω may be constant). Then the *(q, ω)-variation* ($q \geqslant p$) of local approximation is a function $E^{\langle q,\omega \rangle} : \mathcal{K}(\mathbb{R}^n) \times L_p^{\mathrm{loc}}(\mathbb{R}^n) \to \mathbb{R}_+$ given by

$$E^{\langle q,\omega \rangle}(Q\,;\,f) := \sup_{\pi} \left\{ \sum_{K \in \pi} |K| \left(\frac{\mathcal{E}(K\,;\,f)}{\omega(|K|^{\frac{1}{n}})} \right)^q \right\}^{\frac{1}{q}}, \tag{1.9}$$

where π runs over all disjoint families of subcubes in Q.

Clearly, (q,ω)-variation relates to q-average by the inequality

$$E^{\langle q\rangle}(r\,;\,Q\,;\,f) \leqslant \omega(r)E^{\langle q,\omega\rangle}(Q\,;\,f).$$

An upper bound for (q,ω)-variation may be obtained using a kind of the weighted Hardy–Littlewood maximal operator given for $Q \in \mathcal{K}(\mathbb{R}^n)$, $f \in L_p^{\mathrm{loc}}(\mathbb{R}^n)$, and $x \in Q$ by

$$E^{\#,\omega}(x\,;\,Q\,;\,f) := \sup_{Q\supset K\ni x} \frac{\mathcal{E}(K\,;\,f)}{\omega(|K|^{\frac{1}{n}})}. \tag{1.10}$$

In the case $\omega(r) = r^\lambda$, $0 \leqslant \lambda < 1$, this notion was introduced and studied by Calderón and Scott [20]; the case of arbitrary $\lambda \in [0,k]$ was then investigated in [26].

It is the matter of definition to check that

$$E^{\langle q,\omega\rangle}(Q\,;\,f) \leqslant \big\|E^{\#,\omega}(\cdot\,;\,Q\,;\,f)\big\|_{L_q(Q)}.$$

Now, we present the second partial inversion of the inequality (1.6). To this end, we use the following analog of the rearrangement inequality of Theorem 1.9, see [9, Sect. 2, Theorem 2].

Theorem 1.12. *Assume that* $1 \leqslant p \leqslant q \leqslant \infty$*. There exists a constant* $c = c(k,n)$ *such that for* $0 < t \leqslant |Q|^{\frac{1}{n}}$

$$f_Q^*(t^n) \leqslant c\Bigg(\int\limits_t^{|Q|^{\frac{1}{n}}} \frac{\omega(s)}{s^{\frac{n}{q}}}\,\frac{ds}{s}\Bigg)E^{\langle q,\omega\rangle}(Q\,;\,f).$$

Exploiting this result for $\omega := \omega_\lambda$ defined by $\omega_\lambda(t) := r^\lambda$, we obtain the following conversion of the inequality (1.6).

Corollary 1.13. *Assume that* $\lambda \in (0,k]$ *and* $1 \leqslant p < q \leqslant +\infty$ *satisfy the inequality*

$$\frac{\lambda}{n} > \frac{1}{p} - \frac{1}{q}.$$

There exists a constant $c = c(k,n,p,q)$ *such that for any* $Q \in \mathcal{K}(\mathbb{R}^n)$ *and* $f \in L_p^{\mathrm{loc}}(\mathbb{R}^n)$,

$$\mathcal{E}(Q\,;\,f\,;\,L_q) \leqslant c|Q|^{\frac{\lambda}{n}} \sup_{K\subset Q} \frac{\mathcal{E}(K\,;\,f)}{|K|^{\frac{\lambda}{n}}}. \tag{1.11}$$

Remark 1.14. In the case of $q \in (p, +\infty)$ satisfying $\frac{\lambda}{n} = \frac{1}{p} - \frac{1}{q}$, the inequality (1.11) holds if the L_q-norm is replaced by the Marcinkiewicz–Lorentz norm $L_{q\infty}$.

In the remaining case of $\frac{\lambda}{n} = \frac{1}{p}$, i.e., $q = \infty$, the inequality holds with the BMO norm substituted for that of L_∞.

Now, we discuss local analogs of the classical Jackson and Bernstein theorems that relate the degree of approximation of functions to their smoothness properties (see, for example, [43]). Smoothness in these results is measured by the number of derivatives and behavior of moduli of continuity for the higher derivatives. The latter concept is recalled to be introduced for a p-integrable function of f over a *domain* [3] $G \subset \mathbb{R}^n$ and for $0 < t < \frac{\operatorname{diam} G}{k}$ by

$$\omega_k(t\,;\, f)_{L_p(G)} := \sup_{|h| \leqslant t} \|\Delta_h^k f\|_{L_p(G_{kh})}; \tag{1.12}$$

here $|h|$ stands for the standard Euclidean norm of $h \in \mathbb{R}^n$, and for $y \in \mathbb{R}^n$,

$$G_y := \{x \in G\,;\, [x, x+y] \subset G\}.$$

A local analog of this notion, (k,p)-*oscillation*, is a function over $\mathcal{K}(\mathbb{R}^n)$ given by

$$\operatorname{osc}_k(Q\,;\, f\,;\, L_p) := \sup_h \|\Delta_h^k f\|_{L_p(Q_{kh})}, \tag{1.13}$$

where h runs over all vectors of norm at most $\frac{\operatorname{diam} Q}{k}$.

In some applications, the following generalization of this notion is of use. Let μ be a Borel measure on $[0,1]$ whose support consists of a finite number of points and such that

$$\int t^j d\mu = \begin{cases} 0, & 0 \leqslant j \leqslant k-1, \\ \neq 0, & j = k. \end{cases}$$

Then we define a μ-*difference* (or order k) acting on functions f over $\mathbb{R}^n$ by

$$\mu_h f(x) := \int f(x+sh)\,d\mu(s).$$

If, for example, we take $\mu = \sum\limits_{j=0}^{k} (-1)^{k-j} \binom{k}{j} \delta_{\frac{j}{k}}$, then μ_h coincides with the standard k-difference $\Delta^k_{\frac{h}{k}}$.

Now, μ-*oscillation* is defined similarly to that in (1.13):

[3] i.e., an open connected subset

$$\operatorname{osc}_{\mu}(Q\,;\,f\,;\,L_p) := \sup_{|h|\leqslant \operatorname{diam} Q} \|\mu_h f\|_{L_p(Q_h)}. \tag{1.14}$$

The following result relats this characteristic to local approximation.

Theorem 1.15. *There exist constants $c_1, c_2 > 0$ depending only on k, n, p^* such that for all $Q \in \mathcal{K}(\mathbb{R}^n)$ and $f \in L_p^{\mathrm{loc}}(\mathbb{R}^n)$, $1 \leqslant p \leqslant \infty$,*

$$c_1 E_k(Q\,;\,f\,;\,L_p) \leqslant \operatorname{osc}_{\mu}(Q\,;\,f\,;\,L_p) \leqslant c_2 E_k(Q\,;\,f\,;\,L_p). \tag{1.15}$$

For univariate functions in $C[0,1]$ and $\mu = \Delta^k_{\frac{1}{k}}$ the left inequality was due to Whitney [46]. The general result was proed by the author [6, 7]. A nonconstructive proof for $p \geqslant 1$ is presented in [13, Appendix I].

As a consequence, we obtain a generalization of the classical Poincaré–Sobolev inequality.

Corollary 1.16. *Let f belong to the Sobolev space $W_p^\ell(Q)$, $0 \leqslant \ell < k$. Assume that $1 \leqslant p < q \leqslant \infty$ and these exponents and ℓ satisfy*

$$\sigma := \frac{\ell}{n} - \frac{1}{p} + \frac{1}{q} \geqslant 0$$

for all q except $q = \infty$. In the latter case, assume that $\sigma > 0$. There exists a constant $c = c(k, n, p, q)$ such that

$$\mathcal{E}_k(Q\,;\,f\,;\,L_q) \leqslant c|Q|^{\frac{\ell}{n}} \sup_{|\alpha|=\ell} \mathcal{E}_{k-\ell}(Q\,;\,D^\alpha f\,;\,L_p). \tag{1.16}$$

Using the q-average of local approximation (see (1.7)), we obtain a partial conversion of (1.16).

Theorem 1.17. *Let $f \in L_p(Q)$, $1 \leqslant p \leqslant \infty$, and an integer $\ell \in [1, k)$ and exponent $q \in [p, +\infty]$ be given. Assume that*

$$I(t\,;\,f) := \int_0^t \frac{E^{\langle q\rangle}(s\,;Q\,;f\,;L_p)}{s^{\ell+1}} < \infty$$

for some $t > 0$. Then f belongs to the Sobolev space $W_p^\ell(Q)$ and its higher derivatives satisfy for some constant $c = c(k, n, p, q)$

$$\max_{|\alpha|=\ell} E_{k-\ell}(Q\,;\,D^\alpha f\,;\,L_q) \leqslant cI\big(|Q|^{\frac{1}{n}}\,;\,f\big).$$

Remark 1.18. The last two results are true for a wider class of domains than cubes. In particular, they hold for uniform domains, also known as (ε, δ)-domains (see, for example, [29] for the definition).

Further, we present a global version of Theorem 1.15 proved in [9] for $p \geqslant 1$ and in [13, Appendix I] for $0 < p < 1$.

Theorem 1.19. *There are constants $c_1, c_2 > 0$ depending only on k, n, p^* such that for all $Q \in \mathcal{K}(\mathbb{R}^n)$, $f \in L_p^{\mathrm{loc}}(Q)$, and $0 < t \leqslant \operatorname{diam} Q$*

$$c_1 \omega_k(t\,; f)_{L_p(Q)} \leqslant E^{\langle p \rangle}(t\,; Q\,; f\, L_p) \leqslant c_2 \omega_k(t\,; f)_{L_p(Q)}.$$

As above, Q can be replaced by a bounded uniform domain. Moreover, k-modulus of continuity can be replaced by its analog ω_μ defined as in (1.12), but with μ-difference of order k substituted for Δ_h^k.

Finally, we discuss relations between Taylor's and best local approximations. To this end, we recall the definition of the Taylor classes introduced, in essence, by Calderón and Zygmund [21].

Definition 1.20. (a) Let λ be a positive nonintegral number. A function $f \in L_p^{\mathrm{loc}}(\mathbb{R}^n)$ belongs to a Taylor space $T_p^\lambda(x)$ if there exists a polynomial m_x of degree strictly less than λ such that for some constant c and all $0 < r \leqslant 1$

$$\left\{ \frac{1}{r^n} \int\limits_{|y| \leqslant r} \big| f(x+y) - m_x(y) \big|^p dy \right\}^{\frac{1}{p}} \leqslant c r^\lambda. \tag{1.17}$$

The infimum of c in this inequality is called a (quasi)norm of f in $T_p^\lambda(x)$.

(b) Let λ be a natural number. The Taylor space $T_p^\lambda(x)$ is defined as a real interpolation space given for some $\varepsilon \in (0,1)$ by

$$T_p^\lambda(x) := \big(T_p^{\lambda-\varepsilon}(x), T^{\lambda+\varepsilon}(x)\big)_{\frac{1}{2},\infty}$$

It can be shown that up to equivalence of norms the definition of $T_p^\lambda(x)$ with $\lambda \in \mathbb{N}$ is independent of ε.

Remark 1.21. (a) For $\lambda \in \mathbb{N}$ our definition of $T_p^\lambda(x)$ deviates from that in [21]. In fact, the latter exploits the inequality (1.17) for any λ to define such a space. The Calderón–Zygmund space with $\lambda \in \mathbb{N}$ is denoted by $\widehat{T}_p^\lambda(x)$.

Definition 1.22. A function $f \in L_p^{\mathrm{loc}}(\mathbb{R}^n)$ belongs to the Taylor space $t_p^\lambda(x)$ if there exists a polynomial m_x of degree less or equal to λ such that (1.17) holds with $o(r^\lambda)$ as $r \to 0$ substituted for cr^λ.

It can be shown that $t_p^\lambda(x)$ is a separable subspace of $T_p^\lambda(x)$ (or $\widehat{T}_p^\lambda(x)$).

The following two results describe relations between the Taylor and best local approximations (see [6, 10] and Appendix III in [13]).

Theorem 1.23. (a) *If $0 < \lambda < k$, then a function $f \in L_p^{\mathrm{loc}}(\mathbb{R}^n)$ belongs to $T_p^\lambda(x)$ if and only if*

$$\mathcal{E}_k(Q\,;f\,;L_p) = O\big(|Q|^{\frac{\lambda}{n}}\big) \quad as \quad Q \to x. \tag{1.18}$$

(b) *If* $\lambda = k$, *the condition* (1.18) *is necessary and sufficient for* f *to belong to* $\widehat{T}_p^k(x)$.

(c) *If* λ *is an integer and* $0 < \lambda < k$, *then* f *belongs to* $\widehat{T}_p^\lambda(x)$ *if and only if* (1.18) *holds and, moreover,*

$$\overline{\lim_{Q\to x}} \max_{|\alpha|=\lambda} \big|D^\alpha P_k(Q\,;f\,;L_p)\big| < \infty.$$

A variant of this result for the separable Taylor subspaces is as follows.

Theorem 1.24. (a) *If* $0 < \lambda < k$ *and* λ *is nonintegral, then a function* $f \in L_p^{\mathrm{loc}}(\mathbb{R}^n)$ *belongs to the space* $t_p^\lambda(x)$ *if and only if*

$$\mathcal{E}(Q\,;f\,;L_p) = o\big(|Q|^{\frac{\lambda}{n}}\big) \quad as \quad Q \to x. \tag{1.19}$$

(b) *If* $0 < \lambda \leqslant k$ *and* λ *is an integer, then* $f \in t_p^\lambda(x)$ *if and only if* (1.19) *holds and, moreover, the limits*

$$\lim_{Q\to x} D^\alpha P_k(Q\,;f\,L_p)$$

exist for all α *with* $|\alpha| = \lambda$.

In the case $\lambda = k$, the criterion of Theorem 1.24 (b) may be simplified if we are interested in the case for f belonging to $t_p^k(x)$ almost everywhere. Actually, the following is true.

Theorem 1.25. *A function* $f \in L_p^{\mathrm{loc}}(\mathbb{R}^n)$ *belongs to the space* $t_p^n(x)$ *for almost all* $x \in \mathbb{R}^n$ *if and only if for almost all* x

$$\overline{\lim_{Q\to x}} \frac{\mathcal{E}_k(Q\,;f\,L_p)}{|Q|^{\frac{k}{n}}} < \infty.$$

Using the above presented criteria, we give a one-point version of Theorem 1.17. To this end, we define *Peano derivatives* for a function $f \in L_p^{\mathrm{loc}}(\mathbb{R}^n)$.

Let f belong to the Taylor space $t_p^k(x_0)$, and let m_{x_0} be a polynomial of degree k satisfying Definition 1.22, i.e.,

$$\|f - m_{x_0}\|_{L_p(Q)} = o\big(|Q|^{\frac{k}{n}}\big) \quad \text{as} \quad Q \to x_0.$$

This polynomial is, clearly, unique. It is called a *Taylor polynomial* and is written in the form

$$T_{x_0}^k f(x) := \sum_{|\alpha|\leqslant k} \frac{f_\alpha(x_0)}{\alpha!}\,(x - x_0)^\alpha.$$

Then the number $f_\alpha(x_0)$ is said to be the *Peano mixed derivative* for f of order α at the point x_0 denoted by $D^\alpha f(x_0)$.

The notation is motivated by the following fact (see, for example, [21]).

If $f \in L_p^{\rm loc}(\mathbb{R}^n)$ belongs to $t_p^k(x)$ for almost all points of a domain and $p \geqslant 1$, then Peano derivatives for f of order at most k coincide with the corresponding weak L_p-derivatives almost everywhere in the domain.

Using this concept, we introduce Sobolev spaces over L_p with $p < 1$.

Definition 1.26. A function $f \in L_p^{\rm loc}(G)$, where $G \subset \mathbb{R}^n$ is a domain and $0 < p \leqslant \infty$, belongs to a homogeneous Sobolev space $\dot{W}_p^k(G)$ if $f \in t_p^k(x)$ for almost all $x \in G$ and its Peano derivatives of order k satisfy

$$|f|_{W_p^k(G)} := \max_{|\alpha|=k} \|D^\alpha f\|_{L_p(G)} < \infty.$$

According to the formulated fact, the space subject to this definition coincides with the classical space for $1 \leqslant p \leqslant \infty$.

The notion introduced allows us to extend Theorem 1.17 to the case of $0 < p < 1$ as follows.

Theorem 1.27. *Let $f \in L_p(Q)$, where $0 < p < 1$ and an integer $\ell \in [1, k)$ and exponent $q \in [p, +\infty]$, be given. Assume that*

$$I(t\,;f) := \left\{ \int\limits_0^t \left[\frac{E^{\langle q \rangle}(s\,;Q\,;f\,;L_p)}{s^{\ell+1}} \right]^p \frac{ds}{s} \right\}^{\frac{1}{p}} < \infty$$

for some $t > 0$. Then f belongs to $\dot{W}_p^\ell(Q)$ and its higher Peano derivatives satisfy for some constant $c = c(k, n, p, q)$

$$\max_{|\alpha|=\ell} E_{k-\ell}(Q\,;D^\alpha f\,;L_q) \leqslant cI\big(|Q|^{\frac{1}{n}}\,;f\big).$$

2 Local Approximation Spaces

We introduce a class of functional spaces defined by the behavior of local approximation of their members. For this purpose, we need the following definition.

Definition 2.1. Let μ be a Borel measure on $\mathcal{K}(\mathbb{R}^n)$. A functional parameter is a quasi-Banach lattice of μ-measurable (classes of) functions of $\mathcal{K}(\mathbb{R}^n)$ satisfying the Faton property. That is to say, a functional parameter φ meets the following conditions:

(FP_1) If $f \in \varphi$ and $|g| \leqslant |f|$ μ-almost everywhere, then $g \in \varphi$ and

$$|g|_\varphi \leqslant \|f\|_\varphi.$$

$(FP_2$ For some constant $c = c(\varphi) \geqslant 1$ and all $f, g \in \varphi$,

$$\|f + g\|_\varphi \leqslant c\{\|f\|_\varphi + \|f\|_\varphi\}.$$

(FP_3) φ is complete.

(FP_4) Every bounded in φ sequence $\{f_j\}$ converging in measure to a function f satisfies the Faton inequality

$$\|f\|_\varphi \leqslant \varliminf_{j\to\infty} \|f_j\|_\varphi.$$

The following classes of function parameter will be used below.

Example 2.2. Identifying $\mathcal{K}(\mathbb{R}^n)$ with the open half-space $\mathbb{R}^{n+1}_+ := \{(r, x);\ r > 0,\ x \in \mathbb{R}^n\}$ by the bijection $Q_r(x) \mapsto (r, x)$, we introduce a Borel measure on $\mathcal{K}(\mathbb{R}^n)$, denoted by dQ, by setting

$$dQ := dx \otimes r^{-1} dr.$$

Then a functional parameter $L_q^\lambda(L_p)$, where $0 < p, q \leqslant \infty$, $\lambda \in \mathbb{R}$, is defined by the quasinorm

$$\|f\|_{L_q^\lambda(L_p)} := \left\{ \int\limits_{\mathbb{R}_+} \left(\int\limits_{\mathbb{R}^n} \left| \frac{f(r,x)}{r^\lambda} \right|^p dx \right)^{\frac{q}{p}} \frac{dr}{r} \right\}^{\frac{1}{q}}. \tag{2.1}$$

Example 2.3. Changing the order of integration in (2.1), we define the functional parameter $L_p(L_q^\lambda)$:

$$\|f\|_{L_p(L_q^\lambda)} := \left\{ \int\limits_{\mathbb{R}^n} \left(\int\limits_{\mathbb{R}_+} \left| \frac{f(r,x)}{r^\lambda} \right|^q \frac{dr}{r} \right)^{\frac{p}{q}} dx \right\}^{\frac{1}{p}}. \tag{2.2}$$

Example 2.4. We fix a point $x_0 \in \mathbb{R}^n$ and define a Borel measure $\mathcal{K}(\mathbb{R}^n)$ to be $\delta_{x_0} \otimes r^{-1} dr$. A (functional) parameter $L_p^\lambda(x_0)$ is then defined by the quasinorm

$$\|f\|_{L_p^\sigma(x_0)} := \left\{ \int\limits_{\mathbb{R}_+} \left| \frac{f(r,x_0)}{r^\sigma} \right|^p \frac{dr}{r} \right\}^{\frac{1}{p}}. \tag{2.3}$$

For $p = \infty$ we use a separable subspace of $L_\infty^\sigma(x_0)$ denoted by $\ell_\infty^\sigma(x_0)$ defined by the condition

$$\lim_{r\to 0}\frac{f(r,x_0)}{r^\sigma}=0. \tag{2.4}$$

Example 2.5. Let μ be the counting measure on $\mathcal{K}(\mathbb{R}^n)$. A parameter V_p^σ is defined by the quasinorm

$$\|f\|_{V_p^\sigma} := \sup_\pi\left\{\sum_{Q\in\pi}|Q|\left|\frac{f(Q)}{|Q|^{\frac{\sigma}{n}}}\right|^p\right\}^{\frac{1}{p}}, \tag{2.5}$$

where π runs over all disjoint families of cubes from $\mathcal{K}(\mathbb{R}^n)$.

Replacing in (2.5) $|Q|^{\frac{\sigma}{n}}$ by $\omega\big(|Q|^{\frac{1}{n}}\big)$, where $\omega:\mathbb{R}_+\to\mathbb{R}_+$ is a k-majorant, we define the parameter $V_p^{(\omega)}$.

We also use a separable subspace of the space V_p^σ denoted by v_p^σ that is defined by the condition

uniformly in $Q\in\mathcal{K}(\mathbb{R}^n)$,

$$\lim_{|Q|\to 0}\sup_{K\in\pi}\left\{\sum_{K\in\pi}|K|\left|\frac{f(K)}{|K|^{\frac{\sigma}{n}}}\right|^p\right\}^{\frac{1}{p}}=0;$$

here π runs over all disjoint families of cubes in Q.

In the sequel, we extend Definition 2.1 by exploiting functional parameters over the set

$$\mathcal{K}(G) := \big\{Q_r(x)\in\mathcal{K}(\mathbb{R}^n)\,;x\in G\ \text{ and }\ 2r<\operatorname{diam}G\big\}. \tag{2.6}$$

In some situations, one can also use a subset of $\mathcal{K}(G)$ defined by

$$\mathcal{K}_{\text{int}}(G) := \big\{Q\in\mathcal{K}(G)\,;Q\subset G\big\}. \tag{2.7}$$

In the latter case, G is assumed to be the closure of a domain $G\subset\mathbb{R}^n$.

It can be easily seen that all the spaces introduced are functional parameters, i.e., satisfy axioms (PF_1)–(PF_3).

Now, we introduce the main object of this section, the *local approximation space.*

Let φ be a parameter over $\mathcal{K}(\mathbb{R}^n)$, and let $0<p\leqslant\infty$ and $k\in\mathbb{N}$ be fixed.

Definition 2.6. A homogeneous local approximation space $\dot{\mathcal{A}}_\varphi^k(L_p)$ consists of functions $f\in L_p^{\text{loc}}(\mathbb{R}^n)$ such that the seminorm

$$|f|_{\mathcal{A}_\varphi^k(L_p)} := \big\|\mathcal{E}_k(\cdot\,;f\,;L_p)\big\|_\varphi$$

is finite.

A (nonhomogeneous) local approximation space $\mathcal{A}_\varphi^k(L_p)$ is then defined by the quasinorm

$$\|f\|_{\mathcal{A}^k_\varphi(L_p)} := |f|_{\mathcal{A}^k_\varphi(L_p)} + \|f\|_{L_p(Q_0)},$$

where $Q_0 := [0,1]^n$.

In the sequel, we encounter also spaces $\dot{\mathcal{A}}^k_\varphi(X)$ obtained by replacing $L_p^{\mathrm{loc}}(\mathbb{R}^n)$ by other functional spaces of locally integrable functions on $\mathbb{R}^n$, for example, X may be the Lorentz space $L_{pq}^{\mathrm{loc}}(\mathbb{R}^n)$, $BMO(\mathbb{R}^n)$, or $C^{\mathrm{loc}}(\mathbb{R}^n)$ (the space of locally bounded continuous functions on $\mathbb{R}^n$).

We also exploit local approximation spaces over measurable subsets $G \subset \mathbb{R}^n$. For their definition we use the trace space[4] $\varphi|_G$ of a functional parameter φ and modify normed local approximation by setting for $Q \in \mathcal{K}(G)$,

$$\mathcal{E}_k(Q\,;f\,;L_p) := |Q|^{-\frac{1}{p}} E_k(Q\cap G\,;f\,;L_p)$$

i.e., replacing, for an evident reason, the factor $|Q\cap G|^{-\frac{1}{p}}$ of the previous definition.

Then the local approximation space $\dot{\mathcal{A}}^k_\varphi(L_p, G)$ is defined by the seminorm

$$|f|_{\dot{\mathcal{A}}^k_\varphi(L_p,G)} := \big\|\mathcal{E}_k(\cdot\,;f\,;L_p)\big\|_{\varphi|_G}.$$

If G is a domain we also introduce a subspace $\dot{\mathcal{A}}^k_\varphi(L_p\,;G)_{\mathrm{int}}$ replacing $\mathcal{K}(G)$ by the subset $\mathcal{K}_{\mathrm{int}}(G)$.

Finally, we introduce subspaces of local approximation spaces whose members possess Peano derivatives. In fact, given a parameter φ and an integer $\ell \in \mathbb{N}$, a space $W^\ell \dot{\mathcal{A}}^k_\varphi(L_p)$ consists of functions $f \in L_p^{\mathrm{loc}}(\mathbb{R}^n)$ whose Peano derivatives $D^\alpha f$ with $|\alpha| \leqslant \ell$ exist almost everywhere and satisfy

$$|f|_{W^k\mathcal{A}^k_\varphi(L_p)} := \max_{|\alpha|=\ell} \big\|\mathcal{E}_k(\cdot\,;D^\alpha f\,;L_p)\big\|_\varphi < \infty.$$

Theorem 2.7. (a) *The spaces $\dot{\mathcal{A}}^k_\varphi(L_p)/\mathcal{P}_{k-1}(\mathbb{R}^n)$ and $\mathcal{A}^k_\varphi(L_p)$ are quasi-Banach (Banach, if φ and L_p are).*

(b) *Let parameters φ, ψ over $\big(\mathcal{K}(\mathbb{R}^n), \mu\big)$ satisfy $\varphi \hookrightarrow \psi$ (continuous embedding), and $\ell \geqslant k$, $p \geqslant q$. Then the following holds:*

$$\dot{\mathcal{A}}^k_\varphi(L_p) \hookrightarrow \dot{\mathcal{A}}^\ell_\psi(L_q)/\mathcal{P}_{k-1}(\mathbb{R}^n). \tag{2.8}$$

(c) *Let (φ_0, φ_1) be a quasi-Banach couple of parameters over $\big(\mathcal{K}(\mathbb{R}^n), \mu\big)$. Then for the real interpolation space of the corresponding local approximation spaces we have*

$$\big(\dot{\mathcal{A}}^k_{\varphi_0}(L_{p_0}),\, \dot{\mathcal{A}}^k_{\varphi_1}(L_{p_1})\big)_{\theta q} \hookrightarrow \dot{\mathcal{A}}^k_\varphi(L_{p_\theta q}), \tag{2.9}$$

---

[4] Recall that the trace of a normed functional space X over $\mathbb{R}^n$ to a subset G is equipped by the canonical trace norm given for $f \in X|_G$ by $\|f\|_{X|_G} := \inf\{\|\varphi\|_X\,;\varphi|_G = f\}$.

where $\varphi := (\varphi_0, \varphi_1)_{\theta q}$ *and*

$$\frac{1}{p_\theta} := \frac{1-\theta}{p_0} + \frac{\theta}{p_1}.$$

(d) *If the local approximation spaces on the left-hand side of* (2.9) *are Banach spaces, then for the complex interpolation of this couple we have*

$$\left[\dot{\mathcal{A}}^k_{\varphi_0}(L_{p_0}), \dot{\mathcal{A}}^k_{\varphi_1}(L_{p_1})\right]_\theta \hookrightarrow \dot{\mathcal{A}}^k_{\varphi}(L_{p_\theta}), \tag{2.10}$$

where $\varphi := \varphi_0^{1-\theta}\varphi_1^\theta$ *(the Calderón construction).*

We refer the reader to the books [2, 17] for information on the real- and complex interpolation and the Calderón construction.

One more result of this type is true for the variants of local approximation spaces introduced above.

Proposition 2.8. (a) *Let* $G \subset \mathbb{R}^n$ *be a set of positive measure. Then*

$$\dot{\mathcal{A}}^k_\varphi(L_p)/G \hookrightarrow \dot{\mathcal{A}}^k_\varphi(L_p\,; G).$$

If, in addition, G is a domain, then the space on the right-hand side can be replaced by its subspace $\dot{\mathcal{A}}^k_\varphi(L_p\,; G)_{\text{int}}$.

(b) *If* $1 \leqslant p \leqslant \infty$, *then*

$$W^\ell \dot{\mathcal{A}}^k_\varphi(L_p) \hookrightarrow \dot{\mathcal{A}}^{k+\ell}_{\varphi^\ell}(L_p),$$

where φ^ℓ *is a weighted space given by the quasinorm*

$$\|f\|_{\varphi^\ell} := \left\| \frac{f(Q)}{|Q|^{\frac{\ell}{n}}} \right\|_\varphi.$$

The basic problems of the theory of local approximation spaces concern the conversions of the above formulated assertions. In important special cases, such conversions hold under mild restrictions on parameters. Proofs of these results are easily derived from the corresponding properties of local approximation presented in Sect. 1. The results obtained in this way can be, in turn, applied to the study of the "classical" function spaces of smooth functions. The success of this approach is provided by the description of every "classical" space via the local approximation behavior of its members. This, in particular, gives a representation for such a space as a local approximation space determined by one of the functional parameters from Examples 2.2–2.5. In turn, properties of local approximation spaces derived from the results of Sect. 1 can be transferred to those of the spaces of smooth functions.

In accordance with the choice of parameters we divide all local approximation spaces being used into several families.

2.1 Λ-spaces

These spaces are determined by parameters of Example 2.2. Most of them coincide with Lipschitz spaces of different kind; this fact explains the choice of the name.

To introduce spaces of this series, we fix an integer $k \geqslant 1$, real number σ and exponents $0 < p \leqslant q \leqslant \infty$ and $0 < r \leqslant \infty$. Then a *homogeneous Λ-space* with these parameters is given by

$$\dot{\Lambda}_q^{\sigma r}(L_p) := \dot{\mathcal{A}}_\varphi^k(L_q) \quad \text{where} \quad \varphi := L_r^\lambda(L_q).$$

In particular,

$$|f|_{\Lambda_\infty^{\sigma\infty}(L_p)} = \sup_Q \frac{\mathcal{E}_k(Q\,;f\,;L_p)}{|Q|^{\frac{\sigma}{n}}}$$

for $0 < \sigma \leqslant k$ defines the space of functions satisfying the Lipschitz–Zygmund condition

$$\left|\Delta_h^k f(x)\right| = O\big(|h|^\sigma\big)$$

(see [22] for nonintegral σ and [6] for the general case).

The following result explains the lack of index k in the notation.

Proposition 2.9. *If $r < \infty$ and $\sigma \notin (0, k)$ or $r = \infty$ and $\sigma > k$, then*

$$\dot{\Lambda}_q^{\sigma r}(L_p) = \mathcal{P}_{k-1}(\mathbb{R}^n).$$

Remark 2.10. In the remaining case of $r = \infty$ and $\sigma \leqslant 0$, this space, up to factorization by $\mathcal{P}_{k-1}(\mathbb{R}^n)$ and equivalence of (quasi)norms, coincides with the Morrey space $\mathcal{M}^{-\frac{1}{p}-\frac{\sigma}{n}}(\mathbb{R}^n)$ if $-\frac{n}{p} < \sigma < 0$ and with the weighted L_p-space $L_p\big(|x|^{-\sigma-\frac{n}{p}}\,;\mathbb{R}^n\big)$ if $\sigma \leqslant -\frac{n}{p}$.

Let us recall that the Morrey space $\mathcal{M}_p^\lambda(\mathbb{R}^n)$ is defined by the (quasi)norm

$$\|f\|_{\mathcal{M}_p^\lambda(\mathbb{R}^n)} := \sup_Q |Q|^\lambda \left\{ \int_Q |f|^p dx \right\}^{\frac{\ell}{p}}.$$

The following result describes the main properties of Λ-spaces.

Theorem 2.11. (a) *If $0 \leqslant \sigma < \ell < k$, then, up to equivalence of (quasi)norms,*

$$\dot{\Lambda}_q^{\sigma r}(L_p) = \dot{\mathcal{A}}_\varphi^\ell(L_p)/\mathcal{P}_{k-1}(\mathbb{R}^n),$$

where $\varphi := L_r^\sigma(L_q)$.

(b) *If $1 \leqslant \ell < \sigma \leqslant k$, then*

$$\dot{\Lambda}_q^{\sigma r}(L_p) \hookrightarrow W^\ell \dot{\Lambda}_q^{\sigma-\ell}(L_p).$$

If $p \geqslant 1$, the converse continuous embedding is also true.

(c) *If $0 < \sigma < k$ and for some $p \in (0, +\infty]$ the parameters satisfy one of the conditions:*

(i) $\left|\frac{1}{p} - \frac{1}{q}\right| < \frac{\sigma}{n}$ *and* $0 < r \leqslant \infty$;

(ii) $\left|\frac{1}{p} - \frac{1}{q}\right| = \frac{\sigma}{n}$ *and* $r \leqslant \max\{p, q\}$,

then, up to equivalence of (quasi)norms,

$$\dot{\Lambda}_q^{\sigma r}(L_p) = \dot{\Lambda}_q^{\sigma r}(L_q).$$

(d) *Under the conditions of assertion* (c), *the Λ-space on the left-hand side coincides, up to equivalence of (quasi)norm, with the homogeneous Besov space $\dot{B}_q^{\sigma r}(\mathbb{R}^n)$.*

Here, the Besov space is defined by the seminorm

$$|f|_{B_q^{\sigma r}(\mathbb{R}^n)} := \left\{ \int\limits_{\mathbb{R}_+} \left(\frac{\omega_k(t\,; f\,; L_p)}{t^\sigma} \right)^r \frac{dt}{t} \right\}^{\frac{1}{r}}. \tag{2.11}$$

This definition is known to be equivalent, for $q \geqslant 1$, to the standard definition of the Besov space (see, for example, [45]). However, the (Peetre) definition of this space for $q < 1$ [35] differs from that given by k-modulus of continuity. Denoting the Peetre space by $\widehat{B}_q^{\sigma r}(\mathbb{R}^n)$ we have the following embedding:

$$\widehat{B}_q^{\sigma}(\mathbb{R}^n) \hookrightarrow B_q^{\sigma r}(\mathbb{R}^n)$$

which is proper only if $\frac{\sigma}{n} < \frac{1}{q} - 1$.

The family of Λ-spaces may be extended to negative values of σ for $p \geqslant 1$. To introduce the extension we need the notion of atoms,

A measurable function $a : \mathbb{R}^n \to \mathbb{R}$ is recalled to be a (k, σ, p)-*atom* if for some cube Q

(i) supp $a \subset Q$;

(ii) $\displaystyle \left\{ \frac{1}{|Q|} \int\limits_Q |a|^p dx \right\}^{\frac{1}{p}} \leqslant |Q|^{-\frac{\sigma}{n}}$;

(iii) for all indices $|\alpha| < k$

$$\int x^\alpha a(x) dx = 0.$$

Further, we say that a function $m : \mathbb{R}^n \to \mathbb{R}$ is a *(k,σ,p)-molecule of size* $t > 0$ if m is the linear combination

$$m = \sum_{Q \in \pi} c_Q a_Q, \tag{2.12}$$

where a_Q are a (k,σ,p)-atom supported by a cube Q of volume t^n.

Given presentation (2.12), for the molecule m we set

$$[m]_q := \Big\{ \sum_{Q \in \pi} |c_Q|^q \Big\}^{\frac{1}{q}}$$

and then define a space $\dot{\Lambda}_q^{-\sigma r}(L_p)$ for $0 < \sigma < k,\ 1 \leqslant p \leqslant \infty$ and $0 < q,\, r \leqslant \infty$ by the seminorm

$$|f|_{\Lambda_q^{-\sigma r}(L_p)} := \inf \Big\{ \sum_{j \in \mathbb{Z}} \big(2^{j\sigma} [m_j]_q\big)^r \Big\}^{\frac{1}{r}},$$

where the infimum is taken over all decompositions

$$f = \sum_{j \in \mathbb{Z}} m_j \quad \text{(convergence in the space } \mathcal{S}' \text{ of distributions)}$$

into the sum of (k,σ,p)-molecules m_j of size $2^j,\ j \in \mathbb{Z}$.

The construction of the space introduced allows us to check easily the duality result presented below. In its formulation, given $0 < s < \infty$, the conjugate exponent s' is defined by $\frac{1}{s} + \frac{1}{s'} = 1$ if $s \geqslant 1$ and $s' = +\infty$ if $s < 1$.

Theorem 2.12. *Assume that* $1 \leqslant p < \infty,\ 0 < q,\, r < \infty$, *and* $0 < \sigma < k$. *Then, up to equivalence of seminorms,*

$$\dot{\Lambda}_q^{-\sigma r}(L_p)^* = \Lambda_{q'}^{\sigma r'}(L_{p'}).$$

Finally, we formulate problems concerning stability of the Λ-family under the real or complex interpolation.

Problem 2.13. What conditions on parameters does the equality

$$\big(\dot{\Lambda}_{q_0}^{\sigma_0, r_0}(Lp_0),\ \dot{\Lambda}_{q_1}^{\sigma_1, r_1}(L_{p_1})\big)_{\theta q} = \dot{\Lambda}_q^{\sigma r}(L_{pq}) \tag{2.13}$$

provide? Here $\sigma := (1-\theta)\sigma_0 + \theta\sigma_1,\ \frac{1}{q} := \frac{1-\theta}{q_0} + \frac{\theta}{q_1}$ and $\frac{1}{r} := \frac{1-\theta}{r_0} + \frac{\theta}{r_1}$.

Problem 2.14. The same question arises for the complex interpolation of parameter θ for the *Banach* couple in (2.13) (i.e., $p, q, r \geqslant 1$ in this case). The answer is assumed to be the space $\dot{\Lambda}_q^{\sigma r}(L_p)$ where σ, r, q are defined as above and $\frac{1}{p} := \frac{1-\theta}{p_0} + \frac{\theta}{p_1}$.

In several cases, the conjectures are valid. In particular, they hold if the spaces in (2.13) are isomorphic to Besov spaces (see assertions (c) and (d) of Theorem 2.11). Another interesting case of validity of (2.13) is the couple of the BMO space ($\sigma = 0$, $p \geqslant 1$ and $q = r = \infty$) and a Λ-space isomorphic to a Besov space (see [36]). On the other hand, (2.13) is not true if one of the spaces is a Morrey space (when $\sigma < 0$) (see Remark 2.10).

Finally, we present an extension theorem for Λ-spaces. In its formulation, S is an *n-set*, i.e., a measurable subset of $\mathbb{R}^n$ satisfying for some $\lambda = \lambda_S \in (0,1)$, condition (1.5) of extension Theorem 1.6.

Theorem 2.15 ([39]). *Assume that $\theta < p$, $q \leqslant \infty$, and $-\frac{n}{p} \leqslant \sigma < k$ or $0 < p \leqslant \infty$, $q = \infty$, and $\sigma = k$. Then, up to equivalence of (quasi)norms,*

$$\dot{\Lambda}_q^{\sigma r}(L_p)\big|_S = \dot{\Lambda}_q^{\sigma r}(L_p\,; S), \tag{2.14}$$

where the space on the right-hand side is defined as in Proposition 2.8 (a) . *If, in addition, $p \geqslant 1$, there exists a linear operator extending functions of the trace space to the function from $\dot{\Lambda}_q^{\sigma r}(L_p)$ whose norm depends only on λ_S and k.*

Due to yet unpublished Shvartsman's theorem [40], the right-hand side in (2.14) can be replaced by its proper subspace $\dot{\Lambda}_q^{\sigma r}(L_p\,; S)_{\mathrm{int}}$ for S being a uniform domain.

2.2 $\mathcal{M}$-spaces

These spaces are determined by parameters of Example 2.3. The family includes spaces defined as the Calderón maximal function (see (1.10)); this explains the choice of the notation.

As above, we fix an integer $k \geqslant 1$, a real number σ, and exponents $0 < p$, $q \leqslant \infty$ and $0 < r < \infty$. Then a *homogeneous $\mathcal{M}$-space* with these parameters is defined by

$$\dot{\mathcal{M}}_q^{\sigma r}(L_p) := \dot{\mathcal{A}}_\varphi^k(L_p) \quad \text{where} \quad \varphi := L_q(L_r^\sigma).$$

In particular,

$$|f|_{\mathcal{M}_q^{\sigma\infty}(L_p)} = \left\|\mathcal{E}_k^{\#,\omega}(\cdot\,; f\,; L_p)\right\|_{L_q(\mathbb{R}^n)},$$

where $\omega(t) := t^\sigma$ and the Calderón maximal function $\mathcal{E}_k^{\#,\omega}$ is defined by (1.10). The space $\dot{\mathcal{M}}_q^{\sigma\infty}(L_1)$ was introduced and studied in detail by Devore and Sharpley (see [26], where this space was denoted by $\dot{\mathcal{C}}_q^\sigma$).

Most results for $\mathcal{M}$-spaces are similar to those formulated for their "twins," Λ-spaces. Therefore, we only present an analog of assertion (d) for Theorem 2.11 relating $\mathcal{M}$-spaces to the family of Triebel–Lizorkin spaces $\dot{F}_q^{\sigma r}(\mathbb{R}^n)$

(see [45, Chapt. 2] for several equivalent definitions). Since functions from $\mathcal{M}$-spaces may be nonintegrable (for $p < 1$), we use the Marcinkiewicz–Strichartz definition of F-spaces which may be naturally extended to $p < 1$.

This definition exploits the notion of *spherical k-modulus of continuity* given for $f \in L_p^{\mathrm{loc}}(\mathbb{R}^n)$, $t > 0$, and $x \in \mathbb{R}^n$ by

$$\widehat{\omega}_k(t, x\,; f\,; L_p) := \left\{ \int\limits_B \left|\Delta_{ty}^k f(x)\right|^p dy \right\}^{\frac{1}{p}},$$

where $B := \{y \in \mathbb{R}^n\,; |y| \leqslant 1\}$. Then we define a homogeneous space $\dot{F}_q^{\sigma r}(L_p)$ by the seminorm,

$$|f|_{F_q^{\sigma r}(L_p)} := \left\| \left\{ \int\limits_{\mathbb{R}_+} \left(\frac{\widehat{\omega}_k(t\,; f\,; L_p)}{t^\sigma} \right)^r \frac{dt}{t} \right\}^{\frac{1}{r}} \right\|_{L_q(\mathbb{R}^n)}.$$

The relation between this space and the homogeneous Triebel–Lizorkin space $\dot{F}_q^{\sigma r}(\mathbb{R}^n)$ is given by the following result (see [45, Sect. 2.63] and [18]).

Theorem 2.16. *Assume that $0 < \sigma < k$ and one of the following conditions holds*:

(i) $0 < q < \infty$, $0 < r \leqslant \infty$ *and*

$$\left(\frac{1}{\min\{q, r\}} - \frac{\sigma}{n} \right)_+ < \min\left(1, \frac{1}{q}\right);$$

(ii) $0 < q < \infty$, $r = \infty$ *and*

$$\frac{1}{q} < \frac{\sigma}{n}.$$

Then, up to equivalence of seminorms,

$$\dot{F}_q^{\sigma r}(L_q) = \dot{F}_q^{\sigma r}(\mathbb{R}^n).$$

The relation between spaces $\dot{F}_q^{\sigma r}(L_q)$ and $\mathcal{M}$-spaces is given by the following results due to Shvartsman and the author [18].

Theorem 2.17. *Assume that $0 < \sigma < k$, $0 < p \leqslant \infty$, and one of the following conditions holds*:

(i) $0 < q < \infty$, $0 < r \leqslant \infty$ *and*

$$\left(\frac{1}{\min\{q, r\}} - \frac{\sigma}{n} \right)_+ < \frac{1}{p};$$

(ii) $p = \infty$ *and*

$$\frac{\sigma}{n} > \frac{1}{\min\{q, r\}}.$$

Then, up to equivalence of seminorms,

$$\dot{F}_q^{\sigma r}(L_p) = \dot{\mathcal{M}}_q^{\sigma r}(L_p).$$

As a consequence of the previous two results, we obtain the following fact due to Dorronsoro [27] in another way for the special case $1 \leqslant p \leqslant \infty$, $1 < q < \infty$, and $0 < r \leqslant \infty$.

Corollary 2.18. *Assume that* $0 < \sigma < k$ *and one of the conditions below holds:*

(i) $0 < q < \infty$ *and* $\left(\frac{1}{\min\{q,r\}} - \frac{\sigma}{n}\right)_+ < \min\{1, \frac{1}{p}\}$;

(ii) $0 < q < \infty$, $p = \infty$, $0 < r \leqslant \infty$ *and* $\frac{1}{q} < \frac{\sigma}{n}$.

Then, up to equivalence of seminorms,

$$\dot{\mathcal{M}}_q^{\sigma r}(L_p) = \dot{F}_q^{\sigma r}(\mathbb{R}^n).$$

The paper [18] also contains an analog of Theorem 2.16 on extensions from n-sets $S \subset \mathbb{R}^n$. Since its conditions are slightly different from those in Theorem 2.16, we present it now.

Theorem 2.19. *Assume that* $0 < p, q, r \leqslant \infty$ *and* $0 < \sigma < k$. *Then, up to equivalence of (quasi)norms,*

$$\dot{\mathcal{M}}_q^{\sigma r}(L_p)\big|_S = \dot{\mathcal{M}}_q^{\sigma r}(L_p\,; S).$$

If, in addition, $p \geqslant 1$, *there exists a linear operator from the trace space to the space* $\dot{\mathcal{M}}_q^{\sigma r}(L_p)$ *whose norm is bounded by a constant depending only on* λ_S *and* k.

2.3 $\mathcal{T}$-spaces

These spaces are determined by parameters of Example 2.4. The family contains Taylor spaces (see Definition 1.20). This explains the choice of the notation.

Let, as above, $k \in \mathbb{N}$, $\sigma \in \mathbb{R}_+$ and $0 < p, q \leqslant \infty$. Then a *homogeneous space* $\dot{\mathcal{T}}_p^{\sigma q}(x_0)$, $x_0 \in \mathbb{R}^n$, is defined by the (quasi)norm

$$|f|_{\mathcal{T}_p^{\sigma q}(x_0)} := \dot{\mathcal{A}}_\varphi^k(L_p) \quad \text{where} \quad \varphi := L_q^\sigma(x_0).$$

For $q = \infty$ we also introduce a separable subspace of $\mathcal{T}_p^{\sigma\infty}(x_0)$ denoted by $\mathfrak{t}_p^{\sigma\infty}(x_0)$ using the parameter $l_\infty^\sigma(x_0)$. Hence f belongs to the subspace if

$$\lim_{r\to 0} \frac{\mathcal{E}_k\big(Q_r(x_0)\,;f\,;L_p\big)}{r^\sigma} = 0.$$

An analog of Theorem 2.11 for $\mathcal{T}$-spaces is as follows.

Theorem 2.20. (a) *If integers ℓ, k satisfy $0 \leqslant \sigma < \ell < k$, then*

$$\dot{\mathcal{T}}_p^{\sigma q}(x_0) = \dot{\mathcal{A}}_\varphi^\ell(L_p),$$

where $\varphi := L_q^\sigma(x_0)$.

(b) *If $0 < \sigma < k$, then*

$$\mathcal{T}_p^{\sigma\infty}(x_0) = T_p^\sigma(x_0).$$

Moreover, for $0 < \sigma \leqslant k$

$$\mathfrak{t}_p^{\sigma\infty}(x_0) = t_p^\sigma(x_0).$$

All these equalities hold up to equivalence of (quasi)norms.

Recall that the Taylor spaces $T_p^\sigma(x_0)$ and $t_p^\sigma(x_0)$ are introduced by Definitions 1.20 and 1.22.

Theorem 2.21. *If $0 < \sigma_i < k$, $0 < q_i \leqslant \infty$, $i = 0, 1$, then*

$$\big(\mathcal{T}_p^{\sigma_0, q_0}(x), \mathcal{T}_p^{\sigma_1, q_1}(x)\big)_{\theta r} = \mathcal{T}_p^{\sigma r}(x),$$

where $\sigma := (1-\theta)\sigma_0 + \theta\sigma_1$.

The reader is referred to [14] for more information on this family.

2.4 $\mathcal{V}$-spaces

Parameters which determine these spaces are spaces of set-functions of bounded variation introduced in Example 2.5. Hence, given $k \in \mathbb{Z}_+$, $\sigma \in [0, k]$, and $0 < p,\, q \leqslant \infty$, a homogeneous $\mathcal{V}$-space with these parameters is given by

$$\dot{\mathcal{V}}_q^\sigma(L_p) := \dot{\mathcal{A}}_\varphi^k(L_p) \quad \text{where} \quad \varphi := V_q^\sigma.$$

Replacing the parameter V_p^σ by its subspace v_q^σ, we then define the homogeneous space $\mathfrak{v}_q^\sigma(L_p)$.

The properties of $\mathcal{V}$-spaces are surveyed by the author in [15]. Therefore, we present here only few results. The first of them explains the absence of index k in the notation.

Theorem 2.22. (a) *Assume that* $\sigma = 0$ *and* $1 \leqslant q < \infty$. *Then*

$$L_p^{\mathrm{loc}}(\mathbb{R}^n)/\mathcal{P}_{k-1}(\mathbb{R}^n) \hookrightarrow \dot{\mathcal{V}}_q^\sigma(L_p) \hookrightarrow L_{p\infty}^{\mathrm{loc}}(\mathbb{R}^n)/\mathcal{P}_{k-1}(\mathbb{R}^n),$$

provided that $p < q$.

If, moreover, $p = q$, *then*

$$\dot{\mathcal{V}}_q^\sigma(L_p) = L_p^{\mathrm{loc}}(\mathbb{R}^n)/\mathcal{P}_{k-1}(\mathbb{R}^n).$$

(b) *If* $\sigma > k$, *then for* $q \leqslant p$

$$\mathcal{V}_q^\sigma(L_p) = \mathcal{P}_{k-1}(\mathbb{R}^n),$$

and for $p < q$ *this space is infinite dimensional, but*

$$\dot{\mathcal{V}}_q^\sigma(L_p) \cap C^k(\mathbb{R}^n) = \mathcal{P}_{k-1}(\mathbb{R}^n)$$

Remark 2.23. (a) Apparently,

$$\dot{\mathcal{A}}_\varphi^k(L_p) \neq \dot{\mathcal{A}}_\varphi^\ell(L_p)$$

for $\varphi = V_q^\sigma$ and $0 \leqslant \sigma < \ell < k$, cf. Theorem 2.11 (a).

(b) Some examples suggest that the space C^k in the second assertion might be replaced by C^{k-1}.

In the following result, we use $\mathcal{V}$-spaces over a space of locally integrable functions X different from L_p^{loc}, such as Marcienkiewicz–Lopentz space $L_{p\infty}^{\mathrm{loc}}$ or BMO. In this case, we use the notation $\dot{\mathcal{V}}_q^\sigma(X)$.

Further, we set $\frac{1}{q^*} := \frac{1}{q} - \frac{\sigma}{n}$ (note that $-\infty < q^* \leqslant \infty$) and

$$X_{q^*} := \begin{cases} L_{q^*\infty}^{\mathrm{loc}}(\mathbb{R}^n) & \text{if } 0 < q^* < \infty, \\ BMO(\mathbb{R}^n) & \text{if } q^* = \infty, \\ C^{\mathrm{loc}}(\mathbb{R}^n) & \text{if } q^* < 0. \end{cases}$$

Theorem 2.24. (a) *Assume that*

$$1 \leqslant p < q^* \ \ \textit{if} \ \ q^* > 1 \quad \textit{and} \quad 1 \leqslant p \leqslant \infty \ \ \textit{if} \ \ q^* < 0.$$

Then, up to equivalence of norms,

$$\dot{\mathcal{V}}_q^\sigma(L_p) = \dot{\mathcal{V}}_q^\sigma(X_{q^*}).$$

(b) *Assume that integers* ℓ *and* k *satisfy*

$$1 \leqslant \ell < \sigma \leqslant k.$$

Then

$$\dot{\mathcal{V}}_q^{\sigma}(L_p) \hookrightarrow W^{\ell}\dot{\mathcal{V}}_q^{\sigma-\ell}(L_p)$$

and the converse is also true if $p \geqslant 1$.

(c) *The following continuous embeddings hold:*

$$\dot{\mathcal{M}}_q^{\sigma\infty}(L_p) \hookrightarrow \dot{\mathcal{V}}_q^{\sigma}(L_p) \hookrightarrow \dot{\Lambda}_q^{\sigma\infty}(L_p).$$

As a consequence of this and previous structure theorems for Λ- and $\mathcal{M}$-spaces, we derive a result relating $\mathcal{V}$-spaces to the classical spaces of smooth functions.

Theorem 2.25. *Let exponents* $0 < p \leqslant \infty$, $0 < q < \infty$, *and a number* σ *be given. Then the following is true.*

(a) $\dot{\mathcal{V}}_q^{\sigma}(L_p) \hookrightarrow \dot{B}_q^{\sigma\infty}(\mathbb{R}^n)$ *provided that* $k > \sigma > \left(\frac{n}{p} - \frac{n}{q}\right)_+$;

(b) $\dot{B}_q^{\sigma}(\mathbb{R}^n) \hookrightarrow \dot{\mathcal{V}}_q^{\sigma}(L_p)$ *provided that* $k > \sigma > \left(\frac{n}{q} - \frac{n}{p}\right)_+$ *and* $p < \infty$ *or* $p = \infty$ *and* $q \leqslant 1$.

In the case $p = \infty$ *and* $1 < q < \infty$, *the left-hand side should be replaced by* $\dot{B}_q^{\sigma 1}(\mathbb{R}^n)$.

Remark 2.26. (a) The embeddings are sharp in the sense that the Besov spaces here cannot be replaced by $\dot{B}_q^{\sigma r}$ with $r < \infty$ for (a) and with $r > \max\{q, 1\}$ for (b).

(b) The formulation of this result in [15, Theorem 3.5] is spoilt by several misprints.

Finally, we present a result relating $\mathcal{V}$-spaces with the extreme value $\sigma = k$ to Sobolev spaces.

Theorem 2.27. *Assume that* $1 \leqslant q < p \leqslant \infty$ *and*

$$\frac{k}{n} \geqslant \frac{1}{q} - \frac{1}{p} \quad \text{if} \quad p < \infty \quad \text{and} \quad > \frac{1}{q} \quad \text{if} \quad p = \infty.$$

Then, up to equivalence of seminorms,

$$\dot{W}_q^k(\mathbb{R}^n) = \dot{\mathfrak{v}}_q^k(L_p).$$

Moreover,

$$\dot{\mathfrak{v}}_q^k(L_p) = \dot{\mathcal{V}}_q^k(L_p) \quad \text{for} \quad q > 1$$

and

$$\dot{v}_1^k(L_p) \neq \dot{\mathcal{V}}_1^k(L_p) = W^{k-1}BV(\mathbb{R}^n).$$

The last space consists of functions from $L_1^{\text{loc}}(\mathbb{R}^n)$ whose weak L_1-derivatives of order $k - 1$ belong to the (homogeneous) space $BV(\mathbb{R}^n)$.

3 Selected Applications

3.1 Embeddings

Most classical trace, embedding, and compactness theorems for Sobolev, Besov and Triebel–Lizorkin spaces may be derived from the results of the previous sections. For instance, the classical trace theorem

$$\dot{W}_p^k(\mathbb{R}^n)\big|_{\mathbb{R}^{n-1}} = \dot{B}_p^{k-\frac{1}{p}}(\mathbb{R}^{n-1})$$

is a direct consequence of the representation $\dot{B}_p^\sigma(\mathbb{R}^n) = \dot{\Lambda}_p^{\sigma p}(L_p)$ and equivalence of norms on a finite-dimensional space ($\mathcal{P}_k(\mathbb{R}^n)$ in this case). It is also possible to derive the results of this kind for smooth spaces defined over Lorentz, BMO and Morrey spaces with similar objects as target spaces.

We present only one result of this kind for the embedding of Besov spaces in the case of limiting exponent (as above, these spaces are defined via k-modulus of continuity (see the text after Theorem 2.11)). To formulate the result, we introduce a family of spaces $BMO_q(\mathbb{R}^n)$, $0 < q \leqslant \infty$, containing for $q = 1$ the classical John–Nirenberg space BMO.

Let f_Q denote the integral mean of $f \in L^{\text{loc}}(\mathbb{R}^n)$ over a cube Q. Further, f_Q^* stands for the nonincreasing rearrangement of the trace $(f - f_Q)\big|_Q$. Then the space $BMO_q(\mathbb{R}^n)$ with $q < \infty$ is defined by the seminorm

$$|f|_{BMO_q(\mathbb{R}^n)} := \sup_{Q \in \mathcal{K}(\mathbb{R}^n)} \left\{ \int\limits_0^{|Q|} \left(\frac{f_Q^*(t)}{\log \frac{2|Q|}{t}} \right)^q \frac{dt}{t} \right\}^{\frac{1}{q}}.$$

The space $BMO_\infty(\mathbb{R}^n)$ consists, by definition, of all continuous functions f with the finite seminorm

$$|f|_{BMO_\infty(\mathbb{R}^n)} := \sup_{Q \in \mathcal{K}(\mathbb{R}^n)} \max_Q |f - f_Q|.$$

The John–Nirenberg exponential estimate for BMO-functions [30] immediately implies $BMO_1 = BMO$.

Theorem 3.1. *Assume that the parameters p, q, σ, and k satisfy*

$$0 < p, q \leqslant \infty, \ 0 < \sigma < k, \quad \text{and} \quad \frac{\sigma}{n} = \frac{1}{p}.$$

Then for every $1 \leqslant m \leqslant n$

$$\dot{B}_p^{\sigma q}(\mathbb{R}^n)\big|_{\mathbb{R}^m} \hookrightarrow BMO_q(\mathbb{R}^m)/\mathcal{P}_{k-1}(\mathbb{R}^m).$$

As a consequence, we obtain an exponential estimate for the functions from $\dot{B}_p^{\sigma q}$ with $\frac{\sigma}{n} = \frac{1}{p}$. For its formulation we denote by $E_q(Q)$, $1 \leqslant q < \infty$, an Orlicz space $L_N(Q)$ determined by the convex function

$$N(t) := \exp(t^q) - 1, \quad t \in \mathbb{R}_+.$$

Recall that the norm of this space is defined by

$$\|f\|_{L_N(Q)} := \inf\Big\{\lambda > 0\, ; \int\limits_Q N(\lambda|f|)dx \leqslant 1\Big\}.$$

Corollary 3.2. *Let Q^m be an m-dimensional face of a cube $Q \in \mathcal{K}(\mathbb{R}^n)$, $1 \leqslant m \leqslant n$. Under the assumptions of Theorem 3.1,*

$$\big\|f - P_k(Q^m\, ; f)\big\|_{E_{q'}(Q^m)} \leqslant c|f|_{B_p^{\sigma q}(Q^m)}$$

provided that $1 < q \leqslant \infty$.

In the case $q = 1$, the function $f \in C(\mathbb{R}^n)$ and

$$\max_{Q^m}\big|f - P_k(Q^m\, ; f)\big| \leqslant c|f|_{B_p^{\sigma 1}(Q^m)}.$$

Here, $c = c(n, k, p, q)$, $\frac{1}{q} + \frac{1}{q'} = 1$, and $P_k(Q^m\, ; f)$ is a polynomial of best approximation for $f|_{Q^m}$ of degree $k - 1$ in $L_1(Q^m)$.

Embeddings results of another kind follow directly from the representation theorems of Sect. 2. To formulate these theorems, denote by X one of the smooth spaces $\dot{W}_q^k(\mathbb{R}^n)$, $\dot{B}_q^{\sigma r}(\mathbb{R}^n)$, or $\dot{F}_q^{\sigma r}(\mathbb{R}^n)$ and assume that $p \in (0, +\infty]$ is such that $X \hookrightarrow L_p^{\text{loc}}(\mathbb{R}^n)$. Then the image of X, denoted by $[X]_p$ and regarded as a subspace of $L_p^{\text{loc}}(\mathbb{R}^n)$, can be characterized in local approximation terms as follows.

Theorem 3.3. *The following equalities hold up to equivalence of seminorms:*

$$[\dot{B}_q^{\sigma r}(\mathbb{R}^n)]_p = \dot{\Lambda}_q^{\sigma r}(L_p);$$
$$[\dot{F}_q^{\sigma r}(\mathbb{R}^n)]_p = \dot{\mathcal{M}}_q^{\sigma r}(L_p);$$
$$[\dot{W}_q^k(\mathbb{R}^n)]_p = \dot{\mathcal{V}}_q^k(L_p) = \dot{\mathcal{M}}_q^{k\infty}(L_p) \quad \text{if} \quad 1 < q \leqslant \infty.$$

Here, the local approximation spaces are defined by local polynomial approximation of order k and $0 < \sigma < k$.

The first result of this kind is the classical F. Riesz lemma (see, for example, [38]) asserting that

$$[\dot{W}_q^1(\mathbb{R})]_\infty = \dot{\mathcal{V}}_q^1(L_\infty) \quad \text{for} \quad 1 \leqslant q < \infty.$$

The last equality of Theorem 3.3 relating Sobolev and $\mathcal{M}$-spaces is a slight reformulation of the Calderón theorem [19] (see also [26, Theorem 6.2]).

To explain the important feature of these embeddings, we consider the space $\dot{B}_q^{\sigma r}(\mathbb{R}^n)$. Functions of this space have smoothness σ in L_q, but essentially lose their smoothness in L_p if $p > q$. Actually, the set

$$\{f \in \dot{B}_q^{\sigma r}(\mathbb{R}^n)\,;\, \omega_k(t\,;f\,;L_p) = O(t^{\widetilde{\sigma}})\},$$

where $\widetilde{\sigma} > \sigma - \left(\frac{n}{q} - \frac{n}{p}\right)$ is *thin* (of first category) in $\dot{B}_q^{\sigma r}(\mathbb{R}^n)$.

Nevertheless, due to Theorem 3.3, functions of the image $[\dot{B}_q^{\sigma r}(\mathbb{R}^n)]_p$ have a kind of "hidden" smoothness in L_p measured by the same number σ. The approximation, extension, and differentiability results for the classical smooth spaces presented below are remote consequences of this phenomenon.

3.2 Extensions

The extension operator of Theorem 1.6 is universal in the sense that it works for a wide class of smoothness spaces. For instance, a generalized version of Theorem 1.6 due to Shvartsman [40] implies simultaneously the P. Jones extension theorem for $W_p^k(G)$. where $G \subset \mathbb{R}^n$ is a uniform domain (see [29]) and similar results for Besov, BMO, and Morrey spaces and their anisotropic analogs. We present two new extension results exploiting the aforementioned operator, but first describe its construction.

Let $S \subset \mathbb{R}^n$ be a closed subset of positive measure, and let $\mathcal{W}$ be a collection of cubes forming the Whitney decomposition of $\mathbb{R}^n \backslash S$. We associate to every $Q \in \mathcal{W}$ a cube $\widehat{Q}$ centered at a point of S closest to Q and of volume $c|Q|$, where $c > 1$ is a constant depending only on S. Further, given $f \in L_p^{\rm loc}(S)$, $0 < p \leqslant \infty$, and $Q \in \mathcal{W}$, we denote by P_Q the projector (linear if $p \geqslant 1$ and quasilinear if $p < 1$) from $L_p(\widehat{Q} \cap S)$ onto $\mathcal{P}_{k-1}(\mathbb{R}^n)$ (see Theorem 1.4 and Remark 1.5 (c) for its definition). Finally, $\{\varphi_Q\}_{Q \in \mathcal{W}}$ is a smooth partition of unity subordinate to the cover $\mathcal{W}$. Then the extension operator, denoted by Ex_k^S, is given for $f \in L_p^{\rm loc}(S)$ by

$$Ex_k^S f := \sum_{Q \in \mathcal{W}} (P_Q f)\varphi_Q.$$

Though the operator is defined for subsets of positive measure, it can be applied for extension problems from subsets of measure zero. The first application of this kind is due to [5, Theorem 2.11], where a Morrey–Companato space over an ℓ-set of $\mathbb{R}^n$ with $n-1 < \ell \leqslant n$ is described as the trace space of a Lipschitz space to the ℓ-set (see its definition below). We present a version of this result recalling before the notions involved.

Let $S \subset \mathbb{R}^n$ be a measurable subset of positive Hausdorfff ℓ-measure, $0 < \ell \leqslant n$. By $L_q(S)$, $1 \leqslant q \leqslant \infty$ we denote the space of $\mathcal{H}_\ell$-measurable functions over S equipped with the norm

$$\|f\|_q := \left\{ \int_S |f|^q \, d\mathcal{H}_\ell \right\}^{\frac{1}{q}}.$$

Similarly to the case $\ell = n$, we define normed local approximation for $f \in L_q(S)$ by

$$\mathcal{E}_k\big(Q\,; f\,; L_q(S)\big) := \inf\left\{ \frac{1}{\mathcal{H}_\ell(Q \cap S)} \int_{Q \cap S} |f - m|^q \, d\mathcal{H}_\ell \right\}^{\frac{1}{\varepsilon}},$$

where m runs over all polynomials of degree $k - 1$.Extending the notion of $\dot{\mathcal{A}}^k_\varphi(L_q\,; S)$ to subsets of positive Hausdorff ℓ-measure (see Sect. 2), we introduce the space $\dot{\Lambda}^{\sigma r}_q(L_p\,; S)$ by the seminorm

$$|f|_{\dot{\Lambda}^{\sigma r}_q(L_p;S)} := \left\{ \int_0^\infty \left(\frac{\mathcal{E}_k(Q_t(x)\,; f\,; L_q(S))}{t^\sigma} \right)^r \frac{dt}{t} \right\}^{\frac{1}{r}}.$$

The following result connects this object to the trace space $\dot{B}^{\sigma r}_q(\mathbb{R}^n)\big|_S$ for ℓ-sets S with $n - 1 < \ell \leqslant n$. Let us recall that $S \subset \mathbb{R}^n$ is said to be an *ℓ-set* if for some constants $c_1, c_2 > 0$ and all cubes $Q \in \mathcal{K}(S)$

$$c_1|Q|^{\frac{\ell}{n}} \leqslant \mathcal{H}_\ell(Q) \leqslant c_2|Q|^{\frac{\ell}{n}}.$$

The class of ℓ-sets, in particular, contains compact Lipschitz ℓ-manifolds (with $\ell \in \mathbb{N}$), Cantor type sets and other self-similar fractals (with arbitrary $\ell \in (0, n]$).

Theorem 3.4. *Let $S \subset \mathbb{R}^n$ be an ℓ-set where $n - 1 < \ell \leqslant n$. Assume that*

$$\dot{B}^{\sigma r}_q(\mathbb{R}^n) \hookrightarrow L^{\mathrm{loc}}_\infty(\mathbb{R}^n).$$

Then, up to equivalence of seminorms,

$$\dot{B}^{\sigma r}_q(\mathbb{R}^n)\big|_S = \dot{\Lambda}^{\sigma r}_q(L_\infty\,; S).$$

Moreover, there exists a continuous extension operator (linear if $q \geqslant 1$) from the trace space to $\dot{B}^{\sigma r}_q(\mathbb{R}^n)$.

The method developed in [5] also admits a similar extension for the embedding pairs $\dot{F}^{\sigma r}_q \subset L^{\mathrm{loc}}_\infty$ and $\dot{W}^k_q \subset L^{\mathrm{loc}}_\infty$ giving local approximation representations of the corresponding trace spaces of $\dot{F}^{\sigma r}_q\big|_S$ and $\dot{W}^k_q\big|_S$ to ℓ-sets with $n - 1 < \ell \leqslant n$.

The recent results by Shvartsman [41] presented below suggest that the restriction on S may be essentially weaker. To formulate Shvartsman's results, we introduce variants of the spaces $\dot{\mathcal{V}}_q^\sigma(L_\infty\,;S)$ and $\dot{\mathcal{M}}_q^\sigma(L_\infty\,;S)$ defining the former by the seminorm

$$|f|_{\mathcal{V}} := \sup_\pi \left\{ \sum_{Q\in\pi} \left(\frac{\mathcal{E}_k(Q\,;f\,;L_\infty)}{|Q|^{\frac{\sigma}{n}}} \right)^q \right\}^{\frac{1}{q}}, \tag{3.1}$$

where π runs over all disjoint families of cubes Q satisfying $4Q\cap S \neq \varphi$; here $\lambda Q := Q_{\lambda r}(x)$ if $Q := Q_r(x)$.

The latter space is defined by the seminorm

$$|f|_{\mathcal{M}} := \|f^\#\|_{L_q(\mathbb{R}^n)},$$

where the maximal function $f^\#$ is given by

$$f^\#(x) := \sup_{Q\ni x} \frac{\mathcal{E}_k(Q\cap S\,;f\,;L_\infty}{|Q|^{\frac{\sigma}{n}}}, \tag{3.2}$$

where Q runs over all cubes intersecting S and containing $x\in\mathbb{R}^n$.

The original space $\dot{\mathcal{V}}_q^\sigma((L_p\,;S)$ is recalled to be defined by disjoint families of cubes centered at S and of length side $< 2\operatorname{diam} S$. In turn, the original space $\dot{\mathcal{M}}_q^{\sigma r}(L_p\,;S)$ is defined by the smaller maximal function

$$\mathcal{E}_k^{\#,\omega}(x\,;f\,;L_\infty) := \sup_Q \frac{\mathcal{E}_k(Q\,;f\,;L_\infty)}{|Q|^{\frac{\sigma}{n}}},$$

where $\omega(t) := t^\sigma$ and Q runs over all cubes from $\mathcal{K}(S)$ containing x.

Theorem 3.5 ([41]). *Let $S\subset\mathbb{R}^n$ be an arbitrary closed subset of $\mathbb{R}^n$. Assume that $\dot{W}_q^1(\mathbb{R}^n)\hookrightarrow L_\infty^{\text{loc}}(\mathbb{R}^n)$, i.e., $q>n$. Then the trace norm of the space $\dot{W}_q^1(\mathbb{R}^n)\big|_S$ is equivalent to seminorm* (3.1) *or* (3.2) *with $k=\sigma=1$. Moreover, there exists a linear continuous extension operator from $\dot{W}_q^1(\mathbb{R}^n)\big|_S$ to $\dot{W}_q^1(\mathbb{R}^n)$.*

Remark 3.6. Theorem 3.4 apparently remains to be true for $L_p^{\text{loc}}(\mathbb{R}^n)$ with $p<\infty$ substituted for L_∞^{loc}, but for now we can prove this only for ℓ-sets with $\ell = n$. The same fact holds for other classical smooth spaces.

Conjecture 3.7. *Let $\dot{X}$ denote one of the "classical" homogeneous spaces over L_q of smoothness σ embedded into $L_p^{\text{loc}}(\mathbb{R}^n)$ where $0<q\leqslant p\leqslant\infty$. Then $\dot{X} = \dot{\mathcal{A}}_\varphi^k(L_p)$ where φ is the corresponding functional parameter* (*see Theorem* 3.3).

Assume that $S\subset\mathbb{R}^n$ is an ℓ-set, where $\ell>\ell(\sigma,q) := \min\{n-1,(n-\sigma q)_+\}$. Then, up to equivalence of seminorms,

$$\dot{X}|_S = \dot{\mathcal{A}}^k_\varphi(L_p\,; S), \tag{3.3}$$

and there exists a continuous extension operator (linear for $q \geqslant 1$) from $\dot{X}|_S$ into $\dot{X}$.

In particular, $\ell(\sigma, q) = 0$ for the space of Theorem 3.4, so validity of the conjecture implies that the assertion of the theorem is true for every ℓ-set with $\ell > 0$.

The same follows from the validity of the conjecture for the space $\dot{W}^k_q(\mathbb{R}^n)$ where $\frac{k}{n} > \frac{1}{q}$. However, in the special case $k = 1$, Theorem 3.5 gives essentially stronger result. This leads to

Conjecture 3.8. *Assume that, in the settings of Conjecture 3.7, $\frac{\sigma}{n} > \frac{1}{q}$, i.e., $\dot{X} \hookrightarrow C^{\mathrm{loc}}(\mathbb{R}^n)$. Then (3.2) holds for every closed subset $S \subset \mathbb{R}^n$.*

3.3 Pointwise differentiability

Using differentiability criteria of Theorems 1.23–1.25, one can reveal the influence of "hidden" smoothness on pointwise behavior and Luzin approximation of functions from smooth spaces. The result presented below firstly announced in [10]. To illustrate a general situation, let us consider the space of L_1-Lipschitz functions on the real line. The corresponding representation of this space as an local approximation space shows that its functions have hidden smoothness 1 in the space L_∞. The general theorem presented below asserts for this special case that every such a function belongs to $t^1_\infty(x)$ for almost all $x \in \mathbb{R}$, i.e., is differentiable almost everywhere. On the other hand, $L_1(\mathbb{R})$-Lipschitz functions are functions of bounded variation and vice versa (Hardy–Littlewood). Therefore, the previous statement coincides with the classical Lebesgue differentiability theorem. For this reason, the result presented now may be regarded as a far reaching generalization of the Lebesgue theorem.

We formulate this result for functions from $\mathcal{V}$-spaces and then derive from there differentiability results related to the hidden smoothness phenomenon for the classical smoothness spaces.

Theorem 3.9. *Assume that a function f belongs to the space $\dot{\mathcal{V}}^\sigma_q(L_p)$, where $0 < q, p \leqslant \infty$, $0 < \sigma \leqslant k$, and for some $s \in (0, n]$*

$$\mu := \sigma - \frac{n-s}{q} > 0.$$

Then

(a) *If $s = n$ and $\mu = k$, then f belongs to the Taylor space $t^k_p(x)$ for almost all $x \in \mathbb{R}^n$.*

(b) *If* $s < n$ *and* $0 < \mu < k$, *then* $f \in T_p^\mu(x)$ *for* $\mathcal{H}_s$ *almost all* $x \in \mathbb{R}^n$.

(c) *If* f *belongs to the subspace* $\dot{v}_q^\sigma(L_p)$, *then assertion (b) holds with* $t_p^\mu(x)$ *substituted for* $T_p^\mu(x)$.

The second result concerns Lusin type approximation of functions from $\mathcal{V}$-spaces. In its proof, a considerable role plays a version of the extension theorem from [5] (see also [13, Sect. 5.3]). In the formulation, we use the notion "Hausdorff s-capacity" given for a set $G \subset \mathbb{R}^n$ by

$$\mathcal{C}_s(G) := \inf\Big\{\sum_i |Q_i|^{\frac{s}{n}}\ ;\ G \subset \cup Q_i\Big\}.$$

Having in mind applications, we formulate the result for the general space $\dot{\mathcal{V}}_q^{(\omega)}(L_p)$, where $\omega : \mathbb{R}_+ \to \mathbb{R}_+$ is k-majorant (i.e., ω is nondecreasing and $t \mapsto \omega(t)/t^k$ is nonincreasing). Recall that

$$|f|_{\dot{\mathcal{V}}_q^{(\omega)}(L_p)} := \sup_\pi \Big\{\sum_{Q\in\pi} |Q| \Big|\frac{\mathcal{E}_k(Q\,;f\,;L_p)}{\omega(|Q|^{\frac{k}{n}})}\Big|^q\Big\}^{\frac{1}{q}},$$

where π runs over all disjoint families of cubes.

Theorem 3.10. *Let* p, q, s *be chosen as in Theorem* 3.9, *and let* f *belong to the space* $\dot{\mathcal{V}}_q^{(\omega)}(L_p)$. *Assume that for some* $t > 0$

$$\widetilde{\omega}(t) := \int_0^t \frac{\omega(u)}{u^{t+\frac{n-s}{q}}}\, du < \infty.$$

Then for every $\varepsilon > 0$ *there exists a function* $f_\varepsilon \in C^{\mathrm{loc}}(\mathbb{R}^n)$ *such that*

$$\mathcal{C}_s\big\{x \in \mathbb{R}^n\,; f_\varepsilon(x) \neq f(x)\big\} < \varepsilon \tag{3.4}$$

and f_ε *satisfies the Lipschitz condition*

$$\sup_{\mathbb{R}^n}\big|\Delta_h^k f_\varepsilon\big| \leqslant c(\varepsilon)\widetilde{\omega}\big(|h|\big), \quad h \in \mathbb{R}^n.$$

Corollary 3.11. *Let* $\omega(t) := t^k$, $t > 0$, *and* $s = n$. *Then for any* $f \in \dot{\mathcal{V}}_q^{(\omega)}(L_p)$ *and* $\varepsilon > 0$ *there exists a function* f_ε *belonging to* $C^{k,\mathrm{loc}}(\mathbb{R}^n)$ *and satisfying*

$$\big|\big\{x \in \mathbb{R}^n\,; f(x) \neq f_\varepsilon(x)\big\}\big| < \varepsilon. \tag{3.5}$$

One may derive from these theorems a few differentiability results for the "classical" smoothness spaces. We present only two beginning with the Calderón space $\dot{C}_q^\sigma(\mathbb{R}^n)$, $0 < \sigma \leqslant k$, introduced in [26]. This space is recalled to coincide with the Triebel–Lizorkin space $\dot{F}_q^{\sigma\infty}(\mathbb{R}^n)$ for $0 < \sigma < k$ and with the Sobolev space $\dot{W}_q^k(\mathbb{R}^n)$ for $\sigma = k$ and $1 < q \leqslant \infty$.

Corollary 3.12. *Let $p \in [q, +\infty]$ be such that*

$$\dot{C}_q^\sigma(\mathbb{R}^n) \hookrightarrow L_p^{\mathrm{loc}}(\mathbb{R}^n),$$

and let

$$\mu := \sigma - \frac{n-s}{q} > 0$$

for some $s \in (0, n]$. Then for $f \in \dot{C}_q^\sigma(\mathbb{R}^n)$ the following holds:

(a) *If $s = n$ and $\sigma = k$, then f belongs to $t_p^k(x)$ for almost all $x \in \mathbb{R}^n$. Moreover, for every $\varepsilon > 0$ there exists a function $f_\varepsilon \in C^{k,\mathrm{loc}}(\mathbb{R}^n)$ such that* (3.5) *holds.*

(b) *If $0 < \mu < k$, then $f \in T_p^\mu(x)$ for $\mathcal{H}_s$ almost all points $x \in \mathbb{R}^n$. Moreover, for every $\varepsilon > 0$ there exists a function f_ε belonging to the Lipschitz space $\dot{B}_\infty^{\mu\infty}(\mathbb{R}^n)$ such that* (3.4) *holds.*

The final result concerns the Besov space $\dot{B}_q^{\sigma\infty}(\mathbb{R}^n)$; in its formulation,

$$\frac{1}{q^*} := \frac{\sigma}{n} + \frac{1}{q}.$$

Corollary 3.13. *Assume that a function $f \in L_p^{\mathrm{loc}}(\mathbb{R}^n)$, $0 < p < \infty$, satisfies for some $\sigma \in (0, k]$ the condition*

$$\sup_{0 \leqslant t \leqslant 1} \frac{\omega_k(t\,; f\, L_p)}{t^\sigma} < \infty. \tag{3.6}$$

Then for any $\varepsilon > 0$ and $\delta > 0$ there exists a function f_ε satisfying (3.5) *and such that for $0 < t \leqslant 1$*

$$\omega_k(t\,; f_\varepsilon\,; L_\infty) \leqslant C(\varepsilon, \delta) t^\sigma \left(\log \frac{2}{t}\right)^{\frac{1}{q^*}+\delta}. \tag{3.7}$$

Remark 3.14. The result is sharp in the sense that the Luzin approximation with $\delta = 0$ in (3.7) fails to be true. According to the counterexample of Oskolkov [34], there exists a function $f \in L_p(\mathbb{R})$ satisfying (3.6) for $k = 1$ and $0 < \sigma < 1$ which differs, on a set of positive measure, from *every* function satisfying (3.7) with $k = 1$ and $\delta = 0$.

3.4 Nonlinear Approximation

One more example of the hidden smoothness phenomenon concerns nonlinear approximation of smooth functions. The first result of this type was due to Birman and Solomyak [3] who studied approximation of functions from $W_q^k(Q_0)$ in the space $L_p(Q_0)$, where $Q_0 := [0,1]^n$ and $\frac{k}{n} > \frac{1}{q} - \frac{1}{p} > 0$ by piecewise polynomials subordinate to a subdivision of Q_0 into a fixed number of dyadic subcubes. They discovered that, in spite of loss of smoothness in L_p, the order of approximation remains the very same as for a similar approximation in L_q-metric. The only difference is a nonlinear nature of the approximation procedure, a fact lying in the core of the problem. Later, it was proved [11, 12] that a similar effect holds for approximation of univariate functions by nonlinear splines and rational functions, and, in these situations, the converse results are also true. Now, this area, *nonlinear approximation theory*, is being intensively developed; the reader is referred to the mini-monograph [25] and forthcoming survey [37] for a detailed account.

We present only one of the possible results of this kind which is most closely related to and proved by the methods of local approximation theory.

Theorem 3.15. *Given $f \in \dot{v}_q^\sigma(L_p; Q_0)$, where $0 \leqslant q \leqslant p \leqslant \infty$, and an integer $N \geqslant 1$, there exists a collection of N dyadic cubes π and a subordinate to π collection of polynomials $\{m_Q\}_{Q\in\pi}$ of degree $k-1$ such that*

$$\left\| f - \sum_{Q\in\pi} \chi_Q m_Q \right\|_{L_p(Q_0)} \leqslant c(n,p,q) N^{-\frac{\sigma}{n}} |f|_{\dot{\mathcal{V}}_q^\sigma(L_p; Q_0)}. \tag{3.8}$$

To formulate a consequence, denote by X_q^k the homogeneous Sobolev space $\dot{W}_q^k(Q_0)$ if $1 < q < \infty$ and the homogeneous space $W^{k-1}BV(Q_0)$ if $q = 1$.

Corollary 3.16. *Assume that*

$$\frac{k}{n} \geqslant \frac{1}{q} - \frac{1}{p} > 0.$$

Then for any $f \in X_q^k$ and $N \geqslant 1$ there exists a collection of N dyadic cubes π and a subordinate to π collection of polynomials $\{m_Q\}_{Q\in\pi}$ of degree $k-1$ such that

$$\left\| f - \sum_{Q\in\pi} \chi_Q m_Q \right\|_{L_p(\mathbb{R}^n)} \leqslant c N^{-\frac{k}{n}} |f|_{X_q^k},$$

where $c = c(k,n,p,q)$.

Remark 3.17. The approximation method gives some information on the structure of π. In particular, if $\frac{k}{n} > \frac{1}{q} - \frac{1}{p}$, then, as in [3], π is a *partition* of Q_0, and if $q = p$, then π is a uniform partition. In the limiting case, there are subfamilies of embedded cubes in π accumulated near singularities of $f \in W_q^k(Q_0)$, seen from the space $L_p(Q_0)$.

In the special case of the space $BV(\mathbb{R}^2)$ and approximation in L_2-metric, the result of the corollary was done in [24].

Using the relations between $\mathcal{V}$-spaces and F-spaces, one can easily derive from Theorem 3.15 a similar to Corollary 3.16 result for the space $\dot{F}_q^{\sigma\infty}(= \dot{C}_q^{\sigma})$. We leave the formulation to the reader.

Finally, we present two applications of Theorem 3.15 of another kind. First, following the authors of [3], we estimate a "massivity" of the unit ball of $W_q^k(Q_0)$ into $L_p(Q_0)$. If $\frac{k}{n} > \frac{1}{p} - \frac{1}{q}$, the ball is compact and its massivity may be measured by its ε-entropy denoted by $\mathcal{H}_\varepsilon(W_q^k\,; L_p)$. According to [3], this quantity satisfies

$$\mathcal{H}_\varepsilon(W_q^k\,; L_p) \approx \varepsilon^{-\frac{n}{k}} \quad \text{as} \quad \varepsilon \to 0. \tag{3.9}$$

In the limiting case, the ball is not compact and we measure its massivity by intersection with a very huge compact, say with the set

$$\operatorname{Lip}_q^\delta := \left\{ f \in L_p(Q_0)\,;\ \|f\|_p + \sup_{t>0} \frac{\omega_1(t\,; f\,; L_p)}{t^\delta} < \infty \right\},$$

where $\delta > 0$ is arbitrarily small. Then for the ε-entropy of this intersection in $L_p(Q_0)$ the following holds.

Corollary 3.18. *If $\frac{k}{n} = \frac{1}{q} - \frac{1}{p}$, then for $\varepsilon \leqslant 1$*

$$\mathcal{H}_\varepsilon\big(W_q^k \cap \operatorname{Lip}_q^\delta\,; L_p\big) \leqslant c\varepsilon^{-\frac{n}{k}} \left(\log \frac{2}{\varepsilon}\right)^{1+\frac{k}{n}},$$

where $c = c(k, n, p, q, \delta)$.

Hence the asymptotic (3.9) is almost preserved in the limiting case.

The second application concerns the still unsolved problem of the real interpolation of the Banach couple (L_p, W_q^k) over Q_0 for $p \neq q$. For $\frac{k}{n} > \frac{1}{q} - \frac{1}{p}$ and $p < \infty$ Krugljak [32] proved that

$$\big(L_p, W_q^k\big)_{\theta p_\theta} = B_{p_\theta}^{k_\theta}, \tag{3.10}$$

where

$$\frac{1}{p_\theta} := \frac{1-\theta}{p} + \frac{\theta}{q} \quad \text{and} \quad k_\theta = (1-\theta)k.$$

Using an approach based on a variant of Theorem 3.15 with piecewise polynomials replaced by splines, one can prove (3.10) also for $\frac{k}{n} \geqslant \frac{1}{q} - \frac{1}{p}$ and $p < \infty$. In the cases $\frac{k}{n} < \frac{1}{q} - \frac{1}{p}$ and $\frac{k}{n} \geqslant \frac{1}{q}$, $p = \infty$, the problem remains open.

References

1. Artin, M., et al.: The situation in Soviet mathematics. Notices Am. Math. Soc. November, 495–497 (1978)
2. Bergh, J., Löfström, J.: Interpolation Spaces. An Introduction. Springer (1971)
3. Birman, M., Solomyak, M.: *Piecewise-polynomial approximation of functions of the classes* W_p^α (Russian). Mat. Sb. **73**, 331–355, (1967); English transl.: Math. USSR–Sb. **2** (1967)
4. Brown, L., Lucier, B.: Best approximations in L^1 are near better in L_ε^p, $p < 1$. Proc. Am. Math. Soc. **120**, 97-100 (1994)
5. Brudnyi, A., Brudnyi, Yu.: Metric spaces with linear extensions preserving Lipschitz condition. Am. Math. J. **129**, 217-314 (2007)
6. Brudnyi, Yu.: On local best approximation of functions by polynomials (Russian). Dokl. Akad. Nauk SSSR **161**, 746-749 (1965)
7. Brudnyi, Yu.: A multidimensional analog of a theorem of Whitney. Mat. Sb. **82(124)**, 175-191 (1970); English transl.: Math. USSR–Sb. **11**, 157-170 (1970)
8. Brudnyi, Yu.: On an extension theorem (Russian). Funk. Anal. Pril. **4**, 96-97 (1970); English transl.: Funct. Anal. Appl. **4**, 252-253 (1970)
9. Brudnyi, Yu.: Spaces defined by local polynomial approximations (Russian). Tr. Mosk. Mat. Obshch. **24**, 69-132 (1971); English transl.: Trans. Mosc. Math. Soc. **24**, 73-139 (1971)
10. Brudnyi, Yu.: Local approximations and differential properties of functions of several variable (Russian). Uspekhi Mat. Nauk **29**, 163-164 (1974)
11. Brudnyi, Yu.: Spline approximation of functions of bounded variation (Russian). Dokl. Akad. Nauk SSSR **215**, 611-613 (1974); English transl.: Sov. Math., Dokl. **15** (1974)
12. Brudnyi, Yu.: Rational approximation and embedding theorems. Dokl. Akad. Nauk SSSR **247**, 269-272 (1979); English transl.: Sov. Math., Dokl. **20** (1979)
13. Brudnyi, Yu.: Adaptive approximation of functions with singularities. Tr. Mosk. Mat. Obshch. **55**, 149-242 (1994) English transl.: Trans. Mosc. Math. Soc. 123-186 (1994)
14. Brudnyi, Yu.: Taylor spaces – approximation space theory approach In: Function Spaces VI, pp. 100-105. World Science (2003)
15. Brudnyi, Yu.: Multivariate functions of bounded variation In: Banach Spaces and their Applications in Analysis, pp. 37-57. de Guyter (2007)
16. Brudnyi, Yu., Ganzburg, M.: On an extremal problems for polynomials in n-variables. Izv. Akad. Nauk SSSR **37**, 344-355 (1973); English transl.: Math. USSR–Izv. **7**, 345-356 (1973)
17. Brudnyi, Yu., Krugljak, N.: Interpolation Functors and Interpolation Spaces. North Holland (1991)
18. Brudnyi, Yu., Shvartsman, P.: On traces of Triebel–Lizorkin spaces to uniform domains [In preparation]
19. Calderón, A.: Estimates for singular integral operators in terms of maximal functions. Stud. Math. **44**, 563-582 (1972)
20. Calderón, A., Scott, R.: Sobolev type inequalities for $p > 0$. Stud. Math. **62**, 75-92 (1978)
21. Calderón, A., Zygmund, A.: Local properties of solutions of elliptic partial differential equations. Stud. Math. **20**, 171-225 (1961)
22. Campanato, S.: Proprietá di una famiglia di spazi funzionali. Ann. Scuola Norm. Super. Pisa, Cl. Sci. (4) **18**, 137-160 (1964)
23. Carbery, A., Wright, J.: Distributional and L^p norm inequalities for polynomials over convex bodies in $\mathbb{R}^n$. Math. Research Lett. **8**, 233-248 (2001)

24. Cohen, A., DeVore, R., Petrushev, P., Xu, H.: Nonlinear approximation and the space $BV(\mathbb{R}^2)$. Am. J. Math. **114**, 587-628 (1999)
25. DeVore, R.: Nonlinear approximation. In: Acta Numer. Birkhäuser (1998)
26. DeVore, R., Sharpley, R.: Maximal functions measuring smoothness. Mem. Am. Math. Soc. **47**. no. 293, 1-115 (1984)
27. Dorronsoro, J.: Poisson integrals of regular functions. Trans. Am. Mat. Soc. **297**, 669-685 (1986)
28. de Guzman, M.: Differentiation of Integrals in $\mathbb{R}^n$. Springer (1975)
29. Jones, P.: Quasiconformal mappings and extendability of Sobolev spaces. Acta Math. **147**, 71-88 (1981)
30. John, F., Nirenberg, L.: On functions of bounded mean oscillation. Commun. Pure Appl. Math. **14**, 415-426 (1961)
31. Kadets, M., Snobar, M.: Certain functionals on the Minkowski compactum (Russian). Mat. Zametki **10**, 453-457 (1971); English transl.: Math. Notes **10** (1971)
32. Krugljak, N.: Smooth analogs of the Calderón–Zygmund decomposition, quantitative covering theorems and the K-functional of the couple (L_q, W_p^k) (Russian). Algebra Anal. **8**, 110-160 (1996); English transl.: St. Petersbg. Math. J. **8**, 617-649 (1997)
33. Maz'ya, V.G.: Sobolev Spaces. Springer-Verlag, Berlin–Tokyo (1985)
34. Oskolkov, K., Approximation properties of a summable function on sets of full measure (Russian). Mat. Sb. **103**, 563-589 (1977); English transl.: Math. USSR–Sb. **32** (1977)
35. Peetre, J.: Remarques sur les espaces de Besov. Le cas $0 < p < 1$. C. R. Acad. Sci. Paris **277**, 947-949 (1973)
36. Peetre, J., Svenson, E.: On the generalized Hardy inequality of McGehee, Pigno and Smith and the problem of interpolation between BMO and a Besov space. Math. Scand. **54**, 221-241 (1984)
37. Petrushev, P.: Nonlinear approximation. Approxim. Theory [To appear]
38. Riesz, F., Sz-Nagy, B.: Lecons d'Analyse Fonctionelle. Akad Kiadó, Budapest (1952)
39. Shvartsman, P.: Extension of functions preserving order of decreasing for (k,p)-modulus of continuity (Russian). In: Investigations in the Theory of Multivariate Functions, pp. 149-160. Yaroslavl' State Univ., Yaroslavl' (1980)
40. Shvartsman, P.: Extension Theorems Preserving Local Polynomial Approximation. (Russian). Yaroslavl' (1986)
41. Shvartsman, P.: Sobolev W_p^1-spaces on closed subsets of $\mathbb{R}^n$ [To appear]
42. Strömberg, J.-O.: Bounded mean oscillation with Orlicz norms and duality of Hardy spaces. Indiana Univ. Math. J. **28**, 511-544 (1979)
43. Timan, A.F.: Theory of Approximation of Functions of Real Variable. Pergamon Press, Oxford (1963)
44. Triebel, H.: Theory of Function Spaces. Birkhäuser, Basel (1983)
45. Triebel, H.: Theory of Function Spaces II. Birkhäuser, Basel (1992)
46. Whitney, H.: On functions of bounded n-differences. J. Math. Pures Appl. **36**, 67-95 (1957)
47. Ziemer, W.P.: Weakly Differentiable Functions. Springer (1989)

Spectral Stability of Higher Order Uniformly Elliptic Operators

Victor Burenkov and Pier Domenico Lamberti

Abstract We prove estimates for the variation of the eigenvalues of uniformly elliptic operators with homogeneous Dirichlet or Neumann boundary conditions upon variation of the open set on which an operator is defined. We consider operators of arbitrary even order and open sets admitting arbitrary strong degeneration. The main estimate is expressed in terms of a natural and easily computable distance between open sets with continuous boundaries. Another estimate is obtained in terms of the lower Hausdorff–Pompeiu deviation of the boundaries, which in general may be much smaller than the usual Hausdorff–Pompeiu distance. Finally, in the case of diffeomorphic open sets, we obtain an estimate even without the assumption of continuity of the boundaries.

1 Introduction

Currently, it is standard that in papers on many topics in partial differential equations Sobolev spaces are used as a natural language for setting problems, defining solutions, and investigating their properties. This paper on spectral stability is not an exception. The assumption on the compactness of certain embeddings involving various variants of Sobolev spaces is systematically used, as well as some other properties of Sobolev spaces and techniques developed in the theory of Sobolev spaces. In particular, the "shrinking" transformation defined by formula (5.3) plays an important role in our proofs, and

Victor Burenkov
Università degli Studi di Padova, 63 Via Trieste, 35121 Padova, Italy, e-mail: `burenkov@math.unipd.it`

Pier Domenico Lamberti
Universitá degli Studi di Padova, 63 Via Trieste, 35121 Padova, Italy, e-mail: `lamberti@math.unipd.it`

V. Maz'ya (ed.), *Sobolev Spaces in Mathematics II*, *International Mathematical Series*.
doi: 10.1007/978-0-387-85650-6,

it should be noted that the idea of using such transformations arose from exploring proofs of approximation theorems for Sobolev spaces (approximation by infinitely differentiable functions).

In this paper, we consider the eigenvalue problem for the operator

$$Hu = (-1)^m \sum_{|\alpha|=|\beta|=m} D^\alpha \left(A_{\alpha\beta}(x) D^\beta u\right), \quad x \in \Omega, \tag{1.1}$$

subject to homogeneous Dirichlet or Neumann boundary conditions, where $m \in \mathbb{N}$, Ω is a bounded open set in $\mathbb{R}^N$, and the coefficients $A_{\alpha\beta}$ are Lipschitz continuous functions satisfying the uniform ellipticity condition (2.4) below on Ω. For a precise statement of the eigenvalue problem see Definition 2.1 and Theorem 2.1.

We consider open sets Ω for which the spectrum is discrete and can be represented by means of a nondecreasing sequence of nonnegative eigenvalues

$$\lambda_1[\Omega] \leqslant \lambda_2[\Omega] \leqslant \ldots \leqslant \lambda_n[\Omega] \leqslant \ldots ,$$

where each eigenvalue is repeated as many times as its multiplicity.

In this paper, we prove estimates for the variation

$$|\lambda_n[\Omega_1] - \lambda_n[\Omega_2]|$$

of the eigenvalues corresponding to two open sets Ω_1, Ω_2.

There is vast literature on spectral stability problems for elliptic operators (see, for example, [11, 12] for references). However, little attention has been devoted to the problem of spectral stability for higher order operators and, in particular, to the problem of finding explicit qualified estimates for the variation of the eigenvalues. Moreover, most of the existing qualified estimates for second order operators were obtained under certain regularity assumptions on the boundaries.

Our analysis comprehends operators of arbitrary even order, with homogeneous Dirichlet or Neumann boundary conditions, and open sets admitting arbitrarily strong degeneration. In fact, we consider bounded open sets whose boundaries are just locally the subgraphs of continuous functions. We only require that the "atlas" $\mathcal{A}$, with the help of which such open sets are described, is fixed: we denote by $C(\mathcal{A})$ the family of all such open sets (see Definition 3.1). In $C(\mathcal{A})$ we introduce a natural metric $d_{\mathcal{A}}$ (the "atlas distance") which can be easily computed. Given two open sets $\Omega_1, \Omega_2 \in C(\mathcal{A})$, the distance $d_{\mathcal{A}}(\Omega_1, \Omega_2)$ is just the maximum of the sup-norms of the differences of the functions describing locally the boundaries of Ω_1 and Ω_2 (see Definition 5.1).

The first main result of the paper is that for both Dirichlet and Neumann boundary conditions the eigenvalues of (1.1) are locally Lipschitz continuous functions of the open set $\Omega \in C(\mathcal{A})$ with respect to the atlas distance $d_{\mathcal{A}}$. Namely, in Theorems 5.1 and 6.1, we prove that for each $n \in \mathbb{N}$ there exist

$c_n, \varepsilon_n > 0$ such that for both Dirichlet and Neumann boundary conditions the estimate

$$|\lambda_n[\Omega_1] - \lambda_n[\Omega_2]| \leqslant c_n d_{\mathcal{A}}(\Omega_1, \Omega_2) \tag{1.2}$$

holds for all open sets $\Omega_1, \Omega_2 \in C(\mathcal{A})$ satisfying $d_{\mathcal{A}}(\Omega_1, \Omega_2) < \varepsilon_n$.

By the estimate (1.2), we deduce an estimate expressed in terms of the lower Hausdorff–Pompeiu deviation of the boundaries

$$d_{\mathcal{HP}}(\partial\Omega_1, \partial\Omega_2) = \min\left\{ \sup_{x\in\partial\Omega_1} d(x, \partial\Omega_2),\ \sup_{x\in\partial\Omega_2} d(x, \partial\Omega_1) \right\}.$$

To do so, we restrict our attention to smaller families of open sets in $C(\mathcal{A})$. Namely, for a fixed $M > 0$ and $\omega : [0,\infty[\to [0,\infty[$ satisfying very weak natural conditions we consider those open sets Ω in $C(\mathcal{A})$ for which any of the functions $\overline{x} \mapsto g(\overline{x})$ describing locally the boundary of Ω satisfies the condition

$$|g(\overline{x}) - g(\overline{y})| \leqslant M\omega(|\overline{x} - \overline{y}|)$$

for all appropriate $\overline{x}, \overline{y}$. We denote by $C_M^{\omega(\cdot)}(\mathcal{A})$ the family of all such open sets (see Definition 7.2). For instance, if $0 < \alpha \leqslant 1$ and $\omega(t) = t^\alpha$ for all $t \geqslant 0$, then we obtain open sets with Hölder continuous boundaries of exponent α: this class is denoted below by $C_M^{0,\alpha}(\mathcal{A})$. It is possible to choose a function ω going to zero arbitrarily slowly which allows dealing with open sets with arbitrarily sharp cusps.

The second main result of the paper is for open sets $\Omega_1, \Omega_2 \in C_M^{\omega(\cdot)}(\mathcal{A})$. Namely, in Theorem 7.1 we prove that for each $n \in \mathbb{N}$ there exist $c_n, \varepsilon_n > 0$ such that

$$|\lambda_n[\Omega_1] - \lambda_n[\Omega_2]| \leqslant c_n \omega(d_{\mathcal{HP}}(\partial\Omega_1, \partial\Omega_2)) \tag{1.3}$$

for all open sets $\Omega_1, \Omega_2 \in C_M^{\omega(\cdot)}(\mathcal{A})$ satisfying $d_{\mathcal{HP}}(\partial\Omega_1, \partial\Omega_2) < \varepsilon_n$.

In particular, in Corollary 7.1 we deduce that if $\Omega_1, \Omega_2 \in C_M^{\omega(\cdot)}(\mathcal{A})$ satisfy

$$(\Omega_1)_\varepsilon \subset \Omega_2 \subset (\Omega_1)^\varepsilon \quad \text{or} \quad (\Omega_2)_\varepsilon \subset \Omega_1 \subset (\Omega_2)^\varepsilon,$$

where $\varepsilon > 0$ is sufficiently small then

$$|\lambda_n[\Omega_1] - \lambda_n[\Omega_2]| \leqslant c_n \omega(\varepsilon). \tag{1.4}$$

Here, $\Omega_\varepsilon = \{x \in \Omega : d(x, \partial\Omega) > \varepsilon\}$, $\Omega^\varepsilon = \big\{x \in \mathbb{R}^N : d(x, \Omega) < \varepsilon\big\}$, for any set Ω in $\mathbb{R}^N$.

In the case $\Omega_1, \Omega_2 \in C_M^{0,\alpha}(\mathcal{A})$, the estimate (1.4) takes the form

$$|\lambda_n[\Omega_1] - \lambda_n[\Omega_2]| \leqslant c_n \varepsilon^\alpha. \tag{1.5}$$

In the case of the Dirichlet boundary conditions and $m = 1$, some estimates of the form (1.5) were obtained in [8] under the assumption that a certain Hardy type inequality is satisfied on Ω_1 (see also [15]). In the case of the

Dirichlet boundary conditions and $m = 1$, the estimate (1.5) was proved in [5]. In the case of the Dirichlet boundary conditions, $m = 2$, and open sets with sufficiently smooth boundaries, an estimate of the form (1.5) was obtained in [1].

In the case of the Neumann boundary conditions and $m = 1$, the estimate (1.5) was proved in [4] for open sets $\Omega_1, \Omega_2 \in C_M^{0,\alpha}(\mathcal{A})$ satisfying $(\Omega_1)_\varepsilon \subset \Omega_2 \subset \Omega_1$. We remark that the result in [4] concerns only inner deformations of an open set and second order elliptic operators. Moreover, the proof in [4] is based on the ultracontractivity which holds for second order elliptic operators in open sets with Hölder continuous boundaries. Since ultracontractivity is not guaranteed for more general open sets, we had to develop a different method.

The third main result of the paper concerns the case $\Omega_1 = \Omega$ and $\Omega_2 = \varphi(\Omega)$, where φ is a suitable diffeomorphism of class C^m. In this case, we make very weak assumptions on Ω: if $m = 1$ it is just the requirement that H has discrete spectrum. Under such general assumptions, we prove that for both Dirichlet and Neumann boundary conditions there exists a constant $c > 0$ independent of n such that

$$|\lambda_n[\Omega] - \lambda_n[\varphi(\Omega)]| \leqslant c(1 + \lambda_n[\Omega]) \max_{0\leqslant|\alpha|\leqslant m} \|D^\alpha(\varphi - \mathrm{Id})\|_{L^\infty(\Omega)}$$

if $\max_{0\leqslant|\alpha|\leqslant m} \|D^\alpha(\varphi - \mathrm{Id})\|_{L^\infty(\Omega)} < c^{-1}$ (see Theorem 4.1 and Corollary 4.1).

The paper is organized as follows. In Sect. 2, we introduce some notation and formulate the eigenvalue problem for the operator (1.1). In Sect. 3, we define the class of open sets under consideration. In Sect. 4, we consider the case of diffeormorphic open sets. In Sect. 5, we prove the estimate (1.2) for the Dirichlet boundary conditions. In Sect. 6, we prove the estimate (1.2) for the Neumann boundary conditions. In Sect. 7, we prove the estimates (1.3) and (1.4) for both Dirichlet and Neumann boundary conditions. In Appendix, we discuss some properties of the atlas distance $d_{\mathcal{A}}$, the Hausdorff–Pompeiu lower deviation $d_{\mathcal{HP}}$, and the Hausdorff–Pompeiu distance $d^{\mathcal{HP}}$.

2 Preliminaries and Notation

Let $N, m \in \mathbb{N}$, and let Ω be an open set in $\mathbb{R}^N$. We denote by $W^{m,2}(\Omega)$ the Sobolev space of complex-valued functions in $L^2(\Omega)$ which have all distributional derivatives up to order m in $L^2(\Omega)$, endowed with the norm

$$\|u\|_{W^{m,2}(\Omega)} = \sum_{|\alpha|\leqslant m} \|D^\alpha u\|_{L^2(\Omega)}. \tag{2.1}$$

We denote by $W_0^{m,2}(\Omega)$ the closure in $W^{m,2}(\Omega)$ of the space of the C^∞-functions with compact support in Ω.

Lemma 2.1. *Let Ω be an open set in $\mathbb{R}^N$. Let $V(\Omega)$ be a subspace of $W^{m,2}(\Omega)$ such that the embedding $V(\Omega) \subset W^{m-1,2}(\Omega)$ is compact. Then there exists $c > 0$ such that*

$$\|u\|_{W^{m,2}(\Omega)} \leqslant c\Big(\|u\|_{L^2(\Omega)} + \sum_{|\alpha|=m} \|D^\alpha u\|_{L^2(\Omega)}\Big) \tag{2.2}$$

for all $u \in V(\Omega)$.

Proof. Since $(V(\Omega), \|\cdot\|_{m,2})$ is compactly embedded in $W^{m-1,2}(\Omega)$ and $W^{m-1,2}(\Omega)$ is continuously embedded in $L^2(\Omega)$, from the Lions lemma (see, for example, [2, p. 35]) it follows that for all $\varepsilon \in]0,1[$ there exists $c(\varepsilon) > 0$ such that

$$\|u\|_{W^{m-1,2}(\Omega)} \leqslant \varepsilon \|u\|_{W^{m,2}(\Omega)} + c(\varepsilon)\|u\|_{L^2(\Omega)}.$$

Hence

$$\|u\|_{W^{m-1,2}(\Omega)} \leqslant \frac{\varepsilon}{(1-\varepsilon)} \sum_{|\alpha|=m} \|D^\alpha u\|_{L^2(\Omega)} + \frac{c(\varepsilon)}{(1-\varepsilon)}\|u\|_{L^2(\Omega)} \tag{2.3}$$

for all $u \in V(\Omega)$. The inequality (2.2) immediately follows. □

Let $\widehat{m}$ be the number of the multiindices $\alpha = (\alpha_1, \dots, \alpha_N) \in \mathbb{N}_0^N$ with length $|\alpha| = \alpha_1 + \dots + \alpha_N$ equal to m. Here, $\mathbb{N}_0 = \mathbb{N} \cup \{0\}$. For all $\alpha, \beta \in \mathbb{N}_0^N$ such that $|\alpha| = |\beta| = m$, let $A_{\alpha\beta}$ be bounded measurable real-valued functions defined on Ω satisfying $A_{\alpha\beta} = A_{\beta\alpha}$ and the uniform ellipticity condition

$$\sum_{|\alpha|=|\beta|=m} A_{\alpha\beta}(x)\xi_\alpha\xi_\beta \geqslant \theta|\xi|^2 \tag{2.4}$$

for all $x \in \Omega$, $\xi = (\xi_\alpha)_{|\alpha|=m} \in \mathbb{R}^{\widehat{m}}$, where $\theta > 0$ is the ellipticity constant.

Let $V(\Omega)$ be a closed subspace of $W^{m,2}(\Omega)$ containing $W_0^{m,2}(\Omega)$. We consider the following eigenvalue problem:

$$\int_\Omega \sum_{|\alpha|=|\beta|=m} A_{\alpha\beta} D^\alpha u D^\beta \overline{v} dx = \lambda \int_\Omega u\overline{v} dx \tag{2.5}$$

for all test functions $v \in V(\Omega)$, in the unknowns $u \in V(\Omega)$ (the eigenfunctions) and $\lambda \in \mathbb{R}$ (the eigenvalues).

It is clear that the problem (2.5) is the weak formulation of an eigenvalue problem for the operator H in (1.1) subject to suitable homogeneous boundary conditions and the choice of $V(\Omega)$ corresponds to the choice of the boundary conditions (see, for example, [14]).

We set

$$Q_\Omega(u,v) = \int_\Omega \sum_{|\alpha|=|\beta|=m} A_{\alpha\beta} D^\alpha u D^\beta \overline{v} dx, \quad Q_\Omega(u) = Q_\Omega(u,u), \tag{2.6}$$

for all $u, v \in W^{m,2}(\Omega)$.

If the embedding $V(\Omega) \subset W^{m-1,2}(\Omega)$ is compact, then the eigenvalues of Eq. (2.5) coincide with the eigenvalues of a suitable operator $H_{V(\Omega)}$ canonically associated with the restriction of the quadratic form Q_Ω to $V(\Omega)$. In fact, we have the following theorem.

Theorem 2.1. *Let Ω be an open set in $\mathbb{R}^N$. Let $m \in \mathbb{N}$, $\theta > 0$ and, for all $\alpha, \beta \in \mathbb{N}_0^N$ such that $|\alpha| = |\beta| = m$, let $A_{\alpha\beta}$ be bounded measurable real-valued functions defined on Ω, satisfying $A_{\alpha\beta} = A_{\beta\alpha}$ and the condition* (2.4).

Let $V(\Omega)$ be a closed subspace of $W^{m,2}(\Omega)$ containing $W_0^{m,2}(\Omega)$ and such that the embedding $V(\Omega) \subset W^{m-1,2}(\Omega)$ is compact. Then there exists a nonnegative self-adjoint linear operator $H_{V(\Omega)}$ on $L^2(\Omega)$ with compact resolvent, such that $\mathrm{Dom}\,(H_{V(\Omega)}^{1/2}) = V(\Omega)$ *and*

$$\langle H_{V(\Omega)}^{1/2} u, H_{V(\Omega)}^{1/2} v \rangle_{L^2(\Omega)} = Q_\Omega(u,v) \tag{2.7}$$

for all $u, v \in V(\Omega)$. Moreover, the eigenvalues of Eq. (2.5) *coincide with the eigenvalues $\lambda_n[H_{V(\Omega)}]$ of $H_{V(\Omega)}$ and*

$$\lambda_n[H_{V(\Omega)}] = \inf_{\substack{\mathcal{L} \leqslant V(\Omega) \\ \dim \mathcal{L} = n}} \sup_{\substack{u \in \mathcal{L} \\ u \neq 0}} \frac{Q_\Omega(u)}{\|u\|^2_{L^2(\Omega)}}. \tag{2.8}$$

Proof. By Lemma 2.1, the inequality (2.4), and the boundedness of the coefficients $A_{\alpha\beta}$, it follows that the space $V(\Omega)$ endowed with the norm

$$(\|u\|^2_{L^2(\Omega)} + Q_\Omega(u))^{1/2} \tag{2.9}$$

for all $u \in V(\Omega)$, is complete. Indeed, this norm is equivalent on $V(\Omega)$ to the norm defined by (2.1). Thus, the restriction of the quadratic form Q_Ω to $V(\Omega)$ is a closed quadratic form on $V(\Omega)$ (see [7, pp. 81-83]) and there exists a nonnegative self-adjoint operator $H_{V(\Omega)}$ on $L^2(\Omega)$ satisfying $\mathrm{Dom}\,(H_{V(\Omega)}^{1/2}) = V(\Omega)$ and the condition (2.7) (see [7, Theorem 4.4.2]). Since the embedding $V(\Omega) \subset L^2(\Omega)$ is compact, $H_{V(\Omega)}$ has compact resolvent (see [7, Exercise 4.2]). The fact that the eigenvalues of Eq. (2.5) coincide with the eigenvalues of the operator $H_{V(\Omega)}$ is well known. Finally, the variational representation in (2.8) is given by the well-known minmax principle (see [7, Theorem 4.5.3]). □

Definition 2.1. Let Ω be an open set in $\mathbb{R}^N$. Let $m \in \mathbb{N}$, $\theta > 0$ and, for all $\alpha, \beta \in \mathbb{N}_0^N$ such that $|\alpha| = |\beta| = m$, let $A_{\alpha\beta}$ be bounded measurable real-valued functions defined on Ω, satisfying $A_{\alpha\beta} = A_{\beta\alpha}$ and the condition (2.4).

If the embedding $W_0^{m,2}(\Omega) \subset W^{m-1,2}(\Omega)$ is compact, we set

$$\lambda_{n,\mathcal{D}}[\Omega] = \lambda_n[H_{W_0^{m,2}(\Omega)}].$$

If the embedding $W^{m,2}(\Omega) \subset W^{m-1,2}(\Omega)$ is compact, we set

$$\lambda_{n,\mathcal{N}}[\Omega] = \lambda_n[H_{W^{m,2}(\Omega)}].$$

The numbers $\lambda_{n,\mathcal{D}}[\Omega]$, $\lambda_{n,\mathcal{N}}[\Omega]$ are called the *Dirichlet eigenvalues*, *Neumann eigenvalues* respectively, of the operator (1.1).

When we refer to both Dirichlet and Neumann boundary conditions, we write just $\lambda_n[\Omega]$ instead of $\lambda_{n,\mathcal{D}}[\Omega]$ and $\lambda_{n,\mathcal{N}}[\Omega]$.

Remark 2.1. If Ω is such that the embedding $W_0^{1,2}(\Omega) \subset L^2(\Omega)$ is compact (for instance, if Ω is an arbitrary open set with finite Lebesgue measure), then also the embedding $W_0^{m,2}(\Omega) \subset W^{m-1,2}(\Omega)$ is compact and the Dirichlet eigenvalues are well defined.

If Ω is such that the embedding $W^{1,2}(\Omega) \subset L^2(\Omega)$ is compact (for instance, if Ω has a continuous boundary, see Definition 3.1), then the embedding $W^{m,2}(\Omega) \subset W^{m-1,2}(\Omega)$ is compact and the Neumann eigenvalues are well defined.

Example 2.1. Let Ω be an open set in $\mathbb{R}^2$. We consider the biharmonic operator Δ^2 in $\mathbb{R}^2$ and the sesquilinear form

$$Q_\Omega(u,v) = \int\limits_\Omega \left(\frac{\partial^2 u}{\partial x_1^2}\frac{\partial^2 \overline{v}}{\partial x_1^2} + 2\frac{\partial^2 u}{\partial x_1 \partial x_2}\frac{\partial^2 \overline{v}}{\partial x_1 \partial x_2} + \frac{\partial^2 u}{\partial x_2^2}\frac{\partial^2 \overline{v}}{\partial x_2^2} \right) dx, \quad u,v \in V(\Omega),$$

where $V(\Omega)$ is either $W_0^{2,2}(\Omega)$ (Dirichlet boundary conditions) or $W^{2,2}(\Omega)$ (Neumann boundary conditions). Recall that the Euler–Lagrange equation for the minimization of the quadratic form $Q_\Omega(u,u)$ is $\Delta^2 u = 0$. Note that the condition (2.4) is satisfied with $\theta = 1$.

Let $H_{V(\Omega)}$ be the operator associated with the quadratic form Q_Ω as in Theorem 2.1. Consider the eigenvalue problem

$$H_{V(\Omega)} u = \lambda u. \tag{2.10}$$

In the case $V(\Omega) = W_0^{2,2}(\Omega)$, Eq. (2.10) is the weak formulation of the classical eigenvalue problem for the biharmonic operator subject to Dirichlet boundary conditions

$$\begin{aligned} &\Delta^2 u = \lambda u \quad \text{in } \Omega, \\ &u = 0 \quad \text{on } \partial\Omega, \\ &\frac{\partial u}{\partial n} = 0 \quad \text{on } \partial\Omega \end{aligned} \tag{2.11}$$

for bounded domains Ω of class C^2. Here, $n = (n_1, n_2)$ is the unit outer normal to $\partial\Omega$.

In the case $V(\Omega) = W^{2,2}(\Omega)$, Eq. (2.10) is the weak formulation of the classical eigenvalue problem for the biharmonic operator subject to Neumann boundary conditions

$$\begin{aligned} &\Delta^2 u = \lambda u, \quad \text{in } \Omega, \\ &\frac{\partial^2 u}{\partial n^2} = 0, \quad \text{on } \partial\Omega, \\ &\frac{d}{ds}\frac{\partial^2 u}{\partial n \partial t} + \frac{\partial \Delta u}{\partial n} = 0, \quad \text{on } \partial\Omega \end{aligned} \tag{2.12}$$

for bounded domains Ω of class C^2. Here,

$$\frac{\partial^2 u}{\partial n^2} = \sum_{i,j=1}^{2} \frac{\partial^2 u}{\partial x_i \partial x_j} n_i n_j, \quad \frac{\partial^2 u}{\partial n \partial t} = \sum_{i,j=1}^{2} \frac{\partial^2 u}{\partial x_i \partial x_j} n_i t_j,$$

s denotes the arclengh of $\partial\Omega$ (with positive orientation), $t = (t_1, t_2)$ denotes the unit tangent vector to $\partial\Omega$ (oriented in the sense of increasing s). This follows by a standard argument and observing that if $u, v \in C^4(\overline{\Omega})$, then, by the divergence theorem,

$$\begin{aligned} Q_\Omega(u,v) &= \int_\Omega \Delta^2 u \overline{v} dx - \int_{\partial\Omega} \frac{\partial \Delta u}{\partial n} \overline{v} d\sigma + \int_{\partial\Omega} \left(n_1 \nabla \frac{\partial u}{\partial x_1} + n_2 \nabla \frac{\partial u}{\partial x_2} \right) \cdot \nabla \overline{v} d\sigma \\ &= \int_\Omega \Delta^2 u \overline{v} dx + \int_{\partial\Omega} \left(\frac{\partial^2 u}{\partial n^2} \frac{\partial \overline{v}}{\partial n} + \frac{\partial^2 u}{\partial n \partial t} \frac{\partial \overline{v}}{\partial t} - \frac{\partial \Delta u}{\partial n} \overline{v} \right) d\sigma. \end{aligned}$$

One may also consider the sesquilinear form

$$Q_\Omega^{(\nu)}(u,v) = \nu \int_\Omega \Delta u \Delta \overline{v} + (1-\nu) Q_\Omega(u,v), \quad u, v \in V(\Omega).$$

If $0 \leqslant \nu < 1$, then the condition (2.4) is satisfied with $\theta = 1 - \nu$. Note that the Euler–Lagrange equation for the minimization of the quadratic form $Q_\Omega^{(\nu)}(u,u)$ is again $\Delta^2 u = 0$.

Let $H_{V(\Omega)}^{(\nu)}$ be the operator associated with the quadratic form $Q_\Omega^{(\nu)}$ as in Theorem 2.1. Consider the eigenvalue problem

$$H_{V(\Omega)}^{(\nu)} u = \lambda u. \tag{2.13}$$

In the case $V(\Omega) = W_0^{2,2}(\Omega)$, Eq. (2.13) is another weak formulation of the classical eigenvalue problem (2.11).

In the case $V(\Omega) = W^{2,2}(\Omega)$, Eq. (2.13) is the weak formulation of the classical eigenvalue problem for the biharmonic operator subject to Neumann boundary conditions depending on ν

$$\begin{aligned} &\Delta^2 u = \lambda u \quad \text{in } \Omega, \\ &\nu \Delta u + (1-\nu)\frac{\partial^2 u}{\partial n^2} = 0 \quad \text{on } \partial\Omega, \\ &(1-\nu)\frac{d}{ds}\frac{\partial^2 u}{\partial n \partial t} + \frac{\partial \Delta u}{\partial n} = 0 \quad \text{on } \partial\Omega. \end{aligned} \tag{2.14}$$

The biharmonic operator subject to these boundary conditions with $0 < \nu < 1/2$ arises in the study of small deformations of a thin plate under Kirchhoff hypothesis in which case ν is the Poisson ratio of the plate (see, for example, [13] and the references therein).

3 Open Sets with Continuous Boundaries

We recall that for any set V in $\mathbb{R}^N$ and $\delta > 0$ we denote by V_δ the set $\{x \in V : d(x, \partial\Omega) > \delta\}$. Moreover, by a rotation in $\mathbb{R}^N$ we mean a $N \times N$-orthogonal matrix with real entries which we identify with the corresponding linear operator acting in $\mathbb{R}^N$.

Definition 3.1. Let $\rho > 0$, $s, s' \in \mathbb{N}$, $s' \leqslant s$ and $\{V_j\}_{j=1}^s$ be a family of bounded open cuboids and $\{r_j\}_{j=1}^s$ be a family of rotations in $\mathbb{R}^N$.

We say that $\mathcal{A} = (\rho, s, s', \{V_j\}_{j=1}^s, \{r_j\}_{j=1}^s)$ is an *atlas* in $\mathbb{R}^N$ with the parameters $\rho, s, s', \{V_j\}_{j=1}^s, \{r_j\}_{j=1}^s$, briefly an atlas in $\mathbb{R}^N$.

We denote by $C(\mathcal{A})$ the family of all open sets Ω in $\mathbb{R}^N$ satisfying the following properties:

(i) $\Omega \subset \bigcup_{j=1}^{s} (V_j)_\rho$ and $(V_j)_\rho \cap \Omega \neq \varnothing$;

(ii) $V_j \cap \partial\Omega \neq \varnothing$ for $j = 1, \dots s'$, $V_j \cap \partial\Omega = \varnothing$ for $s' < j \leqslant s$;

(iii) for $j = 1, \dots, s$

$$r_j(V_j) = \{\, x \in \mathbb{R}^N : \ a_{ij} < x_i < b_{ij},\ i = 1, \dots, N\},$$

and

$$r_j(\Omega \cap V_j) = \{x \in \mathbb{R}^N : \ a_{Nj} < x_N < g_j(\overline{x}),\ \overline{x} \in W_j\},$$

where $\overline{x} = (x_1, \dots, x_{N-1})$, $W_j = \{\overline{x} \in \mathbb{R}^{N-1} : \ a_{ij} < x_i < b_{ij},\ i = 1, \dots, N-1\}$ and g_j is a continuous function defined on $\overline{W}_j$ (this means if $s' < j \leqslant s$, then $g_j(\overline{x}) = b_{Nj}$ for all $\overline{x} \in \overline{W}_j$); moreover for $j = 1, \dots, s'$

$$a_{Nj} + \rho \leqslant g_j(\overline{x}) \leqslant b_{Nj} - \rho,$$

for all $\overline{x} \in \overline{W}_j$.

We say that an open set Ω in $\mathbb{R}^N$ is an open set with a continuous boundary if Ω is of class $C(\mathcal{A})$ for some atlas $\mathcal{A}$.

We note that for an open set Ω of class $C(\mathcal{A})$ the inequality (2.2) holds for all $u \in W^{m,2}(\Omega)$ with a constant c depending only on $\mathcal{A}$. More precisely, we denote by $\mathcal{D}_\Omega$ the *best constant* for which the inequality (2.2) is satisfied for $V(\Omega) = W_0^{m,2}(\Omega)$. We denote by $\mathcal{N}_\Omega$ the *best constant* for which the inequality (2.2) is satisfied for $V(\Omega) = W^{m,2}(\Omega)$. Then we have the following assertion (for a proof we refer to [3, Theorem 6, p. 160]).

Lemma 3.1. *Let $\mathcal{A}$ be an atlas in $\mathbb{R}^N$, $m \in \mathbb{N}$. There exists $c > 0$ depending only on $N, \mathcal{A}$, and m such that*

$$1 \leqslant \mathcal{D}_\Omega \leqslant \mathcal{N}_\Omega \leqslant c \tag{3.1}$$

for all open sets $\Omega \in C(\mathcal{A})$.

Lemma 3.2. *Let $\mathcal{A}$ be an atlas in $\mathbb{R}^N$. Let $m \in \mathbb{N}$, $L, \theta > 0$ and, for all $\alpha, \beta \in \mathbb{N}_0^N$ with $|\alpha| = |\beta| = m$, let $A_{\alpha\beta} \in L^\infty\big(\bigcup_{j=1}^{s} V_j\big)$ satisfy $A_{\alpha\beta} = A_{\beta\alpha}$, $\|A_{\alpha\beta}\|_{L^\infty\left(\bigcup_{j=1}^{s} V_j\right)} \leqslant L$ and the condition (2.4). Then for each $n \in \mathbb{N}$ there exists $\Lambda_n > 0$ depending only on n, N, $\mathcal{A}$, m, and L such that*

$$\lambda_{n,\mathcal{N}}[\Omega] \leqslant \lambda_{n,\mathcal{D}}[\Omega] \leqslant \Lambda_n \tag{3.2}$$

for all open sets $\Omega \in C(\mathcal{A})$.

Proof. The inequality $\lambda_{n,\mathcal{N}}[\Omega] \leqslant \lambda_{n,\mathcal{D}}[\Omega]$ is well known. Now, we prove the second inequality. It is clear that there exists a ball B of radius $\rho/2$ such that $B \subset \Omega$ for all open sets $\Omega \in C(\mathcal{A})$. By the well-known monotonicity of the Dirichlet eigenvalues with respect to inclusion, it follows that

$$\lambda_{n,\mathcal{D}}[\Omega] \leqslant \lambda_{n,\mathcal{D}}[B].$$

Thus, it suffices to estimate $\lambda_{n,\mathcal{D}}[B]$. It is clear that there exists $c > 0$ depending only on N and m such that

$$Q_B(u) \leqslant cL \int_B |\nabla^m u|^2 dx \tag{3.3}$$

for all $u \in W_0^{m,2}(B)$, where $\nabla^m u = (D^\alpha u)_{|\alpha|=m}$. By (2.8) and (3.3),

$$\lambda_{n,\mathcal{D}}[B] \leqslant \Lambda_n \equiv cL \inf_{\substack{\mathcal{L} \leqslant W_0^{m,2}(B) \\ \dim \mathcal{L}=n}} \sup_{\substack{u \in \mathcal{L} \\ u \neq 0}} \frac{\displaystyle\int_B |\nabla^m u|^2 dx}{\displaystyle\int_B |u|^2 dx} < \infty.$$

It is clear that Λ_n depends only on n, N, ρ, m, and L. □

4 The Case of Diffeomorphic Open Sets

Lemma 4.1. *Let Ω be an open set in $\mathbb{R}^N$. Let $m \in \mathbb{N}$, $B_1, B_2 > 0$ and φ be a diffeomorphism of Ω onto $\varphi(\Omega)$ of class C^m such that*

$$\max_{1 \leqslant |\alpha| \leqslant m} |D^\alpha \varphi(x)| \leqslant B_1, \quad |\det \nabla\varphi(x)| \geqslant B_2 \tag{4.1}$$

for all $x \in \Omega$. Let $B_3 > 0$ and, for all $\alpha, \beta \in \mathbb{N}_0^N$ such that $|\alpha| = |\beta| = m$, let $A_{\alpha\beta}$ be measurable real-valued functions defined on $\Omega \cup \varphi(\Omega)$ satisfying

$$\max_{|\alpha|=|\beta|=m} |A_{\alpha\beta}(x)| \leqslant B_3 \tag{4.2}$$

for almost all $x \in \Omega \cup \varphi(\Omega)$. Then there exists $c > 0$ depending only on N, m, B_1, B_2, and B_3 such that

$$\left| Q_{\varphi(\Omega)}(u \circ \varphi^{(-1)}) - Q_\Omega(u) \right| \leqslant c\mathcal{L}(\varphi) \int_\Omega \sum_{1 \leqslant |\alpha| \leqslant m} |D^\alpha u|^2 dx \tag{4.3}$$

for all $u \in W^{m,2}(\Omega)$, where

$$\mathcal{L}(\varphi) = \max_{1 \leqslant |\alpha| \leqslant m} \|D^\alpha(\varphi - \mathrm{Id})\|_{L^\infty(\Omega)} + \max_{|\alpha|=|\beta|=m} \|A_{\alpha\beta} \circ \varphi - A_{\alpha\beta}\|_{L^\infty(\Omega)}. \tag{4.4}$$

Proof. By changing variables and using a known formula for high derivatives of composite functions (see, for example, [10, Formula B]), we have

$$Q_{\varphi(\Omega)}(u \circ \varphi^{(-1)}) = \int_{\varphi(\Omega)} \sum_{|\alpha|=|\beta|=m} A_{\alpha\beta} D^\alpha(u \circ \varphi^{(-1)}) \overline{D^\beta(u \circ \varphi^{(-1)})} dy$$

$$= \int_\Omega \sum_{|\alpha|=|\beta|=m} \left(A_{\alpha\beta} D^\alpha(u \circ \varphi^{(-1)}) \overline{D^\beta(u \circ \varphi^{(-1)})} \right) \circ \varphi |\det \nabla\varphi| dx$$

$$= \int_{\Omega} \sum_{|\alpha|=|\beta|=m} A_{\alpha\beta} \circ \varphi \sum_{\substack{1 \leqslant |\eta| \leqslant |\alpha| \\ 1 \leqslant |\xi| \leqslant |\beta|}} D^{\eta} u \overline{D^{\xi} u} \, (p_{\alpha\eta}(\varphi^{(-1)}) p_{\beta\xi}(\varphi^{(-1)})) \circ \varphi |\det \nabla\varphi| dx$$

$$= \sum_{\substack{|\alpha|=|\beta|=m \\ 1 \leqslant |\eta| \leqslant |\alpha| \\ 1 \leqslant |\xi| \leqslant |\beta|}} \int_{\Omega} (A_{\alpha\beta} p_{\alpha\eta}(\varphi^{(-1)}) p_{\beta\xi}(\varphi^{(-1)})) \circ \varphi \, D^{\eta} u \overline{D^{\xi} u} \, |\det \nabla\varphi| dx \tag{4.5}$$

for all $u \in W^{m,2}(\Omega)$, where for all α, η with $1 \leqslant |\eta| \leqslant |\alpha| = m$, $p_{\alpha\eta}(\varphi^{(-1)})$ denotes a polynomial of degree $|\eta|$ in derivatives of $\varphi^{(-1)}$ of order between 1 and $|\alpha|$, with coefficients depending only on N, α, η.

We recall that for each α with $1 \leqslant |\alpha| \leqslant m$ there exists a polynomial $p_{\alpha}(\varphi)$ in derivatives of φ of order between 1 and $|\alpha|$, with coefficients depending only on N, α, such that

$$(D^{\alpha} \varphi^{(-1)}) \circ \varphi = \frac{p_{\alpha}(\varphi)}{(\det \nabla\varphi)^{2|\alpha|-1}}. \tag{4.6}$$

In order to estimate $Q_{\varphi(\Omega)}(u \circ \varphi^{(-1)}) - Q_{\Omega}(u)$, it is enough to estimate the expressions

$$(A_{\alpha\beta} p_{\alpha\eta}(\varphi^{(-1)}) p_{\beta\xi}(\varphi^{(-1)})) \circ \varphi \; |\det \nabla\varphi| \\ - (A_{\alpha\beta} p_{\alpha\eta}(\widetilde{\varphi}^{(-1)}) p_{\beta\xi}(\widetilde{\varphi}^{(-1)})) \circ \widetilde{\varphi} \; |\det \nabla\widetilde{\varphi}|,$$

where $\widetilde{\varphi} = \mathrm{Id}$. This can be done by using the triangle inequality and observing that (4.6) implies

$$|(D^{\alpha} \varphi^{(-1)}) \circ \varphi - (D^{\alpha} \widetilde{\varphi}^{(-1)}) \circ \widetilde{\varphi}| \leqslant c \max_{1 \leqslant |\beta| \leqslant |\alpha|} \|D^{\beta}(\varphi - \widetilde{\varphi})\|_{L^{\infty}(\Omega)},$$

where c depends only on N, α, B_1, and B_2. □

Theorem 4.1. *Let U be an open set in $\mathbb{R}^N$. Let $m \in \mathbb{N}$, $B_1, B_2, B_3, \theta > 0$. For all $\alpha, \beta \in \mathbb{N}_0^N$ with $|\alpha| = |\beta| = m$, let $A_{\alpha\beta}$ be measurable real-valued functions defined on U satisfying $A_{\alpha\beta} = A_{\beta\alpha}$ and the conditions (2.4), (4.2) in U. The following statements hold.*

(i) *There exists $c_1 > 0$ depending only on N, m, B_1, B_2, B_3, θ such that for all $n \in \mathbb{N}$, for all open sets $\Omega \subset U$ such that the embedding $W_0^{m,2}(\Omega) \subset W^{m-1,2}(\Omega)$ is compact, and for all diffeomorphisms of Ω onto $\varphi(\Omega)$ of class C^m satisfying (4.1) and such that $\varphi(\Omega) \subset U$ the inequality*

$$|\lambda_{n,\mathcal{D}}[\Omega] - \lambda_{n,\mathcal{D}}[\varphi(\Omega)]| \leqslant c_1 \, \mathcal{D}_{\Omega}^2 (1 + \lambda_{n,\mathcal{D}}[\Omega]) \mathcal{L}(\varphi) \tag{4.7}$$

holds if $\mathcal{L}(\varphi) < (c_1 \, \mathcal{D}_{\Omega}^2)^{-1}$.

(ii) *There exists $c_2 > 0$ depending only on N, m, B_1, B_2, B_3, θ such that for all $n \in \mathbb{N}$, for all open sets $\Omega \subset U$ such that the embedding $W^{m,2}(\Omega) \subset$*

$W^{m-1,2}(\Omega)$ is compact, and for all diffeomorphisms of Ω onto $\varphi(\Omega)$ of class C^m satisfying (4.1) and such that $\varphi(\Omega) \subset U$, the inequality

$$|\lambda_{n,\mathcal{N}}[\Omega] - \lambda_{n,\mathcal{N}}[\varphi(\Omega)]| \leqslant c_2 \mathcal{N}_\Omega^2 (1 + \lambda_{n,\mathcal{N}}[\Omega])\mathcal{L}(\varphi) \tag{4.8}$$

holds if $\mathcal{L}(\varphi) < (c_2 \mathcal{N}_\Omega^2)^{-1}$.

Proof. We prove statement (i). Let $\Omega \subset U$ be an open set such that the embedding $W_0^{m,2}(\Omega) \subset W^{m-1,2}(\Omega)$ is compact and φ be a diffeomorphisms of Ω onto $\varphi(\Omega)$ of class C^m satisfying (4.1) and such that $\varphi(\Omega) \subset U$. By the inequalities (2.2), (2.4), and (4.3), there exists $c_3 > 0$ depending only on N, m, B_1, B_2, B_3, and θ such that

$$\Big|Q_{\varphi(\Omega)}(u \circ \varphi^{(-1)}) - Q_\Omega(u)\Big| \leqslant c_3 \mathcal{D}_\Omega^2 (\|u\|^2_{L^2(\Omega)} + Q_\Omega(u))\mathcal{L}(\varphi). \tag{4.9}$$

Then

$$\left| \frac{Q_{\varphi(\Omega)}(u \circ \varphi^{(-1)})}{\|u \circ \varphi^{(-1)}\|^2_{L^2(\varphi(\Omega))}} - \frac{Q_\Omega(u)}{\|u\|^2_{L^2(\Omega)}} \right|$$

$$\leqslant \frac{|Q_{\varphi(\Omega)}(u \circ \varphi^{(-1)}) - Q_\Omega(u)|}{\int\limits_\Omega |u|^2 |\det \nabla\varphi| dx} + \frac{Q_\Omega(u) \int\limits_\Omega |u|^2 || \det \nabla\varphi| - 1| dx}{\int\limits_\Omega |u|^2 |\det \nabla\varphi| dx \int\limits_\Omega |u|^2 dx}. \tag{4.10}$$

Observing that $\mathcal{D}_\Omega \geqslant 1$ and combining the inequalities (4.9) and (4.10), we see that there exists $c_4 > 0$ depending only on N, m, B_1, B_2, B_3, θ such that for all $u \in W_0^{m,2}(\Omega)$

$$\left| \frac{Q_{\varphi(\Omega)}(u \circ \varphi^{(-1)})}{\|u \circ \varphi^{(-1)}\|^2_{L^2(\varphi(\Omega))}} - \frac{Q_\Omega(u)}{\|u\|^2_{L^2(\Omega)}} \right| \leqslant c_4 \mathcal{D}_\Omega^2 \left(1 + \frac{Q_\Omega(u)}{\|u\|^2_{L^2(\Omega)}}\right) \mathcal{L}(\varphi), \tag{4.11}$$

which can be written as

$$(1 - c_4 \mathcal{D}_\Omega^2 \mathcal{L}(\varphi)) \frac{Q_\Omega(u)}{\|u\|^2_{L^2(\Omega)}} - c_4 \mathcal{D}_\Omega^2 \mathcal{L}(\varphi) \tag{4.12}$$

$$\leqslant \frac{Q_{\varphi(\Omega)}(u \circ \varphi^{(-1)})}{\|u \circ \varphi^{(-1)}\|^2_{L^2(\varphi(\Omega))}} \leqslant (1 + c_4 \mathcal{D}_\Omega^2 \mathcal{L}(\varphi)) \frac{Q_\Omega(u)}{\|u\|^2_{L^2(\Omega)}} + c_4 \mathcal{D}_\Omega^2 \mathcal{L}(\varphi).$$

Assume now that $1 - c_4 \mathcal{D}_\Omega^2 \mathcal{L}(\varphi) > 0$. Note that the map C_φ of $L^2(\Omega)$ to $L^2(\varphi(\Omega))$ which takes $u \in L^2(\Omega)$ to $C_\varphi u = u \circ \varphi^{-1}$ is a linear homeomorphism which restricts to a linear homeomorphism of $W_0^{m,2}(\Omega)$ onto $W_0^{m,2}(\varphi(\Omega))$, and that the embedding $W_0^{m,2}(\varphi(\Omega)) \subset W^{m-1,2}(\varphi(\Omega))$ is compact. Then, applying the minmax principle (2.8) and using the inequality (4.12), it easy to deduce the validity of the inequality (4.7).

The proof of statement (ii) is similar. In this case, observe that the map C_φ defined above restricts to a linear homeomorphism from $W^{m,2}(\Omega)$ onto $W^{m,2}(\varphi(\Omega))$ and if the embedding $W^{m,2}(\Omega) \subset W^{m-1,2}(\Omega)$ is compact, then the embedding $W^{m,2}(\varphi(\Omega)) \subset W^{m-1,2}(\varphi(\Omega))$ is also compact. □

Corollary 4.1. *Let $\mathcal{A}$ be an atlas in $\mathbb{R}^N$. Let $m \in \mathbb{N}$, $B_1, B_2, L, \theta > 0$ and, for all $\alpha, \beta \in \mathbb{N}_0^N$ with $|\alpha| = |\beta| = m$, let $A_{\alpha\beta} \in C^{0,1}\big(\bigcup_{j=1}^s V_j\big)$ satisfy $A_{\alpha\beta} = A_{\beta\alpha}$, $\|A_{\alpha\beta}\|_{C^{0,1}(\bigcup_{j=1}^s V_j)} \leqslant L$ and the condition (2.4). Then there exists $c > 0$ depending only on N, $\mathcal{A}$, m, B_1, B_2, L, θ such that for all $n \in \mathbb{N}$, for all open sets $\Omega \in C(\mathcal{A})$, and for all diffeomorphisms of Ω onto $\varphi(\Omega)$ of class C^m satisfying (4.1) and such that $\varphi(\Omega) \subset \bigcup_{j=1}^s V_j$, the inequality*

$$|\lambda_n[\Omega] - \lambda_n[\varphi(\Omega)]| \leqslant c(1 + \lambda_n[\Omega]) \max_{0 \leqslant |\alpha| \leqslant m} \|D^\alpha(\varphi - \mathrm{Id})\|_{L^\infty(\Omega)} \tag{4.13}$$

holds for both Dirichlet and Neumann boundary conditions if

$$\max_{0 \leqslant |\alpha| \leqslant m} \|D^\alpha(\varphi - \mathrm{Id})\|_{L^\infty(\Omega)} < c^{-1}.$$

Proof. It suffices to apply Lemma 3.1 and Theorem 4.1. □

5 Estimates for Dirichlet Eigenvalues via the Atlas Distance

Definition 5.1. Let $\mathcal{A} = (\rho, s, s', \{V_j\}_{j=1}^s, \{r_j\}_{j=1}^s)$ be an atlas in $\mathbb{R}^N$. For all $\Omega_1, \Omega_2 \in C(\mathcal{A})$ we define the *atlas distance* $d_{\mathcal{A}}$ by

$$d_{\mathcal{A}}(\Omega_1, \Omega_2) = \max_{j=1,\dots,s} \sup_{(\overline{x}, x_N) \in r_j(V_j)} |g_{1j}(\overline{x}) - g_{2j}(\overline{x})|, \tag{5.1}$$

where g_{1j} and g_{2j} are the functions describing the boundaries of Ω_1 and Ω_2 respectively, as in Definition 3.1 (iii).

We observe that the function $d_{\mathcal{A}}(\cdot, \cdot)$ is, in fact, a distance in $C(\mathcal{A})$ (for further properties of $d_{\mathcal{A}}$ see also Appendix).

If $\Omega \in C(\mathcal{A})$, it is useful to set

$$d_j(x, \partial\Omega) = |g_j(\overline{(r_j(x))}) - (r_j(x))_N| \tag{5.2}$$

for all $j = 1, \dots, s$ and $x \in V_j$, where g_j and r_j are as in Definition 3.1.

Let $\mathcal{A} = (\rho, s, s', \{V_j\}_{j=1}^s, \{r_j\}_{j=1}^s)$ be an atlas in $\mathbb{R}^N$. We consider a partition of unity $\{\psi_j\}_{j=1}^s$ such that $\psi_j \in C_c^\infty(\mathbb{R}^N)$, $\operatorname{supp}\psi_j \subset (V_j)_{\frac{3}{4}\rho}$, $0 \leqslant \psi_j(x) \leqslant 1$, $|\nabla\psi_j(x)| \leqslant G$ for all $x \in \mathbb{R}^N$ and $j = 1, \dots, s$, where $G > 0$ depends only on $\mathcal{A}$, and such that $\sum_{j=1}^s \psi_j(x) = 1$ for all $x \in \bigcup_{j=1}^s (V_j)_\rho$.

For $\varepsilon \geqslant 0$ we consider the transformation

$$T_\varepsilon(x) = x - \varepsilon \sum_{j=1}^s \xi_j \psi_j(x), \quad x \in \mathbb{R}^N, \tag{5.3}$$

where $\xi_j = r_j^{(-1)}((0, \dots, 1))$, which was introduced in [4].

Then we have the following variant of Lemma 18 in [4].

Lemma 5.1. *Let $\mathcal{A} = (\rho, s, s', \{V_j\}_{j=1}^s, \{r_j\}_{j=1}^s)$ be an atlas in $\mathbb{R}^N$. Then there exist $A_1, A_2, E_1 > 0$ depending only on N and $\mathcal{A}$ such that*

$$\max_{0 \leqslant |\alpha| \leqslant m} \left\| D^\alpha (T_\varepsilon - \mathrm{Id}) \right\|_{L^\infty(\mathbb{R}^N)} \leqslant A_1 \varepsilon \tag{5.4}$$

and such that

$$\frac{1}{2} \leqslant 1 - A_2\varepsilon \leqslant \det \nabla T_\varepsilon \leqslant 1 + A_2\varepsilon \tag{5.5}$$

for all $0 \leqslant \varepsilon < E_1$. Furthermore,

$$T_\varepsilon(\Omega_1) \subset \Omega_2 \tag{5.6}$$

for all $0 < \varepsilon < E_1$ and $\Omega_1, \Omega_2 \in C(\mathcal{A})$ such that $\Omega_2 \subset \Omega_1$ and

$$d_{\mathcal{A}}(\Omega_1, \Omega_2) < \frac{\varepsilon}{s}. \tag{5.7}$$

Proof. The inequalities (5.4) and (5.5) are obvious. We prove the inclusion (5.6). Let $\Omega_1, \Omega_2 \in C(\mathcal{A})$ satisfy $\Omega_2 \subset \Omega_1$ and (5.7). For all $j = 1, \dots, s$ we denote by g_{1j}, g_{2j} respectively, the functions describing the boundaries of Ω_1, Ω_2 respectively, as in Definition 3.1 (iii). For all $x \in \bigcup_{j=1}^s V_j$ we set $J(x) = \{j \in \{1, \dots, s\} : x \in (V_j)_{\frac{3}{4}\rho}\}$. Let $x \in \Omega_1$. From the proof of Lemma 18 in [4] it follows that $0 < \varepsilon < \frac{\rho}{4}$ implies $T_\varepsilon(x) \in \Omega_1 \cap (V_j)_{\frac{\rho}{2}}$ and $d_j(T_\varepsilon(x), \partial\Omega_1) \geqslant \varepsilon\psi_j(x)$ for all $j \in J(x)$, where $\{\psi_j\}_{j=1}^s$ is an appropriate partition of unity satisfying $\operatorname{supp}\psi_j \subset (V_j)_{\frac{3}{4}\rho}$. Therefore,

$$\sum_{j \in J(x)} d_j(T_\varepsilon(x), \partial\Omega_1) \geqslant \varepsilon \sum_{j \in J(x)} \psi_j(x) = \varepsilon \sum_{j=1}^s \psi_j(x) = \varepsilon.$$

Hence there exists $\tilde{j} \in J(x)$ such that $d_{\tilde{j}}(T_\varepsilon(x), \partial\Omega_1) \geqslant \frac{\varepsilon}{s}$, which implies

$$(r_{\tilde{j}}(T_\varepsilon(x)))_N < g_{2\tilde{j}}(\overline{r_{\tilde{j}}(T_\varepsilon(x))}). \tag{5.8}$$

Indeed, assume to the contrary that $(r_{\tilde{j}}(T_\varepsilon(x)))_N \geqslant g_{2\tilde{j}}(\overline{r_{\tilde{j}}(T_\varepsilon(x))})$. Then we have

$$\begin{aligned} d_{\mathcal{A}}(\Omega_1, \Omega_2) &\geqslant g_{1\tilde{j}}(\overline{r_{\tilde{j}}(T_\varepsilon(x))}) - g_{2\tilde{j}}(\overline{r_{\tilde{j}}(T_\varepsilon(x))}) \\ &= g_{1\tilde{j}}(\overline{r_{\tilde{j}}(T_\varepsilon(x))}) - (r_{\tilde{j}}(T_\varepsilon(x)))_N + (r_{\tilde{j}}(T_\varepsilon(x)))_N - g_{2\tilde{j}}(\overline{r_{\tilde{j}}(T_\varepsilon(x))}) \\ &\geqslant g_{1\tilde{j}}(\overline{r_{\tilde{j}}(T_\varepsilon(x))}) - (r_{\tilde{j}}(T_\varepsilon(x)))_N = d_{\tilde{j}}(T_\varepsilon(x), \partial\Omega_1) \geqslant \frac{\varepsilon}{s}, \end{aligned} \tag{5.9}$$

which contradicts (5.7). Thus, (5.8) holds hence $T_\varepsilon(x) \in \Omega_2$. □

Theorem 5.1. *Let $\mathcal{A}$ be an atlas in $\mathbb{R}^N$. Let $m \in \mathbb{N}$, $L, \theta > 0$ and, for all $\alpha, \beta \in \mathbb{N}_0^N$ with $|\alpha| = |\beta| = m$, let $A_{\alpha\beta} \in C^{0,1}\big(\bigcup_{j=1}^{s} V_j\big)$ satisfy $A_{\alpha\beta} = A_{\beta\alpha}$, $\|A_{\alpha\beta}\|_{C^{0,1}(\bigcup_{j=1}^{s} V_j)} \leqslant L$ and the condition (2.4).*

Then for each $n \in \mathbb{N}$ there exist $c_n, \varepsilon_n > 0$ depending only on n, N, $\mathcal{A}$, m, L, θ such that

$$|\lambda_{n,\mathcal{D}}[\Omega_1] - \lambda_{n,\mathcal{D}}[\Omega_2]| \leqslant c_n d_{\mathcal{A}}(\Omega_1, \Omega_2), \tag{5.10}$$

for all $\Omega_1, \Omega_2 \in C(\mathcal{A})$ satisfying $d_{\mathcal{A}}(\Omega_1, \Omega_2) < \varepsilon_n$.

Proof. Let $0 < \varepsilon < E_1$, where $E_1 > 0$ is as in Lemma 5.1, and let $\Omega_1, \Omega_2 \in C(\mathcal{A})$ satisfy (5.7). We set $\Omega_3 = \Omega_1 \cap \Omega_2$. It is clear that $\Omega_3 \in C(\mathcal{A})$ and $d_{\mathcal{A}}(\Omega_3, \Omega_1), d_{\mathcal{A}}(\Omega_3, \Omega_2) < \varepsilon/s$. By Lemma 5.1 applied to the couples of open sets Ω_i, Ω_3, it follows that $T_\varepsilon(\Omega_i) \subset \Omega_3$, $i = 1, 2$. By the monotonicity of the eigenvalues with respect to inclusion,

$$\lambda_{n,\mathcal{D}}[\Omega_i] \leqslant \lambda_{n,\mathcal{D}}[\Omega_3] \leqslant \lambda_{n,\mathcal{D}}[T_\varepsilon(\Omega_i)], \quad i = 1, 2. \tag{5.11}$$

Since Λ_n in Lemma 3.2 depends only on n, N, $\mathcal{A}$, m, and L, c in Corollary 4.1 depends only on N, $\mathcal{A}$, m, B_1, B_2, L, and θ, whereas E_1 and A_1 in Lemma 5.1 depend only on N and $\mathcal{A}$, from (4.13), (3.2), and (5.4) it follows that there exist $\widetilde{c}_n$ and $\widetilde{\varepsilon}_n > 0$ such that

$$\lambda_{n,\mathcal{D}}[\Omega_3] - \lambda_{n,\mathcal{D}}[\Omega_i] \leqslant \lambda_{n,\mathcal{D}}[T_\varepsilon(\Omega_i)] - \lambda_{n,\mathcal{D}}[\Omega_i] \leqslant \widetilde{c}_n \varepsilon, \quad i = 1, 2,$$

if $0 < \varepsilon < \widetilde{\varepsilon}_n$. Hence

$$|\lambda_{n,\mathcal{D}}[\Omega_1] - \lambda_{n,\mathcal{D}}[\Omega_2]| \leqslant \max_{i=1,2}\{\lambda_{n,\mathcal{D}}[\Omega_3] - \lambda_{n,\mathcal{D}}[\Omega_i]\} \leqslant \widetilde{c}_n \varepsilon.$$

Take here $\varepsilon = 2s d_{\mathcal{A}}(\Omega_1, \Omega_2)$. Then the inequality (5.10) holds with $c_n = 2s\widetilde{c}_n$ if $d_{\mathcal{A}}(\Omega_1, \Omega_2) < \varepsilon_n = \widetilde{\varepsilon}_n/(2s)$. □

Let $\mathcal{A} = (\rho, s, s', \{V_j\}_{j=1}^s, \{r_j\}_{j=1}^s)$ be an atlas in $\mathbb{R}^N$. For all $x \in V' = \bigcup_{j=1}^{s'} V_j$ we set $J'(x) = \{j \in \{1, \ldots, s'\} : x \in V_j\}$. Let $\Omega \in C(\mathcal{A})$. Then we set

$$d_{\mathcal{A}}(x, \partial\Omega) = \max_{j \in J'(x)} d_j(x, \partial\Omega)$$

for all $x \in V'$, where $d_j(x, \partial\Omega)$ is defined in (5.2). Note that if $\Omega \in C(\mathcal{A})$, then $\partial\Omega \subset V'$. Therefore, if $\Omega_1, \Omega_2 \in C(\mathcal{A})$, then

$$d_{\mathcal{A}}(\Omega_1, \Omega_2) = \sup_{x \in \partial\Omega_1} d_{\mathcal{A}}(x, \partial\Omega_2) = \sup_{x \in \partial\Omega_2} d_{\mathcal{A}}(x, \partial\Omega_1). \tag{5.12}$$

For all $\varepsilon > 0$ we set

$$\Omega_{\varepsilon,\mathcal{A}} = \Omega \setminus \{x \in V' : d_{\mathcal{A}}(x, \partial\Omega) \leqslant \varepsilon\},$$

$$\Omega^{\varepsilon,\mathcal{A}} = \Omega \cup \{x \in V' : d_{\mathcal{A}}(x, \partial\Omega) < \varepsilon\}.$$

Lemma 5.2. *Let $\mathcal{A}$ be an atlas in $\mathbb{R}^N$ and $\varepsilon > 0$. If Ω_1 and Ω_2 are two open sets in $C(\mathcal{A})$ satisfying the inclusions*

$$(\Omega_1)_{\varepsilon,\mathcal{A}} \subset \Omega_2 \subset (\Omega_1)^{\varepsilon,\mathcal{A}} \tag{5.13}$$

or

$$(\Omega_2)_{\varepsilon,\mathcal{A}} \subset \Omega_1 \subset (\Omega_2)^{\varepsilon,\mathcal{A}}, \tag{5.14}$$

then

$$d_{\mathcal{A}}(\Omega_1, \Omega_2) \leqslant \varepsilon. \tag{5.15}$$

Proof. Assume that the inclusion (5.13) holds. Let $x \in \partial\Omega_2$. We consider three cases.

Case $x \in \Omega_1$. Since $x \notin \Omega_2$, we have $x \notin (\Omega_1)_{\varepsilon,\mathcal{A}}$. By the definition of $(\Omega_1)_{\varepsilon,\mathcal{A}}$, it follows that $x \in V'$ and $d_{\mathcal{A}}(x, \partial\Omega_1) \leqslant \varepsilon$.

Case $x \in \partial\Omega_1$. It is obvious that $d_{\mathcal{A}}(x, \partial\Omega_1) = 0$.

Case $x \notin \overline{\Omega}_1$. In this case, there exists a sequence $x_n \in \Omega_2 \setminus \overline{\Omega}_1$, $n \in \mathbb{N}$, converging to x. Since $x_n \notin \overline{\Omega}_1$, we have $d_{\mathcal{A}}(x_n, \partial\Omega_1) < \varepsilon$ because $x_n \in (\Omega_1)^{\varepsilon,\mathcal{A}}$. Since $J'(x_n) = J'(x)$ for all n sufficiently large, one can pass to the limit and obtain $d_{\mathcal{A}}(x, \partial\Omega_1) \leqslant \varepsilon$.

Thus, in any case, $d_{\mathcal{A}}(x, \partial\Omega_1) \leqslant \varepsilon$ for all $x \in \partial\Omega_2$, and (5.15) follows by (5.12). The same argument applies when the inclusion (5.14) holds. □

Corollary 5.1. *Let $\mathcal{A}$ be an atlas in $\mathbb{R}^N$. Let $m \in \mathbb{N}$, $L, \theta > 0$ and, for all $\alpha, \beta \in \mathbb{N}_0^N$ with $|\alpha| = |\beta| = m$, let $A_{\alpha\beta} \in C^{0,1}\big(\bigcup_{j=1}^{s} V_j\big)$ satisfy $A_{\alpha\beta} = A_{\beta\alpha}$,*

$\|A_{\alpha\beta}\|_{C^{0,1}(\bigcup_{j=1}^{s} V_j)} \leqslant L$ *and the condition* (2.4). *Then for each* $n \in \mathbb{N}$ *there exist* $c_n, \varepsilon_n > 0$ *depending only on* n, N, $\mathcal{A}$, m, L, θ *such that*

$$|\lambda_{n,\mathcal{D}}[\Omega_1] - \lambda_{n,\mathcal{D}}[\Omega_2]| \leqslant c_n \varepsilon \tag{5.16}$$

for all $0 < \varepsilon < \varepsilon_n$ *and for all* $\Omega_1, \Omega_2 \in C(\mathcal{A})$ *satisfying (5.13) or (5.14).*

Proof. The inequality (5.16) follows from (5.10) and (5.15). □

6 Estimates for Neumann Eigenvalues via the Atlas Distance

In this section, we prove Theorem 6.1. The proof is based on Lemmas 6.1 and 6.3.

Definition 6.1. Let U be an open set in $\mathbb{R}^N$ and ρ a rotation. We say that U is a *ρ-patch* if there exists an open set $G_U \subset \mathbb{R}^{N-1}$ and functions $\varphi_U, \psi_U : G_U \to \mathbb{R}$ such that

$$\rho(U) = \left\{(\overline{x}, x_N) \in \mathbb{R}^N : \ \psi_U(\overline{x}) < x_N < \varphi_U(\overline{x}), \ \overline{x} \in G_U\right\}.$$

The *thickness* of the ρ-patch is defined by

$$R_U = \inf_{\overline{x} \in G_U} (\varphi_U(\overline{x}) - \psi_U(\overline{x}));$$

the *thinness* of the ρ-patch is defined by

$$S_U = \sup_{\overline{x} \in G_U} (\varphi_U(\overline{x}) - \psi_U(\overline{x})).$$

If $\Omega_2 \subset \Omega_1$ and $\Omega_1 \backslash \Omega_2$ is covered by a finite number of ρ-patches contained in Ω_1, then we can estimate $\lambda_{n,\mathcal{N}}[\Omega_2] - \lambda_{n,\mathcal{N}}[\Omega_1]$ via the thinness of the patches.

Lemma 6.1. *Let* $m \in \mathbb{N}$, *and let* Ω_1 *be an open set in* $\mathbb{R}^N$ *such that the embedding* $W^{m,2}(\Omega_1) \subset W^{m-1,2}(\Omega_1)$ *is compact. For all* $\alpha, \beta \in \mathbb{N}_0^N$ *with* $|\alpha| = |\beta| = m$, *let* $A_{\alpha\beta}$ *be bounded measurable real-valued functions defined on* Ω_1 *satisfying* $A_{\alpha\beta} = A_{\beta\alpha}$ *and the condition* (2.4) *in* Ω_1. *Let* $\sigma \in \mathbb{N}$, $R > 0$. *Assume that* $\Omega_2 \subset \Omega_1$ *is such that the embedding* $W^{m,2}(\Omega_2) \subset W^{m-1,2}(\Omega_2)$ *is compact and there exist rotations* $\{\rho_j\}_{j=1}^{\sigma}$ *and two sets* $\{U_j\}_{j=1}^{\sigma}$, $\{\widetilde{U}_j\}_{j=1}^{\sigma}$ *of* ρ_j*-patches* U_j *and* $\widetilde{U}_j$ *satisfying the following properties*

(a) $U_j \subset \widetilde{U}_j \subset \Omega_1$ *for all* $j = 1, \dots, \sigma$;

(b) $G_{U_j} = G_{\widetilde{U}_j}$, $\varphi_{U_j} = \varphi_{\widetilde{U}_j}$ *for all* $j = 1, \dots, \sigma$;

(c) $R_{\widetilde{U}_j} > R$ *for all* $j = 1, \ldots, \sigma$;

(d) $\Omega_1 \setminus \Omega_2 \subset \cup_{j=1}^{\sigma} U_j$.

Then there exists $d > 0$ *depending only on* N, m, R *such that for all* $n \in \mathbb{N}$

$$\lambda_{n,\mathcal{N}}[\Omega_2] \leqslant \lambda_{n,\mathcal{N}}[\Omega_1](1 + d_n \max_{j=1,\ldots,\sigma} S_{U_j}) \tag{6.1}$$

if $\max_{j=1,\ldots,\sigma} S_{U_j} < d_n^{-1}$, *where*

$$d_n = 2\sigma d(1 + \theta^{-1}\lambda_{n,\mathcal{N}}[\Omega_1]). \tag{6.2}$$

Proof. By (a) and (b), $\psi_{\widetilde{U}_j} \leqslant \psi_{U_j}$ for all $j = 1, \ldots, \sigma$. Let $u \in W^{m,2}(\Omega_1)$. By (d), we have

$$\int_{\Omega_1 \setminus \Omega_2} |u|^2 dy \leqslant \sum_{j=1}^{\sigma} \int_{U_j} |u|^2 dy = \sum_{j=1}^{\sigma} \int_{\rho_j(U_j)} |u \circ \rho_j^{(-1)}|^2 dx. \tag{6.3}$$

We set $v_j = u \circ \rho_j^{(-1)}$ for brevity. It is clear that

$$\int_{\rho_j(U_j)} |u \circ \rho_j^{-1}|^2 = \int_{G_{U_j}} \int_{\psi_{U_j}(\overline{x})}^{\varphi_{U_j}(\overline{x})} |v_j(\overline{x}, x_N)|^2 d\overline{x} dx_N. \tag{6.4}$$

Since $v_j \in W^{m,2}(\rho_j(\widetilde{U}_j))$, it follows that for almost all $\overline{x} \in G_{\widetilde{U}_j}$ the function $v_j(\overline{x}, \cdot)$ belongs to the space $W^{m,2}(\psi_{\widetilde{U}_j}(\overline{x}), \varphi_{\widetilde{U}_j}(\overline{x}))$. Moreover, by (c), $\varphi_{\widetilde{U}_j}(\overline{x}) - \psi_{\widetilde{U}_j}(\overline{x}) \geqslant R$. Thus, by [3, Theorem. 2, p.127], there exists $\widetilde{d} > 0$ depending only on m, R such that

$$\begin{aligned} &\|v_j(\overline{x}, \cdot)\|^2_{L^\infty(\psi_{\widetilde{U}_j}(\overline{x}), \varphi_{\widetilde{U}_j}(\overline{x}))} \\ &\leqslant \widetilde{d}\Big(\|v_j(\overline{x}, \cdot)\|^2_{L^2(\psi_{\widetilde{U}_j}(\overline{x}), \varphi_{\widetilde{U}_j}(\overline{x}))} + \Big\| \frac{\partial^m v_j}{\partial x_N^m}(\overline{x}, \cdot) \Big\|^2_{L^2(\psi_{\widetilde{U}_j}(\overline{x}), \varphi_{\widetilde{U}_j}(\overline{x}))} \Big). \end{aligned} \tag{6.5}$$

By the inequality (6.5) and property (b),

$$\begin{aligned} &\int_{G_{U_j}} \int_{\psi_{U_j}(\overline{x})}^{\varphi_{U_j}(\overline{x})} |v_j(\overline{x}, x_N)|^2 d\overline{x} dx_N \\ &\leqslant \int_{G_{U_j}} (\varphi_{U_j}(\overline{x}) - \psi_{U_j}(\overline{x})) \|v_j(\overline{x}, \cdot)\|^2_{L^\infty(\psi_{\widetilde{U}_i}(\overline{x}), \varphi_{U_i}(\overline{x}))} d\overline{x} \end{aligned}$$

$$\leqslant \tilde{d}S_{U_j}\Bigg(\|v_j\|^2_{L^2(\rho_j(\tilde{U}_j))} + \left\|\frac{\partial^m v_j}{\partial x_N^m}\right\|^2_{L^2(\rho_j(\tilde{U}_j))}\Bigg)$$

$$\leqslant dS_{U_j}\Bigg(\|u\|^2_{L^2(\Omega_1)} + \sum_{|\alpha|=m} \|D^\alpha u\|^2_{L^2(\Omega_1)}\Bigg), \tag{6.6}$$

where $d > 0$ depends only on N, m, and R.

Let $\psi_n[\Omega_1]$, $n \in \mathbb{N}$, be an orthonormal sequence of eigenfunctions corresponding to the eigenvalues $\lambda_{n,\mathcal{N}}[\Omega_1]$. We denote by $L_n[\Omega_1]$ the linear subspace of $W^{m,2}(\Omega_1)$ generated by the $\psi_1[\Omega_1], \dots, \psi_n[\Omega_1]$. If $u \in L_n[\Omega_1]$ and $\|u\|_{L^2(\Omega_1)} = 1$, then, by (6.3), (6.4), and (6.6),

$$\int\limits_{\Omega_1\setminus\Omega_2} |u|^2 \leqslant \sigma d \max_{j=1,\dots,\sigma} S_{U_j}(1 + \theta^{-1}Q_{\Omega_1}(u))$$

$$\leqslant \sigma d \max_{j=1,\dots,\sigma} S_{U_j}(1 + \theta^{-1}\lambda_n[\Omega_1]). \tag{6.7}$$

Let $\mathcal{T}_{12}$ be the restriction operator from Ω_1 to Ω_2. It is clear that $\mathcal{T}_{12}$ maps $W^{m,2}(\Omega_1)$ to $W^{m,2}(\Omega_2)$. For all $n \in \mathbb{N}$ and $u \in L_n[\Omega_1]$, $\|u\|_{L^2(\Omega_1)} = 1$, we have

$$\|\mathcal{T}_{12}u\|^2_{L^2(\Omega_2)} = \int\limits_{\Omega_1} |u|^2 - \int\limits_{\Omega_1\setminus\Omega_2} |u|^2$$

$$\geqslant 1 - \sigma d \max_{j=1,\dots,\sigma} S_{U_j}(1 + \theta^{-1}\lambda_n[\Omega_1]) \tag{6.8}$$

and

$$Q_{\Omega_2}(\mathcal{T}_{12}u) \leqslant Q_{\Omega_1}(u) \leqslant \lambda_n[\Omega_1]$$

because $\sum\limits_{|\alpha|=|\beta|=m} A_{\alpha\beta}\xi_\alpha\overline{\xi}_\beta \geqslant 0$ for all $\xi_\alpha, \xi_\beta \in \mathbb{C}$. Thus, in the terminology of [6], $\mathcal{T}_{12}$ is a transition operator from $H_{W^{m,2}(\Omega_1)}$ to $H_{W^{m,2}(\Omega_2)}$ with the measure of vicinity $\delta(H_{W^{m,2}(\Omega_1)}, H_{W^{m,2}(\Omega_2)}) = \max\limits_{j=1,\dots,\sigma} S_{U_j}$ and the parameters $a_n = \sigma d(1 + \theta^{-1}\lambda_n[\Omega_1])$, $b_n = 0$. Thus, by the general spectral stability theorem [6, Theorem 3.2] it follows that

$$\lambda_{n,\mathcal{N}}[\Omega_2] \leqslant \lambda_{n,\mathcal{N}}[\Omega_1] + 2(a_n\lambda_{n,\mathcal{N}}[\Omega_1] + b_n)\delta(H_{W^{m,2}(\Omega_1)}, H_{W^{m,2}(\Omega_2)})$$

if $\delta(H_{W^{m,2}(\Omega_1)}, H_{W^{m,2}(\Omega_2)}) < (2a_n)^{-1}$, which immediately yields (6.1). □

Lemma 6.2. *Let $\mathcal{A}$ be an atlas in $\mathbb{R}^N$. Let $m \in \mathbb{N}$, $L, \theta > 0$ and, for all $\alpha, \beta \in \mathbb{N}_0^N$ with $|\alpha| = |\beta| = m$, let $A_{\alpha\beta} \in L^\infty(\bigcup\limits_{j=1}^s V_j)$ satisfy $A_{\alpha\beta} = A_{\beta\alpha}$, $\|A_{\alpha\beta}\|_{L^\infty(\bigcup\limits_{j=1}^s V_j)} \leqslant L$ and the condition (2.4). Then for each $n \in \mathbb{N}$ there exist*

$c_n, \varepsilon_n > 0$ *depending only on* n, N, $\mathcal{A}$, m, L, θ *such that*

$$\lambda_{n,\mathcal{N}}[\Omega_2] \leqslant \lambda_{n,\mathcal{N}}[\Omega_1] + c_n d_{\mathcal{A}}(\Omega_1, \Omega_2) \tag{6.9}$$

for all $\Omega_1, \Omega_2 \in C(\mathcal{A})$ *satisfying* $\Omega_2 \subset \Omega_1$ *and* $d_{\mathcal{A}}(\Omega_1, \Omega_2) < \varepsilon_n$.

Proof. Suppose that $\mathcal{A} = (\rho, s, s', \{V_j\}_{j=1}^s, \{r_j\}_{j=1}^s)$, $\Omega_1, \Omega_2 \in C(\mathcal{A})$ and $\Omega_2 \subset \Omega_1$. For all $j = 1, \dots, s$ we denote by g_{1j} and g_{2j} the functions describing the boundaries of Ω_1 and Ω_2 respectively, as in Definition 3.1 (iii). We consider two sets of $\{U_j\}_{j=1}^s$, $\{\widetilde{U}_j\}_{j=1}^s$ of r_j-patches U_j, $\widetilde{U}_j$ defined as follows:

$$\widetilde{U}_j = r_j^{(-1)}(\{(\overline{x}, x_N) : \ \overline{x} \in W_j, \ a_{Nj} < x_N < g_{1j}(\overline{x})\}),$$

$$U_j = r_j^{(-1)}(\{(\overline{x}, x_N) : \ \overline{x} \in W_j, \ g_{2j}(\overline{x}) < x_N < g_{1j}(\overline{x})\},$$

where W_j and a_{Nj} are as in Definition 3.1. Note that conditions (a), (b), (c), and (d) of Lemma 6.1 are satisfied with $\sigma = s$ and $R = \rho$. Moreover, $\max_{j=1,\dots,\sigma} S_{U_j} = d_{\mathcal{A}}(\Omega_1, \Omega_2)$. Thus, applying Lemma 6.1 to the open sets Ω_1, Ω_2 and the sets of patches defined above, and using Lemma 3.2, we immediately obtain (6.9). □

Our next goal is to consider the case $\Omega_2 = T_\varepsilon(\Omega_1)$ where T_ε is the map defined in (5.3).

Lemma 6.3. *Let* $\mathcal{A}$ *be an atlas in* $\mathbb{R}^N$. *Then there exist* ε_0, A, $R > 0$ *and* $\sigma \in \mathbb{N}$ *depending only on* N, $\mathcal{A}$ *and for every open set* $\Omega \in C(\mathcal{A})$ *and any* $0 < \varepsilon < \varepsilon_0$ *there exist rotations* $\{\rho_j\}_{j=1}^\sigma$ *and sets* $\{U_j\}_{j=1}^\sigma$, $\{\widetilde{U}_j\}_{j=1}^\sigma$ *of* ρ_j-*patches* U_j, $\widetilde{U}_j$ *satisfying conditions* (a), (b), (c), (d) *in Lemma* 6.1 *with* $\Omega_1 = \Omega$ *and* $\Omega_2 = T_\varepsilon(\Omega)$ *and such that* $\max_{j=1,\dots,\sigma} S_{U_j} < A\varepsilon$.

Proof. In fact, we prove that there exists a family of rotations $\{\rho_j\}_{j=1}^\sigma$, a family $\{G_j\}_{j=1}^\sigma$ of bounded open sets in $\mathbb{R}^{N-1}$, and a family $\{\varphi_j\}_{j=1}^\sigma$ of functions φ_j continuous on $\overline{G}_j$ such that for all $0 < \varepsilon < \varepsilon_0$

$$\Omega \setminus T_\varepsilon(\Omega) \subset \bigcup_{j=1}^{\sigma} U_j^{[\varepsilon]} \tag{6.10}$$

and

$$U_j^{[\varepsilon]} \subset \widetilde{U}_j \subset \Omega, \tag{6.11}$$

where the ρ_j-patches $U_j^{[\varepsilon]}$, $\widetilde{U}_j$ are defined by

$$\rho_j(U_j^{[\varepsilon]}) = \{(\overline{x}, x_N) \in \mathbb{R}^{N-1} : \ \varphi_j(\overline{x}) - A\varepsilon < x_N < \varphi_j(\overline{x}), \ \overline{x} \in G_j\} \tag{6.12}$$

and

$$\rho_j(\widetilde{U}_j) = \left\{(\overline{x}, x_N) \in \mathbb{R}^{N-1} : \ \varphi_j(\overline{x}) - R < x_N < \varphi_j(\overline{x}), \ \overline{x} \in G_j\right\}. \quad (6.13)$$

Let $\mathcal{A} = (\rho, s, s', \{V_j\}_{j=1}^s, \{r_j\}_{j=1}^s)$, and $\Omega \in C(\mathcal{A})$. We split the proof into four steps.

Step 1. Assume that for each nonempty set $J \subset \{1, \ldots, s'\}$, $V_J = \bigcap_{j \in J} (V_j)_{\frac{\rho}{2}}$ and $d = \dim \operatorname{Span}\{\xi_j\}_{j \in J}$. Recall that $\xi_j = r_j^{(-1)}(0, \ldots, 1)$. From the proof of Lemma 19 in [4] it follows that there exist vectors $\xi_J \equiv \xi_{J1}, \xi_{J2}, \ldots, \xi_{Jd}$ and a rotation r_J such that

1) $\xi_J, \xi_{J2}, \ldots, \xi_{Jd}$ is an orthonormal basis for $\operatorname{Span}\{\xi_j\}_{j \in J}$ and $r_J(\xi_J) = e_N$, $r_J(\xi_{J2}) = e_{N-1}, \ldots, r_J(\xi_{Jd}) = e_{N-d+1}$,

2) there exist continuous functions φ_J, ψ_J defined on $\overline{G}_J$, where $G_J = \operatorname{Pr}_{x_N=0} r_J(V_J \cap \Omega)$ ($\operatorname{Pr}_{x_N=0}$ denotes the orthogonal projector onto the hyperplane with the equation $x_N = 0$) such that

$$r_J(V_J \cap \Omega) = \left\{(\overline{x}, x_N) \in \mathbb{R}^{N-1} : \ \psi_J(\overline{x}) < x_N < \varphi_J(\overline{x})\right\} \quad (6.14)$$

and such that

$$\left\{(\overline{x}, y) \in \mathbb{R}^{N-1} : \ y < \varphi_J(\overline{x}), \ \overline{x} \in G_J, \ (\overline{x}, y) \in r_J(V_j), \ \forall j \in J\right\} \subset r_J(\Omega), \quad (6.15)$$

3) the function φ_J satisfies the Lipschitz condition with respect to the variables $v = (x_{N-d+1}, \ldots, x_{N-1})$ uniformly with respect to the variables $u = (x_1, \ldots, x_{N-d})$ on G_J, i.e.,

$$|\varphi_J(u, v) - \varphi_J(u, w)| \leqslant L_J |v - w| \quad \text{for all } (u, v), (u, w) \in G_J,$$

where $L_J > 0$ depends only on $\{V_j\}_{j=1}^s$ and $\{r_j\}_{j=1}^s$.

Note that by (6.14) and (6.15) it follows that

$$\{(\overline{x}, x_N) \in \mathbb{R}^{N-1} : \ \overline{x} \in G_J, \ \psi_J(\overline{x}) - \frac{\rho}{4} < x_N < \varphi_J(\overline{x})\} \subset r_J(\Omega) \quad (6.16)$$

because the distance of $(\overline{x}, \psi_J(\overline{x}))$ to the boundary of $r_J(V_j)$ is greater than $\frac{\rho}{2}$ for all $j \in J$ and $\overline{x} \in G_J$.

Step 2. As in [4], for $x \in \mathbb{R}^N$ we set

$$J(x) = \{j \in \{1, \ldots, s\} : \ x \in (V_j)_{\frac{3}{4}\rho}\}.$$

Note that $J(x) \subset \{1, \ldots, s'\}$ if $x \in \partial\Omega$. The inclusion $\operatorname{supp} \psi_j \subset (V_j)_{\frac{3}{4}\rho}$ implies that $\psi_j(x) = 0$ for $j \notin J(x)$ and

$$T_\varepsilon(x) = x - \varepsilon \sum_{j \in J(x)} \xi_j \psi_j(x), \quad x \in \mathbb{R}^N.$$

For any subset $J \subset \{1, \dots, s\}$ we set

$$\widetilde{V}_J = \{x \in \mathbb{R}^N : \ J(x) = J\}$$

so that $\mathbb{R}^N = \mathring{\bigcup}_{J \subset \{1,\dots,s'\}} \widetilde{V}_J$ and

$$T_\varepsilon(x) = x - \varepsilon \sum_{j \in J} \xi_j \psi_j(x), \quad x \in \widetilde{V}_J.$$

Step 3. Let $x \in \widetilde{V}_J \cap \partial\Omega$. Since $\|T_\varepsilon - \mathrm{Id}\ \|_\infty \leqslant \varepsilon$ and $T_\varepsilon(\Omega) \subset \Omega$, we have $T_\varepsilon(x) \in V_J \cap \Omega$ for all $0 < \varepsilon \leqslant \frac{\rho}{4}$. Let $r_J(x) = (\beta^{(1)}, \beta^{(2)}, \beta_N)$, where $\beta^{(1)} = (\beta_1, \dots, \beta_{N-d})$, $\beta^{(2)} = (\beta_{N-d+1}, \dots, \beta_{N-1})$ and $\beta_N = \varphi_J(\beta^{(1)}, \beta^{(2)})$. Since $T_\varepsilon(x) - x \in \mathrm{Span}\{\xi_j\}_{j \in J}$, it follows that $r_J(T_\varepsilon(x)) = (\beta^{(1)}, \gamma^{(2)}, \gamma_N)$ for some $\gamma^{(2)} = (\gamma_{N-d+1}, \dots, \gamma_{N-1})$ and γ_N. Since $T_\varepsilon(x) \in V_J \cap \Omega$, for the distance $d_J(T_\varepsilon(x))$ from $T_\varepsilon(x)$ to $\partial\Omega$ in the direction of the vector ξ_J we have

$$\begin{aligned} d_J(T_\varepsilon(x)) &= \varphi_J(\beta^{(1)}, \gamma^{(2)}) - \gamma_N \\ &= \varphi_J(\beta^{(1)}, \gamma^{(2)}) - \beta_N + \beta_N - \gamma_N \\ &= \varphi_J(\beta^{(1)}, \gamma^{(2)}) - \varphi_J(\beta^{(1)}, \beta^{(2)}) + \beta_N - \gamma_N \\ &\leqslant L_J|\gamma^{(2)} - \beta^{(2)}| + |\gamma_N - \beta_N| \\ &\leqslant (L_J + 1)|r_J(T_\varepsilon(x)) - r_J(x)| \\ &= (L_J + 1)|T_\varepsilon(x) - x|. \end{aligned} \tag{6.17}$$

Let

$$A = \max_{\substack{J \subset \{1,\dots,s'\} \\ J \neq \varnothing}} (L_J + 1), \tag{6.18}$$

and let $U_J^{[\varepsilon]}$ be defined by

$$\rho_J(U_J^{[\varepsilon]}) = \{(\overline{x}, x_N) \in \mathbb{R}^{N-1} : \ \varphi_J(\overline{x}) - A\varepsilon < x_N < \varphi_J(\overline{x}), \ \overline{x} \in G_J\}. \tag{6.19}$$

Then, by (6.17), it follows that $T_\varepsilon(\widetilde{V}_J \cap \partial\Omega) \subset U_J^{[\varepsilon]}$ and

$$T_\varepsilon(\partial\Omega) = \bigcup_{\substack{J \subset \{1,\dots,s'\} \\ J \neq \varnothing}} T_\varepsilon(\widetilde{V}_J \cap \partial\Omega) \subset \bigcup_{\substack{J \subset \{1,\dots,s'\} \\ J \neq \varnothing}} U_J^{[\varepsilon]}. \tag{6.20}$$

Step 4. Let $y \in \Omega \setminus \bigcup_{\substack{J \subset \{1,\dots,s'\} \\ J \neq \varnothing}} U_J^{[\varepsilon]}$. By the definition of $U_J^{[\varepsilon]}$, it follows that $y \notin U_J^{[\varepsilon']}$ for all $0 < \varepsilon' \leqslant \varepsilon$. Thus, by (6.20), $y \notin T_{\varepsilon'}(\partial\Omega)$. This implies that the topological degree deg $(\Omega, T_{\varepsilon'}, y)$ of the triple $(\Omega, T_{\varepsilon'}, y)$ is well defined (see, for example, [9, Sect. 1]) and, by homotopy invariance, deg $(\Omega, T_{\varepsilon'}, y) =$ deg $(\Omega, T_0, y) = 1$ for all $0 < \varepsilon' \leqslant \varepsilon$. Thus, the equation $T_\varepsilon(x) = y$ has a solution $x \in \Omega$ hence $y \in T_\varepsilon(\Omega)$ (see, for example, [9, Theorem 3.1]). This shows that $\Omega \setminus T_\varepsilon(\Omega) \subset \bigcup\limits_{\substack{J \subset \{1,\dots,s'\} \\ J \neq \varnothing}} U_J^{[\varepsilon]}$.

To complete the proof, it suffices to choose σ to be the number of nonempty subsets of $\{1, \dots, s'\}$, $\varepsilon_0 = \frac{\varrho}{4}$, $R = \frac{\varrho}{4}$ (see (6.16)) and take A as in (6.18). □

Theorem 6.1. *Let $\mathcal{A}$ be an atlas in $\mathbb{R}^N$. Let $m \in \mathbb{N}$, $L, \theta > 0$ and, for all $\alpha, \beta \in \mathbb{N}_0^N$ with $|\alpha| = |\beta| = m$, let $A_{\alpha\beta} \in C^{0,1}(\bigcup\limits_{j=1}^{s} V_j)$ satisfy $A_{\alpha\beta} = A_{\beta\alpha}$, $\|A_{\alpha\beta}\|_{C^{0,1}(\bigcup\limits_{j=1}^{s} V_j)} \leqslant L$ and the condition (2.4). Then for each $n \in \mathbb{N}$ there exist c_n, $\varepsilon_n > 0$ depending only on n, N, $\mathcal{A}$, m, L, θ such that*

$$|\lambda_{n,\mathcal{N}}[\Omega_1] - \lambda_{n,\mathcal{N}}[\Omega_2]| \leqslant c_n d_{\mathcal{A}}(\Omega_1, \Omega_2) \tag{6.21}$$

for all $\Omega_1, \Omega_2 \in C(\mathcal{A})$ satisfying $d_{\mathcal{A}}(\Omega_1, \Omega_2) < \varepsilon_n$.

Proof. In this proof, c_n and ε_n denote positive constants depending only on some of the parameters n, N, $\mathcal{A}$, m, L, θ and their value is not necessarily the same for all the inequalities below.

Let $E_1 > 0$ be as in Lemma 5.1. Let $0 < \varepsilon < E_1$, and let $\Omega_1, \Omega_2 \in C(\mathcal{A})$ satisfy (5.7). We set $\Omega_3 = \Omega_1 \cap \Omega_2$. It is clear that $\Omega_3 \in C(\mathcal{A})$ and $d_{\mathcal{A}}(\Omega_3, \Omega_1), d_{\mathcal{A}}(\Omega_3, \Omega_2) < \varepsilon/s$. By Lemma 5.1 applied to the couple of open sets Ω_1, Ω_3, it follows that $T_\varepsilon(\Omega_1) \subset \Omega_3$. Hence

$$T_\varepsilon(\Omega_3) \subset T_\varepsilon(\Omega_1) \subset \Omega_3. \tag{6.22}$$

We now apply Lemma 6.3 to the set $\Omega = \Omega_3$. It follows that if $0 < \varepsilon < \varepsilon_0$ there exist rotations $\{\rho_j\}_{j=1}^{\sigma}$ and two sets $\{U_j\}_{j=1}^{\sigma}$, $\{\widetilde{U}_j\}_{j=1}^{\sigma}$ of ρ_j-patches U_j, $\widetilde{U}_j$ satisfying conditions (a), (b), (c), (d) in Lemma 6.1 with Ω_1 replaced by Ω_3 and Ω_2 replaced by $T_\varepsilon(\Omega_3)$, and such that $\max\limits_{j=1,\dots,\sigma} S_{U_j} < A\varepsilon$. In particular,

$$\Omega_3 \setminus T_\varepsilon(\Omega_3) \subset \cup_{j=1}^{\sigma} U_j. \tag{6.23}$$

Hence, by (6.22) and (6.23),

$$\Omega_3 \setminus T_\varepsilon(\Omega_1) \subset \cup_{j=1}^{\sigma} U_j. \tag{6.24}$$

Now, we apply Lemma 6.1 to the couple of open sets Ω_3, $T_\varepsilon(\Omega_1)$ by using the sets of patches defined above. Since $\max\limits_{j=1,\dots,\sigma} S_{U_j} < A\varepsilon$, from Lemma 6.1 it follows that $A\varepsilon < d_n^{-1}$ implies

$$\lambda_{n,\mathcal{N}}[T_\varepsilon(\Omega_1)] \leqslant \lambda_{n,\mathcal{N}}[\Omega_3](1 + d_n A\varepsilon), \tag{6.25}$$

where d_n is defined by (6.2). By the inequality (6.25) and Lemma 3.2, there exist $c_n, \varepsilon_n > 0$ such that

$$\lambda_{n,\mathcal{N}}[T_\varepsilon(\Omega_1)] \leqslant \lambda_{n,\mathcal{N}}[\Omega_3] + c_n\varepsilon \tag{6.26}$$

if $0 < \varepsilon < \varepsilon_n$. On the other hand, by Lemma 3.2, Corollary 4.1, and the inequalities (5.4), (5.5), it follows that there exist c_n, $\varepsilon_n > 0$ such that

$$|\lambda_{n,\mathcal{N}}[T_\varepsilon[\Omega_1]] - \lambda_{n,\mathcal{N}}[\Omega_1]| \leqslant c_n\varepsilon \tag{6.27}$$

if $0 < \varepsilon < \varepsilon_n$. Thus, by (6.26) and (6.27), there exist $c_n, \varepsilon_n > 0$ such that

$$\lambda_{n,\mathcal{N}}[\Omega_1] \leqslant \lambda_n[\Omega_3] + c_n\varepsilon \tag{6.28}$$

if $0 < \varepsilon < \varepsilon_n$. By Lemma 6.2 applied to the couple of open sets Ω_1, Ω_3, there exist $c_n, \varepsilon_n > 0$ such that

$$\lambda_{n,\mathcal{N}}[\Omega_3] \leqslant \lambda_{n,\mathcal{N}}[\Omega_1] + c_n\varepsilon \tag{6.29}$$

if $0 < \varepsilon < \varepsilon_n$. Thus, by (6.28), (6.29) it follows that

$$|\lambda_{n,\mathcal{N}}[\Omega_1] - \lambda_{n,\mathcal{N}}[\Omega_3]| \leqslant c_n\varepsilon \tag{6.30}$$

if $0 < \varepsilon < \varepsilon_n$. It is clear that the inequality (6.30) holds also with Ω_2 replacing Ω_1: it is simply enough to interchange the role of Ω_1 and Ω_2 from the beginning this proof. Thus,

$$|\lambda_{n,\mathcal{N}}[\Omega_2] - \lambda_{n,\mathcal{N}}[\Omega_3]| \leqslant c_n\varepsilon \tag{6.31}$$

if $0 < \varepsilon < \varepsilon_n$. By (6.30) and (6.31), we finally deduce that for each $n \in \mathbb{N}$ there exist $c_n, \varepsilon_n > 0$ such that

$$|\lambda_{n,\mathcal{N}}[\Omega_1] - \lambda_{n,\mathcal{N}}[\Omega_2]| \leqslant c_n\varepsilon \tag{6.32}$$

for all $0 < \varepsilon < \varepsilon_n$ and $\Omega_1, \Omega_2 \in C(\mathcal{A})$ satisfying $d_{\mathcal{A}}(\Omega_1, \Omega_2) < \varepsilon$. Finally, arguing in the same way as in the last lines of the proof of Theorem 5.1, we deduce the validity of (6.21). □

In the case of the Dirichlet boundary conditions, we have a version of Theorem 6.1 in terms of ε-neighborhoods with respect to the atlas distance.

Corollary 6.1. *Let $\mathcal{A}$ be an atlas in $\mathbb{R}^N$. Let $m \in \mathbb{N}$, $L, \theta > 0$ and, for all $\alpha, \beta \in \mathbb{N}_0^N$, with $|\alpha| = |\beta| = m$ let $A_{\alpha\beta} \in C^{0,1}\big(\bigcup_{j=1}^{s} V_j\big)$ satisfy $A_{\alpha\beta} = A_{\beta\alpha}$, $\|A_{\alpha\beta}\|_{C^{0,1}\left(\bigcup_{j=1}^{s} V_j\right)} \leqslant L$ and the condition (2.4). Then for each $n \in \mathbb{N}$ there exist $c_n, \varepsilon_n > 0$ depending only on n, N, $\mathcal{A}$, m, L, θ such that*

$$|\lambda_{n,\mathcal{N}}[\Omega_1] - \lambda_{n,\mathcal{N}}[\Omega_2]| \leqslant c_n\varepsilon \tag{6.33}$$

for all $0 < \varepsilon < \varepsilon_n$ *and* $\Omega_1, \Omega_2 \in C(\mathcal{A})$ *satisfying* (5.13) *or* (5.14).

Proof. The inequality (6.33) follows from (6.21) and (5.15). □

7 Estimates via the Lower Hausdorff–Pompeiu Deviation

If $C \subset \mathbb{R}^N$ and $x \in \mathbb{R}^N$ we denote by $d(x, C)$ the euclidean distance of x to C.

Definition 7.1. Let $A, B \subset \mathbb{R}^N$. We define the *lower Hausdorff–Pompeiu deviation* of A from B by

$$d_{\mathcal{HP}}(A, B) = \min\Big\{\sup_{x \in A} d(x, B),\ \sup_{x \in B} d(x, A)\Big\}. \tag{7.1}$$

If the minimum in (7.1) is replaced by the maximum, then the right-hand side becomes the usual Hausdorff–Pompeiu distance $d^{\mathcal{HP}}(A, B)$ from A to B. Note that, in contrast to the Hausdorff–Pompeiu distance $d^{\mathcal{HP}}$ which satisfies the triangle inequality and defines a distance on the family of closed sets, the lower Hausdorff–Pompeiu deviation $d_{\mathcal{HP}}$ is not a distance or a quasi-distance. Indeed, it suffices to note that $d_{\mathcal{HP}}(A, B) = 0$ if and only if $A \subset \overline{B}$ or $B \subset \overline{A}$: thus, if $A \not\subset \overline{B}$ and $B \not\subset \overline{A}$, then $d_{\mathcal{HP}}(A, B) > 0$, but $d_{\mathcal{HP}}(A, A \cup B) + d_{\mathcal{HP}}(A \cup B, A) = 0$.

In this section, we prove an estimate for the variation of the eigenvalues in the terms of the lower Hausdorff–Pompeiu deviation of the boundaries of the open sets.

We introduce a class of open sets for which we can estimate the atlas distance $d_{\mathcal{A}}$ via the lower Hausdorff–Pompeiu deviation of the boundaries.

Definition 7.2. Let $\mathcal{A}$ be an atlas in $\mathbb{R}^N$. Let $\omega : [0, \infty[\to [0, \infty[$ be a continuous nondecreasing function such that $\omega(0) = 0$ and, for some $k > 0$, $\omega(t) \geqslant kt$ for all $0 \leqslant t \leqslant 1$. Let $M > 0$. We denote by $C_M^{\omega(\cdot)}(\mathcal{A})$ the family of all open sets Ω in $\mathbb{R}^N$ belonging to $C(\mathcal{A})$ and such that all the functions g_j in Definition 3.1 (iii) satisfy the condition

$$|g_j(\overline{x}) - g_j(\overline{y})| \leqslant M\omega(|\overline{x} - \overline{y}|) \tag{7.2}$$

for all $\overline{x}, \overline{y} \in \overline{W}_j$.

We also say that an open set is of class $C^{\omega(\cdot)}$ if there exists an atlas $\mathcal{A}$ and $M > 0$ such that $\Omega \in C_M^{\omega(\cdot)}(\mathcal{A})$.

Lemma 7.1. *Let* $\omega : [0, \infty[\to [0, \infty[$ *be a continuous nondecreasing function such that* $\omega(0) = 0$ *and for some* $k > 0$, $\omega(t) \geqslant kt$ *for all* $t \geqslant 0$. *Let* W *be an open set in* $\mathbb{R}^{N-1}$. *Let* $M > 0$, *and let* g *be a function of* $\overline{W}$ *to* $\mathbb{R}$ *such that*

$$|g(\overline{x}) - g(\overline{y})| \leqslant M\omega(|\overline{x} - \overline{y}|)$$

for all $\overline{x}$, $\overline{y} \in \overline{W}$. *Then*

$$|g(\overline{x}) - x_N| \leqslant (M + k^{-1})\omega\left(d\left((\overline{x}, x_N), \mathrm{Graph(g)}\right)\right) \tag{7.3}$$

for all $\overline{x} \in \overline{W}$ *and* $x_N \in \mathbb{R}$.

Proof. For all $\overline{x}, \overline{y} \in W$

$$\begin{aligned} |g(\overline{x}) - x_N| &\leqslant |g(\overline{x}) - g(\overline{y})| + |g(\overline{y}) - x_N| \\ &\leqslant M\omega(|\overline{x} - \overline{y}|) + k^{-1}\omega(|g(\overline{y}) - x_N|) \\ &\leqslant (m + k^{-1})\omega(|(\overline{x}, x_N) - (\overline{y}, g(\overline{y}))|). \end{aligned} \tag{7.4}$$

Hence, by the continuity of ω,

$$\begin{aligned} |g(\overline{x}) - x_N| &\leqslant (m + k^{-1}) \inf_{\overline{y} \in \overline{W}} \omega(|(\overline{x}, x_N) - (\overline{y}, g(\overline{y}))| \\ &\leqslant (m + k^{-1})\omega(\inf_{\overline{y} \in \overline{W}} |(\overline{x}, x_N) - (\overline{y}, g(\overline{y}))| \\ &\leqslant (M + k^{-1})\omega\left(d\left((\overline{x}, x_N), \mathrm{Graph(g)}\right)\right). \end{aligned} \tag{7.5}$$

The proof is complete. □

Lemma 7.2. *Let $\mathcal{A}$ be an atlas in $\mathbb{R}^N$. Let $\omega : [0, \infty[\to [0, \infty[$ be a continuous increasing function satisfying $\omega(0) = 0$ and, for some $k > 0$, $\omega(t) \geqslant kt$ for all $0 \leqslant t \leqslant 1$. Let $M > 0$. Then there exists $c > 0$ depending only on $N, \mathcal{A}, \omega, M$ such that*

$$d_j(x, \partial\Omega) \leqslant c\,\omega(d(x, \partial\Omega)) \tag{7.6}$$

for all open sets $\Omega \in C_M^{\omega(\cdot)}(\mathcal{A})$, $j = 1, \ldots, s$, and $x \in (V_j)_{\frac{\rho}{2}}$.

Proof. Let $\widetilde{\omega}$ be the function of $[0, \infty[$ to itself defined by $\widetilde{\omega}(t) = \omega(t)$ for all $0 \leqslant t \leqslant 1$ and $\widetilde{\omega}(t) = t + \omega(1) - 1$ for all $t > 1$. It is clear that $\widetilde{\omega}$ is continuous and nondecreasing and $\widetilde{\omega}(t) \geqslant \widetilde{k}t$ for all $t \geqslant 0$, where $\widetilde{k} = \min\{k, 1, \omega(1)\}$; moreover,

$$\min\left\{1, \frac{\omega(1)}{\omega(A)}\right\}\omega(a) \leqslant \widetilde{\omega}(a) \leqslant \max\left\{1, \frac{\widetilde{\omega}(A)}{\widetilde{\omega}(1)}\right\}\omega(a) \tag{7.7}$$

for all $A > 0$ and $0 \leqslant a \leqslant A$. By the first inequality in (7.7), $\Omega \in \mathrm{C}_{\widetilde{M}}^{\widetilde{\omega}(\cdot)}(\mathcal{A})$, where $\widetilde{M} = \max\left\{1, \frac{\omega(D)}{\omega(1)}\right\}M$ and D is the diameter of $\bigcup_{j=1}^{s} V_j$. Then, by Lemma 7.1 applied to each function g_j describing the boundary of Ω as in Definition 3.1 (iii) and the second inequality in (7.7), it follows that for each $j = 1, \ldots s$ and all $(\overline{y}, y_N) \in r_j(V_j)$

$$
\begin{aligned}
|g_j(\overline{y}) - y_N| &\leqslant (\widetilde{M} + \widetilde{k}^{-1})\,\widetilde{\omega}\,(d\,((\overline{y}, y_N), \mathrm{Graph}(g_j))) \\
&\leqslant \max\left\{1, \frac{\widetilde{\omega}(D)}{\widetilde{\omega}(1)}\right\}(\widetilde{M} + \widetilde{k}^{-1})\,\omega\,(d\,((\overline{y}, y_N), \mathrm{Graph}(g_j))). \qquad (7.8)
\end{aligned}
$$

Note that if $y \in r_j((V_j)_{\frac{\varrho}{2}})$ and $d\,((\overline{y}, y_N), \mathrm{Graph}(g_j)) < \frac{\varrho}{2}$, then $d(r_j^{(-1)}(y), \partial\Omega)$ equals $d\,((\overline{y}, y_N), \mathrm{Graph}(g_j))$; if $y \in r_j((V_j)_{\frac{\varrho}{2}})$ and $d\,((\overline{y}, y_N), \mathrm{Graph}(g_j)) \geqslant \frac{\varrho}{2}$, then $d\left(r_j^{(-1)}((\overline{y}, y_N)), \partial\Omega\right) \geqslant \frac{\varrho}{2}$. Hence

$$
\omega(d(\overline{y}, y_N), \mathrm{Graph}(g_j))) \leqslant \frac{\omega\,(D)}{\omega(\frac{\varrho}{2})}\omega(d(r_j^{(-1)}((\overline{y}, y_N)), \partial\Omega)). \qquad (7.9)
$$

By (7.8) and (7.9), it follows that if $y \in r_j((V_j)_{\frac{\varrho}{2}})$, then

$$
|g_j(\overline{y}) - y_N| \leqslant c\omega\left(d(r_j^{(-1)}(y), \partial\Omega)\right),
$$

where $c = \max\{1, \widetilde{\omega}(D)/\widetilde{\omega}(1)\}\,(\widetilde{M} + \widetilde{k}^{-1})\omega\,(D)\,/\omega(\rho/2)$. Hence, by (5.2), for $x \in (V_j)_{\frac{\varrho}{2}}$ we have

$$
d_j(x, \partial\Omega) = |g_j(\overline{r_j(x)}) - (r_j(x))_N| \leqslant c\omega(d(x, \partial\Omega)).
$$

The proof is complete. □

Lemma 7.3. *Let $\mathcal{A} = (\rho, s, s', \{(V_j)_{j=1}^s, \{r_j\}_{j=1}^s)$ be an atlas in $\mathbb{R}^N$. Let $\widetilde{\mathcal{A}} = (\rho/2, s, s', \{(V_j)_{\rho/2}\}_{j=1}^s, \{r_j\}_{j=1}^s)$. Let $\omega : [0, \infty[\to [0, \infty[$ be a continuous nondecreasing function satisfying $\omega(0) = 0$ and, for some $k > 0$, $\omega(t) \geqslant kt$ for all $0 \leqslant t \leqslant 1$. Let $M > 0$. Then there exists $c > 0$ depending only on N, $\mathcal{A}$, ω, M such that*

$$
d^{\mathcal{HP}}(\partial\Omega_1, \partial\Omega_2) \leqslant d_{\widetilde{\mathcal{A}}}(\Omega_1, \Omega_2) \leqslant c\,\omega(d_{\mathcal{HP}}(\partial\Omega_1, \partial\Omega_2)) \qquad (7.10)
$$

for all opens sets $\Omega_1, \Omega_2 \in C_M^{\omega(\cdot)}(\mathcal{A})$.

Proof. For each $x \in \partial\Omega_1$ there exists $y \in \partial\Omega_2$ such that $|x - y| \leqslant d_{\widetilde{\mathcal{A}}}(\Omega_1, \Omega_2)$. Indeed, if $r_j(x) = (\overline{x}, x_N)$ for some $j = 1, \dots, s'$, it suffices to consider $y \in \partial\Omega_2$ such that $\overline{r_j(y)} = \overline{x}$. It follows that $d(x, \partial\Omega_2) \leqslant d_{\widetilde{\mathcal{A}}}(\Omega_1, \Omega_2)$ for all $x \in \partial\Omega_1$. In the same way, $d(x, \partial\Omega_1) \leqslant d_{\widetilde{\mathcal{A}}}(\Omega_1, \Omega_2)$ for all $x \in \partial\Omega_2$. Thus, the first inequality in (7.10) follows. The second inequality in (7.10) immediately follows by (7.6), the continuity of ω, the property (5.12), and Definition 7.1. □

Theorem 7.1. *Let $\mathcal{A}$ be an atlas in $\mathbb{R}^N$. Let $m \in \mathbb{N}$, $L, M, \theta > 0$ and, for all $\alpha, \beta \in \mathbb{N}_0^N$ with $|\alpha| = |\beta| = m$, let $A_{\alpha\beta} \in C^{0,1}(\bigcup_{j=1}^s V_j)$ satisfy $A_{\alpha\beta} = A_{\beta\alpha}$,*

$\|A_{\alpha\beta}\|_{C^{0,1}(\bigcup_{j=1}^{s} V_j)} \leqslant L$ *and the condition* (2.4). *Let* $\omega : [0,\infty[\to [0,\infty[$ *be a continuous nondecreasing function satisfying* $\omega(0) = 0$ *and, for some* $k > 0$, $\omega(t) \geqslant kt$ *for all* $0 \leqslant t \leqslant 1$. *Then for each* $n \in \mathbb{N}$ *there exist* $c_n, \varepsilon_n > 0$ *depending only on* n, N, $\mathcal{A}$, m, L, M, θ, ω *such that for both Dirichlet and Neumann boundary conditions*

$$|\lambda_n[\Omega_1] - \lambda_n[\Omega_2]| \leqslant c_n \omega(d_{\mathcal{HP}}(\partial\Omega_1, \partial\Omega_2)) \tag{7.11}$$

for all $\Omega_1, \Omega_2 \in C_M^{\omega(\cdot)}(\mathcal{A})$ *satisfying* $d_{\mathcal{HP}}(\partial\Omega_1, \partial\Omega_2) < \varepsilon_n$.

Proof. Note that if $\Omega_1, \Omega_2 \in C(\mathcal{A})$, then also $\Omega_1, \Omega_2 \in C(\widetilde{\mathcal{A}})$, where $\widetilde{\mathcal{A}} = (\rho/2, s, s', \{(V_j)_{\rho/2}\}_{j=1}^{s}, \{r_j\}_{j=1}^{s})$. Thus, by the inequalities (5.10), (6.21) applied to Ω_1, Ω_2 as open sets in $C(\widetilde{\mathcal{A}})$ and, by the inequality (7.10), we deduce the validity of (7.11). □

Recall that for any Ω we set

$$\Omega^{\varepsilon} = \{x \in \mathbb{R}^N : \ d(x, \Omega) < \varepsilon\}, \quad \Omega_{\varepsilon} = \{x \in \Omega : \ d(x, \partial\Omega) > \varepsilon\}.$$

Lemma 7.4. *If* Ω_1 *and* Ω_2 *are two open sets satisfying the inclusions*

$$(\Omega_1)_{\varepsilon} \subset \Omega_2 \subset (\Omega_1)^{\varepsilon} \tag{7.12}$$

or

$$(\Omega_2)_{\varepsilon} \subset \Omega_1 \subset (\Omega_2)^{\varepsilon}, \tag{7.13}$$

then

$$d_{\mathcal{HP}}(\partial\Omega_2, \partial\Omega_1) \leqslant \varepsilon. \tag{7.14}$$

Proof. As in the proof of Lemma 5.2, the inclusions (7.12) and (7.13) imply $\sup_{x\in\partial\Omega_2} d(x, \partial\Omega_1) \leqslant \varepsilon$ and $\sup_{x\in\partial\Omega_1} d(x, \partial\Omega_2) \leqslant \varepsilon$ respectively. Hence if (7.12) or (7.13) is satisfied, then at least one of these inequalities is satisfied, which implies (7.14). □

Note that if Ω_1 and Ω_2 are two open sets satisfying the inclusion (7.12), then it may happen that they do not satisfy the inclusion (7.13) and

$$\sup_{x\in\partial\Omega_1} d(x, \partial\Omega_2) > \varepsilon \tag{7.15}$$

(see Examples 8.1 and 8.2 in Appendix).

Corollary 7.1. *Under the assumptions of Theorem 7.1, for each* $n \in \mathbb{N}$ *there exist* $c_n, \varepsilon_n > 0$ *depending only on* n, N, $\mathcal{A}$, m, L, M, θ, ω *such that for both Dirichlet and Neumann boundary conditions*

$$|\lambda_n[\Omega_1] - \lambda_n[\Omega_2]| \leqslant c_n \omega(\varepsilon) \tag{7.16}$$

for all $0 < \varepsilon < \varepsilon_n$ *and* $\Omega_1, \Omega_2 \in C_M^{\omega(\cdot)}(\mathcal{A})$ *satisfying* (7.12) *or* (7.13).

Proof. The inequality (7.16) follows from (7.11) and (7.14). □

8 Appendix

8.1 On the atlas distance

Given an atlas $\mathcal{A}$ in $\mathbb{R}^N$, it is easy to prove that the function $d_{\mathcal{A}}$ of $C(\mathcal{A}) \times C(\mathcal{A})$ to $\mathbb{R}$ which takes (Ω_1, Ω_2) to $d_{\mathcal{A}}(\Omega_1, \Omega_2)$ for all $(\Omega_1, \Omega_2) \in C(\mathcal{A}) \times C(\mathcal{A})$ is a metric on the set $C(\mathcal{A})$.

Lemma 8.1. *Let $\mathcal{A}$ be an atlas in $\mathbb{R}^N$. Let Ω_n, $n \in \mathbb{N}$, be a sequence in $C(\mathcal{A})$. For each $n \in \mathbb{N}$ let g_{jn}, $j = 1, \dots, s$, be the functions describing the boundary of Ω_n as in Definition* 3.1 (iii)*. Then the sequence Ω_n, $n \in \mathbb{N}$, is convergent in $(C(\mathcal{A}), d_{\mathcal{A}})$ if and only if for all $j = 1, \dots, s$ the sequences g_{jn}, $n \in \mathbb{N}$, are uniformly convergent on $\overline{W}_j$. Moreover, if g_{jn} converge uniformly to g_j on $\overline{W}_j$ for all $j = 1, \dots, s$, then Ω_n converges in $(C(\mathcal{A}), d_{\mathcal{A}})$ to the open set $\Omega \in C(\mathcal{A})$ whose boundary is described by the functions g_j as in Definition* 3.1 (iii).

Proof. It suffices to prove that if the sequences g_{jn}, $n \in \mathbb{N}$, converge to g_j uniformly on $\overline{W}_j$ for all $j = 1, \dots, s$, then the sequence Ω_n, $n \in \mathbb{N}$, converges in $(C(\mathcal{A}), d_{\mathcal{A}})$ to the open set $\Omega \in C(\mathcal{A})$ whose boundary is described by the functions g_j as in Definition 3.1 (iii) (the rest is obvious). We divide the proof into two steps.

Step 1. We prove that if $x \in V_h \cap V_k$ for $h \neq k$ and $r_h(x) = (\overline{x}, g_h(\overline{x}))$ for some $\overline{x} \in W_h$, then there exists $\overline{y} \in W_k$ such that $r_k(x) = (\overline{y}, g_k(\overline{y}))$. Note that $x = \lim_{n\to\infty} r_h^{(-1)}(\overline{x}, g_{hn}(\overline{x}))$ and there exists $\widetilde{n} \in \mathbb{N}$ such that $r_h^{(-1)}(\overline{x}, g_{hn}(\overline{x})) \in V_h \cap V_k$ for all $n \geqslant \widetilde{n}$. For each $n \geqslant \widetilde{n}$ there exists $\overline{y}_n \in W_k$ such that $r_k(r_h^{(-1)}(\overline{x}, g_{hn}(\overline{x}))) = (\overline{y}_n, g_{kn}(\overline{y}_n))$. It is clear that $\lim_{n\to\infty} r_k(r_h^{(-1)}(\overline{x}, g_{hn}(\overline{x}))) = r_k(x)$. Hence $\lim_{n\to\infty} (\overline{y}_n, g_{kn}(\overline{y}_n)) = r_k(x)$. By the uniform convergence of g_{kn} to g_k on W_k, there exists $\overline{y} \in W_k$ such that $\lim_{n\to\infty} (\overline{y}_n, g_{kn}(\overline{y})) = (\overline{y}, g_k(\overline{y}))$. Thus, $\lim_{n\to\infty} r_k(r_h^{(-1)}(\overline{x}, g_{hn}(\overline{x}))) = (\overline{y}, g_k(\overline{y}))$ and $x = r_k^{(-1)}(\overline{y}, g_k(\overline{y}))$ as required.

Step 2. We prove that if $x \in V_h \cap V_k$ for $h \neq k$, $r_h(x) = (\overline{x}, x_N)$ for some $\overline{x} \in W_h$ and $x_N < g_h(\overline{x})$, then there exists $\overline{y} \in W_k$ such that $r_k(x) = (\overline{y}, y_N)$ and $y_N < g_k(\overline{y})$. Indeed, there exists $\widehat{n} \in \mathbb{N}$ such that $x_N < g_{hn}(\overline{x})$ for all $n \geqslant \widehat{n}$. Thus, $x \in V_h \cap V_k \cap \Omega_n$. Hence

$$(r_k(x))_N < g_{kn}(\overline{r_k(x)}),$$

and, passing to the limit, we find

$$(r_k(x))_N \leqslant g_k(\overline{r_k(x)}).$$

If $(r_k(x))_N = g_k(\overline{r_k(x)})$, then, by Step 1, there exists $\overline{z} \in W_h$ such that $r_h(x) = (\overline{z}, g_h(\overline{z}))$ which implies $\overline{z} = \overline{x}$ and $g_h(\overline{z}) = x_N$ which contradicts the assumption that $x_N < g_h(\overline{x})$. Thus, we have proved that

$$(r_k(x))_N < g_k(\overline{r_k(x)}).$$

In other words, $r_k(x) = (\overline{y}, y_N)$, where $\overline{y} = \overline{r_k(x)}$, $y_N = (r_k(x))_N$, and $y_N < g_k(\overline{y})$ as required.

By Steps 1 and 2, the set

$$\Omega = \bigcup_{j=1}^{s} r_j^{(-1)}\left(\{(\overline{x}, x_N) : \ \overline{x} \in W_j, \ a_{Nj} < x_N < g_j(\overline{x})\}\right)$$

is such that

$$r_j(\Omega \cap V_j) = \{(\overline{x}, x_N) : \ \overline{x} \in W_j, \ a_{Nj} < x_N < g_j(\overline{x})\}.$$

Thus, $\Omega \in C(\mathcal{A})$. It is obvious that $\lim\limits_{n\to\infty} d_{\mathcal{A}}(\Omega_n, \Omega) = 0$. □

Theorem 8.1. *Let $\mathcal{A}$ be an atlas in $\mathbb{R}^N$. Then $(C(\mathcal{A}), d_{\mathcal{A}})$ is a complete metric space. Moreover, for every function ω satisfying the assumptions of Definition 7.2 and for each $M > 0$, $C_M^{\omega(\cdot)}(\mathcal{A})$ is a compact set in $(C(\mathcal{A}), d_{\mathcal{A}})$.*

Proof. The completeness of $(C(\mathcal{A}), d_{\mathcal{A}})$ and the closedness of the set $C_M^{\omega(\cdot)}(\mathcal{A})$ follow directly from Lemma 8.1 (in the second case, one should take into account that the condition (7.2) with fixed ω and M allows one to pass to the limit).

By the definition of $C_M^{\omega(\cdot)}(\mathcal{A})$, the sets $G_j = \{g_j[\Omega]\}_{\Omega \in C_M^{\omega(\cdot)}(\mathcal{A})}$, $j = 1, \dots, s$, of functions $g_j[\Omega]$ entering Definition 3.1, which are defined on the bounded cuboids $\overline{W}_j$, are bounded in the sup-norm and are equicontinuous by the condition (7.2), where ω and M are the same for all $\Omega \in C_M^{\omega(\cdot)}(\mathcal{A})$. By the Ascoli-Arzelà theorem, the closed sets G_j are compact with respect to the sup-norm.

Let $\{\Omega_n\}_{n\in\mathbb{N}}$ be a sequence in $C_M^{\omega(\cdot)}(\mathcal{A})$. Since the sets G_j are compact, it follows that, possibly for a subsequence, $\{g_{jn}\}_{n\in\mathbb{N}}$, where $g_{jn} = g_j[\Omega_n]$, converges uniformly on $\overline{W}_j$ to some continuous functions g_j, $j = 1, \dots, s$. By Lemma 8.1, the sequence $\{\Omega_n\}_{n\in\mathbb{N}}$ converges in $(C(\mathcal{A}), d_{\mathcal{A}})$ to the open set Ω defined by the functions g_j, $j = 1, \dots, s$. Therefore, the set $C_M^{\omega(\cdot)}(\mathcal{A})$ is relatively compact in $(C(\mathcal{A}), d_{\mathcal{A}})$ and, being closed, is compact. □

8.2 Comparison of atlas distance, Hausdorff–Pompeiu distance, and lower Hausdorff–Pompeiu deviation

Note that

$$d_{\mathcal{HP}}(\partial\Omega_1,\partial\Omega_2)\leqslant d^{\mathcal{HP}}(\partial\Omega_1,\partial\Omega_2)\leqslant d_{\mathcal{A}}(\Omega_1,\Omega_2) \tag{8.1}$$

for all $\Omega_1,\Omega_2\in C(\mathcal{A})$. (The first inequality is trivial, the second one can be proved in the same way as in the proof of Lemma 7.3.)

The following examples show that $d_{\mathcal{HP}}(\partial\Omega_1,\partial\Omega_2)$ can be much smaller than $d^{\mathcal{HP}}(\partial\Omega_1,\partial\Omega_2)$ and $d^{\mathcal{HP}}(\partial\Omega_1,\partial\Omega_2)$ can be much smaller than $d_{\mathcal{A}}(\Omega_1,\Omega_2)$.

Example 8.1. Let $N=2$, $c>\sqrt{3}$ and $0<\varepsilon<1/\sqrt{3}$. Let

$$\Omega_1=\big\{(x_1,x_2)\in\mathbb{R}^2:\ c|x_2|<x_1<c\big\}$$

and $\Omega_2=(\Omega_1)_\varepsilon$. Then Ω_1,Ω_2 satisfy the inclusion (7.12), but not (7.13).

Moreover, the lower Hausdorff–Pompeiu deviation of the boundaries can be much smaller than their usual Hausdorff–Pompeiu distance because

$$d_{\mathcal{HP}}(\partial\Omega_1,\partial\Omega_2)=\varepsilon\ \text{ and }\ d^{\mathcal{HP}}(\partial\Omega_1,\partial\Omega_2)=\varepsilon\sqrt{c^2+1}.$$

Example 8.2. Let a function $\psi:[0,\infty[\to[0,\infty[$ be such that $\psi(0)=\psi'(0)=0$ and $\psi'(t)>0$ for all $t>0$. Let $\omega:[0,\infty[\to[0,\infty[$ be the inverse function of ψ. Let $N=2$, and let

$$\Omega_1=\{(x_1,x_2)\in\mathbb{R}^2:\ \omega(|x_2|)<x_1<\omega(1)\}.$$

Let $P=(x_1,0)$ be a point with $x_1>0$ sufficiently small so that $d(P,\partial\Omega_1)=d(P,\{(t,\psi(t)):\ 0\leqslant t\leqslant\omega(1)\})=|P-Q|$ for some point $Q=(\xi,\psi(\xi))$ with $0<\xi<\omega(1)$. We set $d(P,\partial\Omega_1)=\varepsilon$. An elementary consideration shows that

$$x_1=\xi+\psi(\xi)\psi'(\xi)\ \text{ and }\ \varepsilon=\psi(\xi)\sqrt{1+\psi'(\xi)^2}.$$

This implies that $x_1\sim\xi$ and $\varepsilon\sim\psi(\xi)$ as $\xi\to0^+$. Hence $x_1\sim\omega(\varepsilon)$ as $\varepsilon\to0^+$.

Now, let $\Omega_2=(\Omega_1)_\varepsilon$. It is clear that Ω_1 and Ω_2 satisfy (7.12) and

$$\sup_{x\in\partial\Omega_2}d(x,\partial\Omega_1)=\varepsilon.$$

However, since $P\in\partial\Omega_2$, we have

$$\sup_{x\in\partial\Omega_1}d(x,\partial\Omega_2)\geqslant x_1\sim\omega(\varepsilon)$$

as $\varepsilon \to 0^+$. Hence Ω_1, Ω_2 cannot satisfy (7.13) for small values of ε because $\lim_{\varepsilon \to 0^+} \omega(\varepsilon)/\varepsilon = \infty$. Moreover, there exist $c_1, c_2 > 0$ such that for all sufficiently small $\varepsilon > 0$

$$c_1\omega(d_{\mathcal{HP}}(\partial\Omega_1, \partial\Omega_2)) \leqslant d^{\mathcal{HP}}(\partial\Omega_1, \partial\Omega_2) \leqslant c_2\omega(d_{\mathcal{HP}}(\partial\Omega_1, \partial\Omega_2)).$$

(The second inequality follows by (7.10).) This means, in particular, that the usual Hausdorff–Pompeiu distance $d^{\mathcal{HP}}$ between the boundaries may tend to zero arbitrarily slower than their lower Hausdorff–Pompeiu deviation $d_{\mathcal{HP}}$.

Example 8.3. Let $N = 2$. Let $\mathcal{A} = (\rho, s, s', \{V_j\}_{j=1}^s, \{r_j\}_{j=1}^s)$ be an atlas in $\mathbb{R}^2$ with $V_1 =]-2,2[\times]-2,2[$. Let $\omega : [0,\infty[\to [0,\infty[$ be a continuous increasing function such that $\omega(0) = 0$ and, for some $k > 0$, $\omega(t) \geqslant kt$ for all $0 \leqslant t \leqslant 1$. Assume also that $\omega(1) = 1$ and that, for some $M > 0$, $|\omega(x) - \omega(y)| \leqslant M\omega(|x-y|)$ for all $0 \leqslant x, y \leqslant 1$. Let $0 < \varepsilon < 1/2$. Let $\Omega_1, \Omega_2 \in C_M^{\omega(\cdot)}(\mathcal{A})$ with

$$r_1(\Omega_1 \cap V_1) = \{(x_1,x_2) :\ -2 < x_1 < 2,\ -2 < x_2 < g_{11}(x_1)\},$$
$$r_1(\Omega_2 \cap V_1) = \{(x_1,x_2) :\ -2 < x_1 < 2,\ -2 < x_2 < g_{12}(x_1)\},$$

where

$$g_{11}(x_1) = \begin{cases} 1 - \omega(|x_1|) & \text{if } |x_1| \leqslant 1, \\ 0 & \text{if } 1 < |x_1| < 2, \end{cases} \tag{8.2}$$

$$g_{12}(x_1) = \begin{cases} g_{11}(x_1 - \varepsilon) & \text{if } -2+\varepsilon \leqslant x_1 < 2, \\ 0 & \text{if } -2 < x_1 < -2+\varepsilon\,, \end{cases} \tag{8.3}$$

and $\Omega_1 \cap V_j = \Omega_2 \cap V_j$ for all $2 \leqslant j \leqslant s$. It is clear that $d^{\mathcal{HP}}(\partial\Omega_1, \partial\Omega_2) \leqslant \varepsilon$, and $d_{\mathcal{A}}(\Omega_1, \Omega_2) \geqslant g_{11}(0) - g_{12}(0) = \omega(\varepsilon)$. Thus,

$$\omega(d^{\mathcal{HP}}(\partial\Omega_1, \partial\Omega_2)) \leqslant d_{\mathcal{A}}(\Omega_1, \Omega_2). \tag{8.4}$$

Acknowledgement. This research was supported by the research project "Problemi di stabilità per operatori differenziali" of the University of Padova, Italy and partially supported by the grant of INTAS (grant no. 05-1000008-8157). The first author was also supported by the Russian Foundation for Basic Research (grant no. 05-01-01050).

References

1. Barbatis, G.: Boundary decay estimates for solutions of fourth order operators elliptic equations. Ark. Mat. **45**, 197-219 (2007)

2. Berger, M.S.: Nonlinearity and Functional Analysis. Academic Press, New York–London (1977)
3. Burenkov, V.I.: Sobolev spaces on domains. B.G. Teubner, Stuttgart (1998)
4. Burenkov, V.I., Davies, E.B.: Spectral stability of the Neumann Laplacian. J. Differ. Equations **186**, 485-508 (2002)
5. Burenkov, V.I., Lamberti, P.D.: Spectral stability of Dirichlet second order uniformly elliptic operators. J. Differ. Equations **244**, 1712-1740 (2008)
6. Burenkov, V.I., Lamberti, P.D.: Spectral stability of general nonnegative self-adjoint operators with applications to Neumann-type operators. J. Differ. Equations **233**, 345-379 (2007)
7. Davies, E.B.: Spectral Theory and Differential Operators. Cambridge Univ. Press, Cambridge (1995)
8. Davies, E.B.: Sharp boundary estimates for elliptic operators. Math. Proc. Camb. Phil. Soc. **129**, 165-178 (2000)
9. Deimling, K.: Nonlinear Functional Analysis. Springer-Verlag, Berlin (1985)
10. Fraenkel, L.E., Formulae for high derivatives of composite functions. Math. Proc. Camb. Phil. Soc. **83**, 159-165 (1978)
11. Hale, J.K.: Eigenvalues and perturbed domains. In: Ten Mathematical Essays on Approximation in Analysis and Topology, pp. 95-123. Elsevier, Amsterdam (2005)
12. Henry, D.: Perturbation of the Boundary in Boundary-Value Problems of Partial Differential Equations. Cambridge Univ. Press, Cambridge (2005)
13. Nazaret, C.: A system of boundary integral equations for polygonal plates with free edges. Math. Methods Appl. Sci. **21**, 165-185 (1998)
14. Nečas, J.: Les méthodes directes en théorie des équations elliptiques. Masson, Paris (1967)
15. Pang, M.M.H.: Approximation of ground state eigenvalues and eigenfunctions of Dirichlet Laplacians. Bull. London Math. Soc., **29**, 720-730 (1997)

Conductor Inequalities and Criteria for Sobolev-Lorentz Two-Weight Inequalities

Serban Costea and Vladimir Maz'ya

In memory of S.L. Sobolev

Abstract We present integral conductor inequalities connecting the Lorentz p, q-(quasi)norm of a gradient of a function to a one-dimensional integral of the p, q-capacitance of the conductor between two level surfaces of the same function. These inequalities generalize an inequality obtained by the second author in the case of the Sobolev norm. Such conductor inequalities lead to necessary and sufficient conditions for Sobolev–Lorentz type inequalities involving two arbitrary measures.

1 Introduction

During the last decades, Sobolev–Lorentz function spaces, which include classical Sobolev spaces, attracted attention not only as an interesting mathematical object, but also as a tool for a finer tuning of properties of solutions to partial differential equations (see, for example, [7, 8, 9, 10, 14, 15, 16, 19, 28, 30, 43]).

In the present paper, we generalize the inequality

Serban Costea
McMaster University, 1280 Main Street West, Hamilton, Ontario L8S 4K1, Canada; Fields Institute for Research in Mathematical Sciences, 222 College Street, Toronto, Ontario M5T 3J1, Canada, e-mail: secostea@math.mcmaster.ca

Vladimir Maz'ya
Ohio State University, Columbus, OH 43210 USA; University of Liverpool, Liverpool L69 7ZL, UK; Linköping University, Linköping SE-58183, Sweden, e-mail: vlmaz@mai.liu.se, vlmaz@math.ohio-state.edu

V. Maz'ya (ed.), *Sobolev Spaces in Mathematics II*, *International Mathematical Series*.
doi: 10.1007/978-0-387-85650-6,

$$\int_0^\infty \operatorname{cap}_p(\overline{M_{at}}, M_t)d(t^p) \leqslant c(a,p)\int_\Omega |\nabla f|^p\,dx \tag{1.1}$$

to the case of Sobolev–Lorentz spaces. Here, $f \in \operatorname{Lip}_0(\Omega)$, i.e., f is an arbitrary Lipschitz function compactly supported in the open set $\Omega \subset \mathbf{R}^n$, while M_t is the set $\{x \in \Omega : |f(x)| > t\}$ with $t > 0$. The inequality (1.1) was obtained in [34] (see also [36, Chapt. 2]). It has various extensions and applications to the theory of Sobolev type spaces on domains in $\mathbf{R}^n$, Riemannian manifolds, metric and topological spaces, to linear and nonlinear partial differential equations, Dirichlet forms, and Markov processes etc. (see, for example, [1, 2, 3, 4, 5, 6, 12, 17, 18, 21, 22, 23, 24, 25, 26, 27, 29, 31, 32, 33, 34, 35, 37, 38, 39, 40, 41, 42, 45, 46, 47, 48]).

In the sequel, we prove the inequalities

$$\int_0^\infty \operatorname{cap}(\overline{M_{at}}, M_t)d(t^p) \leqslant c(a,p,q)||\nabla f||^p_{L^{p,q}(\Omega,m_n;\mathbf{R}^n)} \text{ when } 1 \leqslant q \leqslant p \tag{1.2}$$

and

$$\int_0^\infty \operatorname{cap}_{p,q}(\overline{M_{at}}, M_t)^{q/p}d(t^q) \leqslant c(a,p,q)||\nabla f||^q_{L^{p,q}(\Omega,m_n;\mathbf{R}^n)} \text{ when } p < q < \infty \tag{1.3}$$

for all $f \in \operatorname{Lip}_0(\Omega)$.

The proof of (1.2) and (1.3) is based on the superadditivity of the p,q-capacitance, also justified in this paper.

From (1.2) and (1.3) we derive necessary and sufficient conditions for certain two-weight inequalities involving Sobolev–Lorentz norms, generalizing results obtained in [37, 38]. Specifically, let μ and ν be two locally finite nonnegative measures on Ω, and let p, q, r, s be real numbers such that $1 < s \leqslant \max(p,q) \leqslant r < \infty$ and $q \geqslant 1$. We characterize the inequality

$$||f||_{L^{r,\max(p,q)}(\Omega,\mu)} \leqslant A\left(||\nabla f||_{L^{p,q}(\Omega,m_n;\mathbf{R}^n)} + ||f||_{L^{s,\max(p,q)}(\Omega,\nu)}\right) \tag{1.4}$$

restricted to functions $f \in \operatorname{Lip}_0(\Omega)$ by requiring the condition

$$\mu(g)^{1/r} \leqslant K(\operatorname{cap}_{p,q}(\overline{g}, G)^{1/p} + \nu(G)^{1/s}) \tag{1.5}$$

to be valid for all open bounded sets g and G subject to $\overline{g} \subset G$, $\overline{G} \subset \Omega$. When $n = 1$, the inequality (1.4) becomes

$$||f||_{L^{r,\max(p,q)}(\Omega,\mu)} \leqslant A\left(||f'||_{L^{p,q}(\Omega,m_1)} + ||f||_{L^{s,\max(p,q)}(\Omega,\nu)}\right). \tag{1.6}$$

It is shown that the requirement that (1.6) is valid for all functions $f \in \operatorname{Lip}_0(\Omega)$ when $n = 1$ is equivalent to the condition

$$\mu(\sigma_d(x))^{1/r} \leqslant K(\tau^{(1-p)/p} + \nu(\sigma_{d+\tau}(x))^{1/s}) \tag{1.7}$$

whenever x, d and τ are such that $\overline{\sigma_{d+\tau}(x)} \subset \Omega$. Here and throughout the paper, $\sigma_d(x)$ denotes the open interval $(x-d, x+d)$ for every $d > 0$.

2 Preliminaries

Denote by Ω a nonempty open subset of $\mathbf{R}^n$ and by m_n the Lebesgue n-measure in $\mathbf{R}^n$, where $n \geqslant 1$ is an integer. For a Lebesgue measurable function $u : \Omega \to \mathbf{R}$ denote by supp u the smallest closed set such that u vanishes outside supp u. We also introduce

$$\begin{aligned} \mathrm{Lip}(\Omega) &= \{\varphi : \Omega \to \mathbf{R} : \varphi \text{ is Lipschitz}\}, \\ \mathrm{Lip}_0(\Omega) &= \{\varphi : \Omega \to \mathbf{R} : \varphi \text{ is Lipschitz and with compact support in } \Omega\}. \end{aligned}$$

If $\varphi \in \mathrm{Lip}(\Omega)$, we write $\nabla\varphi$ for the gradient of φ. This notation makes sense since, by the Rademacher theorem [20, Theorem 3.1.6], every Lipschitz function on Ω is m_n-a.e. differentiable.

Throughout this section, we assume that (Ω, μ) is a measure space. Let $f : \Omega \to \mathbf{R}^n$ be a μ-measurable function. We define $\mu_{[f]}$, the *distribution function* of f as follows (see [11, Definition II.1.1]):

$$\mu_{[f]}(t) = \mu(\{x \in \Omega : |f(x)| > t\}), \qquad t \geqslant 0.$$

We define f^*, the *nonincreasing rearrangement* of f, by

$$f^*(t) = \inf\{v : \mu_{[f]}(v) \leqslant t\}, \quad t \geqslant 0$$

(see [11, Definition II.1.5]). We note that f and f^* have the same distribution function. Moreover, for every positive α we have

$$(|f|^\alpha)^* = (|f|^*)^\alpha$$

and if $|g| \leqslant |f|$ a.e. on Ω, then $g^* \leqslant f^*$ (see [11, Proposition II.1.7]). We also define f^{**}, the *maximal function* of f^*, by

$$f^{**}(t) = m_{f^*}(t) = \frac{1}{t}\int_0^t f^*(s)ds, \quad t > 0$$

(see [11, Definition II.3.1]).

Throughout the paper, we denote by p' the Hölder conjugate of $p \in [1, \infty]$.

The *Lorentz space* $L^{p,q}(\Omega, \mu; \mathbf{R}^n)$, $1 < p < \infty$, $1 \leqslant q \leqslant \infty$, is defined as follows:

$$L^{p,q}(\Omega,\mu;\mathbf{R}^n)=\{f:\Omega\to\mathbf{R}^n : f \text{ is } \mu\text{-measurable}, ||f||_{L^{p,q}(\Omega,\mu;\mathbf{R}^n)}<\infty\},$$

where

$$||f||_{L^{p,q}(\Omega,\mu;\mathbf{R}^n)}=|| \, |f| \, ||_{p,q}=\begin{cases}\left(\int\limits_0^\infty (t^{1/p}f^*(t))^q \, \frac{dt}{t}\right)^{1/q}, & 1\leqslant q<\infty,\\ \sup\limits_{t>0} t\mu_{[f]}(t)^{1/p}=\sup\limits_{s>0} s^{1/p}f^*(s), & q=\infty\end{cases}$$

(see [11, Definition IV.4.1] and [44, p. 191]). We omit $\mathbf{R}^n$ in the notation of function spaces in the scalar case, i.e., for $n=1$.

If $1\leqslant q\leqslant p$, then $||\cdot||_{L^{p,q}(\Omega,\mu;\mathbf{R}^n)}$ represents a norm, but for $p<q\leqslant\infty$ it represents a quasinorm equivalent to the norm $||\cdot||_{L^{(p,q)}(\Omega,\mu;\mathbf{R}^n)}$, where

$$||f||_{L^{(p,q)}(\Omega,\mu;\mathbf{R}^n)}=|| \, |f| \, ||_{(p,q)}=\begin{cases}\left(\int\limits_0^\infty (t^{1/p}f^{**}(t))^q \, \frac{dt}{t}\right)^{1/q}, & 1\leqslant q<\infty,\\ \sup\limits_{t>0} t^{1/p}f^{**}(t), & q=\infty\end{cases}$$

(see [11, Definition IV.4.4]). Namely, from [11, Lemma IV.4.5] we have

$$|| \, |f| \, ||_{L^{p,q}(\Omega,\mu)}\leqslant || \, |f| \, ||_{L^{(p,q)}(\Omega,\mu)}\leqslant p'|| \, |f| \, ||_{L^{p,q}(\Omega,\mu)}$$

for all $q\in[1,\infty]$ and μ-measurable functions $f:\Omega\to\mathbf{R}^n$.

It is known that $(L^{p,q}(\Omega,\mu;\mathbf{R}^n), ||\cdot||_{L^{p,q}(\Omega,\mu;\mathbf{R}^n)})$ is a Banach space for $1\leqslant q\leqslant p$, while $(L^{p,q}(\Omega,\mu;\mathbf{R}^n), ||\cdot||_{L^{(p,q)}(\Omega,\mu;\mathbf{R}^n)})$ is a Banach space for $1<p<\infty$, $1\leqslant q\leqslant\infty$.

Remark 2.1. It is also known (see [11, Proposition IV.4.2]) that for every $p\in(1,\infty)$ and $1\leqslant r<s\leqslant\infty$ there exists a constant $C(p,r,s)$ such that

$$|| \, |f| \, ||_{L^{p,s}(\Omega,\mu)}\leqslant C(p,r,s)|| \, |f| \, ||_{L^{p,r}(\Omega,\mu)} \tag{2.1}$$

for all measurable functions $f\in L^{p,r}(\Omega,\mu;\mathbf{R}^n)$ and integers $n\geqslant 1$. In particular, the embedding $L^{p,r}(\Omega,\mu;\mathbf{R}^n)\hookrightarrow L^{p,s}(\Omega,\mu;\mathbf{R}^n)$ holds.

The subadditivity and superadditivity of the Lorentz quasinorms

In the second part of this paper, we will prove a few results by relying on the superadditivity of the Lorentz p,q-quasinorm. Therefore, we recall the known results and present new results concerning the superadditivity and the subadditivity of the Lorentz p,q-quasinorm.

The superadditivity of the Lorentz p,q-norm in the case $1 \leqslant q \leqslant p$ was stated in [13, Lemma 2.5].

Proposition 2.2 (see [13, Lemma 2.5]). *Let (Ω, μ) be a measure space. Suppose that $1 \leqslant q \leqslant p$. Let $\{E_i\}_{i\geqslant 1}$ be a collection of pairwise disjoint measurable subsets of Ω with $E_0 = \bigcup\limits_{i\geqslant 1} E_i$, and let $f \in L^{p,q}(\Omega,\mu)$. Then*

$$\sum_{i\geqslant 1} ||\chi_{E_i} f||^p_{L^{p,q}(\Omega,\mu)} \leqslant ||\chi_{E_0} f||^p_{L^{p,q}(\Omega,\mu)}.$$

We obtain a similar result concerning the superadditivity in the case $1 < p < q < \infty$.

Proposition 2.3. *Let (Ω, μ) be a measure space. Suppose that $1 < p < q < \infty$. Let $\{E_i\}_{i\geqslant 1}$ be a collection of pairwise disjoint measurable subsets of Ω with $E_0 = \bigcup\limits_{i\geqslant 1} E_i$, and let $f \in L^{p,q}(\Omega,\mu)$. Then*

$$\sum_{i\geqslant 1} ||\chi_{E_i} f||^q_{L^{p,q}(\Omega,\mu)} \leqslant ||\chi_{E_0} f||^q_{L^{p,q}(\Omega,\mu)}.$$

Proof. For every $i = 0, 1, 2, \ldots$ we let $f_i = \chi_{E_i} f$, where χ_{E_i} is the characteristic function of E_i. We can assume without loss of generality that all the functions f_i are nonnegative. We have (see [30, Proposition 2.1])

$$||f_i||^q_{L^{p,q}(\Omega,\mu)} = p \int\limits_0^\infty s^{q-1} \mu_{[f_i]}(s)^{q/p} ds,$$

where $\mu_{[f_i]}$ is the distribution function of $f_i, i = 0, 1, 2, \ldots$. From the definition of f_0 we have

$$\mu_{[f_0]}(s) = \sum_{i\geqslant 1} \mu_{[f_i]}(s) \text{ for every } s > 0, \tag{2.2}$$

which implies, since $1 < p < q < \infty$, that

$$\mu_{[f_0]}(s)^{q/p} \geqslant \sum_{i\geqslant 1} \mu_{[f_i]}(s)^{q/p} \text{ for every } s > 0.$$

This yields

$$||f_0||^q_{L^{p,q}(\Omega,\mu)} = p \int\limits_0^\infty s^{q-1} \mu_{[f_0]}(s)^{q/p} ds \geqslant p \int\limits_0^\infty s^{q-1} (\sum_{i\geqslant 1} \mu_{[f_i]}(s)^{q/p}) ds$$

$$= \sum_{i \geqslant 1} p \int_0^\infty s^{q-1} \mu_{[f_i]}(s)^{q/p} ds = \sum_{i \geqslant 1} ||f_i||^q_{L^{p,q}(\Omega,\mu)}.$$

This completes the proof of the superadditivity for $1 < p < q < \infty$. □

We have a similar result for the subadditivity of the Lorentz p, q-quasinorm. When $1 < p < q \leqslant \infty$ we obtain a result that generalizes [16, Theorem 2.5].

Proposition 2.4. *Let (Ω, μ) be a measure space. Suppose that $1 < p < q \leqslant \infty$. Let $\{E_i\}_{i \geqslant 1}$ be a collection of pairwise disjoint measurable subsets of Ω with $E_0 = \bigcup_{i \geqslant 1} E_i$, and let $f \in L^{p,q}(\Omega, \mu)$. Then*

$$\sum_{i \geqslant 1} ||\chi_{E_i} f||^p_{L^{p,q}(\Omega,\mu)} \geqslant ||\chi_{E_0} f||^p_{L^{p,q}(\Omega,\mu)}.$$

Proof. Without loss of generality we can assume that all the functions $f_i = \chi_{E_i} f$ are nonnegative. We have to consider two cases, depending on whether $p < q < \infty$ or $q = \infty$.

Suppose that $p < q < \infty$. We have (see [30, Proposition 2.1])

$$||f_i||^p_{L^{p,q}(\Omega,\mu)} = \left(p \int_0^\infty s^{q-1} \mu_{[f_i]}(s)^{q/p} ds \right)^{p/q},$$

where $\mu_{[f_i]}$ is the distribution function of f_i for $i = 0, 1, 2, \ldots$. From (2.2) we obtain

$$||f_0||^p_{L^{p,q}(\Omega,\mu)} = \left(p \int_0^\infty s^{q-1} \mu_{[f_0]}(s)^{q/p} ds \right)^{p/q}$$

$$\leqslant \sum_{i \geqslant 1} \left(p \int_0^\infty s^{q-1} \mu_{[f_i]}(s)^{q/p} ds \right)^{p/q} = \sum_{i \geqslant 1} ||f_i||^p_{L^{p,q}(\Omega,\mu)}.$$

Now, suppose that $q = \infty$. From (2.2) we obtain

$$s^p \mu_{[f_0]}(s) = \sum_{i \geqslant 1} (s^p \mu_{[f_i]}(s)) \text{ for every } s > 0,$$

which implies

$$s^p \mu_{[f_0]}(s) \leqslant \sum_{i \geqslant 1} ||f_i||^p_{L^{p,\infty}(\Omega,\mu)} \text{ for every } s > 0. \tag{2.3}$$

Taking the supremum over all $s > 0$ in (2.3), we get the desired conclusion. This completes the proof. □

3 Sobolev–Lorentz p, q-Capacitance

Suppose that $1 < p < \infty$ and $1 \leqslant q \leqslant \infty$. Let $\Omega \subset \mathbf{R}^n$ be an open set, $n \geqslant 1$. Let $K \subset \Omega$ be compact. The Sobolev–Lorentz p, q-capacitance of the conductor (K, Ω) is denoted by

$$\mathrm{cap}_{p,q}(K, \Omega) = \inf \{||\nabla u||^p_{L^{p,q}(\Omega, m_n; \mathbf{R}^n)} : u \in W(K, \Omega)\},$$

where

$$W(K, \Omega) = \{u \in \mathrm{Lip}_0(\Omega) : u \geqslant 1 \ \text{ in a neighborhood of } K\}.$$

We call $W(K, \Omega)$ the *set of admissible functions for the conductor* (K, Ω).

Since $W(K, \Omega)$ is closed under truncations from below by 0 and from above by 1 and since these truncations do not increase the p, q-quasinorm of the grandients whenever $1 < p < \infty$ and $1 \leqslant q \leqslant \infty$, it follows that we can choose only functions $u \in W(K, \Omega)$ that satisfy $0 \leqslant u \leqslant 1$ when computing the p, q-capacitance of the conductor (K, Ω).

Lemma 3.1. *If Ω is bounded, then we get the same p, q-capacitance for the conductor (K, Ω) if we restrict ourselves to a bigger set, namely*

$$W_1(K, \Omega) = \{u \in \mathrm{Lip}(\Omega) \cap C(\overline{\Omega}) : u \geqslant 1 \ \textit{on } K \ \textit{and } u = 0 \ \textit{on } \partial\Omega\}.$$

Proof. Let $u \in W_1(K, \Omega)$. We can assume without loss of generality that $0 \leqslant u \leqslant 1$. Moreover, we can also assume that $u = 1$ in an open neighborhood U of K. Let $\widetilde{U}$ be an open neighborhood of K such that $\widetilde{U} \subset\subset U$. We choose a cutoff Lipschitz function $0 \leqslant \eta \leqslant 1$ such that $\eta = 1$ on $\Omega \setminus U$ and $\eta = 0$ on $\widetilde{U}$. We note that $1 - \eta(1 - u) = u$. We also note that there exists a sequence of functions $\varphi_j \in \mathrm{Lip}_0(\Omega)$ such that

$$\lim_{j \to \infty} (||\varphi_j - u||_{L^{p+1}(\Omega, m_n)} + ||\nabla\varphi_j - \nabla u||_{L^{p+1}(\Omega, m_n; \mathbf{R}^n)}) = 0.$$

Without loss of generality the sequence φ_j can be chosen such that $\varphi_j \to u$ and $\nabla\varphi_j \to \nabla u$ pointwise a.e. in Ω. Then $\psi_j = 1 - \eta(1 - \varphi_j)$ is a sequence belonging to $W(K, \Omega)$ and

$$\lim_{j \to \infty} (||\psi_j - u||_{L^{p+1}(\Omega, m_n)} + ||\nabla\psi_j - \nabla u||_{L^{p+1}(\Omega, m_n; \mathbf{R}^n)}) = 0.$$

This, Hölder's inequality for Lorentz spaces, and the behavior of the Lorentz p, q-quasinorm in q yield

$$\lim_{j \to \infty} (||\psi_j - u||_{L^{p,q}(\Omega, m_n)} + ||\nabla\psi_j - \nabla u||_{L^{p,q}(\Omega, m_n; \mathbf{R}^n)}) = 0.$$

The desired conclusion follows. □

Basic properties of the p, q-capacitance

Usually, a capacitance is a monotone and subadditive set function. The following theorem shows, among other things, that this is true in the case of the p,q-capacitance. We follow [16] for (i)–(vi). In addition, we prove some superadditivity properties of the p,q-capacitance.

Theorem 3.2. *Suppose that $1 < p < \infty$ and $1 \leqslant q \leqslant \infty$. Let $\Omega \subset \mathbf{R}^n$ be open. The set function $K \mapsto \operatorname{cap}_{p,q}(K,\Omega)$, $K \subset \Omega$, K compact, enjoys the following properties:*

(i) *If $K_1 \subset K_2$, then $\operatorname{cap}_{p,q}(K_1,\Omega) \leqslant \operatorname{cap}_{p,q}(K_2,\Omega)$.*

(ii) *If $\Omega_1 \subset \Omega_2$ are open and K is a compact subset of Ω_1, then*

$$\operatorname{cap}_{p,q}(K,\Omega_2) \leqslant \operatorname{cap}_{p,q}(K,\Omega_1).$$

(iii) *If K_i is a decreasing sequence of compact subsets of Ω with $K = \bigcap_{i=1}^{\infty} K_i$, then*

$$\operatorname{cap}_{p,q}(K,\Omega) = \lim_{i\to\infty} \operatorname{cap}_{p,q}(K_i,\Omega).$$

(iv) *If Ω_i is an increasing sequence of open sets with $\bigcup_{i=1}^{\infty} \Omega_i = \Omega$ and K is a compact subset of Ω_1, then*

$$\operatorname{cap}_{p,q}(K,\Omega) = \lim_{i\to\infty} \operatorname{cap}_{p,q}(K,\Omega_i).$$

(v) *Suppose that $p \leqslant q \leqslant \infty$. If $K = \bigcup_{i=1}^{k} K_i \subset \Omega$, then*

$$\operatorname{cap}_{p,q}(K,\Omega) \leqslant \sum_{i=1}^{k} \operatorname{cap}_{p,q}(K_i,\Omega),$$

where $k \geqslant 1$ is a positive integer.

(vi) *Suppose that $1 \leqslant q < p$. If $K = \bigcup_{i=1}^{k} K_i \subset \Omega$, then*

$$\operatorname{cap}_{p,q}(K,\Omega)^{q/p} \leqslant \sum_{i=1}^{k} \operatorname{cap}_{p,q}(K_i,\Omega)^{q/p},$$

where $k \geqslant 1$ is a positive integer.

(vii) *Suppose that $1 \leqslant q \leqslant p$. Suppose that $\Omega_i, \dots, \Omega_k$ are k pairwise disjoint open sets and K_i are compact subsets of Ω_i for $i = 1, \dots, k$. Then*

$$\operatorname{cap}_{p,q}\Big(\bigcup_{i=1}^{k} K_i, \bigcup_{i=1}^{k} \Omega_i\Big) \geqslant \sum_{i=1}^{k} \operatorname{cap}_{p,q}(K_i, \Omega_i).$$

(viii) *Suppose that $p < q < \infty$. Suppose that $\Omega_i, \dots, \Omega_k$ are k pairwise disjoint open sets and K_i are compact subsets of Ω_i for $i = 1, \dots, k$. Then*

$$\operatorname{cap}_{p,q}\Big(\bigcup_{i=1}^{k} K_i, \bigcup_{i=1}^{k} \Omega_i\Big)^{q/p} \geqslant \sum_{i=1}^{k} \operatorname{cap}_{p,q}(K_i, \Omega_i)^{q/p}.$$

(ix) *Suppose that $1 \leqslant q < \infty$. If Ω_1 and Ω_2 are two disjoint open sets and $K \subset \Omega_1$, then*

$$\operatorname{cap}_{p,q}(K, \Omega_1 \cup \Omega_2) = \operatorname{cap}_{p,q}(K, \Omega_1).$$

Proof. Properties (i)-(vi) are proved by duplicating the proof of Theorem 3.2 in [16], so we will prove only (vii)-(ix).

In order to prove (vii) and (viii), it is enough to assume that $k = 2$. A finite induction on k would prove each of these claims. So, we assume that $k = 2$. Let $u \in \operatorname{Lip}_0(\Omega_1 \cup \Omega_2)$, and let $u_i = \chi_{\Omega_i} u, i = 1, 2$. Let v_i be the restriction of u to Ω_i for $i = 1, 2$. Then $v_i \in \operatorname{Lip}_0(\Omega_i)$ for $i = 1, 2$. We note that u_i can be regarded as the extension of v_i by 0 to $\Omega_1 \cup \Omega_2$ for $i = 1, 2$. We see that $u \in W(K_1 \cup K_2, \Omega_1 \cup \Omega_2)$ if and only if $v_i \in W(K_i, \Omega_i)$ for $i = 1, 2$.

First, suppose that $1 \leqslant q \leqslant p$. Since Ω_1 and Ω_2 are disjoint and $u = u_1 + u_2$ with the functions u_i supported in Ω_i for $i = 1, 2$, we obtain with the help of Proposition 2.2

$$\begin{aligned}
&||\nabla u||^p_{L^{p,q}(\Omega_1 \cup \Omega_2, m_n; \mathbf{R}^n)} \\
&\geqslant ||\nabla u_1||^p_{L^{p,q}(\Omega_1 \cup \Omega_2, m_n; \mathbf{R}^n)} + ||\nabla u_2||^p_{L^{p,q}(\Omega_1 \cup \Omega_2, m_n; \mathbf{R}^n)} \\
&= ||\nabla v_1||^p_{L^{p,q}(\Omega_1, m_n; \mathbf{R}^n)} + ||\nabla v_2||^p_{L^{p,q}(\Omega_2, m_n; \mathbf{R}^n)}.
\end{aligned}$$

This proves (vii).

Now, suppose that $p < q < \infty$. Since Ω_1 and Ω_2 are disjoint and $u = u_1 + u_2$ with the functions u_i supported in Ω_i for $i = 1, 2$, we obtain with the help of Proposition 2.3

$$\begin{aligned}
&||\nabla u||^q_{L^{p,q}(\Omega_1 \cup \Omega_2, m_n; \mathbf{R}^n)} \\
&\geqslant ||\nabla u_1||^q_{L^{p,q}(\Omega_1 \cup \Omega_2, m_n; \mathbf{R}^n)} + ||\nabla u_2||^q_{L^{p,q}(\Omega_1 \cup \Omega_2, m_n; \mathbf{R}^n)} \\
&= ||\nabla v_1||^q_{L^{p,q}(\Omega_1, m_n; \mathbf{R}^n)} + ||\nabla v_2||^q_{L^{p,q}(\Omega_2, m_n; \mathbf{R}^n)}.
\end{aligned}$$

This proves (viii).

We see that (ix) follows from (vii) and (ii) when $1 \leqslant q \leqslant p$. (We use (vii) with $k = 2$ by taking $K_1 = K$ and $K_2 = \varnothing$.) When $p < q < \infty$, (ix) follows from (viii) and (ii). (We use (viii) with $k = 2$ by taking $K_1 = K$ and $K_2 = \varnothing$.) This completes the proof of the theorem. □

Remark 3.3. The definition of the p, q-capacitance implies

$$\mathrm{cap}_{p,q}(K, \Omega) = \mathrm{cap}_{p,q}(\partial K, \Omega)$$

whenever K is a compact set in Ω. Moreover, if $n = 1$ and Ω is an open interval of $\mathbf{R}$, then

$$\mathrm{cap}_{p,q}(K, \Omega) = \mathrm{cap}_{p,q}(H, \Omega),$$

where H is the smallest compact interval containing K.

4 Conductor Inequalities

Lemma 4.1. *Suppose that $\Omega \subset \mathbf{R}^n$ is open. Let $f \in \mathrm{Lip}_0(\Omega)$, and let $a > 1$ be a constant. For $t > 0$ we denote $M_t = \{x \in \Omega : |f(x)| > t\}$. Then the function $t \mapsto \mathrm{cap}_{p,q}(\overline{M_{at}}, M_t)$ is upper semicontinuous.*

Proof. Let $t_0 > 0$ and $\varepsilon > 0$. Let $u \in W(\overline{M_{at_0}}, M_{t_0})$ be chosen such that

$$||\nabla u||^p_{L^{p,q}(\Omega, m_n; \mathbf{R}^n)} < \mathrm{cap}_{p,q}(\overline{M_{at_0}}, M_{t_0}) + \varepsilon.$$

Let g be an open neighborhood of $\overline{M_{at_0}}$ such that $u \geqslant 1$ on g. Since g contains the compact set $\overline{M_{at_0}}$, there exists $\delta_1 > 0$ small such that $g \supset \overline{M_{a(t_0-\delta_1)}}$. Let G be an open set such that $\mathrm{supp}\, u \subset G \subset\subset M_{t_0}$. There exists a small $\delta_2 > 0$ such that $\overline{G} \subset M_{t_0+\delta_2}$. Thus, we have $\overline{M_{a(t_0-\delta)}} \subset g$ and $\overline{G} \subset M_{t_0+\delta}$ for every $\delta \in (0, \min\{\delta_1, \delta_2\})$. By the choice of g and G, we have $u \in W(K, \Omega)$, whenever $K \subset g$ and $\overline{G} \subset \Omega$. This and the choice of u imply that

$$\mathrm{cap}_{p,q}(\overline{M_{a(t_0-\delta)}}, M_{t_0+\delta}) \leqslant \mathrm{cap}_{p,q}(\overline{M_{at_0}}, M_{t_0}) + \varepsilon$$

for every $\delta \in (0, \min\{\delta_1, \delta_2\})$. Using the monotonicity of $\mathrm{cap}_{p,q}$, we deduce that

$$\mathrm{cap}_{p,q}(\overline{M_{at}}, M_t) \leqslant \mathrm{cap}_{p,q}(\overline{M_{at_0}}, M_{t_0}) + \varepsilon$$

for every t sufficiently close to t_0. The result follows. □

Theorem 4.2. *Let Φ denote an increasing convex (not necessarily strictly convex) function given on $[0, \infty)$, $\Phi(0) = 0$. Suppose that $a > 1$ is a constant.*

(i) *If $1 \leqslant q \leqslant p$, then*

$$\Phi^{-1}\left(\int_0^\infty \Phi(t^p \mathrm{cap}_{p,q}(\overline{M_{at}}, M_t)) \frac{dt}{t} \right) \leqslant c(a, p, q) ||\nabla \varphi||^p_{L^{p,q}(\Omega, m_n; \mathbf{R}^n)}$$

for every $\varphi \in \mathrm{Lip}_0(\Omega)$.

(ii) *If* $p < q < \infty$, *then*

$$\Phi^{-1}\left(\int_0^\infty \Phi(t^q \mathrm{cap}_{p,q}(\overline{M_{at}}, M_t)^{q/p})\frac{dt}{t}\right) \leqslant c(a,p,q)||\nabla\varphi||^q_{L^{p,q}(\Omega,m_n;\mathbf{R}^n)}$$

for every $\varphi \in \mathrm{Lip}_0(\Omega)$.

Proof. The proof follows [37]. When $p = q$, we are in the case of the p-capacitance and for that case the result was proved in [37, Theorem 1]. So, we can assume without loss of generality that $p \neq q$. Let $\varphi \in \mathrm{Lip}_0(\Omega)$. We set

$$\Lambda_t(\varphi) = \frac{1}{(a-1)t}\min\{(|\varphi| - t)_+, (a-1)t\}.$$

From Lemma 3.1 we note that

$$\Lambda_t(\varphi) \in W_1(\overline{M_{at}}, M_t) \text{ and } |\nabla\Lambda_t(\varphi)| = \frac{1}{(a-1)t}\chi_{M_t\setminus M_{at}}|\nabla\varphi| \; m_n\text{-a.e.} \quad (4.1)$$

The proof splits now, depending on whether $1 \leqslant q < p$ or $p < q < \infty$.

We assume first that $1 \leqslant q < p$. From (4.1) we have

$$t^p \mathrm{cap}_{p,q}(\overline{M_{at}}, M_t) \leqslant \frac{1}{(a-1)^p}||\chi_{M_t\setminus M_{at}}\nabla\varphi||^p_{L^{p,q}(\Omega,m_n;\mathbf{R}^n)}.$$

Hence

$$\int_0^\infty \Phi(t^p \mathrm{cap}_{p,q}(\overline{M_{at}}, M_t))\frac{dt}{t} \leqslant \int_0^\infty \Phi\Big(\frac{1}{(a-1)^p}||\chi_{M_t\setminus M_{at}}\nabla\varphi||^p_{L^{p,q}(\Omega,m_n;\mathbf{R}^n)}\Big)\frac{dt}{t}.$$

Let γ denote a locally integrable function on $(0,\infty)$ such that there exist the limits $\gamma(0)$ and $\gamma(\infty)$. Then

$$\int_0^\infty (\gamma(t) - \gamma(at))\frac{dt}{t} = (\gamma(0) - \gamma(\infty))\log a. \quad (4.2)$$

We set

$$\gamma(t) = \Phi\Big(\frac{1}{(a-1)^p}||\chi_{M_t}\nabla\varphi||^p_{L^{p,q}(\Omega,m_n;\mathbf{R}^n)}\Big).$$

Using the monotonicity and convexity of Φ together with Proposition 2.2 and the definition of γ, we see that

$$\Phi\Big(\frac{1}{(a-1)^p}||\chi_{M_t\setminus M_{at}}\nabla\varphi||^p_{L^{p,q}(\Omega,m_n;\mathbf{R}^n)}\Big) \leqslant \gamma(t) - \gamma(at) \text{ for every } t > 0.$$

Since

$$\gamma(0) = \Phi\Big(\frac{1}{(a-1)^p}||\nabla\varphi||^p_{L^{p,q}(\Omega,m_n;\mathbf{R}^n)}\Big) \text{ and } \gamma(\infty) = 0,$$

we get

$$\int_0^\infty \Phi(t^p \operatorname{cap}_{p,q}(\overline{M_{at}}, M_t))\frac{dt}{t} \leqslant \log a \cdot \Phi\Big(\frac{1}{(a-1)^p}||\nabla\varphi||^p_{L^{p,q}(\Omega,m_n;\mathbf{R}^n)}\Big).$$

This completes the proof of the case $1 \leqslant q < p$.

Now, we assume that $p < q < \infty$. From (4.1) we have

$$t^q \operatorname{cap}_{p,q}(\overline{M_{at}}, M_t)^{q/p} \leqslant \frac{1}{(a-1)^q}||\chi_{M_t \setminus M_{at}} \nabla\varphi||^q_{L^{p,q}(\Omega,m_n;\mathbf{R}^n)}.$$

Hence

$$\int_0^\infty \Phi(t^q \operatorname{cap}_{p,q}(\overline{M_{at}}, M_t)^{q/p})\frac{dt}{t} \leqslant \int_0^\infty \Phi\Big(\frac{1}{(a-1)^q}||\chi_{M_t \setminus M_{at}} \nabla\varphi||^q_{L^{p,q}(\Omega,m_n;\mathbf{R}^n)}\Big)\frac{dt}{t}.$$

As before, we let γ denote a locally integrable function on $(0,\infty)$ such that there exist the limits $\gamma(0)$ and $\gamma(\infty)$. We set

$$\gamma(t) = \Phi\Big(\frac{1}{(a-1)^q}||\chi_{M_t} \nabla\varphi||^q_{L^{p,q}(\Omega,m_n;\mathbf{R}^n)}\Big).$$

Using the monotonicity and convexity of Φ together with Proposition 2.3 and the definition of γ, we see that

$$\Phi\Big(\frac{1}{(a-1)^q}||\chi_{M_t \setminus M_{at}} \nabla\varphi||^q_{L^{p,q}(\Omega,m_n;\mathbf{R}^n)}\Big) \leqslant \gamma(t) - \gamma(at) \text{ for every } t > 0.$$

Since

$$\gamma(0) = \Phi\Big(\frac{1}{(a-1)^q}||\nabla\varphi||^q_{L^{p,q}(\Omega,m_n;\mathbf{R}^n)}\Big) \text{ and } \gamma(\infty) = 0,$$

we get

$$\int_0^\infty \Phi(t^q \operatorname{cap}_{p,q}(\overline{M_{at}}, M_t)^{q/p})\frac{dt}{t} \leqslant \log a \cdot \Phi\Big(\frac{1}{(a-1)^q}||\nabla\varphi||^q_{L^{p,q}(\Omega,m_n;\mathbf{R}^n)}\Big).$$

This completes the proof of the case $p < q < \infty$. The theorem is proved. □

Choosing $\Phi(t) = t$, we arrive at the inequalities mentioned at the beginning of this paper.

Corollary 4.3. *Suppose that* $1 < p < \infty$ *and* $1 \leqslant q < \infty$. *Let* $a > 1$ *be a constant. Then* (1.2) *and* (1.3) *hold for every* $\varphi \in \mathrm{Lip}_0(\Omega)$.

5 Necessary and Sufficient Conditions for Two-Weight Embeddings

Now, we derive necessary and sufficient conditions for Sobolev–Lorentz type inequalities involving two measures, generalizing results obtained in [37, 38].

Theorem 5.1. *Let* p, q, r, s *be chosen such that* $1 < p < \infty$, $1 \leqslant q < \infty$, *and* $1 < s \leqslant \max(p, q) \leqslant r < \infty$. *Let* Ω *be an open set in* $\mathbf{R}^n$, *and let* μ *and* ν *be two nonnegative locally finite measures on* Ω.

(i) *Suppose that* $1 \leqslant q \leqslant p$. *The inequality*

$$||f||_{L^{r,p}(\Omega,\mu)} \leqslant A\left(||\nabla f||_{L^{p,q}(\Omega,m_n;\mathbf{R}^n)} + ||f||_{L^{s,p}(\Omega,\nu)}\right) \tag{5.1}$$

holds for every $f \in \mathrm{Lip}_0(\Omega)$ *if and only if there exists a constant* $K > 0$ *such that the inequality* (1.5) *is valid for all open bounded sets* g *and* G *that are subject to* $\overline{g} \subset G \subset \overline{G} \subset \Omega$.

(ii) *Suppose that* $p < q < \infty$. *The inequality*

$$||f||_{L^{r,q}(\Omega,\mu)} \leqslant A\left(||\nabla f||_{L^{p,q}(\Omega,m_n;\mathbf{R}^n)} + ||f||_{L^{s,q}(\Omega,\nu)}\right) \tag{5.2}$$

holds for every $f \in \mathrm{Lip}_0(\Omega)$ *if and only if there exists a constant* $K > 0$ *such that the inequality* (1.5) *is valid for all open bounded sets* g *and* G *that are subject to* $\overline{g} \subset G \subset \overline{G} \subset \Omega$.

Proof. We suppose first that $1 \leqslant q \leqslant p$. The case $q = p$ was studied in [38]. Without loss of generality we can assume that $q < p$. We choose some bounded open sets g and G such that $\overline{g} \subset G \subset \overline{G} \subset \Omega$ and $f \in W(\overline{g}, G)$ with $0 \leqslant f \leqslant 1$. We have

$$\mu(g) \leqslant C(r,p)\, ||f||^r_{L^{r,p}(\Omega,\mu)}$$

and

$$||f||^s_{L^{s,p}(\Omega,\nu)} \leqslant C(s,p)\, \nu(G)$$

for every $f \in W(\overline{g}, G)$ with $0 \leqslant f \leqslant 1$. The necessity for $1 \leqslant q < p$ is obtained by taking the infimum over all such functions f that are admissible for the conductor $(\overline{g}, G)$.

We prove the sufficiency now when $1 \leqslant q < p$. Let $a \in (1, \infty)$. We have

$$a^p \int\limits_0^\infty \mu(M_{at})^{p/r} d(t^p) \leqslant a^p K_1 \left(\int\limits_0^\infty (\mathrm{cap}_{p,q}(\overline{M_{at}}, M_t) + \nu(M_t)^{p/s}) d(t^p) \right).$$

This and (1.2) yield the sufficiency for the case $1 \leqslant q < p$.

Now, suppose that $p < q < \infty$. We choose some bounded open sets g and G such that $\overline{g} \subset G \subset \overline{G} \subset \Omega$ and $f \in W(\overline{g}, G)$ with $0 \leqslant f \leqslant 1$. We have

$$\mu(g) \leqslant C(r,q)\, ||f||^r_{L^{r,q}(\Omega,\mu)}$$

and

$$||f||^s_{L^{s,q}(\Omega,\nu)} \leqslant C(s,q)\, \nu(G)$$

for every $f \in W(\overline{g}, G)$ with $0 \leqslant f \leqslant 1$. The necessity for $p < q < \infty$ is obtained by taking the infimum over all such functions f that are admissible for the conductor $(\overline{g}, G)$.

We prove the sufficiency now when $p < q < \infty$. Let $a \in (1, \infty)$. We have

$$a^q \int\limits_0^\infty \mu(M_{at})^{q/r} d(t^q) \leqslant a^q K_2 \left(\int\limits_0^\infty (\mathrm{cap}_{p,q}(\overline{M_{at}}, M_t)^{q/p} + \nu(M_t)^{q/s}) d(t^q) \right).$$

This and (1.3) yield the sufficiency for the case $p < q < \infty$. The proof is complete. □

We look for a simplified necessary and sufficient two-weight imbedding condition when $n = 1$. Before we state and prove such a condition for the case $n = 1$, we need to obtain sharp estimates for the p,q-capacitance of conductors $([a,b], (A,B))$ with $A < a < b < B$. This is the goal of the following proposition.

Proposition 5.2. *Suppose that $n = 1$, $1 < p < \infty$, and $1 \leqslant q \leqslant \infty$. There exists a constant $C(p,q) \geqslant 1$ such that*

$$C(p,q)^{-1}(\sigma_1^{1-p} + \sigma_2^{1-p}) \leqslant \mathrm{cap}_{p,q}([a,b], (A,B)) \leqslant C(p,q)(\sigma_1^{1-p} + \sigma_2^{1-p}),$$

where $\sigma_1 = a - A$ and $\sigma_2 = B - b$.

Proof. By the behavior of the Lorentz p,q-quasinorm in q (see, for example, [11, Proposition IV.4.2]), it suffices to find the upper bound for the $p,1$-capacitance and the lower bound for the p,∞-capacitance of the conductor $([a,b], (A,B))$. We start with the upper bound for the $p,1$-capacitance of this conductor.

We use the function $u : (A,B) \to \mathbf{R}$ defined by

$$u(x) = \begin{cases} 1 & \text{if } a \leqslant x \leqslant b, \\ \frac{x-A}{\sigma_1} & \text{if } A < x < a, \\ \frac{B-x}{\sigma_2} & \text{if } b < x < B. \end{cases}$$

Then from Lemma 3.1 it follows that $u \in W_1([a,b], (A,B))$ with

$$|u'(x)| = \begin{cases} 0 & \text{if } a < x < b, \\ {\sigma_1}^{-1} & \text{if } A < x < a, \\ {\sigma_2}^{-1} & \text{if } b < x < B. \end{cases}$$

We want to compute an upper estimate for $||u'||_{L^{p,1}((A,B),m_1)}$. We have

$$\begin{aligned} ||u'||_{L^{p,1}((A,B),m_1)} &\leqslant ||{\sigma_1}^{-1}||_{L^{p,1}((A,a),m_1)} + ||{\sigma_2}^{-1}||_{L^{p,1}((b,B),m_1)} \\ &= p({\sigma_1}^{-1+1/p} + {\sigma_2}^{-1+1/p}). \end{aligned} \tag{5.3}$$

Therefore,

$$\text{cap}_{p,1}([a,b],(A,B)) \leqslant C(p)(\sigma_1^{1-p} + \sigma_2^{1-p}).$$

We try to get lower estimates for the p,∞-capacitance of this conductor. Let $v \in W([a,b],(A,B))$ be an arbitrary admissible function such that $0 \leqslant v \leqslant 1$. We let v_1 be the restriction of v to (A,a) and v_2 be the restriction of v to (b,B) respectively. We note that v' is supported in $(A,a) \cup (b,B)$. Therefore, since v' coincides with v_1' on (A,a) and with v_2' on (b,B), we have

$$||v'||_{L^{p,\infty}((A,B),m_1)} \geqslant \max(||v_1'||_{L^{p,\infty}((A,a),m_1)}, ||v_2'||_{L^{p,\infty}((b,B),m_1)}). \tag{5.4}$$

From ([16, Corollary 2.4]) we have

$$||v_1'||_{L^{p,\infty}((A,a),m_1)} \geqslant 1/p' \cdot {\sigma_1^{-1/p'}} ||v_1'||_{L^1((A,a),m_1)}$$

and

$$||v_2'||_{L^{p,\infty}((b,B),m_1)} \geqslant 1/p' \cdot {\sigma_2^{-1/p'}} ||v_2'||_{L^1((b,B),m_1)}.$$

Since

$$||v_1'||_{L^1((A,a),m_1)} = \int_A^a |v_1'(x)|dx \geqslant 1,$$

we obtain

$$||v_1'||_{L^{p,\infty}((A,a),m_1)} \geqslant 1/p' \cdot {\sigma_1^{-1/p'}}. \tag{5.5}$$

Similarly, since

$$||v_2'||_{L^1((b,B),m_1)} = \int_b^B |v_2'(x)|dx \geqslant 1,$$

we obtain

$$||v_2'||_{L^{p,\infty}((b,B),m_1)} \geqslant 1/p' \cdot {\sigma_2^{-1/p'}}. \tag{5.6}$$

From (5.4), (5.5), and (5.6) we get the desired lower bound for the p,∞-capacitance. This completes the proof. □

Now, we state and prove a necessary and sufficient two-weight imbedding condition for the case $n = 1$.

Theorem 5.3. *Suppose that $n = 1$. Let p, q, r, s be chosen such that $1 < p < \infty$, $1 \leqslant q < \infty$ and $1 < s \leqslant \max(p, q) \leqslant r < \infty$. Let Ω be an open set in* $\mathbf{R}$, *and let μ and ν be two nonnegative locally finite measures on Ω.*

(i) *Suppose that $1 \leqslant q \leqslant p$. The inequality*

$$||f||_{L^{r,p}(\Omega,\mu)} \leqslant A\left(||f'||_{L^{p,q}(\Omega,m_1)} + ||f||_{L^{s,p}(\Omega,\nu)}\right) \tag{5.7}$$

holds for every $f \in \mathrm{Lip}_0(\Omega)$ if and only if there exists a constant $K > 0$ such that the inequality (1.7) *is valid whenever x, d and τ are such that $\overline{\sigma_{d+\tau}(x)} \subset \Omega$.*

(ii) *Suppose that $p < q < \infty$. The inequality*

$$||f||_{L^{r,q}(\Omega,\mu)} \leqslant A\left(||f'||_{L^{p,q}(\Omega,m_1)} + ||f||_{L^{s,q}(\Omega,\nu)}\right) \tag{5.8}$$

holds for every $f \in \mathrm{Lip}_0(\Omega)$ if and only if there exists a constant $K > 0$ such that the inequality (1.7) *is valid whenever x, d and τ are such that $\overline{\sigma_{d+\tau}(x)} \subset \Omega$.*

Proof. We only have to prove that the sufficiency condition for intervals implies the sufficiency condition for general bounded and open sets g and G with $\overline{g} \subset G \subset \overline{G} \subset \Omega$. Let G be the union of nonoverlapping intervals G_i, and let $g_i = G \cap g_i$. We denote by h_i the smallest interval containing g_i and by τ_i the minimal distance from h_i to $\mathbf{R} \setminus G_i$. We also denote by H_i the open interval concentric with h_i such that the minimal distance from h_i to $\mathbf{R}\setminus H_i$ is τ_i. Then $H_i \subset G_i$. From Remark 3.3 we have $\mathrm{cap}_{p,q}(\overline{g_i}, G_i) = \mathrm{cap}_{p,q}(\overline{h_i}, G_i)$. Moreover, from Theorem 3.2 (ii) and Proposition 5.2 we have

$$C(p,q)^{-1}\tau_i^{1-p} \leqslant \mathrm{cap}_{p,q}(\overline{h_i}, G_i) \leqslant \mathrm{cap}_{p,q}(\overline{h_i}, H_i) \leqslant 2\,C(p,q)\tau_i^{1-p}$$

for some constant $C(p,q) \geqslant 1$. Since $\overline{g}$ is a compact set lying in $\bigcup\limits_{i\geqslant 1} G_i$, it follows that $\overline{g}$ is covered by only finitely many of the sets G_i. This and Theorem 3.2 (ix) allow us to assume that G is in fact written as a finite union of disjoint intervals G_i. Now the proof splits, depending on whether $1 \leqslant q \leqslant p$ or $p < q < \infty$.

We assume that $1 \leqslant q \leqslant p$. Then

$$\mathrm{cap}_{p,q}(\overline{g}, G) \geqslant \sum_i \mathrm{cap}_{p,q}(\overline{g_i}, G_i) = \sum_i \mathrm{cap}_{p,q}(\overline{h_i}, G_i). \tag{5.9}$$

Using (1.7), we obtain

$$\begin{aligned}\mu(g_i)^{p/r} &\leqslant \mu(h_i)^{p/r} \leqslant K_1(\tau_i^{1-p} + \nu(H_i)^{p/s})\\ &\leqslant K_1\,C(p,q)(\mathrm{cap}_{p,q}(\overline{g_i}, G_i) + \nu(G_i)^{p/s}),\end{aligned}$$

where K_1 is a positive constant independent of g and G. Since $s \leqslant p \leqslant r < \infty$, we have

$$\mu(g)^{p/r} \leqslant \sum_i \mu(g_i)^{p/r}$$

and

$$\sum_i \nu(G_i)^{p/s} \leqslant \nu(G)^{p/s}.$$

This and (5.9) prove the claim when $1 \leqslant q \leqslant p$.

We assume that $p < q < \infty$. Then

$$\mathrm{cap}_{p,q}(\overline{g}, G)^{q/p} \geqslant \sum_i \mathrm{cap}_{p,q}(\overline{g_i}, G_i)^{q/p} = \sum_i \mathrm{cap}_{p,q}(\overline{h_i}, G_i)^{q/p}. \tag{5.10}$$

Using (1.7), we obtain

$$\begin{aligned}\mu(g_i)^{q/r} &\leqslant \mu(h_i)^{q/r} \leqslant K_2(\tau_i^{q(1-p)/p} + \nu(H_i)^{q/s}) \\ &\leqslant K_2\, C(p,q)^{q/p}(\mathrm{cap}_{p,q}(\overline{g_i}, G_i)^{q/p} + \nu(G_i)^{q/s}),\end{aligned}$$

where K_2 is a positive constant independent of g and G. Since $s \leqslant q \leqslant r < \infty$, we have

$$\mu(g)^{q/r} \leqslant \sum_i \mu(g_i)^{q/r}$$

and

$$\sum_i \nu(G_i)^{q/s} \leqslant \nu(G)^{q/s}.$$

This and (5.10) prove the claim when $p < q < \infty$. The theorem is proved. □

Acknowledgement. S. Costea was supported by NSERC and the Fields Institute. V. Maz'ya was supported by NSF (grant no. DMS 0500029).

References

1. Adams, D.R.: On the existence of capacitary strong type estimates in $\mathbb{R}^n$. Ark. Mat. **14**, 125-140 (1976)
2. Adams, D.R., Hedberg, L.I.: Function Spaces and Potential Theory. Springer Verlag (1996)
3. Adams, D.R., Pierre, M.: Capacitary strong type estimates in semilinear problems. Ann. Inst. Fourier (Grenoble) **41**, 117-135 (1991)
4. Adams, D.R., Xiao, J.: Strong type estimates for homogeneous Besov capacities. Math. Ann. **325**, no. 4, 695-709 (2003)
5. Adams, D.R., Xiao, J.: Nonlinear potential analysis on Morrey spaces and their capacities. Indiana Univ. Math. J. **53**, no. 6, 1631-1666 (2004)

6. Aikawa, H.: Capacity and Hausdorff content of certain enlarged sets. Mem. Fac. Sci. Eng. Shimane. Univ. Series B: Math. Sci. **30**, 1-21 (1997)
7. Alberico, A.: Moser type inequalities for higher order derivatives in Lorentz spaces. Potential Anal. [To appear]
8. Alvino, A., Ferone, V., Trombetti, G.: Moser type inequalities in Lorentz spaces. Potential Anal. **5**, no. 3, 273-299 (1996)
9. Alvino, A., Ferone, V., Trombetti, G.: Estimates for the gradient of solutions of nonlinear elliptic equations with L^1 data. Ann. Mat. Pura Appl. **178**, no. 4, 129-142 (2000)
10. Bénilan, P., Boccardo, L., Gallouët, T., Gariepy, R., Pierre, M., Vázquez, J.L.: An L^1-theory of existence and uniqueness of solutions of nonlinear elliptic equations. Ann. Scuola Norm. Sup. Pisa Cl. Sci. (4) **22**, no. 2, 241-273 (1995)
11. Bennett, C., Sharpley, R.: Interpolation of Operators. Academic Press, Boston (1988)
12. Chen, Z.-Q., Song, R.: Conditional gauge theorem for nonlocal Feynman–Kac transforms. Probab. Theory Related Fields. **125**, no. 1, 45-72, (2003)
13. Chung, H.-M., Hunt, R.A., Kurtz, D.S.: The Hardy-Littlewood maximal function on $L(p,q)$ spaces with weights. Indiana Univ. Math. J. **31**, no. 1, 109-120 (1982)
14. Cianchi, A.: Moser–Trudinger inequalities without boundary conditions and isoperimetric problems. Indiana Univ. Math. J. **54**, no. 3, 669-705 (2005)
15. Cianchi, A., Pick, L.: Sobolev embeddings into BMO, VMO, and L_∞. Ark. Mat. **36**, 317-340 (1998)
16. Costea, S.: Scaling invariant Sobolev–Lorentz capacity on $\mathbb{R}^n$. Indiana Univ. Math. J. **56**, no. 6, 2641-2669 (2007)
17. Dafni, G., Karadzhov, G., Xiao, J.: Classes of Carleson type measures generated by capacities. Math. Z. **258**, no. 4, 827-844 (2008)
18. Dahlberg, B.: Regularity properties of Riesz potentials. Indiana Univ. Math. J. **28**, no. 2, 257-268 (1979)
19. Dolzmann, G., Hungerbühler, N., Müller, S.: Uniqueness and maximal regularity for nonlinear elliptic systems of n-Laplace type with measure valued right hand side. J. Reine Angew. Math. **520**, 1-35 (2000)
20. Federer. H.: Geometric Measure Theory. Springer-Verlag (1969)
21. Fitzsimmons, P.J.: Hardy's inequality for Dirichlet forms. J. Math. Anal. Appl. **250**, 548-560 (2000)
22. Fukushima, M., Uemura, T.: On Sobolev and capacitary inequalities for contractive Besov spaces over d-sets. Potential Anal. **18**, no. 1, 59-77 (2003)
23. Fukushima, M., Uemura, T.: Capacitary bounds of measures and ultracontractivity of time changed processes. J. Math. Pures Appl. **82**, no. 5, 553-572 (2003)
24. Grigor'yan, A.: Isoperimetric inequalities and capacities on Riemannian manifolds. In: The Maz'ya Anniversary Collection. Vol. 1, pp. 139-153. Birkhäuser (1999)
25. Hajłasz, P.: Sobolev inequalities, truncation method, and John domains. Rep. Univ. J. Dep. Math. Stat. **83** (2001)
26. Hansson, K.: Embedding theorems of Sobolev type in potential theory, Math. Scand. **45**, 77-102 (1979)
27. Hansson, K., Maz'ya, V., Verbitsky, I.E.: Criteria of solvability for multi-dimensional Riccati's equation. Ark. Mat. **37**, no. 1, 87-120 (1999)
28. Hudson, S., Leckband, M.: A sharp exponential inequality for Lorentz-Sobolev spaces on bounded domains. Proc. Am. Math. Soc. **127**, no. 7, 2029-2033 (1999)
29. Kaimanovich, V.: Dirichlet norms, capacities and generalized isoperimetric inequalities for Markov operators. Potential Anal. **1**, no. 1, 61-82 (1992)
30. Kauhanen, J.., Koskela, P., Malý, J.: On functions with derivatives in a Lorentz space. Manuscr. Math. **100**, no. 1, 87-101 (1999)

31. Kolsrud, T.: Condenser capacities and removable sets in $W^{1,p}$. Ann. Acad. Sci. Fenn. Ser. A I Math. **8**, no. 2, 343-348 (1983)
32. Kolsrud, T.: Capacitary integrals in Dirichlet spaces. Math. Scand. **55**, 95-120 (1984)
33. Malý, J.: Sufficient conditions for change of variables in integral. In: Proc. Anal. Geom. (Novosibirsk Akademgorodok, 1999), pp. 370-386. Izdat. Ross. Akad. Nauk. Sib. Otdel. Inst. Mat., Novosibirsk (2000)
34. Maz'ya, V.G.: On certain integral inequalities for functions of many variables (Russian). Probl. Mat. Anal. **3**, 33-68 (1972); English transl.: J. Sov. Math. **1**, 205-234 (1973)
35. Maz'ya, V.G.: Summability with respect to an arbitrary measure of functions from S.L. Sobolev-L.N. Slobodetsky (Russian). Zap. Nauchn. Semin. Leningr. Otd. Mat. Inst. Steklova **92**, 192-202 (1979)
36. Maz'ya, V.G.: Sobolev Spaces. Springer-Verlag, Berlin-Tokyo (1985)
37. Maz'ya, V.: Conductor and capacitary inequalities for functions on topological spaces and their applications to Sobolev type imbeddings. J. Funct. Anal. **224**, no. 2, 408-430 (2005)
38. Maz'ya, V.: Conductor inequalities and criteria for Sobolev type two-weight imbeddings. J. Comput. Appl. Math. **194**, no. 11, 94-114 (2006)
39. Maz'ya, V., Netrusov, Yu.: Some counterexamples for the theory of Sobolev spaces on bad domains. Potential Anal. **4**, 47-65 (1995)
40. Maz'ya, V., Poborchi, S.: Differentiable Functions on Bad Domains. World Scientific (1997)
41. Netrusov, Yu.V.: Sets of singularities of functions in spaces of Besov and Lizorkin–Triebel type (Russian). Tr. Mat. Inst. Steklova **187**, 162-177 (1989); English transl.: Proc. Steklov Inst. Math. **187**, 185-203 (1990)
42. Rao, M.: Capacitary inequalities for energy. Israel J. Math. **61**, no. 1, 179-191 (1988)
43. Serfaty, S., Tice, I.: Lorentz space estimates for the Ginzburg–Landau energy. J. Funct. Anal. **254**, no. 3, 773-825 (2008)
44. Stein, E., Weiss, G.: Introduction to Fourier Analysis on Euclidean Spaces. Princeton Univ. Press, Princeton (1975)
45. Takeda, M.: L^p-independence of the spectral radius of symmetric Markov semigroups. Canad. Math. Soc. Conf. Proc. **29**, 613-623 (2000)
46. Verbitsky, I.E.: Superlinear equations, potential theory and weighted norm inequalities. In: Proc. the Spring School VI (Prague, May 31-June 6, 1998)
47. Verbitsky, I.E.: Nonlinear potentials and trace inequalities. In: The Maz'ya Anniversary Collection, Vol. 2, pp. 323-343. Birkhäuser (1999)
48. Vondraček, Z.: An estimate for the L^2-norm of a quasi continuous function with respect to smooth measure. Arch. Math. **67**, 408-414 (1996)

Besov Regularity for the Poisson Equation in Smooth and Polyhedral Cones

Stephan Dahlke and Winfried Sickel

Abstract The regularity of solutions to the Dirichlet and Neumann problems in smooth and polyhedral cones contained in $\mathbf{R}^3$ is studied with particular attention to the specific scale $B^s_\tau(L_\tau)$, $1/\tau = s/3 + 1/2$, of Besov spaces. The regularity of the solution in these Besov spaces determines the order of approximation that can be achieved by adaptive and nonlinear numerical schemes. We show that the solutions are much smoother in the specific Besov scale than in the usual L_2-Sobolev scale, which justifies the use of adaptive schemes. The proofs are performed by combining weighted Sobolev estimates with characterizations of Besov spaces by wavelet expansions.

1 Introduction

We study the regularity of solutions of the Poisson equation in smooth and polyhedral cones $\mathcal{K} \subset \mathbf{R}^3$ respectively, within Besov spaces $B^s_\tau(L_\tau(\mathcal{K}))$ with $0 < \tau < 2$. The motivation can be explained as follows.

Recent years, the numerical treatment of operator equations by adaptive numerical algorithms became a field of an increasing importance, with many applications in science and engineering. In particular, adaptive finite element schemes have been successfully developed and implemented, and innumerable numerical experiments impressively confine their excellent performance. Complementary to this, also adaptive algorithms based on wavelets have become more and more at the center of attraction during the last years for the

Stephan Dahlke
Philipps–Universität Marburg, Hans Meerwein Str., Lahnberge, 35032 Marburg, Germany, e-mail: `dahlke@mathematik.uni-marburg.de`

Winfried Sickel
Friedrich-Schiller-Universität Jena, Mathematisches Institut, Ernst–Abbe–Platz 2, D-07740 Jena, Germany, e-mail: `sickel@minet.uni-jena.de`

V. Maz'ya (ed.), *Sobolev Spaces in Mathematics II, International Mathematical Series.*
doi: 10.1007/978-0-387-85650-6,

following reason. The strong analytical properties of wavelets can be used to derive adaptive strategies which are guaranteed to converge for a huge class of elliptic operator equations, involving operators of negative order [4, 10]. Moreover, these algorithms are optimal in the sense that they asymptotically realize the convergence order of the optimal (but not directly implementable) approximation scheme, i.e., the order of the best n-term wavelet approximation. Moreover, the number of arithmetic operations that is needed stays proportional to the number of degrees of freedom [4]. By now, various generalizations to nonelliptic equations [5], saddle point problems [11], and also nonlinear operator equations [6] exist. For finite element schemes, rigorous statements of these forms have been rather rare although inspired by the results for wavelet schemes, the situation has changed during the last years [2, 16]. Although the above mentioned results are quite impressive, in the realm of adaptivity one is always faced with the following question: Does adaptivity really pay for the problem under consideration, i.e., does our favorite adaptive scheme really provide a substantial gain of efficiency compared to more conventional nonadaptive schemes which are usually much easier to implement? At least, in the case of adaptive wavelet schemes, it is possible to give a quite rigorous answer. A reasonable comparison would be to compare the performance of wavelet algorithms with classical, nonadaptive schemes which consist of approximations by linear spaces that are generated by uniform grid refinements. It is well known that, under natural assumptions, the approximation order that can be achieved by such a uniform method depends on the smoothness of the exact solution as measured in the classical L_2-Sobolev scale [9] (called *Sobolev regularity* below). On the other hand, as already outlined above, for adaptive wavelet methods the best n-term approximation serves as the benchmark scheme. It is well known that the convergence order that can be achieved by the best n-term approximations also depends on the smoothness of the object we want to approximate, but now the smoothness has to be measured in specific Besov spaces, usually corresponding to L_τ-spaces with $0 < \tau < 2$. Therefore, we can make the following statement: The use of adaptive wavelet schemes is completely justified if the Besov smoothness of the unknown solution of our operator equation is higher compared to its regularity in the Sobolev scale.

At this point, the shape of a domain comes into play. As the classical model problem of elliptic operator equations, let us discuss the Poisson equation. If a domain Ω is smooth, for example, C^∞, then the problem is completely regular, i.e., if the right-hand side is contained in $H^s(\Omega)$, $s \geqslant -1$, the solution is contained in $H^{s+2}(\Omega)$ [1, 18], and there is no reason why the Besov smoothness should be higher. However, in a nonsmooth domain, the situation is completely different. In this case, singularities near the boundary occur which significantly diminish the Sobolev regularity [19] and, consequently, the order of convergence of uniform methods drops down. Fortunately, in recent studies, it was shown that these singularities do not influence the Besov smoothness too much [9, 12], so that for certain nonsmooth domains the use of adaptive schemes is completely justified. In the specific case of polygonal domains contained in $\mathbf{R}^2$, even more can be said. Then, the Besov smoothness depends only on the smoothness of the right-hand side, so that for arbitrary smooth right-hand sides one gets arbitrary high order of convergence, at least in principle [7]. The proof of this result relies on the fact that for polygonal domains the exact solution can be decomposed into a regular part and a singular part corresponding to reentrant corners [17]. By these results, it is quite natural to try to generalize them to the very important case of polyhedral domains in $\mathbf{R}^3$, and this is exactly the task we are concerned with in this paper. In the polyhedral case, the solution can also be decomposed into a singular part and a regular part [17]; however, the situation is much more complicated since edge singularities, as well as vertex singularities occur which have to be treated separately. For edge singularities the first positive result was obtained in [8]. Therefore, in this paper, we concentrate on vertex singularities.

For vertex singularities in 3D the situation is much more unclear compared to the 2D-setting since the singularity functions are not given explicitly, but depend in a somewhat complicated way on the shape of the domain in the vicinity of the vertex [17]. A quite promising way to handle this difficulty is the following: reduce the problem to the case of a smooth cone or a polyhedral cone and treat the cone case by using *weighted* smoothness spaces [21, 26]. The weight takes into account the distance to the vertex or, more general, the distance to parts of the boundary of the cone. Although the problem is not regular in the classical Sobolev spaces, one has regularity in these weighted spaces in the following sense: if the right-hand side has smoothness $l-2$ in the weighted scale, then the solution has smoothness l in the same scale (see [21, 26] and Appendix A for details). In this paper, we show that this regularity of the solution in weighted Sobolev spaces is sufficient to establish Besov smoothness (in the original unweighted sense). Consequently, the use of adaptive wavelet schemes for problems in polyhedral domains is also justified.

In the context of adaptive approximation for elliptic problems, also the recent work of Nitsche [29] should be mentioned. In his pioneering studies, Nitsche is primary concerned with approximations of singularity functions by anisotropic tensor product refinements, whereas, in this paper, we focus on isotropic wavelet approximations.

This paper is organized as follows. In Sect. 2, we first of all discuss the case of a smooth cone. We show that the regularity results in weighted Sobolev spaces are indeed sufficient to establish Besov regularity. The proof is based on the fact that smoothness norms such as Besov norms are equivalent to weighted sequence norms of wavelet expansion coefficients, and we use the weighted regularity results to estimate wavelet coefficients. In Sect. 3, we study polyhedral cones. In this case, the situation is more difficult since we deal with weights that include the distance to the vertex, as well as the distance to the edges. Nevertheless, the wavelet coefficients can again be estimated and Besov smoothness can be established. An additional information is presented in Appendices A and B. In Appendix A, we collect relevant facts concerning the regularity theory for elliptic partial differential equations as far as they are needed for our purposes. Finally, in Appendix B, we recall the definition of Besov spaces and introduce their characterizations by wavelet expansions.

2 Regularity Result for a Smooth Cone

Let $\mathcal{K} \subset \mathbf{R}^3$ be an infinite cone with vertex at the origin, i.e.,

$$\mathcal{K} := \{x \in \mathbf{R}^3 : x = \rho\omega,\ 0 < \rho < \infty,\ \omega \in \Omega\}, \tag{2.1}$$

where Ω is a domain on the unit sphere S^2 with smooth boundary $\partial\Omega$ and ρ and ω are the spherical coordinates of x. For integer $l \geqslant 0$ and real β we define the weighted Sobolev spaces $V^l_{2,\beta}(\mathcal{K})$ as the closure of $C_0^\infty(\overline{\mathcal{K}}\backslash\{0\})$ with respect to the norm

$$\|u\|_{V^l_{2,\beta}(\mathcal{K})} := \left(\int_{\mathcal{K}} \sum_{|\alpha| \leqslant l} \rho^{2(\beta - l + |\alpha|)} \, |D^\alpha u(x)|^2 \, dx \right)^{1/2} . \tag{2.2}$$

If $l \geqslant 1$, then $V^{l-1/2}_{2,\beta}(\partial\mathcal{K})$ denotes the space of traces of functions from $V^l_{2,\beta}(\mathcal{K})$ on the boundary equipped with the norm

$$\| u \|_{V^{l-1/2}_{2,\beta}(\partial\mathcal{K})} := \inf \left\{ \| v \|_{V^l_{2,\beta}(\mathcal{K})} : v \in V^l_{2,\beta}(\mathcal{K}) \, , v_{|\partial\mathcal{K}} = u \right\} .$$

A more explicit description of these trace classes, using differences and derivatives, is given in [21, Lemma 6.1.2]. Let us consider the Poisson equation

$$\begin{aligned} -\Delta u &= f \qquad \text{in } \mathcal{K}, \\ u|_{\partial\mathcal{K}} &= g \, . \end{aligned} \tag{2.3}$$

Denote by $\mathcal{K}_0$ an arbitrary truncated cone, i.e., there exists a positive real number r_0 such that

$$\mathcal{K}_0 = \{x \in \mathcal{K} : |x| < r_0\} \, . \tag{2.4}$$

Theorem 2.1. *Suppose that the right-hand side f is contained in $V^{l-2}_{2,\beta}(\mathcal{K}) \cap L_2(\mathcal{K}_0)$, where $l \geqslant 2$ is a natural number. Further, we assume that $g \in V^{l-1/2}_{2,\beta}(\partial\mathcal{K})$. Let $\alpha_0 = \alpha_0(\mathcal{K})$ be the number defined in Remark 4.1 below. Then there exists a countable set E of complex numbers such that the following holds. If a real number β is chosen such that*

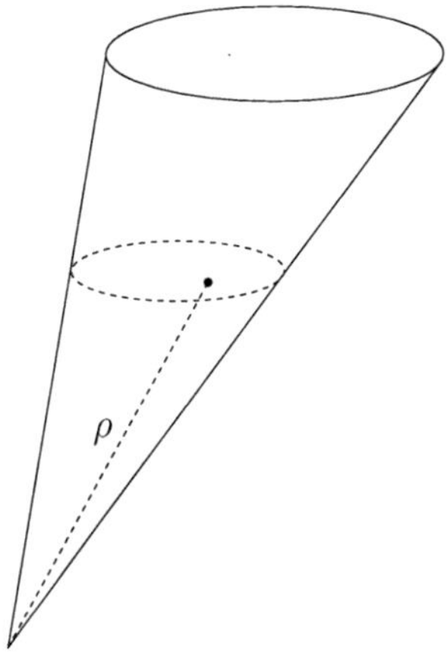

Fig. 1 A smooth cone.

$$\operatorname{Re} \lambda \neq -\beta + l - 3/2 \qquad \textit{for all} \quad \lambda \in E\,, \tag{2.5}$$

then the solution u *of* (2.3) *satisfies*

$$u \in B^s_\tau(L_\tau(\mathcal{K}_0)), \quad \frac{1}{\tau} = \frac{s}{3} + \frac{1}{2}, \quad s < \min\left(l,\, \frac{3}{2}\alpha_0\right). \tag{2.6}$$

Remark 2.1. (i) Our set-up for the partial differential equation is taken from [21, Sect. 6.1]. It turns out that the exceptional set E coincides with the collection of the eigenvalues of the operator pencil associated to (2.3). In particular situations, there are explicit formulas for E, we refer to [21, Lemma 6.6.3]. Furthermore, under the given restrictions, there is an a priori estimate for u within the scale $V^l_{2,\beta}(\mathcal{K})$ (see [21, Theorem 6.1.1] or Proposition 4.3). However, for adaptive wavelet methods we need to know the regularity within unweighted Besov spaces $B^s_\tau(L_\tau(\mathcal{K}_0))$ with s as large as possible, compare with (ii). Curiously, we cannot use the regularity theory for (2.3) within unweighted Sobolev spaces (see, for example, [14]) for deriving the above regularity result. For us the investigations of (2.3) in weighted Sobolev spaces seemingly started by Kondrat'ev [20] and continued by Maz'ya and Plamenevskij [24, 25], Koslov, Maz'ya, and Rossmann [21], and Maz'ya and Rossmann [26], to mention at least a few, were most helpful.

(ii) *Best n-term approximation.* It is well known that the order of convergence of the best n-term wavelet approximation in $\mathbf{R}^3$ is determined by the regularity of the object one wants to approximate as measured in the specific Besov scale $B^s_\tau(L_\tau)$, $\frac{1}{\tau} = \frac{s}{3} + \frac{1}{2}$ introduced in (2.6) (see again [9, 15] for details). As an immediate consequence of (2.6), we conclude that for the solution u of (2.3) there exist subsets $\Gamma \subset \mathbf{Z}^3$ and $\Lambda \subset \{1, \ldots, 7\} \times \mathbf{N}_0 \times \mathbf{Z}^3$ such that $|\Gamma| + |\Lambda| \leqslant n$ (here, $|\Gamma|$ and $|\Lambda|$ denotes the cardinality of the sets

Γ and Λ respectively) and

$$S_n u := \sum_{k\in\Gamma} \langle u, \widetilde{\varphi}_k\rangle\, \varphi_k + \sum_{(i,j,k)\in\Lambda} \langle u, \widetilde{\psi}_{i,j,k}\rangle\, \psi_{i,j,k} \tag{2.7}$$

satisfies

$$\| u - S_n u \|_{L_2(\mathcal{K}_0)} \lesssim \| u \|_{B^s_\tau(L_\tau(\mathcal{K}_0))}\, n^{-s/3}, \qquad s < \min\Big(l, \frac{3}{2}\alpha_0\Big), \tag{2.8}$$

and s and τ are coupled as in (2.6). (We refer to Appendix B for the definition of φ_k, $\widetilde{\varphi}_k$, $\psi_{i,j.k}$, and $\widetilde{\psi}_{i,j,k}$. In this paper, "$a \lesssim b$" always means that there exists a constant c such that $a \leqslant cb$, independent of all context relevant parameters a and b may depend on.) In contrary to this, the order of convergence of uniform methods is determined by the regularity in the L_2-Sobolev scale H^s. Therefore, since the critical Sobolev index α_0 is multiplied by $3/2$, Theorem 2.1 implies that for l large enough the Besov smoothness is always higher compared to the Sobolev smoothness, so that the use of adaptive wavelet schemes is completely justified. In Fig. 2, we plotted the situation where $l \geqslant 3\alpha_0/2$ and $3\alpha_0/2 = 3(\frac{1}{\tau_0} - \frac{1}{2})$.

(iii) So far, we have discussed the best n-term approximation in L_2. However, it is well known that adaptive wavelet methods realize the order of the best n-term approximation with respect to the energy norm, i.e., the H^1-norm would be more natural. Theorem 2.1 also implies a result in this direction. We refer to [9, 13], where similar arguments were used. Indeed, the following estimate for the best n-term approximation in the H^1-norm holds. For all $u \in B^s_{\tau_1}(L_{\tau_1})$, $\frac{1}{\tau_1} = \frac{(s-1)}{3} + \frac{1}{2}$ and all $n \in \mathbf{N}$ there exist subsets $\Gamma \subset \mathbf{Z}^3$ and $\Lambda \subset \{1,\dots,7\} \times \mathbf{N}_0 \times \mathbf{Z}^3$ such that $|\Gamma| + |\Lambda| \leqslant n$ and $S_n u$ (defined as in (2.7)) satisfies

$$\| u - S_n u \|_{H^1(\mathcal{K}_0)} \lesssim \| u \|_{B^s_{\tau_1}(L_{\tau_1}(\mathcal{K}_0))}\, n^{-(s-1)/3}, \qquad \frac{1}{\tau_1} = \frac{(s-1)}{3} + \frac{1}{2}. \tag{2.9}$$

We therefore have to estimate the Besov norm $B^s_{\tau_1}(L_{\tau_1}(\mathcal{K}_0))$ of u. For simplicity, assume that $l \geqslant \frac{3}{2}\alpha_0$. We know that the solution is contained in the Sobolev space $H^\alpha(\mathcal{K}_0)$, $\alpha < \alpha_0$, as well as in the Besov space $B^{\overline{\alpha}}_{\tau_0}(L_{\tau_0}(\mathcal{K}_0))$, $\frac{1}{\tau} = \frac{\overline{\alpha}}{3} + \frac{1}{2}$, $\overline{\alpha} < 3\alpha_0/2$. We continue by real interpolation

$$\Big(H^{s_0}(\mathcal{K}_0)), B^{s_1}_{\tau_0}(L_{\tau_0}(\mathcal{K}_0))\Big)_{\Theta,\tau_1} = B^s_{\tau_1}(L_{\tau_1}(\mathcal{K}_0)),$$

where $0 < \Theta < 1$,

$$\frac{1}{\tau_1} = \frac{1-\Theta}{2} + \frac{\Theta}{\tau_0} \qquad \text{and} \qquad s = (1-\Theta)\, s_0 + \Theta\, s_1\,,$$

see [37]. This shows that

$$u \in B^s_{\tau_1}(L_{\tau_1}(\mathcal{K}_0)), \quad s < \frac{3}{2}\alpha_0 - \frac{1}{2}, \quad \frac{1}{\tau_1} = \frac{(s-1)}{3} + \frac{1}{2}$$

(see Fig. 2). There we plotted the situation, where $l \geqslant 3\alpha_0/2$, $s_0 = \alpha_0$, and $s_1 = 3\alpha_0/2$.

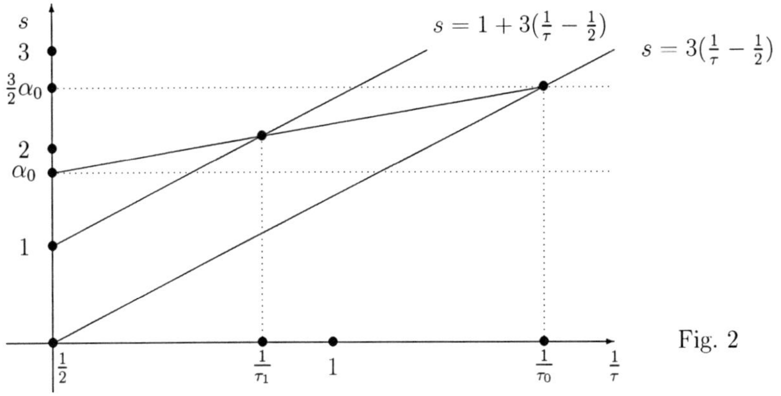

Fig. 2

(iv) Let $f \in C^\infty(\overline{\mathcal{K}})$ such that supp $f \subset \mathcal{K}_0$. Then $f \in V^l_{2,\beta}(\mathcal{K})$ for all pairs (l, β) such that $\beta > l - 3/2$. Hence we can apply Theorem 2.1 with $s < 3\alpha_0/2$.

(v) At first sight, the condition (2.5) looks restrictive. However, it is well known that the set E consists of a countable number of isolated points (see again [21] for details). Therefore, by a minor modification of β, the condition (2.5) is satisfied, and this minor modification does not change the arguments outlined below. This argument also shows that an explicit knowledge of E in our context is not necessary.

Proof of Theorem 2.1. The proof is based on the characterizations of Besov spaces by wavelet expansions (see Proposition 5.1 below). Therefore, we estimate the wavelet coefficients of the solution u to (2.3) and show that they are contained in the weighted sequence spaces that are related to the scale $B^s_\tau(L_\tau(\mathcal{K}_0))$, $\frac{1}{\tau} = \frac{s}{3} + \frac{1}{2}$.

Step 1. Preparations. First of all, we make the following agreement concerning the wavelet characterization of Besov spaces on $\mathbf{R}^3$ (see again Proposition 5.1): to each dyadic cube $I := 2^{-j}k + 2^{-j}[0,1]^3$ we associate the functions

$$\eta_I := \widetilde{\psi}_{i,j,k}\,, \qquad j \in \mathbf{N}\,, \quad k \in \mathbf{Z}^3, \quad i = 1, \dots, 7\,,$$

by ignoring the dependence on i. In the case $I = k + [0,1]^3$, i.e., $j = 0$, we use $\widetilde{\varphi}_k$ instead of $\widetilde{\psi}_{i,0,k}$, $k \in \mathbf{Z}^3$, $i = 1, \ldots, 7$. We denote by η_I^* the corresponding element of the dual basis. Since the wavelet basis is assumed to be compactly supported, there exists a cube Q, centered at the origin, such that $Q(I) := 2^{-j}k + 2^{-j}Q$ contains the support of η_I and of η_I^* for all I.

Step 2. Since $f \in L_2(\mathcal{K}_0)$, we a priori know $u \in H^s(\mathcal{K}_0)$ for some $s > 0$ (see Proposition 4.2 below). We start by estimating the coefficients corresponding to interior wavelets, i.e., we estimate those coefficients $\langle u, \eta_I \rangle$, where $\operatorname{supp} \eta_I \subset \mathcal{K}_0$. Let ρ_I denote the distance of the cube $Q(I)$ to the vertex. We fix a refinement level j and introduce the sets

$$\Lambda_j := \{\, I \mid \operatorname{supp} \eta_I \subset \mathcal{K}_0\,,\ 2^{-3j} \leqslant |I| \leqslant 2^{-3j+2}\},$$

$$\Lambda_{j,k} := \{\, I \in \Lambda_j \mid k2^{-j} \leqslant \rho_I < (k+1)2^{-j}\}\,, \quad j \in \mathbf{N}_0\,, \quad k \in \mathbf{N}_0\,.$$

At this first step, we deal with $k \geqslant 1$ only. Further, we put

$$|u|_{W^l(L_2(Q(I)))} := \Big(\int\limits_{Q(I)} |\nabla^l u(x)|^2\, dx \Big)^{1/2}.$$

Let P_I denote the polynomial of order at most l such that

$$\|u - P_I\|_{L_2(Q(I))} = \inf \Big\{ \|u - P\|_{L_2(Q(I))} :\ P \text{ is a polynomial of degree } \leqslant l \Big\}.$$

Employing the vanishing moment properties of wavelets (see Sect. 5.3.) and the classical Whitney estimate, we find

$$\begin{aligned} |\langle u, \eta_I \rangle| &\leqslant \|u - P_I\|_{L_2} \|\eta_I\|_{L_2} \\ &\lesssim |I|^{l/3} |u|_{W^l(L_2(Q(I)))} \\ &\lesssim 2^{-lj} |u|_{W^l(L_2(Q(I)))} \end{aligned} \tag{2.10}$$

if $I \in \Lambda_j$. Let $0 < \tau < 2$. Summing up over $I \in \Lambda_{j,k}$, we find

$$\begin{aligned} \sum_{I \in \Lambda_{j,k}} |\langle u, \eta_I \rangle|^\tau &\lesssim \sum_{I \in \Lambda_{j,k}} 2^{-lj\tau} \Big(\int\limits_{Q(I)} |\nabla^l u|^2 dx \Big)^{\tau/2} \\ &\lesssim \sum_{I \in \Lambda_{j,k}} 2^{-lj\tau} \rho_I^{-\beta\tau} \Big(\int\limits_{Q(I)} |\rho^\beta |\nabla^l u||^2 dx \Big)^{\tau/2} \end{aligned}$$

$$\lesssim (k\,2^{-j})^{-\beta\tau} \sum_{I\in\Lambda_{j,k}} 2^{-lj\tau}\Big(\int_{Q(I)} |\rho^\beta|\nabla^l u||^2 dx\Big)^{\tau/2}.$$

The next step consists in the use of the Hölder inequality with $p = 2/(2-\tau)$ and $q = 2/\tau$. This yields

$$\sum_{I\in\Lambda_{j,k}} |\langle u,\eta_I\rangle|^\tau \lesssim (k\,2^{-j})^{-\beta\tau}\Big(\sum_{I\in\Lambda_{j,k}} 2^{-\frac{2lj\tau}{2-\tau}}\Big)^{\frac{2-\tau}{2}}\Big(\sum_{I\in\Lambda_{j,k}}\int_{Q(I)} |\rho^\beta|\nabla^l u||^2 dx\Big)^{\frac{\tau}{2}}.$$

Observe that for the cardinality $|\Lambda_{j,k}|$ of $\Lambda_{j,k}$ we have

$$|\Lambda_{j,k}| \lesssim k^2, \qquad k\in\mathbf{N},$$

where the constant is independent of j, but depends on the shape of the domain Ω. Therefore, we further obtain

$$\sum_{I\in\Lambda_{j,k}} |\langle u,\eta_I\rangle|^\tau \lesssim (k\,2^{-j})^{-\beta\tau}\Big(k^2\,2^{-\frac{2lj\tau}{2-\tau}}\Big)^{\frac{2-\tau}{2}}\Big(\sum_{I\in\Lambda_{j,k}}\int_{Q(I)} |\rho^\beta|\nabla^l u||^2 dx\Big)^{\frac{\tau}{2}}$$

$$\lesssim k^{2-\tau-\beta\tau}2^{(\beta-l)j\tau}\Big(\sum_{I\in\Lambda_{j,k}}\int_{Q(I)} |\rho^\beta|\nabla^l u||^2 dx\Big)^{\frac{\tau}{2}}.$$

Now we have to sum over the set Λ_j. Since we are restricting to a truncated cone, there is a general number C such that

$$I\cap\mathcal{K}_0 = \varnothing \qquad \text{if} \quad I\in\Lambda_{j,k}\,, \quad k > C\,2^j\,. \tag{2.11}$$

Using the Hölder inequality once again and invoking (4.1), we find

$$\sum_{k=1}^{C\,2^j}\sum_{I\in\Lambda_{j,k}} |\langle u,\eta_I\rangle|^\tau$$

$$\lesssim \Big(\sum_{k=1}^{C2^j} k^{(2-\tau-\beta\tau)\frac{2}{2-\tau}}\Big)^{\frac{2-\tau}{2}} 2^{-j(l-\beta)\tau}\Big(\sum_{I\in\Lambda_j}\int_{Q(I)} |\rho^\beta|\nabla^l u||^2 dx\Big)^{\frac{\tau}{2}}$$

$$\lesssim 2^{-j(l-\beta)\tau}\,\|\,u\,\|^\tau_{V^l_{2,\beta}(\mathcal{K})}\begin{cases} 2^{j(3-\frac{3}{2}\tau-\beta\tau)} & \text{if } 3\Big(\frac{1}{\tau}-\frac{1}{2}\Big) > \beta\,,\\ (1+j)^{\frac{2-\tau}{2}} & \text{if } 3\Big(\frac{1}{\tau}-\frac{1}{2}\Big) = \beta\,,\\ 1 & \text{if } 3\Big(\frac{1}{\tau}-\frac{1}{2}\Big) < \beta\,,\end{cases}$$

$$\lesssim \| f \|^{\tau}_{V^{l-2}_{2,\beta}(\mathcal{K})} \begin{cases} 2^{j(3-\frac{3}{2}\tau - l\tau)} & \text{if } 3\left(\frac{1}{\tau} - \frac{1}{2}\right) > \beta\,, \\ (1+j)^{\frac{2-\tau}{2}}\, 2^{-j(l-\beta)\tau} & \text{if } 3\left(\frac{1}{\tau} - \frac{1}{2}\right) = \beta\,, \\ 2^{-j(l-\beta)\tau} & \text{if } 3\left(\frac{1}{\tau} - \frac{1}{2}\right) < \beta. \end{cases}$$

This implies that the function

$$u^* := \sum_{j=0}^{\infty} \sum_{k=1}^{C2^j} \sum_{I \in \Lambda_{j,k}} \langle u, \eta_I \rangle\, \eta_I^* \tag{2.12}$$

belongs to

$$\begin{cases} B^l_\infty(L_\tau(\mathbf{R}^3)) & \text{if } 3\left(\frac{1}{\tau} - \frac{1}{2}\right) > \beta\,, \\ B^{l-\beta-\delta-3(\frac{1}{2}-\frac{1}{\tau})}_\tau(L_\tau(\mathbf{R}^3)) & \text{if } 3\left(\frac{1}{\tau} - \frac{1}{2}\right) = \beta\,, \quad \delta > 0\,, \\ B^{l-\beta-3(\frac{1}{2}-\frac{1}{\tau})}_\infty(L_\tau(\mathbf{R}^3)) & \text{if } 3\left(\frac{1}{\tau} - \frac{1}{2}\right) < \beta\,. \end{cases} \tag{2.13}$$

Now we consider the cases $\beta < l$ and $\beta \geqslant l$ separately.

In the first case, we choose s (τ respectively) such that $\beta < s < l$ and s sufficiently close to l. Then, because of $s = 3\left(\frac{1}{\tau} - \frac{1}{2}\right)$ we may use the first line in (2.13) and the continuous embedding $B^l_\infty(L_\tau(\mathbf{R}^3)) \hookrightarrow B^s_\tau(L_\tau(\mathbf{R}^3))$.

In the second case, we choose $s < \beta$ sufficiently close to β and argue by using the third line in (2.13). With

$$\beta - s = \beta - 3\Big(\frac{1}{\tau} - \frac{1}{2}\Big) = \varepsilon > 0\,, \qquad \varepsilon < l - s\,,$$

we obtain $B^{l-\varepsilon}_\infty(L_\tau(\mathbf{R}^3)) \hookrightarrow B^s_\tau(L_\tau(\mathbf{R}^3))$, as in the first case.

Step 3. Estimate of the boundary layer. We recall the argument from Theorem 3.2 in [12]. The set $\Lambda_{j,0}$ can be empty (depending on the cone and on the wavelet system). If it is the case, then nothing is to do. If not, then we argue as follows. From the Lipschitz character of $\mathcal{K}_0$ it follows

$$|\Lambda_{j,0}| \lesssim 2^{2j}\,, \qquad j \in \mathbf{N}_0\,.$$

Let $0 < p < 2$. Using the Hölder inequality, we find

$$\sum_{I \in \Lambda_{j,0}} |\langle u, \eta_I \rangle|^p \lesssim 2^{j2(1-p/2)} \Big(\sum_{I \in \Lambda_{j,0}} |\langle u, \eta_I \rangle|^2 \Big)^{\frac{p}{2}}\,.$$

Summing up over $j \in \mathbf{N}_0$, we finally obtain

$$\sum_{j=0}^{\infty} 2^{j(s+3(\frac{1}{2}-\frac{1}{p}))p} \sum_{I\in\Lambda_{j,0}} |\langle u,\eta_I\rangle|^p$$

$$\lesssim \sum_{j=0}^{\infty} 2^{j(s+3(\frac{1}{2}-\frac{1}{p}))p}\, 2^{j(\frac{2}{p}-1)p} \Big(\sum_{I\in\Lambda_{j,0}} |\langle u,\eta_I\rangle|^2\Big)^{\frac{p}{2}}$$

$$\lesssim \| u \|^p_{B_p^{s+\frac{1}{2}-\frac{1}{p}}(L_2(\mathbf{R}^3))}$$

$$\lesssim \| u \|^p_{B_2^{s+\frac{1}{2}-\frac{1}{p}}(L_2(\mathbf{R}^3))},$$

since $p < 2$ (see Appendix B for the last step). Choosing s and p such that

$$s := \frac{3\,\alpha}{2} \qquad \text{and} \qquad \frac{1}{p} := \frac{s}{3} + \frac{1}{2}, \qquad \text{i.e.,} \quad s = 3\Big(\frac{1}{p} - \frac{1}{2}\Big),$$

we get $\alpha = \frac{2}{p} - 1$, as well as $\alpha = s + \frac{1}{2} - \frac{1}{p}$. This means that we have proved that

$$u^{**} := \sum_{j=0}^{\infty} \sum_{I\in\Lambda_{j,0}} \langle u,\eta_I\rangle\, \eta_I^* \tag{2.14}$$

belongs to $B_p^{3\alpha/2}(L_p(\mathbf{R}^3))$ for all $\alpha < \alpha_0$.

Step 4. Finally we need to deal with those wavelets for which the support intersects the boundary of the truncated cone. We put

$$\Lambda_j^{\#} := \{\, I \mid \operatorname{supp} \eta_I \cap \partial\mathcal{K}_0 \neq \varnothing\,,\ 2^{-3j} \leqslant |I| \leqslant 2^{-3j+2}\}\,, \qquad j \in \mathbf{N}_0\,.$$

Furthermore, since $\mathcal{K}_0$ is a bounded Lipschitz domain, there exists a linear and bounded extension operator

$$\mathcal{E}:\ H^\alpha(\mathcal{K}_0) \to H^\alpha(\mathbf{R}^3)$$

which is simultaneously a bounded operator belonging to

$$\mathcal{L}(B_q^s(L_p(\mathcal{K}_0)), B_q^s(L_p(\mathbf{R}^3)))$$

for all s, p, and q (see, for example, [32]). Defining

$$u^{\#} := \sum_{j=0}^{\infty} \sum_{I\in\Lambda_j^{\#}} \langle \mathcal{E}u,\eta_I\rangle\, \eta_I^*, \tag{2.15}$$

we can argue as at Step 3 since

$$|\Lambda_j^{\#}| \lesssim 2^{2j}\,, \qquad j \in \mathbf{N}_0\,.$$

This implies

$$\| u^{\#} \|_{B_p^{3\alpha/2}(L_p(\mathbf{R}^3))} \lesssim \| \mathcal{E}u \|^p_{B_2^{\alpha}(L_2(\mathbf{R}^3))} \lesssim \| u \|^p_{H^{\alpha}(\mathcal{K}_0))}$$

(see Appendix B for the last step). Adding up the finitely many functions of type u^*, u^{**}, and $u^{\#}$ (see Step 1), we end up with a function which belongs to $B^s_\tau(L_\tau(\mathbf{R}^3))$ (where s satisfies the restrictions in (2.6)) and which coincides with u on $\mathcal{K}_0$. Hence $u \in B^s_\tau(L_\tau(\mathcal{K}_0))$. □

Remark 2.2. Observe that the estimates of the parts u^{**}, see (2.14), and $u^{\#}$, see (2.15), depend only on the Lipschitz character of the cone $\mathcal{K}_0$ and the number α_0 associated via Proposition 4.1 below to the cone $\mathcal{K}$.

3 Besov Regularity for the Neumann Problem

Let

$$\mathcal{K} = \{x \in \mathbf{R}^3 \ : \ x = \rho\,\omega\,, \quad 0 < \rho < \infty\,, \omega \in \Omega\} \tag{3.1}$$

be a polyhedral cone with faces $\Gamma_j = \{x \ : \ x/|x| \in \gamma_j\}$ and edges M_j, $j = 1, \ldots, n$. Here, Ω is a curvilinear polygon on the unit sphere bounded by the arcs $\gamma_1, \ldots, \gamma_n$. The angle at the edge M_j is denoted by θ_j. We consider the problem

$$-\triangle u = f \quad \text{in } \mathcal{K}, \qquad \frac{\partial u}{\partial n} = g_j \text{ on } \Gamma_j\,,\ j = 1, \ldots, n. \tag{3.2}$$

We denote by $\rho(x) = |x|$ the distance to the vertex of the cone and by $r_j(x)$ the distance to the edge M_j. Let $\beta \in \mathbf{R}$, and let $\boldsymbol{\delta} = (\delta_1, \ldots, \delta_n) \in \mathbf{R}^n$ be such that $\delta_j > -1$ for all j. We use the abbreviation $|\boldsymbol{\delta}| := \delta_1 + \ldots + \delta_n$ without assuming that the components δ_j of $\boldsymbol{\delta}$ are positive. Then the weighted Sobolev space $W^{l,2}_{\beta,\boldsymbol{\delta}}(\mathcal{K})$ is defined as the collection of all functions $u \in H^{l,\ell oc}(\mathcal{K})$ such that

$$\|u\|_{W^{l,2}_{\beta,\boldsymbol{\delta}}(\mathcal{K})} := \Big(\int_{\mathcal{K}} \sum_{|\alpha| \leqslant l} \rho^{2(\beta - l + |\alpha|)} \Big(\prod_{j=1}^{n} (\frac{r_j}{\rho})^{\delta_j} \Big)^2 |D^\alpha u(x)|^2 \, dx \Big)^{1/2} < \infty\,. \tag{3.3}$$

If $\boldsymbol{\delta} = \mathbf{0}$, then we are back in case of (2.2). If $l \geqslant 1$, then $W^{l-1/2,2}_{\beta,\boldsymbol{\delta}}(\Gamma_j)$ denotes the space of traces of functions from $W^{l,2}_{\beta,\boldsymbol{\delta}}(\mathcal{K})$ on the face Γ_j equipped with the norm

$$\| u \|_{W^{l-1/2,2}_{\beta,\boldsymbol{\delta}}(\Gamma_j)} := \inf \big\{ \| v \|_{W^{l,2}_{\beta,\boldsymbol{\delta}}(\mathcal{K})} : v \in W^{l,2}_{\beta,\boldsymbol{\delta}}(\mathcal{K})\,, v_{|\Gamma_j} = u \big\}\,.$$

As in the previous section, $\mathcal{K}_0$ denotes an arbitrary truncated cone (see (2.4)).

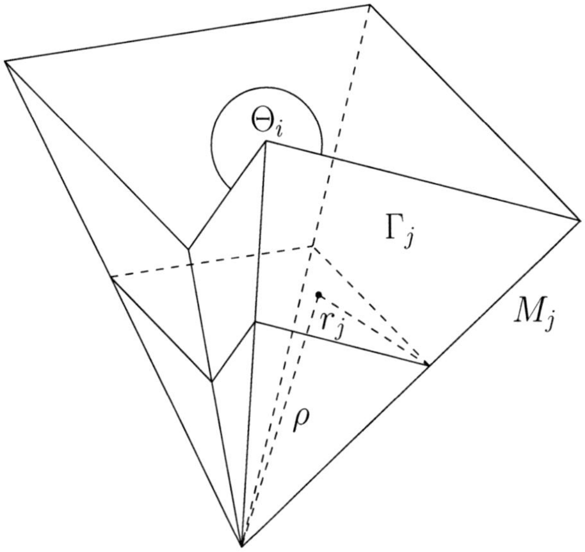

Fig. 2 A polyhedral cone.

Theorem 3.1. *Suppose that $f \in W^{l-2,2}_{\beta,\boldsymbol{\delta}}(\mathcal{K}) \cap L_2(\mathcal{K})$, where $l \geqslant 2$ is a natural number. Further, we assume that $g_j \in W^{l-3/2,2}_{\beta,\boldsymbol{\delta}}(\Gamma_j)$, $j = 1, \dots, n$. Let $\alpha_0 = \alpha_0(\mathcal{K})$ be the number defined in Proposition 4.1 below. Then there exists a countable set E of complex numbers such that the following holds. If a real number β and a vector $\boldsymbol{\delta}$ are chosen such that $\beta < l$,*

$$\lambda \neq l - \beta - \frac{3}{2} \qquad \textit{for all} \quad \lambda \in E\,,$$

and

$$\max\left(l - \frac{\pi}{\theta_j}, 0\right) < \delta_j + 1 < l, \quad j = 1, \dots, n\,, \tag{3.4}$$

then the solution u of (3.2) satisfies

$$u \in B^s_\tau(L_\tau(\mathcal{K}_0)), \quad \frac{1}{\tau} = \frac{s}{3} + \frac{1}{2}, \quad s < \min\left(l, \frac{3}{2}\alpha_0, 3\,(l - |\boldsymbol{\delta}|)\right). \tag{3.5}$$

Remark 3.1. (i) In contrary to Sect. 2, here we formulate the main result for the Neumann problem and not for the Dirichlet problem. The reason is that the analysis in this section heavily relies on the results in [26]. In that paper, the weighted Sobolov estimates are in particular tuned to the Neumann problem (compare with Proposition 4.3 below). However, by suitable modifications, also similar results for the Dirichlet problem can be shown [31].

(ii) Since again the critical Sobolev index α_0 is multiplied by $3/2$, it turns out that also for the Neumann problem the use of adaptive schemes is completely justified.

(iii) By using real interpolation arguments as outlined in Remark 2.1, again a result for the best n-term approximation in H^1 can be derived.

(iv) We comment on the additional restriction $s < 3(l - |\boldsymbol{\delta}|)$ for s in (3.5) compared with (2.6). This restriction comes into play if $|\boldsymbol{\delta}| > 2l/3$. We will be forced to take such a vector $\boldsymbol{\delta}$ if there are some large angles θ_j (see (3.4)). However, also the relation between l and the number of faces n plays a role. To see this, we suppose that $|\boldsymbol{\delta}| > 2l/3$ and choose all δ_j as small as possible in (3.4). Further, we denote by $k \in \{1, \ldots, n\}$ the number of angles θ_j such that $\theta_j > \pi/l$. Observe that $k = 0$ is impossible. Without loss of generality, we assume that $\theta_1 \geqslant \theta_2 \geqslant \ldots \geqslant \theta_n$. As a consequence, we find

$$\frac{2}{3}\, l \leqslant -n + \sum_{j=1}^{k} \Big(l - \frac{\pi}{\theta_j} \Big) \qquad (\text{see } (3.4))\,.$$

This implies

$$l \geqslant \frac{n + \pi \sum\limits_{j=1}^{k} \frac{1}{\theta_j}}{k - 2/3}\,.$$

Using the trivial inequality $\theta_j < 2\pi$, we conclude

$$l > \frac{2n + k}{2(k - 2/3)}\,. \tag{3.6}$$

This inequality allows us different interpretations. For example, if there is only one large angle (i.e., $k = 1$), then (3.6) implies $l > 10$ (since n must be at least 3). However, on this way, the geometry of the polyhedral cone enters once again but we do not know whether this is caused by our method.

(v) Both, the Poisson equation with Dirichlet boundary conditions (2.3) and the Poisson equation with Neumann boundary conditions (3.2), are to understand as model cases. Since we did not use any specific property besides the existence, uniqueness, and regularity of the solution, both Theorems 2.1 and 3.1 extend to much more general classes of elliptic differential equations (see [21, Theorem 6.1.1] and [26] for details).

Proof of Theorem 3.1. The proof is organized as that of Theorem 2.1. We use the same agreements concerning the wavelets as in the proof of this theorem. By Remark 2.2, it suffices to concentrate on the estimate of the interior wavelets.

Let ρ_I denote the distance of the cube $Q(I)$ to the vertex, and let

$$r_I := \min_{j=1,\ldots,n} \; \min_{x \in Q(I)} \; r_j(x)\,.$$

As above, we work with the following decomposition of the set of interior wavelets:

$$\Lambda_j := \{\, I \mid \operatorname{supp} \eta_I \subset \mathcal{K}_0\,,\ 2^{-3j} \leqslant |I| \leqslant 2^{-3j+2}\},$$

$$\Lambda_{j,k} := \{\, I \in \Lambda_j \mid k2^{-j} \leqslant \rho_I < (k+1)2^{-j}\}\,, \quad j \in \mathbf{N}_0\,, \quad k \in \mathbf{N}\,,$$

$$\Lambda_{j,k,m} := \{\, I \in \Lambda_{j,k} \mid 2^{-j}m \leqslant r_I < 2^{-j}(m+1)\}\,, \quad m \in \mathbf{N}\,.$$

Elementary arguments yield

$$|\Lambda_{j,k}| \lesssim k^2 \qquad \text{and} \qquad |\Lambda_{j,k,m}| \lesssim m \tag{3.7}$$

independent of j, k, and m. Let $0 < \tau < 2$. Using the Whitney estimate (2.10), we obtain

$$\sum_{I\in\Lambda_{j,k}} |\langle u,\eta_I\rangle|^\tau \lesssim \sum_{I\in\Lambda_{j,k}} 2^{-lj\tau}\Big(\int_{Q(I)} |\nabla^l u|^2 dx\Big)^{\tau/2}$$

$$\lesssim \sum_{I\in\Lambda_{j,k}} 2^{-lj\tau}\, r_I^{-\tau|\boldsymbol{\delta}|}\, \rho_I^{-\tau(\beta-|\boldsymbol{\delta}|)}\Big(\int_{Q(I)} \rho^{2(\beta-|\boldsymbol{\delta}|)}\Big(\prod_{t=1}^n r_t^{\delta_j}\Big)^2 |\nabla^l u|^2 dx\Big)^{\tau/2}.$$

We put

$$u_I := \int_{Q(I)} \rho^{2(\beta-|\boldsymbol{\delta}|)}\Big(\prod_{t=1}^n r_t^{\delta_j}\Big)^2 |\nabla^l u|^2 dx\,.$$

To continue our estimate, we concentrate first on the set $\Lambda_{j,k,m}$. We use the Hölder inequality with $p = 2/\tau$, $q = 2/(2-\tau)$ and the fact that the layer $\Lambda_{j,k,m}$ contains of order m cubes (see (3.7)). This yields

$$\sum_{I\in\Lambda_{j,k,m}} |\langle u,\eta_I\rangle|^\tau$$

$$\lesssim 2^{-l\tau j}(k2^{-j})^{-\tau(\beta-|\boldsymbol{\delta}|)}\Big(\sum_{I\in\Lambda_{j,k,m}} r_I^{-\tau|\boldsymbol{\delta}|\frac{2}{2-\tau}}\Big)^{\frac{2-\tau}{2}}\Big(\sum_{I\in\Lambda_{j,k,m}} u_I\Big)^{\tau/2}$$

$$\lesssim 2^{-l\tau j}(k2^{-j})^{-\tau(\beta-|\boldsymbol{\delta}|)}\Big(\sum_{I\in\Lambda_{j,k,m}} (m2^{-j})^{-\tau|\boldsymbol{\delta}|\frac{2}{2-\tau}}\Big)^{\frac{2-\tau}{2}}\Big(\sum_{I\in\Lambda_{j,k,m}} u_I\Big)^{\tau/2}$$

$$\lesssim 2^{\tau j(\beta-l)}\, k^{-\tau(\beta-|\boldsymbol{\delta}|)}\, m^{-\tau|\boldsymbol{\delta}|+\frac{2-\tau}{2}}\Big(\sum_{I\in\Lambda_{j,k,m}} u_I\Big)^{\tau/2}.$$

The next step is to exploit the fact that there are of order k sets $\Lambda_{j,k,m}$ in each layer $\Lambda_{j,k}$ (the distance of a point in $\mathcal{K}_0$ to the edges cannot be much larger than the distance to the vertex). Together with the Hölder inequality, this leads us to

$$\sum_{I\in\Lambda_{j,k}} |\langle u,\eta_I\rangle|^\tau \lesssim 2^{j\tau(\beta-l)}\, k^{-\tau(\beta-|\boldsymbol{\delta}|)} \times \Big(\sum_{m=1}^{Ck} m^{-\tau|\boldsymbol{\delta}|\frac{2}{2-\tau}+1}\Big)^{\frac{2-\tau}{2}} \Big(\sum_{I\in\Lambda_{j,k}} u_I\Big)^{\tau/2}, \qquad (3.8)$$

where C is an appropriate constant depending on $\mathcal{K}_0$ only. Observe that

$$\Big(\sum_{m=1}^{Ck} m^{-\tau|\boldsymbol{\delta}|\frac{2}{2-\tau}+1}\Big)^{\frac{2-\tau}{2}} \lesssim \begin{cases} k^{-\tau|\boldsymbol{\delta}|+2-\tau} & \text{if } \ 2>\tau(1+|\boldsymbol{\delta}|)\,,\\ (\log(1+k))^{\frac{2-\tau}{2}} & \text{if } \ 2=\tau(1+|\boldsymbol{\delta}|)\,,\\ 1 & \text{if } \ 2<\tau(1+|\boldsymbol{\delta}|)\,. \end{cases}$$

Inserting this into (3.8) we obtain

$$\sum_{I\in\Lambda_{j,k}} |\langle u,\eta_I\rangle|^\tau \lesssim 2^{j\tau(\beta-l)}\Big(\sum_{I\in\Lambda_{j,k}} u_I\Big)^{\tau/2} \times \begin{cases} k^{-\tau(\beta+1)+2} & \text{if } \ 2>\tau(1+|\boldsymbol{\delta}|)\,,\\ k^{-\tau(\beta-|\boldsymbol{\delta}|)}\,(\log(1+k))^{\frac{2-\tau}{2}} & \text{if } \ 2=\tau(1+|\boldsymbol{\delta}|)\,,\\ k^{-\tau(\beta-|\boldsymbol{\delta}|)} & \text{if } \ 2<\tau(1+|\boldsymbol{\delta}|)\,. \end{cases}$$

To simplify notation, we denote these functions of k in the second line by a_k. For each refinement level j we have to take $C2^j$ layers $\Lambda_{j,k}$ into account (see (2.11)). Therefore, using the Hölder inequality for another time and Proposition 4.4, we finally get

$$\begin{aligned}\sum_{I\in\Lambda_j} |\langle u,\eta_I\rangle|^\tau &\lesssim 2^{j\tau(\beta-l)}\Big(\sum_{k=1}^{C2^j} a_k^{\frac{2}{2-\tau}}\Big)^{\frac{2-\tau}{2}} \Big(\sum_{I\in\Lambda_j} u_I\Big)^{\tau/2}\\ &\lesssim 2^{j\tau(\beta-l)}\Big(\sum_{k=1}^{C2^j} a_k^{\frac{2}{2-\tau}}\Big)^{\frac{2-\tau}{2}} \|u\|^\tau_{W^{l,2}_{\beta,\boldsymbol{\delta}}(\mathcal{K})}\\ &\lesssim 2^{j\tau(\beta-l)}\Big(\sum_{k=1}^{C2^j} a_k^{\frac{2}{2-\tau}}\Big)^{\frac{2-\tau}{2}} \Big(\|f\|_{W^{l-2,2}_{\beta,\boldsymbol{\delta}}(\mathcal{K})} + \sum_{j=1}^{n} \|g_j\|_{W^{l-1/2,2}_{\beta,\boldsymbol{\delta}}(\Gamma_j)}\Big)^\tau.\end{aligned}$$

To complete the estimate, we have to sum with respect to $j\in\mathbf{N}_0$. Formally, the discussion of this splits into nine cases. However, by using the abbreviation from (2.12), we end up with

$$\| u^* \|_{B^s_\tau(L_\tau(\mathbf{R}^3))} \lesssim \|u\|_{W^{l,2}_{\beta,\delta}(\mathcal{K})}\,, \qquad \frac{1}{\tau} = \frac{s}{3} + \frac{1}{2}\,, \tag{3.9}$$

if one of the following conditions is satisfied:

$$\begin{array}{rll} 3\left(\frac{1}{\tau} - \frac{1}{2}\right) < l & \text{if} \quad \tau\,(1+|\boldsymbol{\delta}|) < 2 & \text{and} \quad \beta < 3\left(\frac{1}{\tau} - \frac{1}{2}\right), \\ \beta < l & \text{if} \quad \tau\,(1+|\boldsymbol{\delta}|) < 2 & \text{and} \quad \beta \geqslant 3\left(\frac{1}{\tau} - \frac{1}{2}\right), \\ \frac{3}{2}\,|\boldsymbol{\delta}| < l & \text{if} \quad \tau\,(1+|\boldsymbol{\delta}|) = 2 & \text{and} \quad \beta < \frac{3}{2}\,|\boldsymbol{\delta}|\,, \\ \beta < l & \text{if} \quad \tau\,(1+|\boldsymbol{\delta}|) = 2 & \text{and} \quad \beta \geqslant \frac{3}{2}\,|\boldsymbol{\delta}|\,, \\ \frac{1}{\tau} - \frac{1}{2} < l - |\boldsymbol{\delta}| & \text{if} \quad \tau\,(1+|\boldsymbol{\delta}|) > 2 & \text{and} \quad \frac{1}{\tau} - \frac{1}{2} > \beta - |\boldsymbol{\delta}|\,, \\ \beta < l & \text{if} \quad \tau\,(1+|\boldsymbol{\delta}|) > 2 & \text{and} \quad \frac{1}{\tau} - \frac{1}{2} \leqslant \beta - |\boldsymbol{\delta}|\,. \end{array}$$

Observe that $\beta < l$ is necessary in all six cases. If $\beta < l$ and $|\boldsymbol{\delta}| < 2l/3$, then, according to the first case, we can choose $\beta < s < l$, s arbitrary close to l. Now, let $|\boldsymbol{\delta}| \geqslant 2l/3$. We employ case five in our list of sufficient conditions above. Using $s = 3(\frac{1}{\tau} - \frac{1}{2})$, we can reformulate this as follows:

$$\beta - |\boldsymbol{\delta}| < s < 3\,(l - |\boldsymbol{\delta}|) \qquad \text{and} \qquad s < \frac{3}{2}\,|\boldsymbol{\delta}|\,.$$

Since $|\boldsymbol{\delta}| \geqslant 2l/3$ implies $3\,(l - |\boldsymbol{\delta}|) \leqslant 3\,|\boldsymbol{\delta}|/2$, we have found the third restriction for s in (3.5). But the second originates from the estimates of those terms connected with the boundary (see Remark 2.2). This proves the theorem. □

4 Appendix A. Regularity of Solutions of the Poisson Equation

First of all, we recall a result of Grisvard [17, Corollary 2.6.7].

Proposition 4.1. *Let Ω be any bounded polyhedral open subset of $\mathbf{R}^3$. Then there exists a number $\alpha_0 > 3/2$ such that for every $f \in L_2(\Omega)$ the variational solution u of the Poisson equation (with either the Dirichlet boundary conditions or the Neumann boundary conditions) belongs to $H^s(\Omega)$ for every $s < \alpha_0$.*

The second result which is of certain use for us is taken from [19, Theorems 0.5 and 5.1].

Proposition 4.2. *Let Ω be any bounded Lipschitz domain in $\mathbf{R}^3$. Let $1/2 < \alpha < 3/2$. Suppose that $f \in H^{\alpha-2}(\Omega)$ and $g \in H^{\alpha-1/2}(\partial\Omega)$. Then the Poisson problem* (2.3) *has a unique solution $u \in H^{\alpha}(\Omega)$.*

Remark 4.1. Summarizing, for bounded smooth and polyhedral cones there exists a number $\alpha_0 \geqslant 3/2$ such that for all

$$(f,g) \in L_2(\mathcal{K}) \times H^{\alpha-1/2}(\partial\mathcal{K})$$

the solution u of (2.3) belongs to $H^{\alpha}(\mathcal{K})$, as long as $\alpha < \alpha_0$.

Next, we quote an a priori estimate from [21, Theorem 6.1.1]. It is the basis of our treatment in Sect. 2.

Proposition 4.3. *Let $\mathcal{K}$ be a smooth cone as defined in* (2.1)*. Suppose that the right-hand side f is contained in $V^{l-2}_{2,\beta}(\mathcal{K})$, where $l \geqslant 2$ is a natural number. Further, we assume that $g \in V^{l-1/2}_{2,\beta}(\partial\mathcal{K})$. Then there exists a countable set E of complex numbers such that the following holds. If a real number β is chosen such that*

$$\operatorname{Re}\lambda \neq -\beta + l - 3/2 \qquad \textit{for all} \quad \lambda \in E\,,$$

then the solution u of (2.3) *satisfies*

$$\|u\|_{V^{l}_{2,\beta}(\mathcal{K})} \lesssim \left(\|f\|_{V^{l-2}_{2,\beta}(\mathcal{K})} + \|g\|_{V^{l-1/2}_{2,\beta}(\partial\mathcal{K})}\right). \tag{4.1}$$

Finally, the following result of Maz'ya and Rossmann [26] plays a fundamental role in Sect. 3.

Proposition 4.4. *Let $\mathcal{K}$ be a polyhedral cone as defined in* (3.1)*. Suppose that the right-hand side $f \in W^{l-2,2}_{\beta,\boldsymbol{\delta}}(\mathcal{K}) \cap L_2(\mathcal{K})$, where $l \geqslant 2$ is a natural number. Further, we assume that $g_j \in W^{l-3/2,2}_{\beta,\boldsymbol{\delta}}(\Gamma_j)$, $j = 1,\ldots,n$. Then there exists a countable set E of complex numbers such that the following holds. If a real number β and a vector $\boldsymbol{\delta}$ are chosen such that*

$$\lambda \neq l - \beta - \frac{3}{2} \qquad \textit{for all} \quad \lambda \in E$$

and

$$\max\left(l - \frac{\pi}{\theta_j}, 0\right) < \delta_j + 1 < l, \quad j = 1,\ldots,n\,,$$

then the solution u of (3.2) *satisfies*

$$\|u\|_{W^{l,2}_{\beta,\boldsymbol{\delta}}(\mathcal{K})} \lesssim \left(\|f\|_{W^{l-2,2}_{\beta,\boldsymbol{\delta}}(\mathcal{K})} + \sum_{j=1}^{n} \|g_j\|_{W^{l-1/2,2}_{\beta,\boldsymbol{\delta}}(\Gamma_j)}\right). \tag{4.2}$$

5 Appendix B. Function Spaces

We take it for granted that the reader is familiar with Sobolev and Besov spaces on $\mathbf{R}^d$. There are different approaches to spaces defined on domains. We make a few remarks in this direction.

5.1 *Besov spaces on domains*

Let $\Omega \subset \mathbf{R}^d$ be a bounded open nonempty set. Then we define $B^s_q(L_p(\Omega))$ to be the collection of all distributions $f \in D'(\Omega)$ such that there exists a tempered distribution $g \in B^s_q(L_p(\mathbf{R}^d))$ satisfying

$$f(\varphi) = g(\varphi) \qquad \text{for all} \quad \varphi \in D(\Omega)\,,$$

i.e., $g|_\Omega = f$ in $D'(\Omega)$. We put

$$\|\, f \,|B^s_q(L_p(\Omega))\| := \inf \|\, g \,|B^s_q(L_p(\mathbf{R}^d))\|\,,$$

where the infimum is taken with respect to all distributions g as above.

5.2 *Sobolev spaces on domains*

Let Ω be a bounded Lipschitz domain. Let $m \in \mathbf{N}$. As usual, $H^m(\Omega)$ denotes the collection of all functions f such that the distributional derivatives $D^\alpha f$ of order $|\alpha| \leqslant m$ belong to $L_2(\Omega)$. The norm is defined by

$$\|\, f \,|H^m(\Omega)\| := \sum_{|\alpha| \leqslant m} \|\, D^\alpha f \,|L_2(\Omega)\|\,.$$

It is well known that $H^m(\mathbf{R}^d) = B^m_2(L_2(\mathbf{R}^d))$ in the sense of equivalent norms (see, for example, [35]). As a consequence of the existence of a bounded linear extension operator for Sobolev spaces on bounded Lipschitz domains (see [33, p. 181] or [32]), it follows that

$$H^m(\Omega) = B^m_2(L_2(\Omega)) \qquad \text{(equivalent norms)}$$

for such domains. For fractional $s > 0$ we introduce the classes by complex interpolation. Let $0 < s < m$, $s \notin \mathbf{N}$. Then, following [23, 9.1], we define

$$H^s(\Omega) := \Big[H^m(\Omega), L_2(\Omega)\Big]_\Theta\,, \qquad \Theta = 1 - \frac{s}{m}\,.$$

This definition does not depend on m in the sense of equivalent norms (see [37]). The outcome $H^s(\Omega)$ coincides with $B^s_2(L_2(\Omega))$ (see [37, 38] for further details).

5.3 Besov spaces and wavelets

Here, we collect some properties of Besov spaces which were used above. For a general information on Besov spaces we refer to the monographs [28, 30, 34, 35, 36, 38].

For the construction of biorthogonal wavelet bases as considered below we refer to the recent monograph by Cohen [3, Chapt. 2]. Let φ be a compactly supported scaling function of sufficiently high regularity, and let ψ_i, $i = 1, \ldots 2^d - 1$, be the corresponding wavelets. More exactly, we suppose that for some $N > 0$ and $r \in \mathbf{N}$

$$\begin{aligned} &\operatorname{supp} \varphi, \operatorname{supp} \psi_i \quad \subset \quad [-N,N]^d, \qquad i = 1,\ldots,2^d-1,\\ &\varphi, \psi_i \in C^r(\mathbf{R}^d), \qquad i = 1,\ldots,2^d-1,\\ &\int x^\alpha\, \psi_i(x)\, dx = 0 \qquad \text{for all} \quad |\alpha| \leqslant r, \qquad i = 1,\ldots,2^d-1, \end{aligned}$$

and

$$\varphi(x-k),\ 2^{jd/2}\,\psi_i(2^j x - k), \qquad j \in \mathbf{N}_0, \quad k \in \mathbf{Z}^d, \quad i = 1,\ldots,2^d-1,$$

is a Riesz basis for $L_2(\mathbf{R}^d)$. We use the standard abbreviations

$$\psi_{i,j,k}(x) = 2^{jd/2}\,\psi_i(2^j x - k) \qquad \text{and} \qquad \varphi_k(x) = \varphi(x-k).$$

Further, the dual Riesz basis should fulfill the same requirements, i.e., there exist functions $\widetilde{\varphi}$ and $\widetilde{\psi}_i$, $i = 1, \ldots, 2^d - 1$, such that

$$\begin{aligned} \langle \widetilde{\varphi}_k, \psi_{i,j,k}\rangle &= \langle \widetilde{\psi}_{i,j,k}, \varphi_k\rangle = 0,\\ \langle \widetilde{\varphi}_k, \varphi_\ell\rangle &= \delta_{k,\ell} \qquad \text{(Kronecker symbol)},\\ \langle \widetilde{\psi}_{i,j,k}, \psi_{u,v,\ell}\rangle &= \delta_{i,u}\,\delta_{j,v}\,\delta_{k,\ell},\\ \operatorname{supp} \widetilde{\varphi}, &\quad \operatorname{supp} \widetilde{\psi}_i \quad \subset \quad [-N,N]^d, \qquad i = 1,\ldots,2^d-1,\\ \widetilde{\varphi}, \widetilde{\psi}_i &\in C^r(\mathbf{R}^d), \qquad i = 1,\ldots,2^d-1,\\ \int x^\alpha\, \widetilde{\psi}_i(x)\, dx &= 0 \qquad \text{for all} \quad |\alpha| \leqslant r, \qquad i = 1,\ldots,2^d-1. \end{aligned}$$

For $f \in S'(\mathbf{R}^d)$ we put

$$\langle f, \psi_{i,j,k}\rangle = f(\overline{\psi_{i,j,k}}) \qquad \text{and} \qquad \langle f, \varphi_k\rangle = f(\overline{\varphi_k}), \tag{5.1}$$

whenever this makes sense.

Proposition 5.1. *Let $s \in \mathbf{R}$, and let $0 < p, q \leqslant \infty$. Suppose that*

$$r > \max\Big(s, d\,\max\Big(0, \frac{1}{p} - 1\Big) - s\Big). \tag{5.2}$$

Then $B^s_q(L_p(\mathbf{R}^d))$ is the collection of all tempered distributions f such that f is representable as

$$f = \sum_{k \in \mathbf{Z}^d} a_k\, \varphi_k + \sum_{i=1}^{2^d-1} \sum_{j=0}^{\infty} \sum_{k \in \mathbf{Z}^d} a_{i,j,k}\, \psi_{i,j,k} \qquad (convergence\ in\ \ S')$$

with

$$\| f\, |B^s_q(L_p(\mathbf{R}^d))\|^* := \Big(\sum_{k \in \mathbf{Z}^d} |a_k|^p\Big)^{1/p}$$

$$+ \Big(\sum_{i=1}^{2^d-1} \sum_{j=0}^{\infty} 2^{j(s+d(\frac{1}{2}-\frac{1}{p}))q} \Big(\sum_{k \in \mathbf{Z}^d} |a_{i,j,k}|^p\Big)^{q/p}\Big)^{1/q} < \infty$$

if $q < \infty$ and

$$\| f\, |B^s_\infty(L_p(\mathbf{R}^d))\|^* := \Big(\sum_{k \in \mathbf{Z}^d} |a_k|^p\Big)^{1/p}$$

$$+ \sup_{i=1,\dots,2^d-1}\ \sup_{j=0,\dots}\ 2^{j(s+d(\frac{1}{2}-\frac{1}{p}))} \Big(\sum_{k \in \mathbf{Z}^d} |a_{i,j,k}|^p\Big)^{1/p} < \infty\,.$$

The representation is unique and

$$a_{i,j,k} = \langle f, \widetilde{\psi}_{i,j,k}\rangle \qquad and \qquad a_k = \langle f, \widetilde{\varphi}_k\rangle.$$

Further, $J : f \mapsto \{\langle f, \widetilde{\varphi}_k\rangle, \langle f, \widetilde{\psi}_{i,j,k}\rangle\}$ is an isomorphic map of $B^s_q(L_p(\mathbf{R}^d))$ onto the sequence space (equipped with the quasinorm $\|\cdot|B^s_q(L_p(\mathbf{R}^d))\|^*$), i.e., $\|\cdot|B^s_q(L_p(\mathbf{R}^d))\|^*$ may serve as an equivalent quasinorm on $B^s_q(L_p(\mathbf{R}^d))$.

Proposition 5.1 was proved in [39] (see also [22] for a homogeneous version). A different proof, but restricted to $s > d(\frac{1}{p} - 1)_+$, is given in [3, Theorem 3.7.7]. However, there are many fore-runners with some restrictions on s, p, and q.

References

1. Agmon, S., Douglis, A., Nierenberg, L.: Estimates near the boundary for solutions of elliptic partial differential equations satisfying general boundary conditions I. Commun. Pure Appl. Math. **12**, 623-727 (1959)
2. Binev, P., Dahmen, W., DeVore, R.: Adaptive finite element methods with convergence rates. Numer. Math. **97**, no. 2, 219-268 (2004)
3. Cohen, A.: Numerical Analysis of Wavelet Methods. Elsevier Science, Amsterdam (2003)
4. Cohen, A., Dahmen, W., DeVore, R.: Adaptive wavelet methods for elliptic operator equations - Convergence rates. Math. Comp. **70**, 21-75 (2001)
5. Cohen, A., Dahmen, W., DeVore, R.: Adaptive wavelet methods II - Beyond the elliptic case. Found. Comput. Math. **2**, 203-245 (2002)
6. Cohen, A., Dahmen, W., DeVore, R.: Adaptive wavelet methods for nonlinear variational problems. SIAM J. Numer. Anal. **41**, no. 5, 1785-1823 (2003)
7. Dahlke, S., Besov regularity for elliptic boundary value problems on polygonal domains. Appl. Math. Lett. **12**, no. 6, 31-38 (1999)
8. Dahlke, S., Besov regularity of edge singularities for the Poisson equation in polyhedral domains. Numer. Linear Algebra Appl. **9**, no. 6-7, 457–466 (2002)
9. Dahlke, S., Dahmen, W., DeVore, R.: Nonlinear approximation and adaptive techniques for solving elliptic operator equations. In: Dahmen, W., Kurdila, A., Oswald, P. Eds., Multiscale Wavelet Methods for Partial Differential Equations. Academic Press, San Diego, pp. 237–283 (1997)
10. Dahlke, S., Dahmen, W., Hochmuth, R., Schneider, R.: Stable multiscale bases and local error estimation for elliptic problems. Appl. Numer. Math. **23**, no. 1, 21–48 (1997)
11. Dahlke, S., Dahmen, W., Urban, K.: Adaptive wavelet methods for saddle point problems - Convergence rates. SIAM J. Numer. Anal. **40**, no. 4, 1230–1262 (2002)
12. Dahlke, S., DeVore, R.: Besov regularity for elliptic boundary value problems. Commun. Partial Differ. Equ. **22**, no. 1-2, 1–16 (1997)
13. Dahlke, S., Novak, E., Sickel, W.: Optimal approximation of elliptic problems by linear and nonlinear mappings I. J. Complexity **22**, 29–49 (2006)
14. Dauge, M.: Neumann and mixed problems on curvilinear polyhedra. Integral Equ. Operator Theory **15**, 227–261 (1992)
15. DeVore, R.: Nonlinear approximation. Acta Numerica **7**, 51–150 (1998)
16. Dörfler, W.: A convergent adaptive algorithm for Poisson's equation. SIAM J. Numer. Anal. **33**, no. 3, 1106–1124 (1995)
17. Grisvard, P.: Singularities in Boundary Value Problems. Springer, Berlin (1992)
18. Hackbusch, W.: Elliptic Differential Equations, Springer-Verlag, Berlin–Heidelberg (1992)
19. Jerison, D., Kenig, C.E.: The inhomogeneous Dirichlet problem in Lipschitz domains. J. Funct. Anal. **130**, 161–219 (1995)
20. Kondrat'ev, V.A.: Boundary value problems for elliptic equations in domains with conical or angular points (Russian). Tr. Mosk. Mat. Obshch. **16**, 209–292 (1967)
21. Kozlov, V.A., Mazya, V.G., Rossmann, J.: Elliptic Boundary Value Problems in Domains with Point Singularities. Am. Math. Soc., Providence, RI (1997)
22. Kyriazis, G.: Decomposition systems for function spaces. Studia Math. **157**, 133–169 (2003)
23. Lions, J.L., Magenes, E.: Non-Homogeneous Boundary Value Problems and Applications I. Springer, Berlin (1972)
24. Mazya, V.G., Plamenevskii, B.A.: On the coefficients in the asymptotics of solutions of elliptic boundary value problems in domains with conical points. Math. Nachr. **76**, 29-60 (1977)

25. Mazya, V.G., Plamenevskii,B.A.: Estimates in L_p and Hölder classes and the Miranda–Agmon maximum principle for solutions of elliptic boundary value problems in domains with singular points on the boundary. Math. Nachr. **81**, 25-82 (1977)
26. Mazya, V.G., Roßmann, J.: Weighted L_p estimates of solutions to boundary value problems for second order elliptic systems in poyhedral domains. Z. Angew. Math. Mech. **83**, 435–467 (2003)
27. Meyer, Y.: Wavelets and Operators. Cambridge Univ. Press, Cambridge (1992)
28. Nikol'skij, S.M.: Approximation of Functions of Several Variables and Embedding Theorems (Russian). Nauka, Moscow (1969); Second ed. (1977); English transl.: Springer, Berlin (1975)
29. Nitsche, P.-A.: Sparse approximation of singularity functions. Constr. Approx. **21**, no. 1, 63–81 (2005)
30. Peetre, J.: New Thoughts on Besov Spaces. Duke Univ. Math. Series. Durham (1976)
31. Roßmann. J.: [Privat communication] (2007)
32. Rychkov, V.S.: On restrictions and extensions of the Besov and Triebel-Lizorkin spaces with respect to Lipschitz domains. J. London Math. Soc. **60**, 237–257 (1999)
33. Stein, E.M.: Singular Integrals and Differentiability Properties of Functions. Princeton University Press, Princeton (1970)
34. Triebel, H.: Interpolation Theory, Function Spaces, Differential Operators. North-Holland, Amsterdam (1978); Second ed. Barth, Heidelberg (1995)
35. Triebel, H.: Theory of Function Spaces. Birkhäuser, Basel (1983)
36. Triebel, H.: Theory of Function Spaces II. Birkhäuser, Basel (1992)
37. Triebel, H.: Function spaces in Lipschitz domains and on Lipschitz manifolds. Characteristic functions as pointwise multipliers. Rev. Mat. Complutense **15**, 475-524 (2002)
38. Triebel, H.: Theory of Function Spaces III. Birkhäuser, Basel (2006)
39. Triebel, H.: Local means and wavelets in function spaces. Banach Center Publ. **79**, pp. 215-234. Inst. Math. Polish Acad. Sci., Warsaw (2008)

Variational Approach to Complicated Similarity Solutions of Higher Order Nonlinear Evolution Partial Differential Equations

Victor Galaktionov, Enzo Mitidieri, and Stanislav Pokhozhaev

Dedicated to the memory of S.L. Sobolev
on the occasion of his centenary

Abstract We consider the Cauchy problem for three higher order degenerate quasilinear partial differential equations, as basic models,

$$u_t = (-1)^{m+1}\Delta^m(|u|^n u) + |u|^n u,$$
$$u_{tt} = (-1)^{m+1}\Delta^m(|u|^n u) + |u|^n u,$$
$$u_t = (-1)^{m+1}[\Delta^m(|u|^n u)]_{x_1} + (|u|^n u)_{x_1},$$

where $(x,t) \in \mathbb{R}^N \times \mathbb{R}_+$, $n > 0$, and Δ^m is the $(m \geqslant 1)$th iteration of the Laplacian. Based on the blow-up similarity and travelling wave solutions, we investigate general local, global, and blow-up properties of such equations. The nonexistence of global in time solutions is established by different methods. In particular, for $m = 2$ and $m = 3$ such similarity patterns lead to the semilinear 4th and 6th order elliptic partial differential equations with noncoercive operators and non-Lipschitz nonlinearities

$$-\Delta^2 F + F - |F|^{-\frac{n}{n+1}} F = 0 \text{ and } \Delta^3 F + F - |F|^{-\frac{n}{n+1}} F = 0 \text{ in } \mathbb{R}^N, \qquad (1)$$

which were not addressed in the mathematical literature. Using analytic variational, qualitative, and numerical methods, we prove that Eqs. (1) admit an infinite at least countable set of countable families of compactly supported solutions that are oscillatory near finite interfaces. This shows typical properties of a set of solutions of chaotic structure.

Victor Galaktionov
University of Bath, Bath, BA2 7AY, UK, e-mail: `vag@maths.bath.ac.uk`

Enzo Mitidieri
Università di Trieste, Via Valerio 12/1, 34127 Trieste, Italy, e-mail: `mitidier@units.it`

Stanislav Pokhozhaev
Steklov Mathematical Institute, Gubkina St. 8, 119991 Moscow, Russia, e-mail: `pokhozhaev@mi.ras.ru`

V. Maz'ya (ed.), *Sobolev Spaces in Mathematics II*, *International Mathematical Series*.
doi: 10.1007/978-0-387-85650-6,

1 Introduction. Higher-Order Models and Blow-up or Compacton Solutions

It is difficult to exaggerate the role of Sobolev's celebrated book [35] of 1950 which initiated a massive research in the second half of the twentieth century not only in the area of applications of functional analysis on the basis of the classical concepts of Sobolev spaces, but also in the core problems of weak (generalized) solutions of linear and nonlinear partial differential equations (PDEs). It seems it is impossible to imagine nowadays a complicated nonlinear PDE from modern application that can be studied without Sobolev's ideas and methods proposed more than a half of the century ago.

In this paper, in a unified manner, we study the local existence, blow-up, and singularity formation phenomena for three classes of nonlinear degenerate PDEs which were not that much treated in the existing mathematical literature. Concepts associated with Sobolev spaces and weak solutions will play a key role.

1.1 Three types of nonlinear PDEs under consideration

We study common local and global properties of weak compactly solutions of three classes of quasilinear PDEs of parabolic, hyperbolic, and nonlinear dispersion type, which, in general, look like having nothing in common. Studying and better understanding of nonlinear degenerate PDEs of higher order including a new class of less developed *nonlinear dispersion equations* (NDEs) from the compacton theory are striking features of the modern general PDE theory at the beginning of the twenty first century. It is worth noting and realizing that several key theoretical demands of modern mathematics are already associated and connected with some common local and global features of nonlinear evolution PDEs of different types and orders, including higher order parabolic, hyperbolic, nonlinear dispersion, etc., as typical representatives.

Regardless the great progress of the PDE theory achieved in the twentieth century for many key classes of nonlinear equations [6], the transition process to higher order degenerate PDEs with more and more complicated nonmonotone, nonpotential, and nonsymmetric nonlinear operators will require different new methods and mathematical approaches. Moreover, it seems that for some types of such nonlinear higher order problems the entirely rigorous "exhaustive" goal of developing a complete description of solutions, their properties, functional settings of problems of interest, etc., cannot be achieved in principle in view of an extremal variety of singular, bifurcation, and branching phenomena contained in such multi-dimensional evolution. In many cases, the main results should be extracted by a combination of methods of various analytic, qualitative, and numerical origins. Of course, this is not a novelty in modern mathematics, where several fundamental rigorous results have been already justified with the aid of reliable numerical experiments.

In the present paper, we deal with very complicated pattern sets, where for the elliptic problems in $\mathbb{R}^N$ and even for the corresponding one-dimensional ordinary differential equation (ODE) reductions, the use of proposed *analytic-numerical* approaches is necessary and unavoidable. As the first illustration of such features, let us mention that, according to our current experience, for such classes of second order C^1 variational problems, distinguishing the classical Lusternik–Schnirelman (LS) countable sequence of critical values and points is not possible without refined numerical methods in view of huge complicated multiplicity of other admitted solutions. It is essential that the arising problems do not admit, as customary for other classes of elliptic equations, any homotopic classification of solutions (say, on the hodograph plane) since all compactly supported solutions are infinitely oscillatory, which makes the homotopy rotational parameter infinite and hence nonapplicable.

Now, we introduce three classes of PDEs to be studied.

1.2 (I) *Combustion type models: regional blow-up, global stability, main goals, and first discussion*

We begin with the following quasilinear degenerate $2m$th order parabolic equation of reaction-diffusion (combustion) type:

$$u_t = (-1)^{m+1} \Delta^m (|u|^n u) + |u|^n u \quad \text{in} \quad \mathbb{R}^N \times \mathbb{R}_+, \tag{1.1}$$

where $n > 0$ is a fixed exponent, $m \geqslant 2$ is an integer, and Δ denotes the Laplace operator in $\mathbb{R}$.

Globally asymptotically stable exact blow-up solutions of S-regime

In the simplest case $m = 1$ and $N = 1$, (1.1) is nowadays the canonical *quasilinear heat equation*

$$u_t = (u^{n+1})_{xx} + u^{n+1} \quad \text{in} \quad \mathbb{R} \times \mathbb{R}_+ \quad (u \geqslant 0) \tag{1.2}$$

which occurs in the combustion theory. The reaction-diffusion equation (1.2), playing a key role in the blow-up PDE theory, was under scrutiny since the middle of the 1970s. In 1976, Kurdyumov, together with his former PhD students Mikhailov and Zmitrenko (see [33] and [32, Chapt. 4] for history), discovered the phenomenon of *heat* and *combustion localization* by studying the blow-up separate variable *Zmitrenko–Kurdyumov solution* of (1.2):

$$u_S(x,t) = (T-t)^{-\frac{1}{n}} f(x) \quad \text{in} \quad \mathbb{R} \times (0,T), \tag{1.3}$$

where $T > 0$ is the blow-up time and f satisfies the ordinary differential equation

$$\frac{1}{n} f = (f^{n+1})'' + f^{n+1} \quad \text{for} \quad x \in \mathbb{R}. \tag{1.4}$$

It turned out that (1.4) possesses the explicit compactly supported solution

$$f(x) = \begin{cases} \Big[\frac{2(n+1)}{n(n+2)} \cos^2\Big(\frac{nx}{2(n+1)}\Big)\Big]^{\frac{1}{n}}, & |x| \leqslant \frac{n+1}{n}\pi, \\ 0, & |x| > \frac{n+1}{n}\pi. \end{cases} \tag{1.5}$$

This explicit integration of (1.4) was amazing and rather surprising in the middle of the 1970s and led then to the foundation of the blow-up and heat localization theory. In dimension $N > 1$, the blow-up solution (1.3) does indeed exist [32, p. 183], but not in an explicit form (so that it seems that (1.5) is a unique available elegant form).

Blow-up S-regime for higher order parabolic PDEs

One can see that the $2m$th order counterpart of (1.1) admits a regional blow-up solution of the same form (1.3), but the profile $f = f(y)$ solves the more complicated ODE

$$(-1)^{m+1} \Delta^m (|f|^n f) + |f|^n f = \frac{1}{n} f \quad \text{in} \quad \mathbb{R}^N. \tag{1.6}$$

Under a natural change, this gives the following equation with non-Lipschitz nonlinearity:

$$F = |f|^n f \quad \Longrightarrow \quad (-1)^{m+1} \Delta^m F + F - \frac{1}{n} |F|^{-\frac{n}{n+1}} F = 0 \quad \text{in} \quad \mathbb{R}^N.$$

We scale out the multiplier $\frac{1}{n}$ in the nonlinear term:

$$F \mapsto n^{-\frac{n+1}{n}} F \quad \Longrightarrow \quad \boxed{(-1)^{m+1} \Delta^m F + F - |F|^{-\frac{n}{n+1}} F = 0 \text{ in } \mathbb{R}^N.} \tag{1.7}$$

In the one-dimensional case $N = 1$, we obtain the simpler ordinary differential equation

$$F \mapsto n^{-\frac{n+1}{n}} F \quad \Longrightarrow \quad \boxed{(-1)^{m+1} F^{(2m)} + F - |F|^{-\frac{n}{n+1}} F = 0 \text{ in } \mathbb{R}.} \tag{1.8}$$

Thus, according to (1.3), the elliptic problem (1.7) and the ODE (1.8) for $N = 1$ are responsible for possible "geometric shapes" of regional blow-up described by the higher order combustion model (1.1).

Plan and main goals of the paper related to parabolic PDEs

Unlike the second order case (1.5), *explicit* compactly supported solutions $F(x)$ of the ODE (1.8) for any $m \geqslant 2$ are not available. Moreover, such profiles $F(x)$ have rather complicated local and global structure. We are not aware of any rigorous or even formal qualitative results concerning the existence, multiplicity, or global structure of solutions of ODEs like (1.8).

Our main goal is four-fold:

(ii) PROBLEM "BLOW-UP": proving finite-time blow-up in the parabolic (and hyperbolic) PDEs under consideration (Sect. 2)

(ii) PROBLEM "MULTIPLICITY": existence and multiplicity for the elliptic PDE (1.7) and the ODE (1.8) (Sect. 3)

(iii) PROBLEM "OSCILLATIONS": the generic structure of oscillatory solutions of the ODE (1.8) near interfaces (Sect. 4)

(iv) PROBLEM "NUMERICS": numerical study of various families of $F(x)$ (Sect. 5)

In particular, we show that the ODE (1.8), as well as the PDE (1.7), for any $m \geqslant 2$ admits infinitely many countable families of compactly supported solutions in $\mathbb{R}$, and the set of all solutions exhibits certain *chaotic* properties. Our analysis is based on a combination of analytic (variational and others), numerical, and various formal techniques. Discussing the existence, multiplicity, and asymptotics for the nonlinear problems under consideration, we formulate several open mathematical problems. Some of these problems are extremely difficult in the case of higher order equations.

1.3 (II) *Regional blow-up in quasilinear hyperbolic equations*

We consider the $2m$th order hyperbolic counterpart of (1.1)

$$u_{tt} = (-1)^{m+1} \Delta^m(|u|^n u) + |u|^n u \quad \text{in} \quad \mathbb{R}^N \times \mathbb{R}_+. \tag{1.9}$$

We begin the discussion of blow-up solutions of (1.9) with the case 1D, i.e.,

$$u_{tt} = (u^{n+1})_{xx} + u^{n+1} \quad \text{in} \quad \mathbb{R} \times \mathbb{R}_+ \quad (u \geqslant 0). \tag{1.10}$$

Then the blow-up solutions and ODE take the form

$$u_{\rm S}(x,t) = (T-t)^{-\frac{2}{n}} \widetilde{f}(x) \quad \Longrightarrow \quad \frac{2}{n}\Big(\frac{2}{n}+1\Big)\widetilde{f} = (\widetilde{f}^{n+1})'' + \widetilde{f}^{n+1}. \tag{1.11}$$

Using extra scaling, from

$$\widetilde{f}(x) = \Big[\frac{2(n+2)}{n}\Big]^{\frac{1}{n}} f(x) \tag{1.12}$$

we obtain the same ODE (1.4) and hence the exact localized solution (1.5).

For the N-dimensional PDE (1.9), looking for the same solution (1.11) we obtain, after scaling, the elliptic equation (1.7).

1.4 (III) *Nonlinear dispersion equations and compactons*

In a general setting, these rather unusual PDEs take the form

$$u_t = (-1)^{m+1}[\Delta^m(|u|^n u)]_{x_1} + (|u|^n u)_{x_1} \quad \text{in} \quad \mathbb{R}^N \times \mathbb{R}_+, \tag{1.13}$$

where the right-hand side is just the derivation D_{x_1} of that in the parabolic counterpart of (1.1). Then the elliptic problem (1.7) arises in the study of *travelling wave* solutions of (1.13). As usual, we begin with the simple 1D case.

Setting $N = 1$ and $m = 1$ in (1.13), we obtain the third order *Rosenau–Hyman equation*

$$u_t = (u^2)_{xxx} + (u^2)_x \tag{1.14}$$

simulating the effect of *nonlinear dispersion* in the pattern formation in liquid drops [31]. It is the $K(2,2)$ equation from the general $K(m,n)$ family of *nonlinear dispersion equations*

$$u_t = (u^n)_{xxx} + (u^m)_x \quad (u \geqslant 0) \tag{1.15}$$

describing phenomena of compact pattern formation [28, 29]. Furthermore, such PDEs appear in curve motion and shortening flows [30]. As in the above models, the $K(m,n)$ equation (1.15) with $n > 1$ is degenerated at $u = 0$ and, therefore, may exhibit a finite speed of propagation and admit solutions with finite interfaces. A permanent source of NDEs is the integrable equation theory, for example, the integrable fifth order *Kawamoto equation* [18] (see [15, Chapt. 4] for other models)

$$u_t = u^5 u_{xxxxx} + 5\,u^4 u_x u_{xxxx} + 10\,u^5 u_{xx} u_{xxx}. \tag{1.16}$$

The existence, uniqueness, regularity, shock and rarefaction wave formation, finite propagation and interfaces for degenerate higher order models under consideration were discussed in [14] (see also comments in [15, Chapt. 4.2]).

We study particular continuous solutions of NDEs that give insight on several generic properties of such nonlinear PDEs. The crucial advantage of the Rosenau–Hyman equation (1.14) is that it possesses *explicit* moving compactly supported soliton type solutions, called *compactons* [31], which are *travelling wave* solutions.

Compactons: manifolds of travelling waves and blow-up S-regime solutions coincide

Let us show that such compactons are directly related to the blow-up patterns presented above. Actually, explicit travelling wave compactons exist for the nonlinear dispersion Korteweg-de Vries type equations with arbitrary power nonlinearities

$$u_t = (u^{n+1})_{xxx} + (u^{n+1})_x \quad \text{in} \quad \mathbb{R} \times \mathbb{R}_+. \tag{1.17}$$

This is the $K(1+n, 1+n)$ model [31].

Thus, compactons, regarded as solutions of Eq. (1.17), have the travelling wave structure

$$u_{\rm c}(x,t) = f(y), \quad y = x - \lambda t, \tag{1.18}$$

so that, by substitution, f satisfies the ODE

$$-\lambda f' = (f^{n+1})''' + (f^{n+1})' \tag{1.19}$$

and, by integration,

$$-\lambda f = (f^{n+1})'' + f^{n+1} + D, \tag{1.20}$$

where $D \in \mathbb{R}$ is a constant of integration. Setting $D = 0$, which means the physical condition of zero flux at the interfaces, we obtain the blow-up ODE (1.4), so that the compacton equation (1.20) *coincides* with the blow-up equation (1.4) if

$$-\lambda = \frac{1}{n} \quad \Big(\text{or } -\lambda = \tfrac{2}{n}\Big(\tfrac{2}{n}+1\Big) \text{ to match (1.11)}\Big). \tag{1.21}$$

This yields the compacton solution (1.18) with the same compactly supported profile (1.5) with translation $x \mapsto y = x - \lambda t$.

Therefore, in 1D, the blow-up solutions (1.3), (1.11) of the parabolic and hyperbolic PDEs and the compacton solution (1.18) of the NDE (1.17) are of similar *mathematical* (both ODE and PDE) *nature*. This fact reflects the *universality principle* of compact structure formation in nonlinear evolution

PDEs. The stability property of the travelling wave compacton (1.18) for the PDE setting (1.17), as well as for the higher order counterparts considered below, is unknown.

In the N-dimensional geometry, i.e., for the PDE (1.13), looking for a travelling wave moving only in the x_1-direction

$$u_{\rm c}(x,t) = f(x_1 - \lambda t, x_2, \dots, x_N) \quad \Big(\lambda = -\frac{1}{n}\Big), \tag{1.22}$$

we obtain, by integrating with respect to x_1, the elliptic problem (1.7).

Thus, we have introduced three classes **(I), (II), and (III)** of nonlinear higher order PDEs in $\mathbb{R}^N \times \mathbb{R}_+$ which, being representatives of very different three equations types, nevertheless have quite similar evolution features (possibly, up to replacing blow-up by travelling wave moving); moreover, the complicated countable sets of evolution patterns coincide. These common features reveal a concept of a certain unified principle of singularity formation phenomena in the general nonlinear PDE theory, which we just begin to touch and study in the twenty first century. Some classical mathematical concepts and techniques successfully developed in the twentieth century (including Sobolev legacy) continue to be key tools, but also new ideas from different ranges of various rigorous and qualitative natures are required for tackling such fundamental difficulties and open problems.

2 Blow-up Problem: General Blow-up Analysis of Parabolic and Hyperbolic PDEs

2.1 *Global existence and blow-up in higher order parabolic equations*

We begin with the parabolic model (1.1). Bearing in mind the compactly supported nature of the solutions under consideration, we consider (1.1) in a bounded domain $\Omega \subset \mathbb{R}^N$ with smooth boundary $\partial\Omega$, with the Dirichlet boundary conditions

$$u = Du = \ldots = D^{m-1}u = 0 \quad \text{on} \quad \partial\Omega \times \mathbb{R}_+, \tag{2.1}$$

and a sufficiently smooth and bounded initial function

$$u(x,0) = u_0(x) \quad \text{in} \quad \Omega. \tag{2.2}$$

We show that the blow-up phenomenon essentially depends on the size of domain. Beforehand, we note that the diffusion operator on the right-hand side of (1.1) is a monotone operator in $H^{-m}(\Omega)$, so that the unique local solvability of the problem in suitable Sobolev spaces follow by the classical theory of monotone operators (see [21, Chapt. 2]). We show that, under certain conditions, some of these solutions are global in time, whereas some ones cannot be globally extended and blow-up in finite time.

For the sake of convenience, we use the natural substitution

$$v = |u|^n u \quad \Longrightarrow \quad v_0(x) = |u_0(x)|^n u_0(x) \tag{2.3}$$

which leads to the following parabolic equation with a standard linear operator on the right-hand side:

$$(\psi(v))_t = (-1)^{m+1} \Delta^m v + v \quad \text{in} \quad \mathbb{R}^N \times \mathbb{R}_+, \quad \psi(v) = |v|^{-\frac{n}{n+1}} v, \tag{2.4}$$

where v satisfies the Dirichlet boundary conditions (2.1).

Multiplying (2.4) by v in $L^2(\Omega)$ and integrating by parts with the help of (2.1), we find

$$\frac{n+1}{n+2} \frac{\mathrm{d}}{\mathrm{d}t} \int\limits_{\Omega} |v|^{\frac{n+2}{n+1}} \mathrm{d}x = -\int\limits_{\Omega} |\widetilde{D}^m v|^2 \mathrm{d}x + \int\limits_{\Omega} v^2 \mathrm{d}x \equiv E(v), \tag{2.5}$$

where we used the notation: $\widetilde{D}^m = \Delta^{\frac{m}{2}}$ for even m and $\widetilde{D}^m = \nabla \Delta^{\frac{m-1}{2}}$ for odd m. By Sobolev's embedding theorem, $H^m(\Omega) \subset L^2(\Omega)$ compactly; moreover, the following sharp estimate holds:

$$\int\limits_{\Omega} v^2 \mathrm{d}x \leqslant \frac{1}{\lambda_1} \int\limits_{\Omega} |\widetilde{D}^m v|^2 \mathrm{d}x \quad \text{in} \quad H_0^m(\Omega), \tag{2.6}$$

where $\lambda_1 = \lambda_1(\Omega) > 0$ is the first eigenvalue of the polyharmonic operator $(-\Delta)^m$ with the Dirichlet boundary conditions (2.1):

$$(-\Delta)^m e_1 = \lambda_1 e_1 \quad \text{in} \quad \Omega, \quad e_1 \in H_0^{2m}(\Omega). \tag{2.7}$$

In the case $m = 1$, since $(-\Delta) > 0$ is strictly positive in the $L^2(\Omega)$-metric in view of the classical Jentzsch theorem (1912) about the positivity of the first eigenfunction for linear integral operators with positive kernels, we have

$$e_1(x) > 0 \quad \text{in} \quad \Omega. \tag{2.8}$$

In the case $m \geqslant 2$, the inequality (2.8) remains valid, for example, for the unit ball $\Omega = B_1$. Indeed, if $\Omega = B_1$, then the Green function of the polyharmonic operator $(-\Delta)^m$ with the Dirichlet boundary conditions is positive (see the first results dues to Boggio (1901-1905) [4, 5]). Using again the Jentzsch theorem, we conclude that (2.8) holds.

From (2.5) and (2.6) it follows that

$$\frac{n+1}{n+2}\frac{\mathrm{d}}{\mathrm{d}t}\int\limits_{\Omega}|v|^{\frac{n+2}{n+1}}\mathrm{d}x \leqslant \Big(\frac{1}{\lambda_1}-1\Big)\int\limits_{\Omega}|\widetilde{D}^m v|^2\mathrm{d}x.$$

Global existence for $\lambda_1 > 1$

Thus, we obtain the inequality

$$\frac{n+1}{n+2}\frac{\mathrm{d}}{\mathrm{d}t}\int\limits_{\Omega}|v|^{\frac{n+2}{n+1}}\mathrm{d}x + \Big(1-\frac{1}{\lambda_1}\Big)\int\limits_{\Omega}|\widetilde{D}^m v|^2\mathrm{d}x \leqslant 0. \tag{2.9}$$

Consequently, for

$$\lambda_1(\Omega) > 1 \tag{2.10}$$

(2.9) yields good *a priori* estimates of solutions in $\Omega \times (0,T)$ for arbitrarily large $T > 0$. By the standard Galerkin method [21, Chapt. 1], we get the global existence of solutions of the initial-boundary value problem (2.4), (2.1), (2.2). This means that there is no finite-time blow-up for the initial-boundary value problem provided that (2.10) holds, meaning that the size of domain is sufficiently small.

Global existence for $\lambda_1 = 1$

For $\lambda_1 = 1$ the inequality (2.9) also yields an *a priori* uniform bound. However, the proof of global existence becomes more tricky and requires extra scaling (this is not related to our discussion here and we omit details). In this case, we have the conservation law

$$\int\limits_{\Omega}\psi(v(t))e_1\,\mathrm{d}x = c_0 = \int\limits_{\Omega}\psi(v_0)e_1\,\mathrm{d}x \quad \text{for all} \quad t > 0. \tag{2.11}$$

By the gradient system property (see below), the global bounded orbit must stabilize to a unique stationary solution which is characterized as follows (recall that λ_1 is a simple eigenvalue, so the eigenspace is 1D):

$$v(x,t) \to C_0e_1(x) \quad \text{as} \quad t \to +\infty, \quad \text{where} \quad \int\limits_{\Omega}\psi(C_0e_1)e_1\,\mathrm{d}x = c_0. \tag{2.12}$$

Blow-up for $\lambda_1 < 1$

We show that, in the case of the opposite inequality

$$\lambda_1(\Omega) < 1, \tag{2.13}$$

the solutions blow-up in finite time.

Blow-up of nonnegative solutions in the case $m = 1$

We begin with the simple case $m = 1$. By the maximum principle, we can restrict ourselves to the class of nonnegative solutions

$$v = v(x,t) \geqslant 0, \quad \text{i.e., assuming } u_0(x) \geqslant 0. \tag{2.14}$$

In this case, we can directly study the evolution of the first Fourier coefficient of the function $\psi(v(\cdot,t))$. Multiplying (2.4) by the positive eigenfunction e_1 in $L^2(\Omega)$, we obtain

$$\frac{\mathrm{d}}{\mathrm{d}t} \int\limits_{\Omega} \psi(v) e_1 \mathrm{d}x = (1-\lambda_1) \int\limits_{\Omega} v e_1 \mathrm{d}x. \tag{2.15}$$

Taking into account (2.14) and using the Hölder inequality in the right-hand side of (2.15), we derive the following ordinary differential inequality for the Fourier coefficient:

$$\begin{gathered} \frac{\mathrm{d}J}{\mathrm{d}t} \geqslant (1-\lambda_1) c_2 J^{n+1}, \\ \text{where} \quad J(t) = \int\limits_{\Omega} v^{\frac{1}{n+1}}(x,t) e_1(x)\, \mathrm{d}x, \quad c_2 = \Big(\int\limits_{\Omega} e_1 \, \mathrm{d}x \Big)^{-n}. \end{gathered} \tag{2.16}$$

Hence for any nontrivial nonnegative initial data

$$u_0(x) \not\equiv 0 \quad \Longrightarrow \quad J_0 = \int\limits_{\Omega} v_0 e_1 \, \mathrm{d}x > 0$$

we have finite-time blow-up of the solution with the following lower estimate of the Fourier coefficient:

$$\begin{gathered} J(t) \geqslant A(T-t)^{-\frac{1}{n}}, \\ \text{where} \quad A = \Big(\frac{1}{nc_2(1-\lambda_1)} \Big)^{\frac{1}{n}}, \quad T = \frac{J_0^{-n}}{nc_2(1-\lambda_1)}. \end{gathered} \tag{2.17}$$

Unbounded orbits and blow-up in the case $m \geqslant 2$

It is curious that we do not know a similar simple proof of blow-up for the higher order equations with $m \geqslant 2$. The main technical difficulty is that the set of nonnegative solutions (2.14) is not an invariant of the parabolic flow, so we have to deal with solutions $v(x,t)$ of changing sign. Then (2.16) cannot be derived from (2.15) by the Hölder inequality. Nevertheless, we easily obtain the following result as the first step to blow-up of orbits.

Proposition 2.1. *Assume that* $m \geqslant 2$, *the inequality* (2.13) *holds, and*

$$E(v_0) > 0. \tag{2.18}$$

Then the solution of the initial-boundary value problem (2.4), (2.1), (2.2) *is not uniformly bounded for* $t > 0$.

Proof. We use the obvious fact that (2.4) is a gradient system in $H_0^m(\Omega)$. Indeed, multiplying (2.4) by v_t, on sufficiently smooth local solutions we have

$$\frac{1}{2}\frac{\mathrm{d}}{\mathrm{d}t}E(v(t)) = \frac{1}{n+1}\int\limits_{\Omega} |v|^{-\frac{n}{n+1}}(v_t)^2\,\mathrm{d}x \geqslant 0. \tag{2.19}$$

Therefore, under the assumption (2.18), from (2.5) we obtain

$$E(v(t)) \geqslant E(v_0) > 0 \implies \frac{n+1}{n+2}\frac{\mathrm{d}}{\mathrm{d}t}\int\limits_{\Omega} |v|^{\frac{n+2}{n+1}}\,\mathrm{d}x = E(v) \geqslant E(v_0) > 0, \tag{2.20}$$

i.e.,

$$\int\limits_{\Omega} |v(t)|^{\frac{n+2}{n+1}}\,\mathrm{d}x \geqslant \frac{n+2}{n+1}E(v_0)t \to +\infty \quad \text{as} \quad t \to +\infty. \tag{2.21}$$

The proof is complete. □

Concerning the assumption (2.18), we recall that, by the classical theory of dynamical systems [16], the ω-limit set of bounded orbits of gradient systems consists only of equilibria, i.e.,

$$\omega(v_0) \subseteq \mathcal{S} = \{V \in H_0^m(\Omega) : -(-\Delta)^m V + V = 0\}. \tag{2.22}$$

Therefore, stabilization to a nontrivial equilibrium is possible if

$$\lambda_l = 1 \quad \text{for some} \quad l \geqslant 2;$$

otherwise,

$$\mathcal{S} = \{0\} \quad (\lambda_l \neq 1 \quad \text{for any} \quad l \geqslant 1). \tag{2.23}$$

By the gradient structure of (2.4), one should take into account solutions that decay to 0 as $t \to +\infty$. One can check (at least formally, the necessary functional framework could take some time) that the trivial solution 0 has the empty stable manifold. Hence, under the assumption (2.23), the assertion of Proposition 2.1 is naturally expected to be true for any nontrivial solution.

Thus, in the case (2.18), i.e., for a sufficiently large domain Ω, solutions become arbitrarily large in any suitable metric (including the $H_0^m(\Omega)$- or uniform $C_0(\Omega)$-metric). Then it is a technical matter to show that such large solutions must blow-up in finite time.

Blow-up for a similar modified model in the case $m \geqslant 2$

The above arguments can be easily adapted to the slightly modified equation (2.4):

$$(\psi(v))_t = (-1)^{m+1}\Delta^m v + |v|, \tag{2.24}$$

where the source term is replaced by $|v|$. For "positively dominant" solutions (i.e., for solutions with a nonzero integral $\int u(x,t)\,\mathrm{d}x$) the argument is similar. The most of our self-similar patterns exist for (2.24) and the oscillatory properties of solutions near interfaces remain practically the same (since the source term plays no role there).

Let $\Omega = B_1$. Then (2.8) holds. Instead of (2.15), we obtain the following similar inequality:

$$\frac{\mathrm{d}}{\mathrm{d}t}\int_\Omega \psi(v)e_1\mathrm{d}x = \int_\Omega |v|e_1\mathrm{d}x - \lambda_1\int_\Omega ve_1\mathrm{d}x \geqslant (1-\lambda_1)\int_\Omega |v|e_1\mathrm{d}x > 0, \tag{2.25}$$

where $J(t)$ is defined without the positivity sign restriction;

$$J(t) = \int_\Omega (|v|^{-\frac{n}{n+1}}v)(x,t)e_1(x)\,\mathrm{d}x. \tag{2.26}$$

From (2.25) it follows that for $\lambda_1 < 1$

$$J(0) > 0 \quad \Longrightarrow \quad J(t) > 0 \quad \text{for} \quad t > 0. \tag{2.27}$$

By the Hölder inequality,

$$\begin{aligned}\int |v|e_1\,\mathrm{d}x &\geqslant c_2\Big(\int |v|^{\frac{1}{n+1}}e_1\,\mathrm{d}x\Big)^{n+1} \\ &\geqslant c_2\Big(\int |v|^{-\frac{n}{n+1}}ve_1\,\mathrm{d}x\Big)^{n+1} \equiv c_2J^{n+1}.\end{aligned} \tag{2.28}$$

Hence we can obtain the inequality (2.16) for the function (2.26) and the blow-up estimate (2.17) is valid. The same result holds in a general domain Ω in the case of the Navier boundary conditions

$$u = \Delta u = \cdots = \Delta^{m-1}u = 0 \quad \text{on} \quad \partial\Omega \times \mathbb{R}_+.$$

2.2 Blow-up data for higher order parabolic and hyperbolic PDEs

We have seen that blow-up can occur for some initial data since small data can lead to globally existing sufficiently small solutions (if 0 has a nontrivial stable manifold).

Now, we introduce classes of "blow-up data," i.e., initial functions generating finite-time blow-up of solutions. To deal with such crucial data, we need to study the corresponding elliptic systems with non-Lipschitz nonlinearities in detail.

Parabolic equations

We begin with the transformed parabolic equation (2.4) and consider the separate variable solutions

$$v(x,t) = (T-t)^{-\frac{n+1}{n}} F(x). \tag{2.29}$$

Then $F(x)$ solves the elliptic equation (1.7) in Ω, i.e.,

$$\begin{aligned} &(-1)^{m+1} \Delta^m F + F - \frac{1}{n} |F|^{-\frac{n}{n+1}} F = 0 \quad \text{in} \quad \Omega, \\ &F = DF = \ldots = D^{m-1} F = 0 \quad \text{on} \quad \partial\Omega. \end{aligned} \tag{2.30}$$

Let $F(x) \not\equiv 0$ be a solution of the problem (2.30). From (2.29) it follows that the initial data

$$v_0(x) = CF(x), \tag{2.31}$$

where $C \neq 0$ is an arbitrary constant to be scaled out, generate blow-up of the solution of (2.4) according to (2.29).

Hyperbolic equations

For the hyperbolic counterpart of (2.4)

$$(\psi(v))_{tt} = (-1)^{m+1} \Delta^m v + v \tag{2.32}$$

we consider the initial data

$$v(x,0) = cF(x) \quad \text{and} \quad v_t(x,0) = c_1 F(x), \tag{2.33}$$

where c and c_1 are constants such that $cc_1 > 0$. Then the solution blows up in finite time. In particular, taking

$$c > 0 \quad \text{and} \quad c_1 = \frac{2(n+1)}{n} B^{\frac{1}{\beta}} c^{1-\frac{1}{\beta}},$$

with $\beta = -\frac{2(n+1)}{n}$ and $B = \left(\frac{2(n+2)}{n^2}\right)^{\frac{n+1}{n}}$, we obtain the blow-up solution of (2.32) in the separable form

$$v(x,t) = (T-t)^{\beta} B F(x), \quad \text{where} \quad T = \left(\frac{c}{B}\right)^{\frac{1}{\beta}}.$$

2.3 Blow-up rescaled equation as a gradient system: towards the generic blow-up behavior for parabolic PDEs

Let us briefly discuss another important issue associated with the scaling (2.29). We consider a general solution $v(x,t)$ of the initial-boundary value problem for (2.4) which blows up first time at $t = T$. Introducing the rescaled variables

$$v(x,t) = (T-t)^{-\frac{n+1}{n}} w(x,\tau), \quad \tau = -\ln(T-t) \to +\infty \quad \text{as} \quad t \to T^-, \tag{2.34}$$

we see that $w(x,\tau)$ solves the rescaled equation

$$(\psi(w))_\tau = (-1)^{m+1} \Delta^m w + w - \frac{1}{n} |w|^{-\frac{n}{n+1}} w, \tag{2.35}$$

where the right-hand side contains the same operator with non-Lipschitz nonlinearity as in (1.7) or (2.30). By an analogous argument, (2.35) is a gradient system and admits a Lyapunov function that is strictly monotone on non-equilibrium orbits:

$$\begin{aligned} &\frac{\mathrm{d}}{\mathrm{d}\tau}\Big(-\frac{1}{2}\int |\tilde{D}w|^2 + \frac{1}{2}\int w^2 - \frac{n+1}{n(n+2)}\int |w|^{\frac{n+2}{n+1}}\Big) \\ &= \frac{1}{n+1}\int |w|^{-\frac{n}{n+1}} |w_t|^2 > 0. \end{aligned} \tag{2.36}$$

Therefore, an analog of (2.22) holds, i.e., all bounded orbits can approach only stationary solutions:

$$\omega(w_0) \subseteq \mathcal{S} = \Big\{ F \in H_0^m(\Omega) : (-1)^{m+1} \Delta^m F + F - \frac{1}{n} |F|^{-\frac{n}{n+1}} F = 0 \Big\}. \tag{2.37}$$

Moreover, since under natural smoothness parabolic properties, $\omega(w_0)$ is connected and invariant [16], the omega-limit set reduces to a single equilibrium provided that $\mathcal{S}$ consists of isolated points. Here, the structure of the stationary rescaled set $\mathcal{S}$ is very important for understanding the blow-up behavior of general solutions of the higher order parabolic flow (1.1).

Thus, the above analysis again shows that the "stationary" elliptic problems (1.7) and (2.30) are crucial for revealing various local and global evolution properties of all three classes of PDEs involved. We begin this study by applying the classical variational techniques.

3 Existence Problem: Variational Approach and Countable Families of Solutions by Lusternik–Schnirelman Category and Fibering Theory

3.1 Variational setting and compactly supported solutions

Thus, we need to study, in a general multi-dimensional geometry, the existence and multiplicity of compactly supported solutions of the elliptic problem (1.7).

Since all the operators in (1.7) are potential, the problem admits a variational setting in L^2. Hence solutions can be obtained as critical points of the C^1 functional

$$E(F) = -\frac{1}{2}\int |\widetilde{D}^m F|^2 + \frac{1}{2}\int F^2 - \frac{1}{\beta}\int |F|^\beta, \quad \beta = \frac{n+2}{n+1} \in (1,2), \quad (3.1)$$

where $\widetilde{D}^m = \Delta^{\frac{m}{2}}$ for even m and $\widetilde{D}^m = \nabla \Delta^{\frac{m-1}{2}}$ for odd m. In general, we have to look for critical points in $W^2_m(\mathbb{R}^N) \cap L^2(\mathbb{R}^N) \cap L^\beta(\mathbb{R}^N)$. Bearing in mind compactly supported solutions, we choose a sufficiently large radius $R > 0$ of the ball B_R and consider the variational problem for (3.1) in $W^2_{m,0}(B_R)$ with the Dirichlet boundary conditions on $S_R = \partial B_R$. Then both spaces $L^2(B_R)$ and $L^{p+1}(B_R)$ are compactly embedded into $W^2_{m,0}(B_R)$ in the subcritical Sobolev range

$$1 < p < p_S = \frac{N+2m}{N-2m} \quad (\beta < p_S). \quad (3.2)$$

In general, we have to use the following preliminary result.

Proposition 3.1. *Let F be a continuous weak solution of Eq.* (1.7) *such that*

$$F(y) \to 0 \quad \text{as} \quad |y| \to \infty. \quad (3.3)$$

Then F is compactly supported in $\mathbb{R}^N$.

Note that the continuity of F is guaranteed by the Sobolev embedding $H^m(\mathbb{R}^N) \subset C(\mathbb{R}^N)$ for $N < 2m$ and the local elliptic regularity theory for the whole range (3.2) (necessary information about embeddings of function spaces can be found in [22, Chapt. 1]).

Proof. We consider the parabolic equation with the same elliptic operator

$$w_t = (-1)^{m+1} \Delta^m w + w - |w|^{-\frac{n}{n+1}} w \quad \text{in} \quad \mathbb{R}^N \times \mathbb{R}_+ \tag{3.4}$$

and the initial data $F(y)$. Setting $w = \mathrm{e}^t \widehat{w}$, we obtain the equation

$$\widehat{w}_t = (-1)^{m+1} \Delta^m \widehat{w} - \mathrm{e}^{-\frac{n}{n+1} t} |\widehat{w}|^{p-1} \widehat{w}, \quad p = \frac{1}{n+1} \in (0,1),$$

where the operator is monotone in $L^2(\mathbb{R}^N)$. Therefore, the Cauchy problem with initial data F has a unique weak solution [21, Chapt. 2]. Thus, (3.4) has a unique solution $w(y,t) \equiv F(y)$ which must be compactly supported for arbitrarily small $t > 0$. Such nonstationary instant compactification phenomena for quasilinear absorption-diffusion equations with singular absorption $-|u|^{p-1}u$, $p < 1$, were known since the 1970s. These phenomena, called the *instantaneous shrinking* of the support of solutions, were proved for quasilinear higher order parabolic equations with non-Lipschitz absorption terms [34]. □

Thus, to provide compactly supported patterns $F(y)$, we can consider the problem in sufficiently large bounded balls since, by (3.1), nontrivial solutions are impossible in small domains.

3.2 The Lusternik–Schnirelman theory and direct application of fibering method

Since the functional (3.1) is of class C^1, uniformly differentiable, and weakly continuous, we can use the classical Lusternik–Schnirelman (LS) theory of calculus of variations [20, Sect. 57] in the form of the fibering method [25, 26], which can be regarded as a convenient generalization of the previous versions [7, 27] of variational approaches.

Following the LS theory and fibering approach [26], the number of critical points of the functional (3.1) depends on the *category* (or *genus*) of the functional subset where the fibering takes place. Critical points of $E(F)$ are obtained by *spherical fibering*

$$F = r(v) v \quad (r \geqslant 0), \tag{3.5}$$

where $r(v)$ is a scalar functional and v belongs to a subset of $W^2_{m,0}(B_R)$ defined by the formula

$$\mathcal{H}_0 = \Big\{ v \in W^2_{m,0}(B_R) : H_0(v) \equiv -\int |\widetilde{D}^m v|^2 + \int v^2 = 1 \Big\}. \tag{3.6}$$

The new functional

$$H(r,v) = \frac{1}{2}\, r^2 - \frac{1}{\beta}\, r^\beta \int |v|^\beta \tag{3.7}$$

has an absolute minimum point, where

$$H'_r \equiv r - r^{\beta-1} \int |v|^\beta = 0 \quad \Longrightarrow \quad r_0(v) = \Big(\int |v|^\beta \Big)^{\frac{1}{2-\beta}}. \tag{3.8}$$

Then we obtain the functional

$$\widetilde{H}(v) = H(r_0(v), v) = -\frac{2-\beta}{2\beta}\, r_0^2(v) \equiv -\frac{2-\beta}{2\beta} \Big(\int |v|^\beta \Big)^{\frac{2}{2-\beta}}. \tag{3.9}$$

The critical points of the functional (3.9) on the set (3.6) coincide with the critical point of the functional

$$\widetilde{H}(v) = \int |v|^\beta. \tag{3.10}$$

Hence we obtain an even, nonnegative, convex, and uniformly differentiable functional, to which the LS theory can be applied [20, Sect. 57] (see also [9, p. 353]). Following [26], to find critical points of $\widetilde{H}$ on the set $\mathcal{H}_0$, it is necessary to estimate the category ρ of the set $\mathcal{H}_0$, see the notation and basic results in [3, p. 378]. Note that the Morse index q of the quadratic form Q in Theorem 6.7.9 therein is precisely the dimension of the space where the corresponding form is negative definite. This includes all the multiplicities of eigenfunctions involved in the corresponding subspace (see definitions of genus and cogenus, as well as applications to variational problems in [1] and [2]).

By the above variational construction, F is an eigenfunction,

$$(-1)^{m+1} \Delta^m F + F - \mu |F|^{-\frac{n}{n+1}} F = 0,$$

where $\mu > 0$ is the Lagrange multiplier. Then, scaling $F \mapsto \mu^{(n+1)/n} F$, we obtain the original equation in (1.7).

For further discussion of geometric shapes of patterns we recall that, by Berger's version [3, p. 368] of this minimax analysis of the LS category theory [20, p. 387], the critical values $\{c_k\}$ and the corresponding critical points $\{v_k\}$ are determined by the formula

$$c_k = \inf_{\mathcal{F} \in \mathcal{M}_k} \sup_{v \in \mathcal{F}} \widetilde{H}(v). \tag{3.11}$$

Here, $\mathcal{F} \subset \mathcal{H}_0$ is a closed set, $\mathcal{M}_k$ is the set of all subsets of the form $BS^{k-1} \subset \mathcal{H}_0$, where S^{k-1} is a sufficiently smooth $(k-1)$-dimensional manifold (say, a sphere) in $\mathcal{H}_0$, and B is an odd continuous map. Then each member of $\mathcal{M}_k$ is of genus at least k (available in $\mathcal{H}_0$). It is also important to note that the definition of genus [20, p. 385] assumes that $\rho(\mathcal{F}) = 1$ if no *component* of $\mathcal{F} \cup \mathcal{F}^*$, where $\mathcal{F}^* = \{v : v^* = -v \in \mathcal{F}\}$ is the *reflection* of $\mathcal{F}$ relative to 0, contains a pair of antipodal points v and $v^* = -v$. Furthermore, $\rho(\mathcal{F}) = n$ if each compact subset of $\mathcal{F}$ can be covered by at least n sets of genus one. According to (3.11),

$$c_1 \leqslant c_2 \leqslant \ldots \leqslant c_{l_0},$$

where $l_0 = l_0(R)$ is the category of $\mathcal{H}_0$ (see an estimate below) such that

$$l_0(R) \to +\infty \quad \text{as} \quad R \to \infty. \tag{3.12}$$

Roughly speaking, since the dimension of $\mathcal{F}$ involved in the construction of $\mathcal{M}_k$ increases with k, the critical points delivering critical values (3.11) are all different. By (3.6), the category $l_0 = \rho(\mathcal{H}_0)$ of the set $\mathcal{H}_0$ is equal to the number (with multiplicities) of the eigenvalues $\lambda_k < 1$,

$$l_0 = \rho(\mathcal{H}_0) = \sharp\{\lambda_k < 1\} \tag{3.13}$$

of the linear polyharmonic operator $(-1)^m \Delta^m > 0$,

$$(-1)^m \Delta^m \psi_k = \lambda_k \psi_k, \quad \psi_k \in W^2_{m,0}(B_R) \tag{3.14}$$

(see [3, p. 368]). Since the dependence of the spectrum on R is expressed as

$$\lambda_k(R) = R^{-2m} \lambda_k(1), \quad k = 0, 1, 2, \ldots, \tag{3.15}$$

the category $\rho(\mathcal{H}_0)$ can be arbitrarily large for $R \gg 1$, and (3.12) holds. We formulate this as the following assertion.

Proposition 3.2. *The elliptic problem* (1.7) *has at least a countable set* $\{F_l,\, l \geqslant 0\}$ *of different solutions obtained as critical points of the functional* (3.1) *in* $W^2_{m,0}(B_R)$ *with sufficiently large* $R = R(l) > 0$.

Indeed, in view of Proposition 3.1, we choose $R \gg 1$ such that $\operatorname{supp} F_l \subset B_R$.

3.3 On a model with an explicit description of the Lusternik–Schnirelman sequence

As we will see below, detecting the LS sequence of critical values for the original functional (3.1) is a hard problem. To this end, the numerical estimates of the functional play a key role. However, for some similar models this can be done much easier. Now, we slightly modify (3.1) and consider the functional

$$E_1(F) = -\frac{1}{2}\int |\widetilde{D}^m F|^2 + \frac{1}{2}\int F^2 - \frac{1}{\beta}\Big(\int F^2\Big)^{\frac{\beta}{2}} \Big(\beta = \frac{n+2}{n+1} \in (1,2)\Big) \tag{3.16}$$

corresponding to the nonlocal elliptic problem

$$-(-\Delta)^m F + F - F\Big(\int F^2\Big)^{\frac{\beta}{2}-1} = 0 \quad (\text{in} \quad B_R, \text{ etc.}) \tag{3.17}$$

Denoting by $\{\lambda_k\}$ the spectrum in (3.14) and by $\{\psi_k\}$ the corresponding eigenfunction set, we can solve the problem (3.17) explicitly, looking for solutions

$$F = \sum_{(k\geqslant 1)} c_k \psi_k \implies c_k\Big[-\lambda_k + 1 - \Big(\sum_{(j\geqslant 1)} c_j^2\Big)^{\frac{\beta}{2}-1}\Big] = 0, \;\; k = 1, 2, \ldots . \tag{3.18}$$

The algebraic system in (3.18) is easy and yields precisely the number (3.13) of various nontrivial basic solutions F_l of the form

$$F_l(y) = c_l \psi_l(y), \quad \text{where} \quad |c_l|^{\beta-2} = -\lambda_l + 1 > 0, \quad l = 1, 2, \ldots, l_0. \tag{3.19}$$

3.4 Preliminary analysis of geometric shapes of patterns

The forthcoming discussions and conclusions should be understood in conjunction with the results obtained in Sect. 5 by numerical and other analytic and formal methods. In particular, we use here the concepts of the index and Sturm classification of patterns.

Thus, we now discuss key questions of the spatial structure of patterns constructed by the LS method. Namely, we would like to know how the genus k of subsets involved in the minimax procedure (3.11) can be attributed to the "geometry" of the critical point $v_k(y)$ obtained in such a manner. In this discussion, we assume to explain how to merge the LS genus variational aspects with the actual practical structure of "essential zeros and extrema" of basic patterns $\{F_l\}$. Recall that, in the second order case $m = 1$, $N = 1$, this is easy: by the Sturm theorem, the genus l, which can be formally "attributed" to the function F_l, is equal to the number of zeros (sign changes) $l - 1$ or the number l of isolated local extremum points. Though, even for $m = 1$, this is not that univalent: there are other structures that do not obey the Sturmian order (think about the solution via gluing $\{F_0, F_0, \ldots, F_0\}$ without sign changes); see more comments below.

For $m \geqslant 2$ this question is more difficult, and seems does not admit a clear rigorous treatment. Nevertheless, we will try to clarify some aspects.

Given a solution F of (1.7) (a critical point of (3.1)), let us calculate the corresponding critical value c_F of (3.10) on the set (3.6) by taking

$$\begin{aligned} & v = CF \in \mathcal{H}_0 \quad \Longrightarrow \quad C = \frac{1}{\sqrt{-\int |\widetilde{D}^m F|^2 + \int F^2}} \\ & \text{so that} \quad c_F \equiv \widetilde{H}(v) = \frac{\int |F|^\beta}{(-\int |\widetilde{D}^m F|^2 + \int F^2)^{\beta/2}} \quad \Big(\beta = \frac{n+2}{n+1}\Big). \end{aligned} \tag{3.20}$$

This formula is important in what follows.

Genus one

As usual in many variational elliptic problems, the first pattern F_0 (typically, called a *ground state*) is always of the simplest geometric shape, is radially symmetric, and is a localized profile such as those in Fig. 8. Indeed, this simple shape with a single dominant maximum is associated with the variational formulation for F_0:

$$F_0 = r(v_0)v_0, \quad \text{with } v_0: \quad \inf \widetilde{H}(v) \equiv \inf \int |v|^\beta, \quad v \in \mathcal{H}_0. \tag{3.21}$$

This is (3.11) with the simplest choice of closed sets as points, $\mathcal{F} = \{v\}$.

Let us illustrate why a localized pattern like F_0 delivers the minimum to $\widetilde{H}$ in (3.21). Take, for example, a two-hump structure

$$\widehat{v}(y) = C[v_0(y) + v_0(y+a)], \quad C \in \mathbb{R},$$

with sufficiently large $|a| \geqslant \operatorname{diam} \operatorname{supp} F_0$, so that supports of these two functions do not overlap. Then, evidently, $\widehat{v} \in \mathcal{H}_0$ implies that $C = \frac{1}{\sqrt{2}}$ and, since $\beta \in (1,2)$,

$$\widetilde{H}(\widehat{v}) = 2^{\frac{2-\beta}{2}} \widetilde{H}(v_0) > \widetilde{H}(v_0).$$

By a similar reason, $F_0(y)$ and $v_0(y)$ cannot have "strong nonlinear oscillations" (see next sections for related concepts developed in this direction), i.e., the positive part $(F_0)_+$ must be dominant, so that the negative part $(F_0)_-$ cannot be considered as a separate dominant 1-hump structure. Otherwise, deleting it will diminish $\widetilde{H}(v)$ as above. In other words, essentially nonmonotone patterns such as in Fig. 10 or 11 cannot correspond to the variational problem (3.21), i.e., the genus of the functional sets involved is $\rho = 1$.

Radial symmetry of v_0 is pretty standard in elliptic theory, though not straightforward in view of the lack of the maximum principle and moving planes/spheres tools. We just note that small nonradial deformations of this structure, $v_0 \mapsto \widehat{v}_0$ will more essentially affect (increase) the first differential term $\int |\widetilde{D}^m \widehat{v}_0|^2$ rather than the second one in the formula for C in (3.20).

Therefore, a standard scaling to keep this function in $\mathcal{H}_0$ would mean taking $C\widehat{v}_0$ with a constant $C > 1$. Hence

$$\widetilde{H}(C\widehat{v}_0) = C^\beta \widetilde{H}(\widehat{v}_0) \approx C^\beta \widetilde{H}(v_0) > \widetilde{H}(v_0),$$

so infinitesimal nonradial perturbations do not provide us with critical points of (3.21).

For $N = 1$ this shows that c_1 cannot be attained at another "positively dominant" pattern F_{+4}, with a shape shown in Fig. 17(a). See Tabl. 1 below, where for $n = 1$

$$c_{F_{+4}} = 1.9488\ldots > c_2 = c_{F_1} = 1.8855\ldots > c_1 = c_{F_0} = 1.6203\ldots.$$

Genus two

Let $N = 1$ for the sake of simplicity, and let F_0 obtained above for the genus $\rho = 1$ be a simple compactly supported pattern as in Fig. 8. We denote by $v_0(y)$ the corresponding critical point given by (3.21). We take the function corresponding to the difference (5.7),

$$\widehat{v}_2(y) = \frac{1}{\sqrt{2}} \left[-v_0(y - y_0) + v(y + y_0)\right] \in \mathcal{H}_0 \quad (\operatorname{supp} v_0 = [-y_0, y_0]) \tag{3.22}$$

which approximates the basic profile F_1 given in Fig. 10. One can see that

$$\widetilde{H}(\widehat{v}_2) = 2^{\frac{2-\beta}{2}} \widetilde{H}(v_0) = 2^{\frac{2-\beta}{2}} c_1, \tag{3.23}$$

so that, by (3.11) with $k = 2$

$$c_1 < c_2 \leqslant 2^{\frac{2-\beta}{2}} c_1. \tag{3.24}$$

On the other hand, the sum as in (5.6) (see Fig. 11)

$$\widetilde{v}_2(y) = \frac{1}{\sqrt{2}} \left[v_0(y - y_0) + v(y + y_0)\right] \in \mathcal{H}_0 \tag{3.25}$$

also delivers the same value (3.23) to the functional $\widetilde{H}$. It is easy to see that the patterns F_1 and $F_{+2,2,+2}$, as well as F_{+4} and many others with two dominant extrema, can be embedded into a 1D subset of genus two on $\mathcal{H}_0$.

It seems that, with such a huge, at least, countable variety of similar patterns, we first distinguish a profile that delivers the critical value c_2 given by (3.11) by comparing the values (3.20) for each pattern. The results are presented in Tabl. 1 for $n = 1$, for which the critical values (3.20) are

$$c_F = \widetilde{H}(CF) = \frac{\int |F|^{3/2}}{\left(-\int |\widetilde{D}^m F|^2 + \int F^2\right)^{3/4}} \quad \left(\beta = \frac{3}{2}\right). \tag{3.26}$$

The corresponding profiles are shown in Fig. 1. Calculations have been performed with the enhanced values Tols$= \varepsilon = 10^{-4}$. Comparing the critical values in Tabl. 1, we thus arrive at the following conclusion based on this analytical–numerical evidence:

$$\begin{aligned}&\text{for genus } k = 2, \text{ the LS critical}\\ &\text{value } c_2 = 1.855\ldots \text{ is delivered by } F_1.\end{aligned} \tag{3.27}$$

Note that the critical values c_F for F_1 and $F_{+2,2,+2}$ are close by just two percents.

Table 1 Critical values of $\widetilde{H}(v)$; genus two

F	c_F
F_0	$1.6203\ldots = c_1$
F_1	$1.8855\ldots = c_2$
$F_{+2,2,+2}$	$1.9255\ldots$
$F_{-2,3,+2}$	$1.9268\ldots$
$F_{+2,4,+2}$	$1.9269\ldots$
$F_{+2,\infty,+2}$	$1.9269\ldots$
F_{+4}	$1.9488\ldots$

Thus, according to Tabl. 1, the second critical value c_2 is achieved at the 1-dipole solution $F_1(y)$ having the transversal zero at $y = 0$, i.e., without any part of the oscillatory tail for $y \approx 0$. Therefore, the neighboring profile $F_{-2,3,+2}$ (see the dotted line in Fig. 1) which has a small remnant of the oscillatory tail with just 3 extra zeros, delivers another, worse value (see Sect. 4)

$$c_F = 1.9268\ldots \quad \text{for} \quad F = F_{-2,3,+2}.$$

In addition, the lines from second to fifth in Tabl. 1 clearly show how $\widetilde{H}$ increases with the number of zeros in between the $\pm F_0$-structures involved.

Remark 3.1. Even for $m = 1$ profiles are not variationally recognizable.

Recall that for $m = 1$, i.e., for the ODE (5.5), the $F_0(y)$ profile is not oscillatory at the interface, so that the future rule (3.31) fails. This does not explain the difference between $F_1(y)$ and, say, $F_{+2,0,+2}$, which, obviously deliver the same critical LS values by (3.11). This is the case where we should conventionally attribute the LS critical point to F_1. Of course, for $m = 1$ the existence of profiles $F_l(y)$ with precisely l zeros (sign changes) and $l + 1$ extrema follows from the Sturm theorem.

Checking the accuracy of numerics and using (3.23), we take the critical values in the first and fifth lines in Tabl. 1 to get for the profile $F_{+2,\infty,+2}$, consisting of two independent F_0's, to within 10^{-4},

$$c_F = 2^{\frac{2-\beta}{2}} \widetilde{H}(v_0) = 2^{\frac{1}{4}} c_1 = 1.1892\ldots \times 1.6203\ldots = 1.9269\ldots .$$

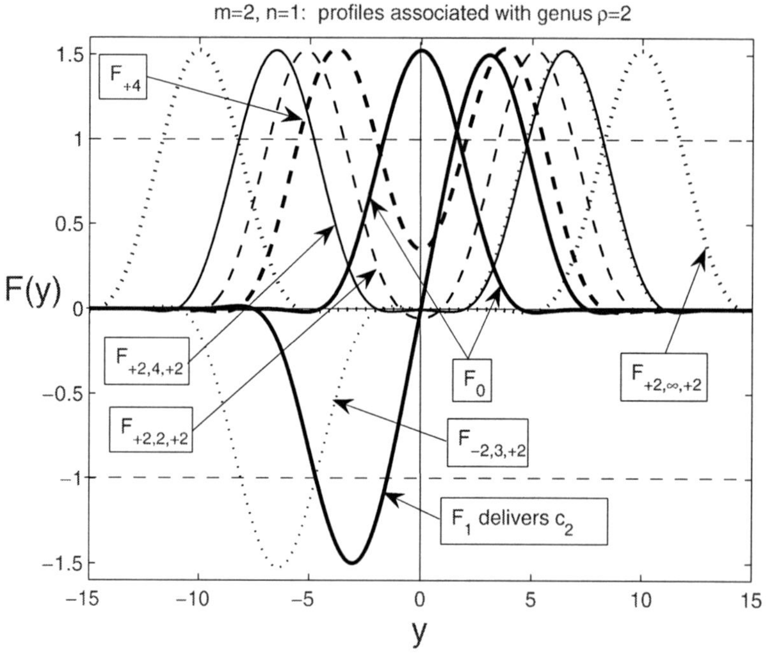

Fig. 1 Seven patterns $F(y)$ indicated in Tabl. 1; $m = 2$ and $n = 1$.

Genus three

Similarly, for $k = 3$ (genus $\rho = 3$), there are also several patterns that can pretend to deliver the LS critical value c_3 (see Fig. 2).

The corresponding critical values (3.26) for $n = 1$ are shown in Tabl. 2, which allows us to conclude:

$$\begin{aligned} k = 3: \ & \text{the LS critical value } c_3 = 2.0710\ldots \\ & \text{is again delivered by the basic } F_2. \end{aligned} \tag{3.28}$$

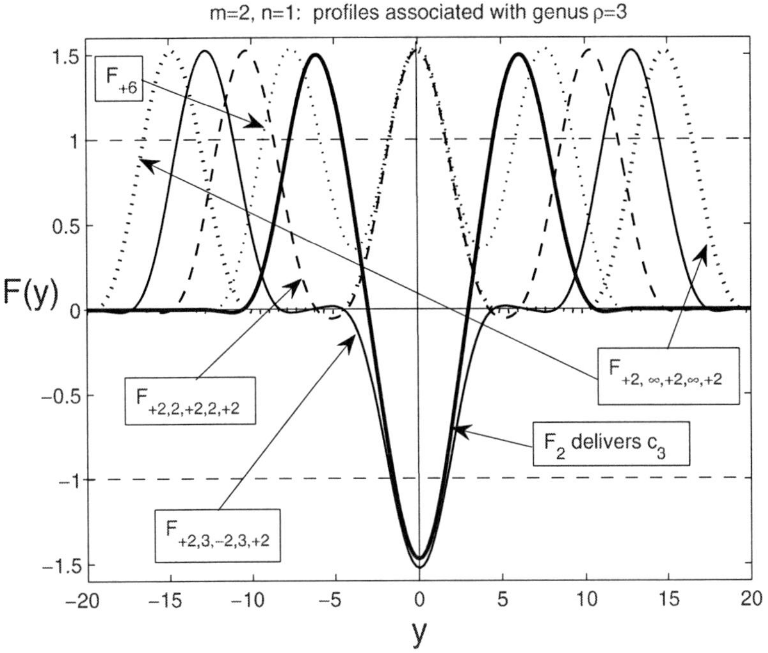

Fig. 2 Five patterns $F(y)$ indicated in Tabl. 2; $m = 2$ and $n = 1$.

All the critical values in Tabl. 2 are very close to each other. Again, checking the accuracy of numerics and taking the critical values c_1 in Tabl. 1 and c_F for $F_{+2,\infty,+2,\infty,+2}$ in Tabl. 2, consisting of three independent F_0's, we obtain, to within 10^{-4},

$$c_F = 3^{\frac{2-\beta}{2}} \widetilde{H}(v_0) = 3^{\frac{1}{4}} c_1 = 1.31607\ldots \times 1.6203\ldots = 2.1324\ldots.$$

Table 2 Critical values of $\widetilde{H}(v)$; genus three

F	c_F
F_2	$2.0710\ldots = c_3$
$F_{+2,2,+2,2,+2}$	$2.1305\ldots$
$F_{+2,3,-2,3,+2}$	$2.1322\ldots$
$F_{+2,\infty,+2,\infty,+2}$	$2.1324\ldots$
F_{+6}	$2.1647\ldots$

Note that the LS category-genus construction (3.11) itself guarantees that all solutions $\{v_k\}$ as critical points will be (geometrically) distinct (see [20, p. 381]). Here we stress upon two important conclusions:

(I) First, what is key for us, is that closed subsets in $\mathcal{H}_0$ of functions of the sum type in (3.25) *do not deliver LS critical values in* (3.11);

(II) On the other hand, patterns of the $\{F_0, F_0\}$-interaction, i.e., those associated with the sum structure as in (3.25) *do exist* (see Fig. 11 for $m = 2$), and

(III) Hence these patterns (different from the basic ones $\{F_l\}$) as well as many others *are not obtainable by a direct LS approach.* Therefore, we will need another version of the LS theory and fibering approach, with different type of decomposition of function spaces to be introduced below.

Genus k

Similarly taking a proper sum of shifted and reflected functions $\pm v_0(y \pm l y_0)$, we obtain from (3.11) that

$$c_{k-1} < c_k \leqslant k^{\frac{2-\beta}{2}} c_1. \tag{3.29}$$

Conclusions: conjecture and open problem

In spite of the closeness of the critical values c_F, the above numerics confirm that there is a geometrical–algebraic way to distinguish the LS patterns delivering (3.11). It can be seen from (3.26) that if we destroy the internal oscillatory "tail" or even any two-three zeros between two F_0-like patterns in the complicated pattern $F(y)$ then

$$\text{two main terms } -\int |\widetilde{D}^m F|^2 \text{ and } \int |F|^{\frac{3}{2}} \text{ in } c_F \text{ in (3.20) decrease.} \tag{3.30}$$

Recall that precisely these terms in the ODE

$$F^{(4)} = -|F|^{-\frac{1}{2}} F + \dots \quad \text{(see (5.2) for } n = 1)$$

are responsible for the formation of a tail, as shown in Sect. 4, while the F-term, giving $\int F^2$, is negligible in the tail. Decreasing both terms, i.e., eliminating the tail between F_0's, we decrease c_F since, in (3.20), the numerator becomes less and the denominator becomes larger. Therefore, composing a complicated pattern $F_l(y)$ from several elementary profiles like $F_0(y)$ and using $(k-1)$-dimensional manifolds of genus k, we follow

Formal rule for composing patterns. *By maximazing $\widetilde{H}(v)$ of any $(k-1)$-dimensional manifold $\mathcal{F} \in \mathcal{M}_k$,*

$$\begin{aligned}&\textit{the LS point } F_{k-1}(y) \textit{ is obtained by}\\&\textit{minimizing all the internal tails and zeros,}\end{aligned} \tag{3.31}$$

i.e., making the minimal number of the internal transversal zeros between single structures.

Regardless such a simple variational-oscillatory meaning (3.30) of this formal rule, we do not know how to make it more rigorous.

Concerning the actual critical LS points, we formulate a conjecture which well corresponds to (3.31):

Conjecture 1. *For $N = 1$ and any $m \geqslant 2$ the critical* LS *value* (3.11), *$k \geqslant 1$, is delivered by the basic pattern F_{k-1} obtained by the minimization over the corresponding $(k-1)$-dimensional manifold $\mathcal{F} \in \mathcal{M}_k$, which is the interaction*

$$F_{k-1} = (-1)^{k-1}\{+F_0, -F_0, +F_0, \dots, (-1)^{k-1}F_0\}, \tag{3.32}$$

where each neighboring pair $\{F_0, -F_0\}$ or $\{-F_0, F_0\}$ has a single transversal zero between the structures.

We also formulate an assertion (as an open problem) which is associated with the specific structure of the LS construction (3.11) over suitable subsets $\mathcal{F}$ as smooth $(k-1)$-dimensional manifolds of genus k:

Open problem. *For $N = 1$ and $m \geqslant 2$ there are no purely geometrical–topological arguments confirming the validity of Conjecture* 1. *The same is true in $\mathbb{R}^N$.*

In other words, the metric analysis of "tails' for the functionals involved in (3.31) cannot be provided by any geometrical type arguments. On the other hand, the geometrical analysis is nicely applied to the case $m = 1$, and this is perfectly covered by the Sturm theorem about zeros for second order ODEs. If the assertion of the Sturm theorem does not hold for the variational problem under consideration, this means that the problem is not of geometrical-topological (or purely homotopical if the tails are oscillatory) nature.

4 Oscillation Problem: Local Oscillatory Structure of Solutions Close to Interfaces and Periodic Connections with Singularities

As we have seen, the first principal feature of the ODEs (1.8) (and the elliptic counterparts) is that these admit compactly supported solutions. This was proved in Proposition 3.1 in a general elliptic setting.

Therefore, we will study the typical local behavior of solutions of the ODE (1.8) close to singular points, i.e., to finite interfaces. We will reveal the extremely oscillatory structure with that well corresponds to the global oscillatory behavior of solutions obtained above by variational techniques.

The phenomenon of the oscillatory changing sign behavior of solutions of the Cauchy problem was detected for various classes of evolution PDEs (see a general view in [15, Chapts. 3–5] and various results for different PDEs in [10, 11, 12]). For the present $2m$th order equations the oscillatory behavior exhibits special features to be revealed.

4.1 Autonomous ODEs for oscillatory components

Assume that a finite interface of $F(y)$ is situated at the origin $y = 0$. Then we can use the trivial extension $F(y) \equiv 0$ to $y < 0$. Since we are interested in describing the behavior of solutions as $y \to 0^+$, we consider the ODE (1.8) written in the form

$$\begin{aligned} &F^{(2m)} = (-1)^{m+1}|F|^{-\alpha}F + (-1)^m F \quad \text{for} \quad y > 0, \\ &F(0) = 0 \quad \Big(\alpha = \frac{n}{n+1} \in (0,1)\Big). \end{aligned} \tag{4.1}$$

In view of the scaling structure of the first two terms, we make extra rescaling and introduce the *oscillatory component* $\varphi(s)$ of F by the formula

$$F(y) = y^\gamma \varphi(s), \quad \text{where} \quad s = \ln y \quad \text{and} \quad \gamma = \frac{2m}{\alpha} \equiv \frac{2m(n+1)}{n}. \tag{4.2}$$

Since $s \to -\infty$ (the new interface position) as $y \to 0^-$, the monotone function y^γ in (4.2) plays the role of the *envelope* to the oscillatory function $F(y)$. Substituting (4.2) into (4.1), we obtain the following equation for φ:

$$P_{2m}(\varphi) = (-1)^{m+1}|\varphi|^{-\alpha}\varphi + (-1)^m \mathrm{e}^{2ms}\varphi. \tag{4.3}$$

Here, P_k, $k \geqslant 0$, are linear differential operators defined by the recursion

$$P_{k+1}(\varphi) = (P_k(\varphi))' + (\gamma - k)P_k(\varphi) \quad \text{for} \quad k = 0, 1, \dots, \quad P_0(\varphi) = \varphi. \tag{4.4}$$

We write out the first four operators, which is sufficient for our purposes:

$$\begin{aligned} &P_1(\varphi) = \varphi' + \gamma\varphi; \\ &P_2(\varphi) = \varphi'' + (2\gamma - 1)\varphi' + \gamma(\gamma - 1)\varphi; \\ &P_3(\varphi) = \varphi''' + 3(\gamma - 1)\varphi'' + (3\gamma^2 - 6\gamma + 2)\varphi' + \gamma(\gamma - 1)(\gamma - 2)\varphi; \end{aligned}$$

$$P_4(\varphi) = \varphi^{(4)} + 2(2\gamma - 3)\varphi''' + (6\gamma^2 - 18\gamma + 11)\varphi'' \\ + 2(2\gamma^3 - 9\gamma^2 + 11\gamma - 3)\varphi' + \gamma(\gamma - 1)(\gamma - 2)(\gamma - 3)\varphi.$$

According to (4.2), the interface at $y = 0$ corresponds to $s = -\infty$, so that (4.3) is an exponentially (as $s \to -\infty$) perturbed autonomous ODE

$$P_{2m}(\varphi) = (-1)^{m+1}|\varphi|^{-\alpha}\varphi \quad \text{in} \quad \mathbb{R} \quad \Big(\alpha = \frac{n}{n+1}\Big). \tag{4.5}$$

By the classical ODE theory [8], one can expect that for $s \ll -1$ the typical (generic) solutions of (4.3) and (4.5) asymptotically differ by exponentially small factors. Of course, we must admit that (4.5) is a singular ODE with a non-Lipschitz term, so the results on continuous dependence need extra justification in general.

Thus, in two principal cases, the ODEs for the oscillatory component $\varphi(s)$ are as follows:

$$m = 2: \quad P_4(\varphi) = -|\varphi|^{-\alpha}\varphi, \tag{4.6}$$

$$m = 3: \quad P_6(\varphi) = +|\varphi|^{-\alpha}\varphi, \tag{4.7}$$

which show rather different properties because comprise even and odd m. For instance, (4.5) for any odd $m \geqslant 1$ (including (4.7)) has two constant equilibria since

$$\gamma(\gamma-1)\dots(\gamma-(2m-1))\varphi = |\varphi|^{-\alpha}\varphi \quad \Longrightarrow \\ \varphi(s) = \pm\varphi_0 \equiv \pm[\gamma(\gamma-1)\dots(\gamma-(2m-1))]^{-\frac{1}{\alpha}} \quad \text{for all} \quad n > 0. \tag{4.8}$$

For even m including (4.6) such equilibria for (4.5) do not exist, at least for $n \in (0, 1]$. We show how this fact affects the oscillatory properties of solutions for odd and even m.

4.2 Periodic oscillatory components

Now, we look for *periodic* solutions of (4.5) which are the simplest nontrivial bounded solutions admitting the continuation up to the interface at $s = -\infty$. Periodic solutions, together with their stable manifolds, are simple *connections* with the interface, as a singular point of the ODE (1.8).

Note that (4.5) does not admit variational setting, so we cannot apply the well-developed potential theory (see [23, Chapt. 8] for existence–nonexistence results and further references therein) or the degree theory [19, 20]. For $m = 2$ the existence of φ_* can be proved by shooting (see [10, Sect. 7.1]), which can be extended to the case $m = 3$ as well. Nevertheless, the uniqueness of a periodic orbit is still an open question. Therefore we formulate the following assertion supported by various numerical and analytical results [15, Sect. 3.7] as a conjecture.

Conjecture 2. *For any* $m \geqslant 2$ *and* $\alpha \in (0,1]$ *the* ODE (4.5) *admits a unique nontrivial periodic solution* $\varphi_*(s)$ *of changing sign.*

4.3 Numerical construction of periodic orbits; $m = 2$

Numerical results clearly suggest that (4.6) possesses a unique periodic solution $\varphi_*(s)$ and this solution is stable in the direction opposite to the interface, i.e., as $s \to +\infty$ (see Fig. 3). The exponential stability and hyperbolicity of φ_* is proved by estimating the eigenvalues of the linearized operator. This agrees with the obviously correct similar result for $n = 0$; namely, for the linear equation (4.1) with $\alpha = 0$

$$F^{(4)} = -F \quad \text{as} \quad y \to -\infty. \tag{4.9}$$

Here, the interface is infinite, so its position corresponds to $y = -\infty$. Indeed, setting $F(y) = \mathrm{e}^{\mu y}$, we obtain the characteristic equation and a unique exponentially decaying pattern

$$\mu^4 = -1 \quad \Longrightarrow \quad F(y) \sim \mathrm{e}^{y/\sqrt{2}} \Big[A \cos\Big(\frac{y}{\sqrt{2}}\Big) + B \sin\Big(\frac{y}{\sqrt{2}}\Big)\Big]. \tag{4.10}$$

The continuous dependence on $n \geqslant 0$ of typical solutions of (4.5) with "transversal" zeros only will be the key of our analysis. In fact, this property means the existence of a "homotopic" connection between the nonlinear equation and the linear ($n = 0$) one. The passage to the limit as $n \to 0$ in similar degenerate ODEs from the thin film equation theory is discussed in [10, Sect. 7.6].

The oscillation amplitude becomes very small for $n \approx 0$, so we perform extra scaling.

Limit $n \to 0$

This scaling is

$$\varphi(s) = \Big(\frac{n}{4}\Big)^{\frac{4}{n}} \varPhi(\eta), \quad \text{where} \quad \eta = \frac{4s}{n}, \tag{4.11}$$

where $\varPhi$ solves a simpler *limit* binomial ODE

$$\mathrm{e}^{-\eta}(\mathrm{e}^{\eta}\varPhi)^{(4)} \equiv \varPhi^{(4)} + 4\varPhi''' + 6\varPhi'' + 4\varPhi' + \varPhi = -|\varPhi|^{-\frac{n}{n+1}}\varPhi. \tag{4.12}$$

The stable oscillatory patterns of (4.12) are shown in Fig. 4. For such small n in Fig. 4(a) and (b), by scaling (4.11), the periodic components φ_* become really small, for example,

$$\max|\varphi_*(s)| \sim 3\cdot 10^{-4}\left(\tfrac{n}{4}\right)^{\frac{4}{n}} \sim 3\cdot 10^{-30} \quad \text{for} \quad n=0.2 \text{ in (a)},$$
$$\text{and} \quad \max|\varphi_*(s)| \sim 10^{-93} \quad \text{for} \quad n=0.08 \text{ in (b)}.$$

Limit $n \to \infty$

Then $\alpha \to 1$, so the original ODE (4.6) approaches the following equation with a discontinuous sign-nonlinearity:

$$\varphi_\infty^{(4)} + 10\varphi_\infty''' + 35\varphi_\infty'' + 50\varphi_\infty' + 24\varphi_\infty = -\operatorname{sign}\varphi_\infty. \tag{4.13}$$

This also admits a stable periodic solution, as shown in Fig. 5.

4.4 Numerical construction of periodic orbits; $m = 3$

We consider Eq. (4.7) which admits constant equilibria (4.8) existing for all $n > 0$. It is easy to check that the equilibria $\pm\varphi_0$ are asymptotically stable as $s \to +\infty$. Then the necessary periodic orbit is situated between of these stable equilibria, so it is unstable as $s \to +\infty$.

Such an unstable periodic solution of (4.7) is shown in Fig. 6 for $n = 15$. It was obtained by shooting from $s = 0$ with prescribed Cauchy data.

As for $m = 2$, in order to reveal periodic oscillations for smaller n (actually, numerical difficulties arise already in the case $n \leqslant 4$), we apply the scaling

$$\varphi(s) = \left(\frac{n}{6}\right)^{\frac{6}{n}} \Phi(\eta), \quad \text{where} \quad \eta = \frac{6s}{n}. \tag{4.14}$$

This gives in the limit a simplified ODE with the binomial linear operator

$$\mathrm{e}^{-\eta}(\mathrm{e}^{\eta}\Phi)^{(6)} \equiv \Phi^{(6)} + 6\Phi^{(5)} + 15\Phi^{(4)} + 20\Phi''' + 15\Phi'' + 6\Phi' + \Phi = |\Phi|^{-\frac{n}{n+1}}\Phi. \tag{4.15}$$

Figure 7 shows the trace of the periodic behavior for Eq. (4.15) with $n = \frac{1}{2}$.

According to scaling (4.14), the periodic oscillatory component $\varphi_*(s)$ becomes very small,

$$\max|\varphi_*| \sim 1.1 \times 10^{-18} \quad \text{for} \quad n = 0.5.$$

A more detailed study of the behavior of the oscillatory component as $n \to 0$ was done in [11, Sect. 12].

(a)

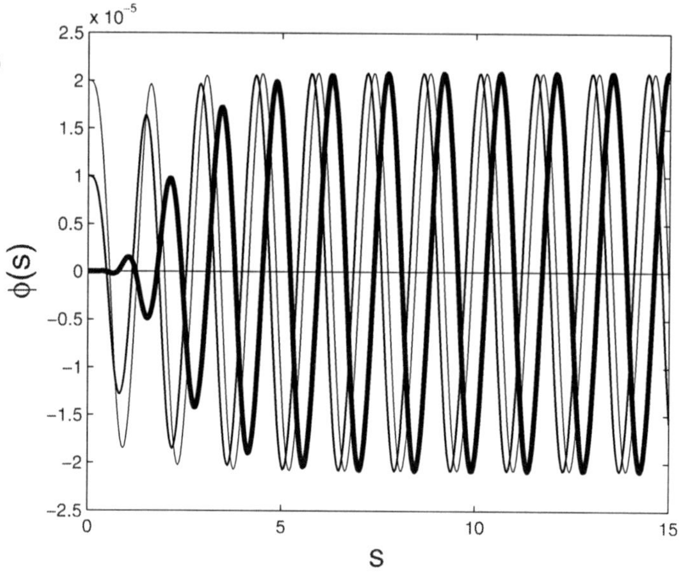

$n = 2$

(b)

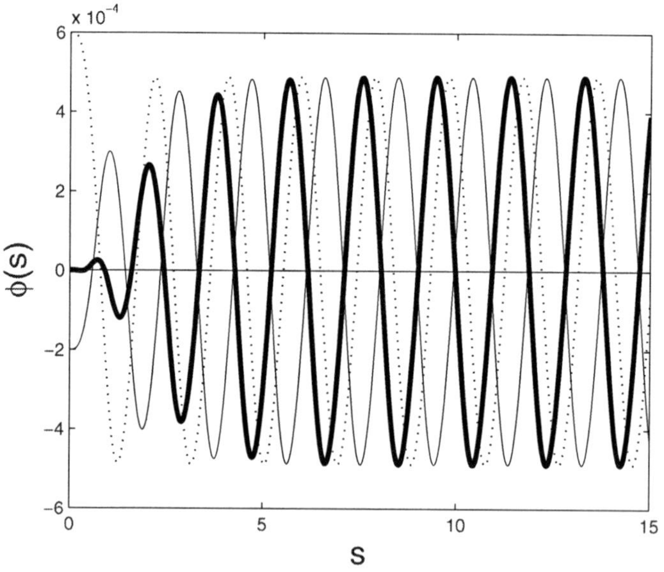

$n = 4$

Fig. 3 Convergence to the stable periodic solution of (4.6) for $n = 2$ (a) and $n = 4$ (b).

(a)

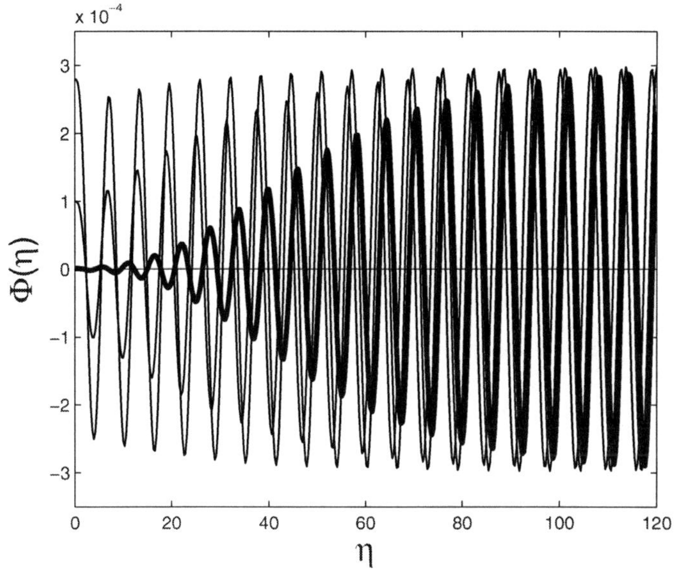

$n = 0.2$

(b)

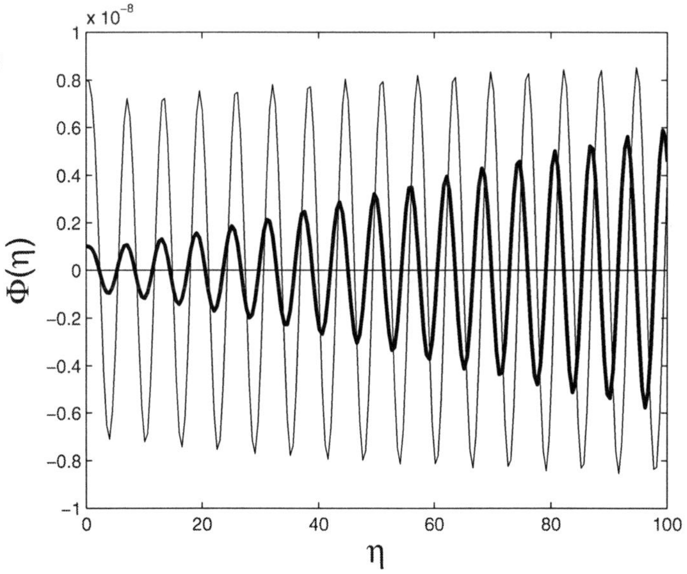

$n = 0.08$

Fig. 4 Stable periodic oscillations in the ODE (4.12) for $n = 0.2$ (a) and $n = 0.08$ (b).

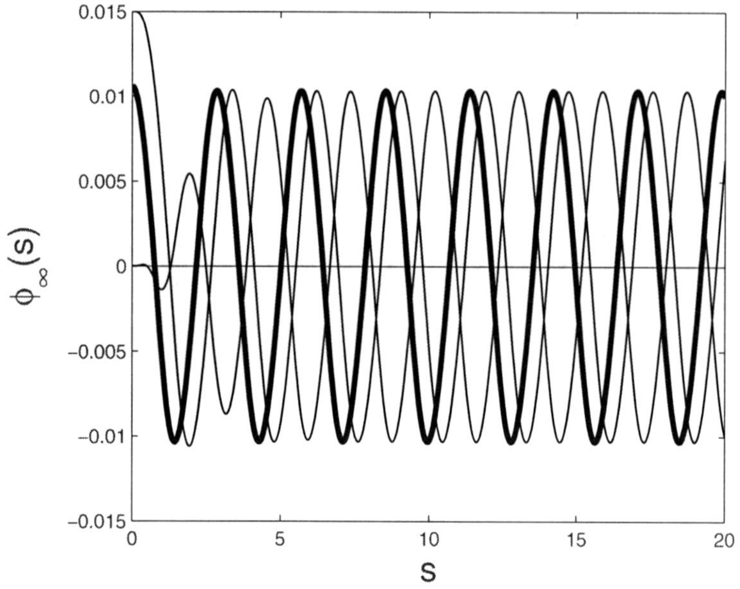

Fig. 5 Convergence to the stable periodic solution of (4.13) ($n = +\infty$).

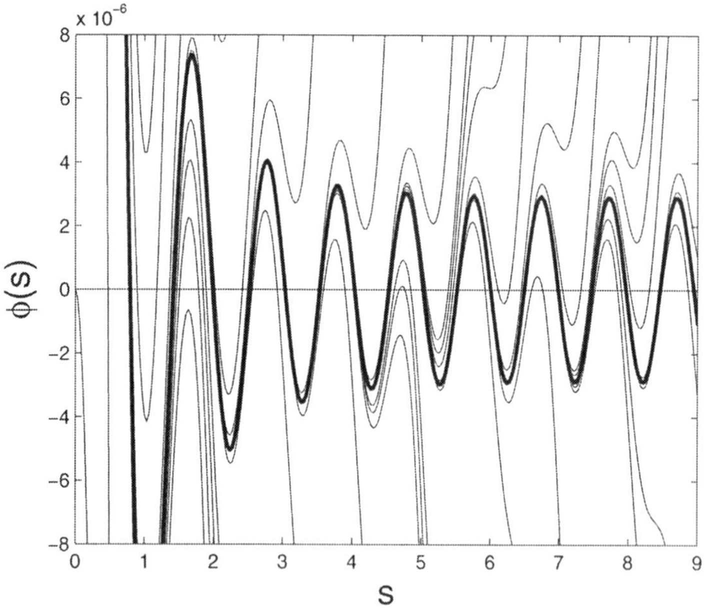

Fig. 6 Unstable periodic behavior of the ODE (4.7) for $n = 15$. Cauchy data are given by $\varphi(0) = 10^{-4}$, $\varphi'(0) = \varphi'''(0) = \ldots = \varphi^{(5)}(0) = 0$, and $\varphi''(0) = -5.0680839826093907\ldots \times 10^{-4}$.

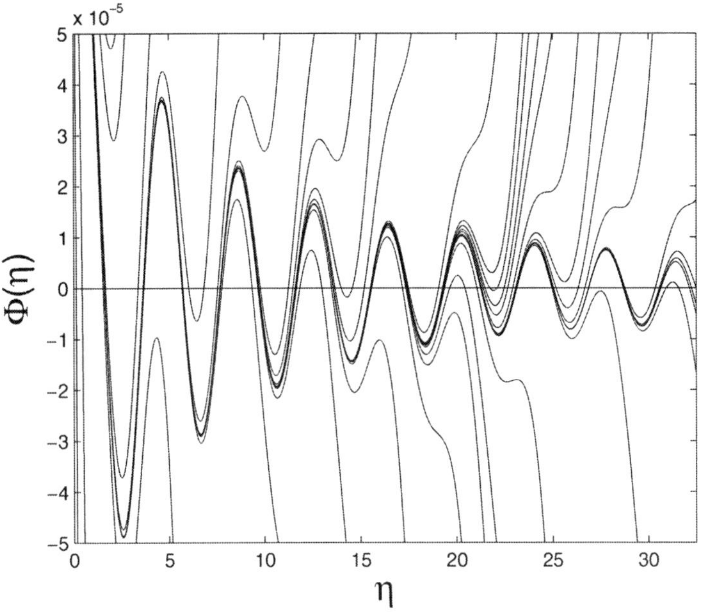

Fig. 7 Unstable periodic behavior of the ODE (4.15) for $n = \frac{1}{2}$. Cauchy data are given by $\varphi(0) = 10^{-4}$, $\varphi'(0) = \varphi'''(0) = \ldots = \varphi^{(5)}(0) = 0$, and $\varphi''(0) = -9.456770333415\ldots \times 10^{-4}$.

The passage to the limit $n \to +\infty$ leads to an equation with discontinuous nonlinearity which is easily obtained from (4.7). This admits a periodic solution, which is rather close to the periodic orbit in Fig. 6 obtained for $n = 15$.

We claim that the above two cases $m = 2$ (even) and $m = 3$ (odd) exhaust all key types of periodic behaviors in ordinary differential equations like the ODE (1.8). Namely, periodic orbits are stable for even m and are unstable for odd m, with typical stable and unstable manifolds as $s \to \pm\infty$. So, we observe a dichotomy relative to all orders $2m$ of the ODEs under consideration.

5 Numeric Problem: Numerical Construction and First Classification of Basic Types of Localized Blow-up or Compacton Patterns

We need a careful numerical description of various families of solutions of the ODEs (1.8). In practical computations, we have to use the regularized version of equations

$$(-1)^m F^{(2m)} = F - (\varepsilon^2 + F^2)^{-\frac{n}{2(n+1)}} F \quad \text{in} \quad \mathbb{R}, \tag{5.1}$$

which, for $\varepsilon > 0$, have smooth analytic nonlinearities. In numerical analysis, it is typical to take $\varepsilon = 10^{-4}$ or, at least, 10^{-3} which is sufficient to revealing global structures.

It is worth mentioning that detecting in Sect. 4 a highly oscillatory structure of solutions close to interfaces makes it impossible to use the well-developed *homotopy* theory [17, 36] which was successfully applied to other classes of fourth order ODEs with coercive operators (see also [24]).

Roughly speaking, our nonsmooth problem cannot be used in homotopic classifications since the oscillatory behavior close to interfaces destroys standard homotopy parameters, for example, the number of rotations on the hodograph plane $\{F, F'\}$. Indeed, for any solution of the ODE (1.8) the rotation number about the origin is always infinite. Then, as $F \to 0$, i.e., as $y \to \pm\infty$, the linearized equation is (4.5) which admits the oscillatory behavior (4.2).

5.1 Fourth order equation: $m = 2$

We describe the main families of solutions.

First basic pattern and structure of zeros

For $m = 2$ the ODE (1.8) takes the form

$$F^{(4)} = F - |F|^{-\frac{n}{n+1}}F \quad \text{in} \quad \mathbb{R}. \tag{5.2}$$

We are looking for compactly supported patterns F (see Proposition 3.1) such that

$$\begin{aligned} &\operatorname{meas}\operatorname{supp} F > 2R_*, \quad \text{where } R_* > \tfrac{\pi}{2} \text{ is the first} \\ &\text{positive root of the equation } \tanh R = -\tan R. \end{aligned} \tag{5.3}$$

Figure 8 presents the first basic pattern denoted by $F_0(y)$ for various $n \in [\frac{1}{10}, 100]$. Concerning the last profile $n = 100$, we note that (5.2) admits a natural passage to the limit as $n \to +\infty$, which leads to an ODE with a discontinuous nonlinearity

$$F^{(4)} = F - \operatorname{sign} F \equiv \begin{cases} F - 1 & \text{for} \quad F > 0, \\ F + 1 & \text{for} \quad F < 0. \end{cases} \tag{5.4}$$

A unique oscillatory solution of (5.4) can be treated by an algebraic approach (see [10, Sect. 7.4]). For $n = 1000$ and $n = +\infty$ the profiles are close to that for $n = 100$ in Fig. 8.

The profiles in Fig. 8 are constructed by MATLAB with extra accuracy, where ε in (5.1) and both tolerances in the `bvp4c` solver have been enhanced and took the values $\varepsilon = 10^{-7}$ and Tols $= 10^{-7}$. This allows us also to check the refined local structure of multiple zeros at the interfaces. Figure 9 corresponding to $n = 1$ explains how the zero structure repeats itself from one zero to another in the usual linear scale.

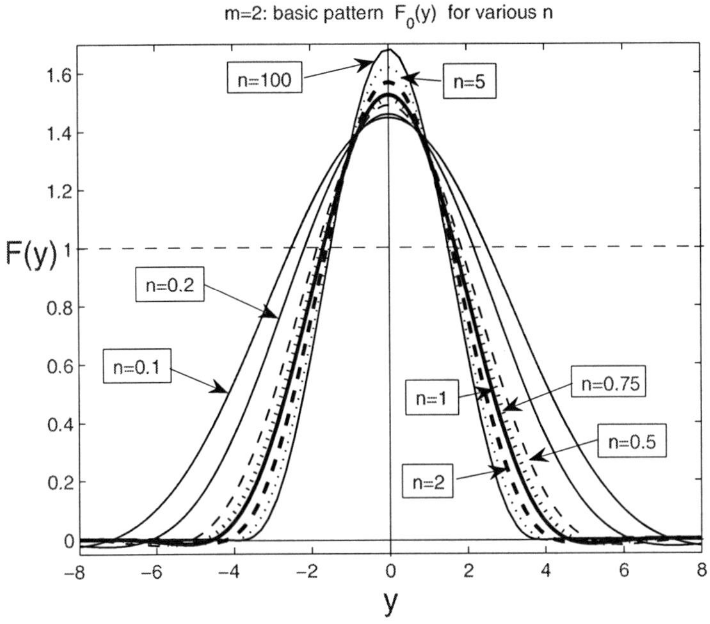

Fig. 8 The first (stable) solution $F_0(y)$ of (5.2) for various n.

Basic countable family: approximate Sturm property

Figure 10 presents the basic family $\{F_l,\ l = 0, 1, 2, \ldots\}$ of solutions of (5.2) for $n = 1$. Each profile $F_l(y)$ has precisely $l + 1$ "dominant" extrema and l "transversal" zeros (see the further discussion below and [13, Sect. 4] for details). It is important that *all the internal zeros of* $F_l(y)$ *are transversal* (obviously, excluding the oscillatory endpoints of the support). In other words, each profile F_l is approximately obtained by a simple "interaction" (gluing together) of $l + 1$ copies of the first pattern $\pm F_0$ taking with suitable signs (see the further development below).

Actually, if we forget for a moment about the complicated oscillatory structure of solutions near interfaces, where an infinite number of extrema and zeros occur, the dominant geometry of profiles in Fig. 10 looks like it

(a)

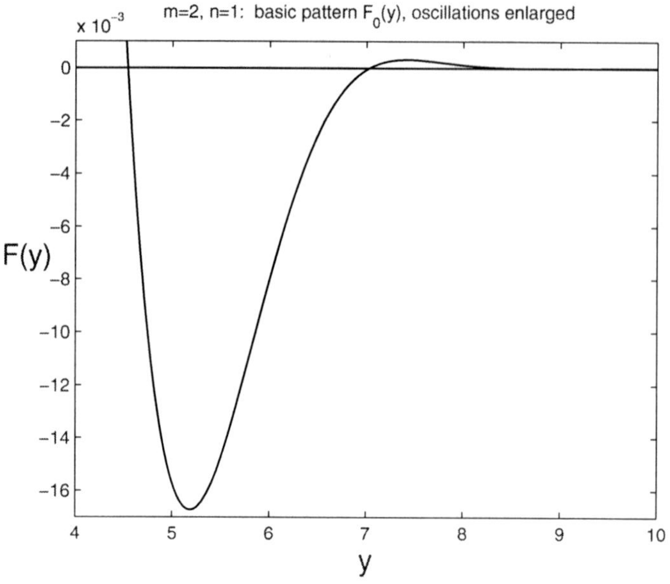

scale 10^{-3}

(b)

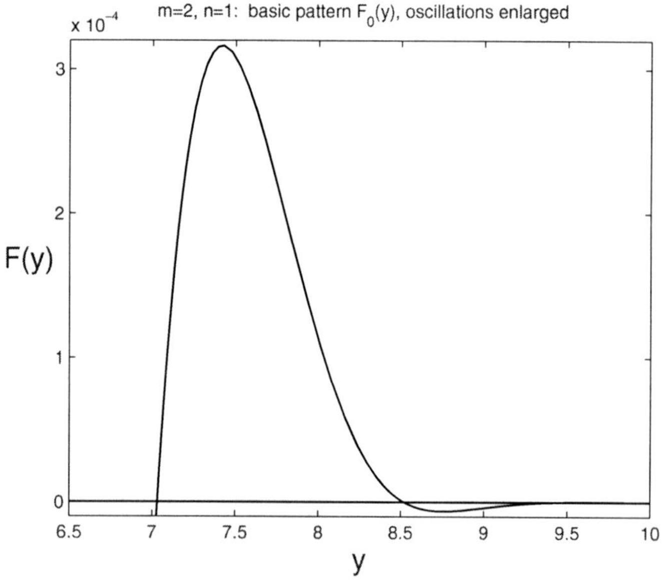

scale 10^{-4}

Fig. 9 Enlarged zero structure of the profile $F_0(y)$ for $n = 1$ in the linear scale.

approximately obeys the Sturm classical zero set property, which is true only for $m = 1$, i.e., for the second order ODE

$$F'' = -F + |F|^{-\frac{n}{n+1}} F \quad \text{in} \quad \mathbb{R}. \tag{5.5}$$

For (5.5) the basic family $\{F_l\}$ is constructed by direct gluing together the explicit patterns (1.5), i.e., $\pm F_0$. Therefore, each F_l consists of precisely $l+1$ patterns (1.5) (with signs $\pm F_0$), so that the Sturm property is true. In Sect. 3, we present some analytic evidence showing that precisely this basic family $\{F_l\}$ is obtained by a direct application of the LS category theory.

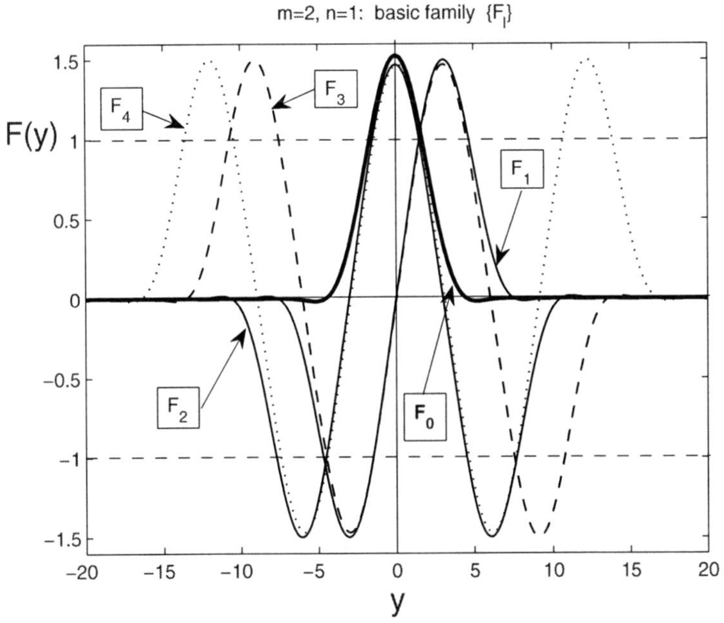

Fig. 10 The first five patterns of the basic family $\{F_l\}$ of the ODE (5.2) for $n = 1$.

5.2 Countable family of $\{F_0, F_0\}$-interactions

We show that the actual nonlinear interaction of the two first patterns $+F_0(y)$ leads to a new family of profiles.

Figure 11, $n = 1$, shows the first profiles from this family denoted by $\{F_{+2,k,+2}\}$, where in each function $F_{+2,k,+2}$ the multiindex $\sigma = \{+2, k, +2\}$ means, from left to right, $+2$ intersections with the equilibrium $+1$, then next k intersections with zero, and final $+2$ stands again for 2 intersections

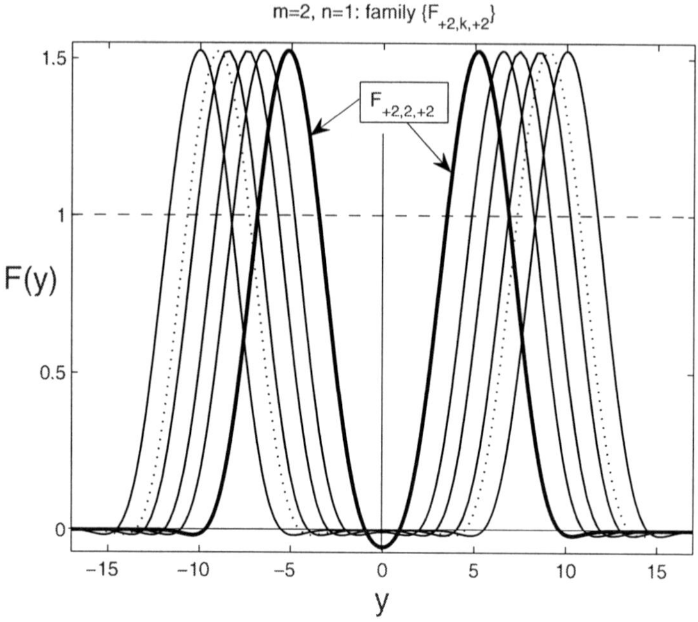

Fig. 11 First patterns from the family $\{F_{+2,k,+2}\}$ of the $\{F_0, F_0\}$-interaction; $n = 1$.

with $+1$. Later on, we will use such a multiindex notation to classify other patterns obtained.

In Fig. 12, we present the enlarged behavior of zeros explaining the structure of the interior layer of connection of two profiles $\sim +F_0(y)$. In particular, (b) shows that there exist *two* profiles $F_{+2,6,+2}$, these are given by the dashed line and the previous one, both having two zeros on $[-1, 1]$. Therefore, the identification and classification of profiles just by the successive number of intersections with equilibria 0 and ± 1 is not always possible (in view of a nonhomotopical nature of the problem), and some extra geometry of curves near intersections should be taken into account. In fact, precisely this proves that a standard homotopic classification of patterns is not consistent for such noncoercive and oscillatory equations. Anyway, whenever possible without confusion, we will continue to use such a multiindex classification, though now meaning that in general a profile F_σ with a given multiindex σ may denote actually a *class* of profiles with the given geometric characteristics. Note that the last profile in Fig. 11 is $F_{+2,6,+2}$, where the last two zeros are seen in the scale $\sim 10^{-6}$ in Fig. 13. Observe here a clear nonsmoothness of two last profiles as a numerical discrete mesh phenomenon, which nevertheless does nor spoil at all this differential presentation.

In view of the oscillatory character of $F_0(y)$ at the interfaces, we expect that the family $\{F_{+2,k,+2}\}$ is countable, and such functions exist for any even $k = 0, 2, 4, \ldots$. Then $k = +\infty$ corresponds to the noninteracting pair

(a)

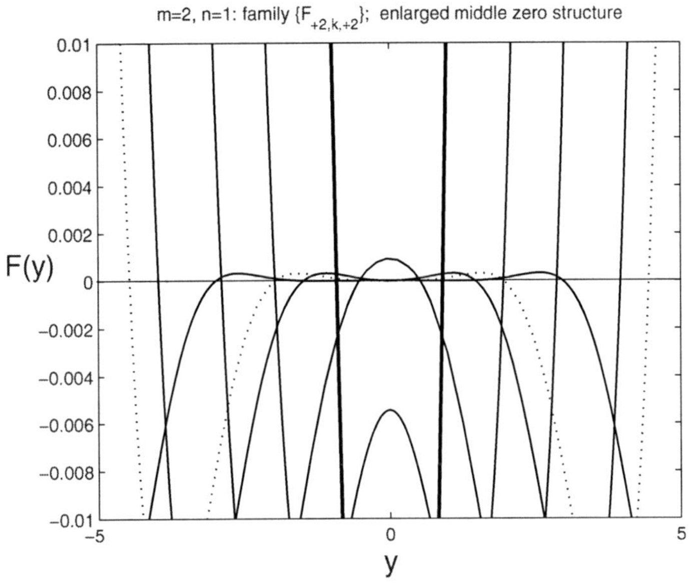

zeros: scale 10^{-2}

(b)

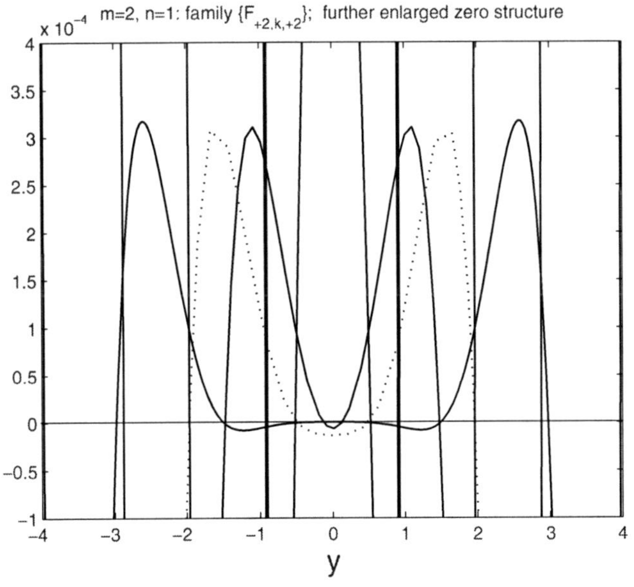

zeros: scale 10^{-4}

Fig. 12 Enlarged middle zero structure of the profiles $F_{+2,k,+2}$ from Fig. 11.

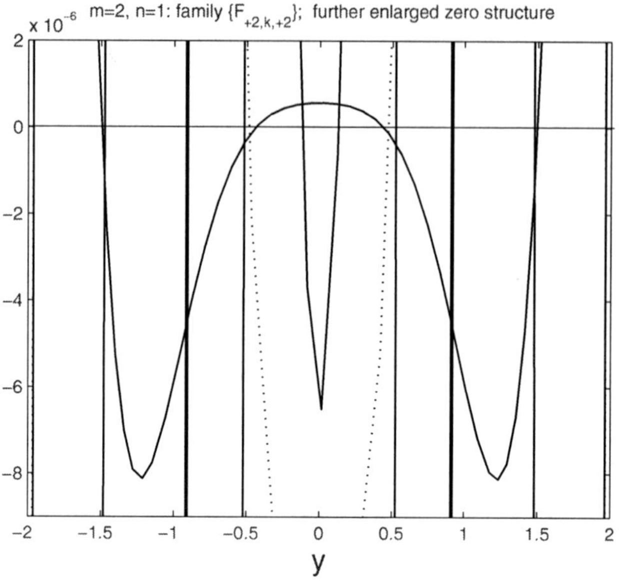

Fig. 13 Enlarged middle zero structure of the profiles $F_{+2,6,+2}$ from Fig. 11.

$$F_0(y+y_0)+F_0(y-y_0), \quad \text{where} \quad \operatorname{supp} F_0(y)=[-y_0,y_0]. \tag{5.6}$$

Of course, there exist various triple $\{F_0, F_0, F_0\}$ and any multiple interactions $\{F_0, \ldots, F_0\}$ of k single profiles, with different distributions of zeros between any pair of neighbors.

5.3 Countable family of $\{-F_0, F_0\}$*-interactions*

We describe the interaction of $-F_0(y)$ with $F_0(y)$. In Fig. 14, $n = 1$, we show the first profiles from this family denoted by $\{F_{-2,k,+2}\}$, where for the multiindex $\sigma = \{-2, k, +2\}$, the first number -2 means 2 intersections with the equilibrium -1, etc. The zero structure close to $y = 0$ is presented in Fig. 15. It follows from (b) that the first two profiles belong to the class $F_{-2,1,2}$, i.e., both have a single zero for $y \approx 0$. The last solution shown is $F_{-2,5,+2}$. Again, we expect that the family $\{F_{-2,k,+2}\}$ is countable, and such functions exist for any odd $k = 1, 3, 5, \ldots$, and $k = +\infty$ corresponds to the noninteracting pair

$$-F_0(y+y_0)+F_0(y-y_0) \quad (\operatorname{supp} F_0(y)=[-y_0,y_0]). \tag{5.7}$$

There exist families of arbitrarily many interactions such as $\{\pm F_0, \pm F_0, \ldots, \pm F_0\}$ consisting of any $k \geqslant 2$ members.

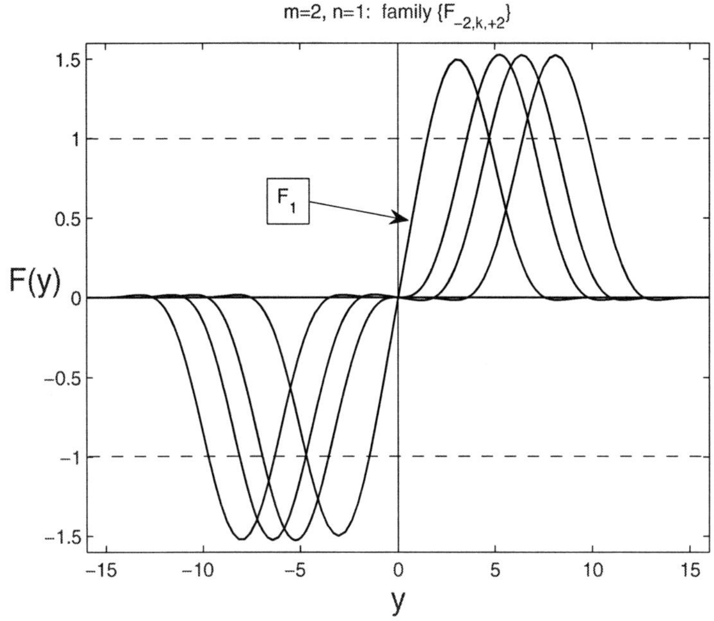

Fig. 14 First four patterns from the family $\{F_{-2,k,+2}\}$ of the $\{-F_0, F_0\}$-interactions; $n = 1$.

5.4 Periodic solutions in $\mathbb{R}$

Before introducing new types of patterns, we need to describe other noncompactly supported solutions in $\mathbb{R}$. As a variational problem, Eq. (5.2) admits an infinite number of periodic solutions (see, for example, [23, Chapt. 8]). In Fig. 16 for $n = 1$, we present a special unstable periodic solution obtained by shooting from the origin with conditions

$$F(0) = 1.5, \quad F'(0) = F'''(0) = 0, \quad F''(0) = -0.3787329255\ldots.$$

We will show next that precisely the periodic orbit $F_*(y)$ with

$$F_*(0) \approx 1.535\ldots \tag{5.8}$$

plays an important part in the construction of other families of compactly supported patterns. Namely, all the variety of solutions of (5.2) that have oscillations about equilibria ± 1 are close to $\pm F_*(y)$ there.

(a)

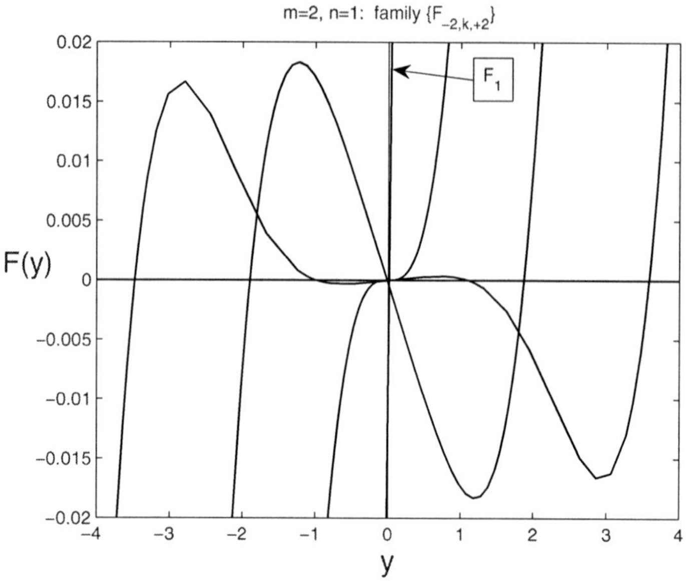

zeros: scale 10^{-2}

(b)

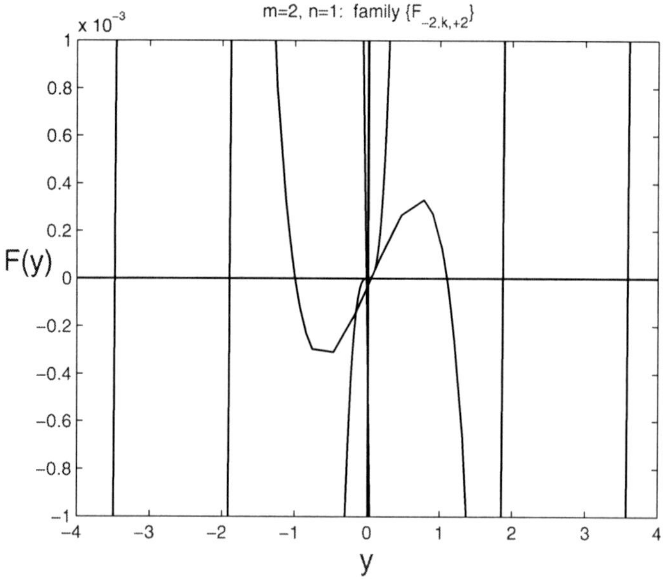

zeros: scale 10^{-4}

Fig. 15 Enlarged middle zero structure of the profiles $F_{-2,k,+2}$ from Fig. 14.

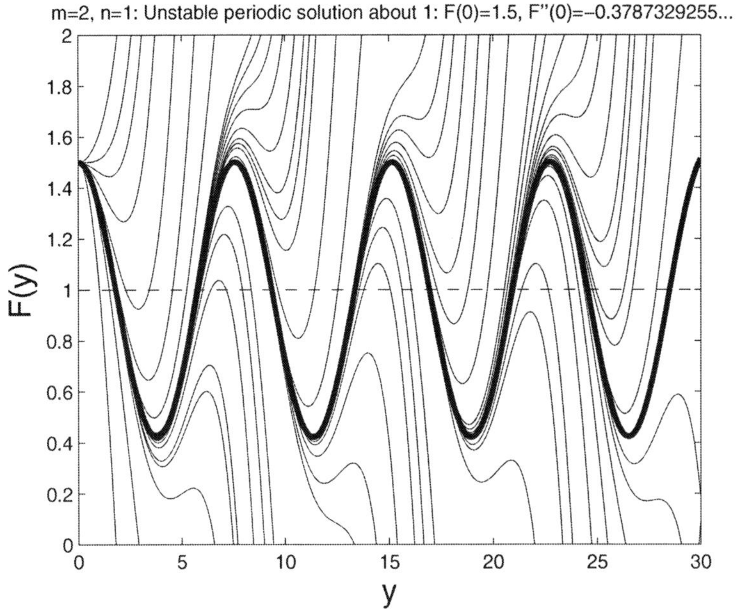

Fig. 16 An example of a periodic solution of the ODE (5.2) for $n = 1$.

5.5 Family $\{F_{+2k}\}$

Such functions F_{+2k} for $k \geqslant 1$ have $2k$ intersection with the single equilibrium $+1$ only and have a clear "almost" periodic structure of oscillations about (see Fig. 17(a)). The number of intersections denoted by $+2k$ gives an extra Strum index to such a pattern. In this notation, $F_{+2} = F_0$.

5.6 More complicated patterns: towards chaotic structures

Using the above rather simple families of patterns, we claim that a pattern (possibly, a class of patterns) with an arbitrary multiindex of any length

$$\sigma = \{\pm\sigma_1, \sigma_2, \pm\sigma_3, \sigma_4, \ldots, \pm\sigma_l\} \qquad (5.9)$$

can be constructed. Figure 17(b) shows several profiles from the family with the index $\sigma = \{+k, l, -m, l, +k\}$. In Fig. 18, we show further four different patterns, while in Fig. 19, a single most complicated pattern is presented, for which

$$\sigma = \{-8, 1, +4, 1, -10, 1, +8, 1, 3, -2, 2, -8, 2, 2, -2\}. \qquad (5.10)$$

(a)

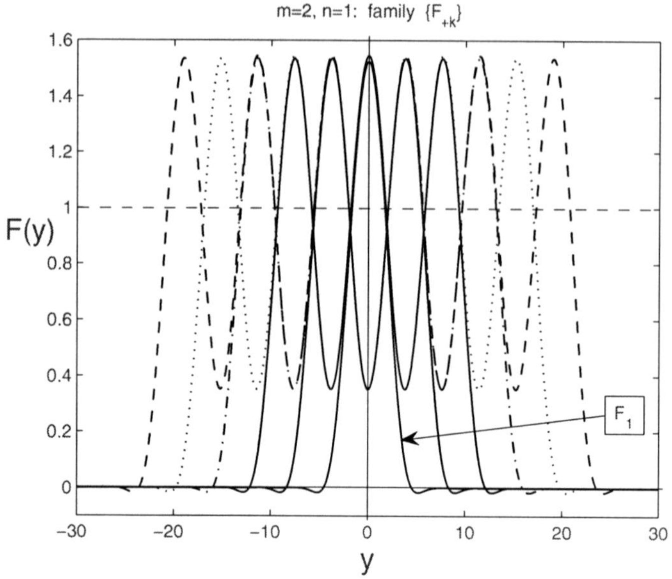

$F_{+2k}(y)$

(b)

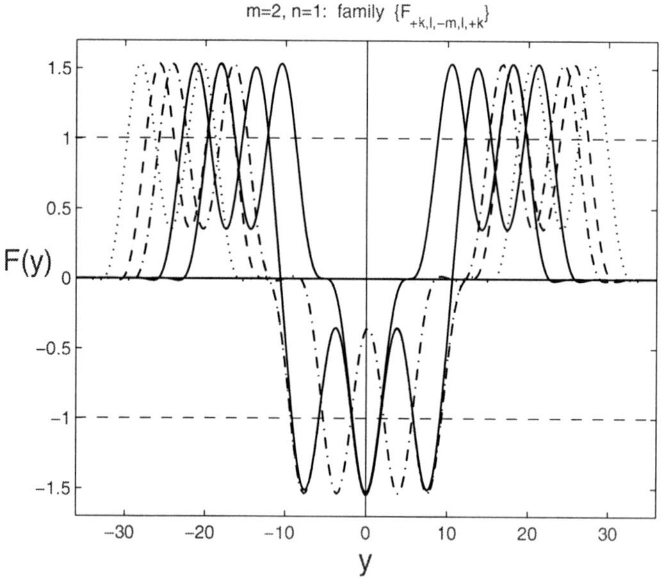

$F_{+k,l,-m,l,+k}$

Fig. 17 Two families of solutions of (5.2) for $n = 1$; $F_{+2k}(y)$ (a) and $F_{+k,l,-m,l,+k}$ (b).

(a)

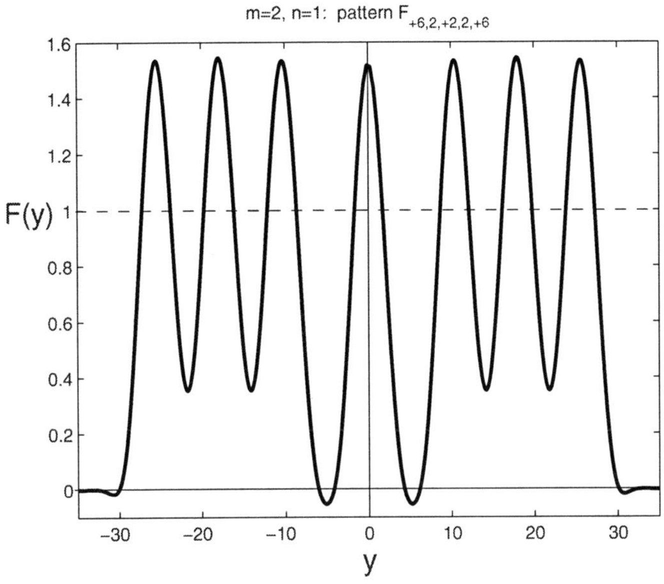

$\sigma = \{+6, 2, +2, 2, +6\}$

(b)

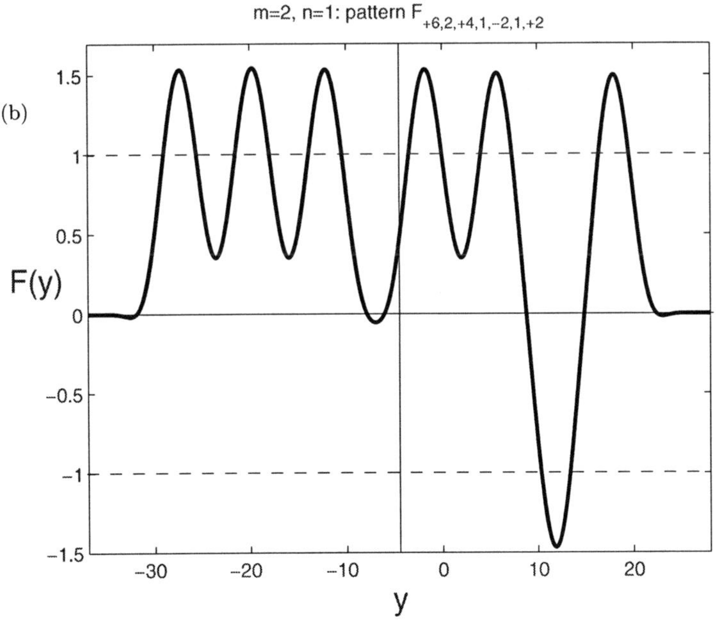

$\sigma = \{+6, 2, +4, 1, -2, 1 + 2\}$

(c)

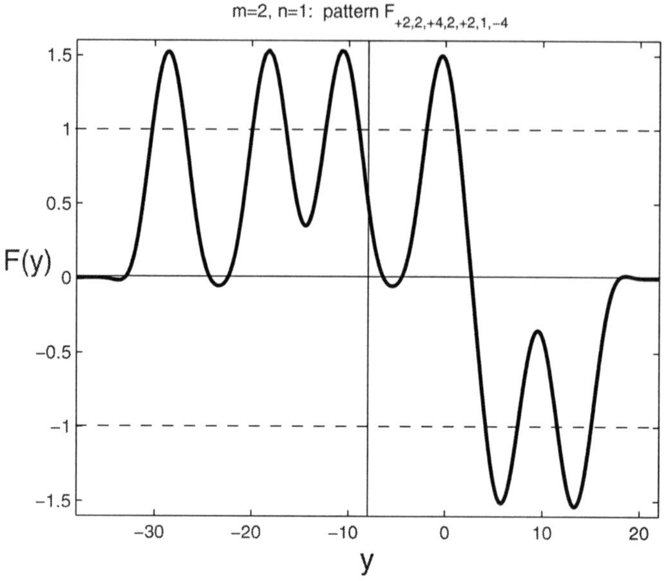

$\sigma = \{+2, 2, +4, 2, +2, 1, -4\}$

(d)

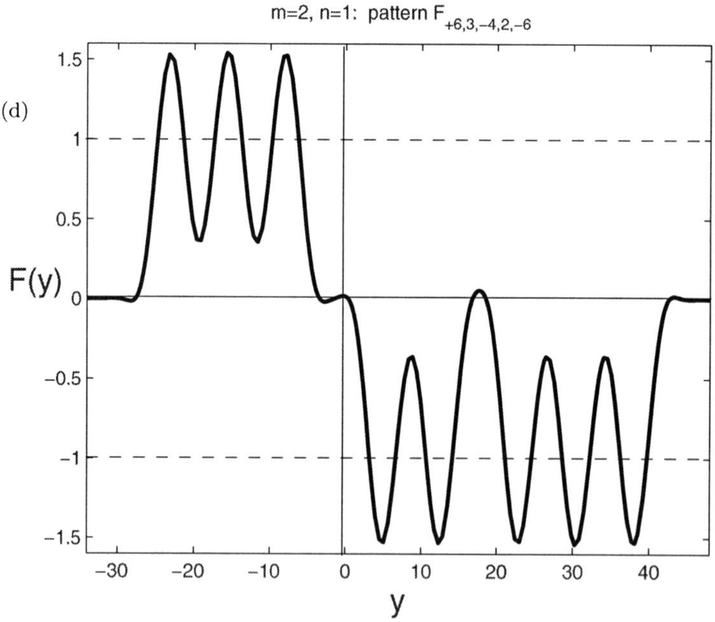

$\sigma = \{+6, 3, -4, 2, -6\}$

Fig. 18 Various patterns for (5.2) for $n = 1$.

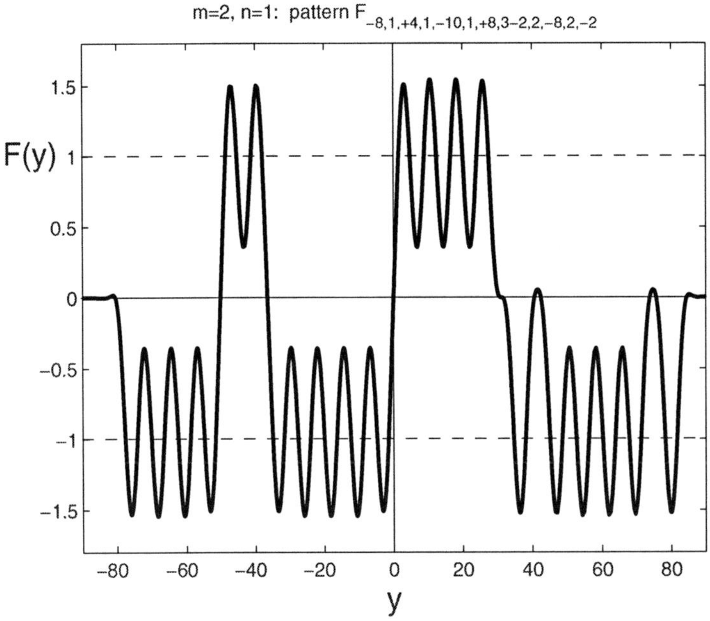

Fig. 19 A complicated pattern $F_\sigma(y)$ for (5.2) for $n = 1$.

All computations are performed for $n = 1$ as usual. Actually, we claim that the multiindex (5.9) can be arbitrary and takes any finite part of any nonperiodic fraction. Actually, this means *chaotic features* of the whole family of solutions $\{F_\sigma\}$. These chaotic types of behavior are known for other fourth order ODEs with coercive operators, [24, p. 198].

Acknowledgement. E. Mitidieri and S. Pokhozhaev were supported by INTAS: Investigation of Global Catastrophes for Nonlinear Processes in Continuum Mechanics (no. 05-1000008-7921).

References

1. Bahri, A., Berestycki, H.: A perturbation method in critical point theory and applications. Trans. Am. Math. Soc. **267**, 1–32 (1981)
2. Bahri, A., Lions, P.L.: Morse index in some min–max critical points. I. Application to multiplicity results. Commun. Pure Appl. Math. **41**, no. 8, 1027–1037 (1988)
3. Berger, M.: Nonlinearity and Functional Analysis. Acad. Press, New York (1977)
4. Boggio, T.: Sull'equilibrio delle piastre elastiche incastrate. Rend. Acc. Lincei **10**, 197–205 (1901)
5. Boggio, T.: Sulle funzioni di Green d'ordine m. Rend. Circ. Mat. Palermo **20**, 97–135 (1905)

6. Brezis, H., Brawder, F.: Partial Differential Equations in the 20th Century. Adv. Math., **135**, 76–144 (1998)
7. Clark, D.C.: A variant of Lusternik–Schnirelman theory. Indiana Univ. Math. J. **22**, 65–74 (1972)
8. Coddington, E.A.: Levinson, N.: Theory of Ordinary Differential Equations. McGraw-Hill Book Company, Inc., New York–London (1955)
9. Deimling, K.: Nonlinear Functional Analysis. Springer-Verlag, Berlin–Tokyo (1985)
10. Evans, J.D., Galaktionov, V.A., King, J.R.: Source-type solutions of the fourth order unstable thin film equation. Euro J. Appl. Math. **18**, 273–321 (2007)
11. Evans, J.D., Galaktionov, V.A., King, J.R.: Unstable sixth order thin film equation. I. Blow-up similarity solutions; II. Global similarity patterns. Nonlinearity **20**, 1799–1841, 1843–1881 (2007)
12. Galaktionov, V.A.: On interfaces and oscillatory solutions of higher order semilinear parabolic equations with nonlipschitz nonlinearities. Stud. Appl. Math. **117**, 353–389 (2006)
13. Galaktionov, V.A., Harwin, P.J.: Non-uniqueness and global similarity solutions for a higher order semilinear parabolic equation. Nonlinearity **18**, 717–746 (2005)
14. Galaktionov, V.A., Pohozaev (Pokhozhaev), S.I.: On shock, rarefaction waves, and compactons for odd order nonlinear dispersion PDEs. Comput. Math. Math. Phys. [To appear]
15. Galaktionov, V.A., Svirshchevskii, S.R.: Exact Solutions and Invariant Subspaces of Nonlinear Partial Differential Equations in Mechanics and Physics. Chapman & Hall/CRC, Boca Raton, Florida (2007)
16. Hale, J.K.: Asymptotic Behavior of Dissipative Systems. Am. Math. Soc., Providence, RI (1988)
17. Kalies, W.D., Kwapisz, J., VandenBerg, J.B., VanderVorst, R.C.A.M.: Homotopy classes for stable periodic and chaotic patterns in fourth order Hamiltonian systems. Commun. Math. Phys. **214**, 573–592 (2000)
18. Kawamoto, S.: An exact transformation from the Harry Dym equation to the modified KdV equation. J. Phys. Soc. Japan **54**, 2055–2056 (1985)
19. Krasnosel'skii, M. A.: Topological Methods in the Theory of Nonlinear Integral Equations. Pergamon Press, Oxford–Paris (1964)
20. Krasnosel'skii, M. A., Zabreiko, P.P.: Geometrical Methods of Nonlinear Analysis. Springer-Verlag, Berlin–Tokyo (1984)
21. Lions, J.L.: Quelques Méthodes de Résolution des Problèmes aux Limites non Linéaires. Dunod, Gauthier–Villars, Paris (1969)
22. Maz'ya, V.G.: Sobolev Spaces. Springer-Verlag, Berlin–Tokyo (1985)
23. Mitidieri, E., Pohozaev (Pokhozhaev), S.I.: Apriori Estimates and Blow-up of Solutions to Nonlinear Partial Differential Equations and Inequalities. Proc. Steklov Inst. Math. **234**, Intern. Acad. Publ. Comp. Nauka/Interperiodica, Moscow (2001)
24. Peletier, L.A., Troy, W.C.: Spatial Patterns: Higher Order Models in Physics and Mechanics. Birkhäuser, Boston–Berlin (2001)
25. Pohozaev (Pokhozhaev), S.I.: On an approach to nonlinear equations. Sov. Math., Dokl. **20**, 912–916 (1979)
26. Pohozaev (Pokhozhaev), S.I.: The fibering method in nonlinear variational problems. Pitman Research Notes in Math. **365**, 35–88 (1997)
27. Rabinowitz, P.: Variational methods for nonlinear eigenvalue problems. In: Eigenvalue of Nonlinear Problems, pp. 141–195. Edizioni Cremonese, Rome (1974)
28. Rosenau, P.: Nonlinear dispertion and compact structures. Phys. Rev. Lett. **73**, 1737–1741 (1994)
29. Rosenau, P.: On solitons, compactons, and Lagrange maps. Phys. Lett. A **211**, 265–275 (1996)

30. Rosenau, P.: Compact and noncompact dispersive patterns. Phys. Lett. A, **275**, 193–203 (2000)
31. Rosenau, P., Hyman, J.M.: Compactons: solitons with finite wavelength. Phys. Rev. Lett. **70**, 564–567 (1993)
32. Samarskii, A.A., Galaktionov, V.A., Kurdyumov, S.P., Mikhailov, A.P.: Blow-up in Quasilinear Parabolic Equations. Walter de Gruyter, Berlin–New York (1995)
33. Samarskii, A.A., Zmitrenko, N.V., Kurdyumov, S.P., Mikhailov, A.P.: Thermal structures and fundamental length in a medium with nonlinear heat conduction and volumetric heat sources. Sov. Phys., Dokl. **21**, 141–143 (1976)
34. Shishkov, A.E.: Dead cores and instantaneous compactification of the supports of energy solutions of quasilinear parabolic equations of arbitrary order. Sbornik: Math. **190**, 1843–1869 (1999)
35. Sobolev, S.L.: Some Applications of Functional Analysis in Mathematical Physics (Russian). 1st ed. Leningrad State Univ., Leningrad (1950); 3rd ed. Nauka, Moscow (1988); English transl. of the 1st Ed.: Am. Math. Soc., Providence, RI (1963); English transl. of the 3rd Ed. with comments by V. P. Palamodov: Am. Math. Soc., Providence, RI (1991)
36. Van Den Berg J.B., Vandervorst, R.C.: Stable patterns for fourth order parabolic equations. Duke Math. J. **115**, 513–558 (2002)

$L_{q,p}$-Cohomology of Riemannian Manifolds with Negative Curvature

Vladimir Gol'dshtein and Marc Troyanov

Dedicated to the memory of Sergey L'vovich Sobolev

Abstract We consider the $L_{q,p}$-cohomology of a complete simply connected Riemannian manifold (M, g) with pinched negative curvature. The connection between the $L_{q,p}$-cohomology of (M, g) and Sobolev inequalities for differential forms on (M, g) was established by the authors in the previous publications.

1 Introduction

In [5], we established a connection between Sobolev inequalities for differential forms on a Riemannian manifold (M, g) and an invariant, called the *$L_{q,p}$-cohomology* $\left(H^k_{q,p}(M)\right)$ of the manifold (M, g). In this paper, we prove nonvanishing results for the $L_{q,p}$-cohomology of simply connected complete manifolds with negative curvature.

1.1 $L_{q,p}$-cohomology and Sobolev inequalities

To define the $L_{q,p}$-cohomology of a Riemannian manifold (M, g), we first recall the notion of the *weak exterior differential* of a locally integrable

Vladimir Gol'dshtein
Ben Gurion University of the Negev, P.O.B. 653, Beer Sheva 84105, Israel, e-mail: `vladimir@bgu.ac.il`

Marc Troyanov
Institute of Geometry, Algebra, and Topology, École Polytechnique Fédérale de Lausanne, 1015 Lausanne, Switzerland, e-mail: `marc.troyanov@epfl.ch`

V. Maz'ya (ed.), *Sobolev Spaces in Mathematics II*, *International Mathematical Series*.
doi: 10.1007/978-0-387-85650-6,

differential form. Denote by $C_c^\infty(M, \Lambda^k)$ the space of smooth differential forms of degree k with compact support on M.

Definition. A form $\theta \in L^1_{loc}(M, \Lambda^k)$ is the *weak exterior differential* of a form $\varphi \in L^1_{loc}(M, \Lambda^{k-1})$ and one writes $d\varphi = \theta$ if for every $\omega \in C_c^\infty(M, \Lambda^{n-k})$

$$\int_M \theta \wedge \omega = (-1)^k \int_M \varphi \wedge d\omega.$$

The Sobolev space $W^{1,p}(M, \Lambda^k)$ of differential k-forms is defined as the space of k-forms φ in $L^p(M)$ such that $d\varphi \in L^p(M)$ and $d(*\varphi) \in L^p(M)$, where $* : \Lambda^k \to \Lambda^{n-k}$ is the Hodge star homomorphism. In this paper, we are interested in a different "Sobolev type" space of differential forms, denoted by $\Omega^k_{q,p}(M)$; namely, the space of all k-forms φ in $L^q(M)$ such that $d\varphi \in L^p(M)$ $(1 \leqslant q, p \leqslant \infty)$. It is a Banach space relative to the graph norm

$$\|\omega\|_{\Omega^k_{q,p}} := \|\omega\|_{L^q} + \|d\omega\|_{L^p}\,. \tag{1.1}$$

When $k = 0$ and $q = p$, the space $\Omega^0_{p,p}(M)$ coincides with the classical Sobolev space $W^1_p(M)$ of functions in L^p with gradient in L^p. Note that a more general space $\Omega^0_{q,p}(M)$ was considered in [10] in the context of embedding theorems and Sobolev inequalities.

To define the $L_{q,p}$–cohomology of (M, g), we introduce the space of weakly closed forms

$$Z^k_p(M) = \{\omega \in L^p(M, \Lambda^k) \mid d\omega = 0\}$$

and the space of differential forms in $L^p(M)$ having a primitive in $L^q(M)$

$$B^k_{q,p}(M) = d(\Omega^{k-1}_{q,p}).$$

Note that $Z^k_p(M) \subset L^p(M, \Lambda^k)$ is always a closed subspace, but this is, in general, not the case of $B^k_{q,p}(M)$, and we denote by $\overline{B}^k_{q,p}(M)$ its closure in the L^p-topology. We also note that $\overline{B}^k_{q,p}(M) \subset Z^k_p(M)$ (by continuity and $d \circ d = 0$). Thus,

$$B^k_{q,p}(M) \subset \overline{B}^k_{q,p}(M) \subset Z^k_p(M) = \overline{Z}^k_p(M) \subset L^p(M, \Lambda^k).$$

Definition. The *$L_{q,p}$-cohomology* of (M, g) (where $1 \leqslant p, q \leqslant \infty$) is

$$H^k_{q,p}(M) := Z^k_p(M) / B^k_{q,p}(M)$$

and the *reduced $L_{q,p}$-cohomology* of (M, g) is

$$\overline{H}^k_{q,p}(M) := Z^k_p(M) / \overline{B}^k_{q,p}(M)\,.$$

The reduced cohomology is a Banach space, whereas the unreduced cohomology is a Banach space if and only if it coincides with the reduced one.

In [5, Theorem 6.1], we established the following connection between Sobolev inequalities for differential forms on a Riemannian manifold (M, g) and the $L_{q,p}$–cohomology of (M, g).

Theorem 1.1. *$H^k_{q,p}(M,g) = 0$ if and only if there exists a constant $C < \infty$ such that for any closed p-integrable differential form ω of degree k there exists a differential form θ of degree $k-1$ such that $d\theta = \omega$ and*

$$\|\theta\|_{L^q} \leqslant C\, \|\omega\|_{L^p}\,.$$

Suppose that $k = 1$. If M is simply connected (or, more generally, $H^1_{\text{deRham}}(M) = 0$), then any $\omega \in Z^1_p(M)$ has a primitive locally integrable function f, $df = \omega$. This means that for simply connected manifolds the space $Z^1_p(M)$ coincides with the seminormed Sobolev space $L^1_p(M)$, $\|f\|_{L^1_p(M)} := \|df\|_{L^p(M)}$. Then Theorem 1.1 reads as follows.

Corollary 1.2. *Suppose that (M, g) is a simply connected Riemannian manifold. Then $H^1_{q,p}(M, g) = 0$ if and only if there exists a constant $C < \infty$ depending only on M, (q, p), and a constant $a_f < \infty$ depending also on $f \in L^1_p(M, g)$ such that*

$$\|f - a_f\|_{L^q} \leqslant C\, \|df\|_{L^p}$$

for any $f \in L^1_p(M, g)$.

In this paper, we prove nonvanishing results for the $L_{q,p}$-cohomology of simply connected complete manifolds with negative curvature, which concerns the nonexistence of Sobolev inequalities for such pairs (q, p).

1.2 Statement of the main result

The main goal of the present paper is to prove the following nonvanishing result for the $L_{q,p}$-cohomology of simply connected complete manifolds with negative curvature.

Theorem 1.3. *Let (M, g) be an n-dimensional Cartan–Hadamard manifold*[1] *with sectional curvature $K \leqslant -1$ and Ricci curvature* $\text{Ric} \geqslant -(1+\varepsilon)^2(n-1)$.

(A) *Assume that*

$$\frac{1+\varepsilon}{p} < \frac{k}{n-1} \quad \textit{and} \quad \frac{k-1}{n-1} + \varepsilon < \frac{1+\varepsilon}{q}.$$

[1] Recall that a *Cartan-Hadamard* manifold is a complete simply connected Riemannian manifold of nonpositive sectional curvature.

Then $H^k_{q,p}(M) \neq 0$.

(B) *If, in addition,*

$$\frac{1+\varepsilon}{p} < \frac{k}{n-1} \quad \textit{and} \quad \frac{k-1}{n-1} + \varepsilon < \min\left\{\frac{1+\varepsilon}{q}, \frac{1+\varepsilon}{p}\right\},$$

then $\overline{H}^k_{q,p}(M) \neq 0$.

Theorem 1.3, together with Theorem 1.1, has the following (negative) consequence regarding Sobolev inequalities for differential forms.

Corollary 1.4. *Let* (M, g) *be a Cartan–Hadamard manifold as above. If* q *and* p *satisfy assumption* (A) *of Theorem* 1.3, *then there is no finite constant* C *such that any smooth closed* k*-form* ω *on* M *admits a primitive* θ *such that* $d\theta = \omega$ *and*

$$\|\theta\|_{L^q(M)} \leqslant C\, \|\omega\|_{L^p(M)}.$$

The proof of Theorem 1.3 is based on the duality principle established in [5] and a comparison argument inspired by Chapt. 8 of the book [8] by Gromov. Now, we discuss some particular cases.

- If M is the hyperbolic plane $\mathbb{H}^2$ ($n = 2, \varepsilon = 0$), Theorem 1.3 says that $\overline{H}^1_{q,p}(\mathbb{H}^2) \neq 0$ for any $q, p \in (1, \infty)$; and another proof can be found in [5, Theorem 10.1].

- For $q = p$ Theorem 1.3 says that $\overline{H}^k_{p,p}(M) \neq 0$ provided that

$$\frac{k-1}{n-1} + \varepsilon < \frac{1+\varepsilon}{p} < \frac{k}{n-1}; \tag{1.2}$$

  this result was already known [8, p. 244]. The inequalities (1.2) can also be written in terms of k as follows:

$$\frac{n-\tau}{p} < k < \frac{n-\tau}{p} + \tau$$

  with $\tau = 1 - \varepsilon(n-1)$.

- By contrast, Pansu [11, Theorem A] proved that $H^k_{p,p}(M) = 0$ if the sectional curvature satisfies

$$-(1+\varepsilon)^2 \leqslant K \leqslant -1 \quad \text{and} \quad (1+\varepsilon)p \leqslant \frac{n-1}{k} + \varepsilon.$$

- A Poincaré duality for the reduced L^p-cohomology was proved in [2], it says that for a complete Riemannian manifold

$$\overline{H}^k_{p,p}(M) = \overline{H}^{n-k}_{p',p'}(M)$$

with $p' = p/(p-1)$. This duality, together with the result of Pansu and some algebraic computations, implies that for a manifold M as in Theorem 1.3 we also have $\overline{H}^k_{p,p}(M) = 0$ if

$$p \geqslant \frac{(n-1) + \varepsilon(n-k)}{k-1}.$$

- Consider, for example, the hyperbolic space $\mathbb{H}^n$. This space is a Cartan–Hadamard manifold with constant sectional curvature $K \equiv -1$, and the reduced cohomology is known. Indeed, we have $\varepsilon = 0$ and, by the above three inequalities, $\overline{H}^k_{p,p}(\mathbb{H}^n) \neq 0$ if and only if $p \in (\frac{n-1}{k}, \frac{n-1}{k-1})$ (or, equivalently, $\frac{n-1}{p} < k < \frac{n-1}{p} + 1$). This assertion also follows from the computation of the L_p-cohomology of warped cylinders in [3, 4].

For $\varepsilon > 0$ there is still a gap between vanishing and nonvanishing results for the $L_{p,p}$-cohomology. When $\varepsilon \geqslant \frac{1}{n-1}$, the estimate (1.2) no longer gives any information about the $L_{p,p}$-cohomology. Note, by contrast, that Theorem 1.3 always produces some nonvanishing $L_{q,p}$-cohomology.

2 Manifolds with Contraction onto the Closed Unit Ball

Theorem 2.1 below, inspired by [8], can be regarded as an application of the concept of almost duality in [5]. It will be used in the proof of Theorem 1.3. Recall that, by the Rademacher theorem, a Lipshitz map $f : M \to N$ is differentiable for almost all $x \in M$ and its differential df_x defines a homomorphism

$$\Lambda^k f_x : \Lambda^k(T_{fx}N) \to \Lambda^k(T_x M).$$

Denote by $|\Lambda^k f_x|$ the norm of this homomorphism.

Theorem 2.1. *Let (M, g) be a complete Riemannian manifold, and let $f : M \to \overline{\mathbb{B}}^n$ be a Lipschitz map such that*

$$|\Lambda^k f| \in L^p(M) \quad \textit{and} \quad |\Lambda^{n-k} f| \in L^{q'}(M),$$

where $\overline{\mathbb{B}}^n$ is the closed unit ball in $\mathbb{R}^n$ and $q' = q/(q-1)$. Assume that

$$f^*\omega \in L^1(M) \quad \textit{and} \quad \int\limits_M f^*\omega \neq 0,$$

where $\omega = dx_1 \wedge dx_2 \wedge \cdots \wedge dx_n$ is the standard volume form on $\overline{\mathbb{B}}^n$. Then $H^k_{q,p}(M) \neq 0$.

Furthermore, if $|\Lambda^{n-k} f| \in L^{p'}(M)$ for $p' = \frac{p}{p-1}$, then $\overline{H}^k_{q,p}(M) \neq 0$.

The proof of this theorem uses the following "almost duality" result.

Proposition 2.2. *Let (M, g) be a complete Riemannian manifold, and let $\alpha \in Z_p^k(M)$. Assume that there exists a closed $(n-k)$-form $\gamma \in Z_{q'}^{n-k}(M)$ for $q' = \frac{q}{q-1}$ such that $\gamma \wedge \alpha \in L^1(M)$ and*

$$\int_M \gamma \wedge \alpha \neq 0.$$

Then $H_{q,p}^k(M) \neq 0$. Furthermore, if $\gamma \in Z_{p'}^{n-k}(M) \cap Z_{q'}^{n-k}(M)$ for $p' = \frac{p}{p-1}$ and $q' = \frac{q}{q-1}$, then $\overline{H}_{q,p}^k(M) \neq 0$.

This assertion is contained in [5, Propositions 8.4 and 8.5].

We also need some facts about locally Lipschitz differential forms.

Lemma 2.3. *For any locally Lipschitz functions $g, h_1, \ldots, h_k : M \to \mathbb{R}$*

$$d(g\, dh_1 \wedge dh_2 \wedge \ldots \wedge dh_k) = dg \wedge dh_1 \wedge dh_2 \wedge \ldots \wedge dh_k$$

in the weak sense.

Denote by $\mathrm{Lip}^*(M)$ the algebra generated by locally Lipschitz functions and the wedge product. By Lemma 2.3, $\mathrm{Lip}^*(M)$ is a graded differential algebra. Elements of $\mathrm{Lip}^*(M)$ are referred to as *locally Lipschitz forms.*

Proposition 2.4. *For any locally Lipschitz map $f : M \to N$ between two Riemannian manifolds the pullback $f^*(\omega)$ of any locally Lipschitz form ω is a locally Lipschitz form and $d(f^*(\omega)) = f^*(d\omega)$.*

A proof of Lemma 2.3 and Proposition 2.4 can be found in [1] (see also [6] for some related results).

Proof of Theorem 2.1. We set $\omega' = dx_1 \wedge dx_2 \wedge \cdots \wedge dx_k$ and $\omega'' = dx_{k+1} \wedge dx_2 \wedge \cdots \wedge dx_n$. Using the inequality

$$|(f^*\omega)_x| \leqslant |\Lambda^k f| \cdot \left|\omega_{f(x)}\right|,$$

we find that

$$\|f^*\omega'\|_{L^p(M,\Lambda^k)} = \left(\int_M |(f^*\omega')_x|^p\, dx\right)^{\frac{1}{p}} \leqslant \left(\int_M \left(|\Lambda^k f|^p \cdot \left|\omega'_{f(x)}\right|^p\right) dx\right)^{\frac{1}{p}}$$
$$\leqslant \left\|\Lambda^k f\right\|_{L^p(M)} \|\omega'\|_{L^\infty(M,\Lambda^k)} < \infty.$$

We set $\alpha = f^*\omega'$. Since f is a Lipschitz map, α is a Lipshitz form and, by Proposition 2.4, we have $d\alpha = f^*d\omega' = 0$. The previous inequality says that $\alpha \in L^p(M, \Lambda^k)$. Thus, $\alpha \in Z_p^k(M)$. The same argument shows that $\gamma =\in Z_{q'}^{n-k}(M)$, where $\gamma = f^*\omega''$.

By assumption, $\alpha \wedge \gamma = f^*(\omega' \wedge \omega'') = f^*(\omega) \in L^1(M)$ and

$$\int\limits_M \gamma \wedge \alpha = \int\limits_M f^*\omega \neq 0.$$

Hence, by Proposition 2.2, we have $H_{q,p}^k(M) \neq 0$.

If, additionally, we assume that $\Lambda^{n-k} f_x \in L^{p'}(M)$ for $p' = \frac{p}{p-1}$, then $\gamma \in Z_{p'}^{n-k}(M)$ and, by the second part of Proposition 2.2, $\overline{H}_{q,p}^k(M) \neq 0$. □

The paper [7] contains other results relating the $L_{q,p}$-cohomology and classes of mappings.

3 Proof of the Main Result

Let (M, g) be a complete simply connected manifold of negative sectional curvature of dimension n. Fix a base point $o \in M$ and identify T_oM with $\mathbb{R}^n$ by a linear isometry. Then the exponential map $\exp_o : \mathbb{R}^n = T_oM \to M$ is a diffeomorphism, and we define a map $f : M \to \overline{\mathbb{B}}^n$, where $\overline{\mathbb{B}}^n \subset \mathbb{R}^n$ is the closed Euclidean unit ball, by the formula

$$f(x) = \begin{cases} \exp_o^{-1}(x) & \text{if } |\exp_o^{-1}(x)| \leqslant 1, \\[2ex] \dfrac{\exp_o^{-1}(x)}{|\exp_o^{-1}(x)|} & \text{if } |\exp_o^{-1}(x)| \geqslant 1. \end{cases}$$

Using the polar coordinates (r, u) on M, i.e., writing a point $x \in M$ as $x = \exp_o(r \cdot u)$ with $u \in \mathbb{S}^{n-1}$ and $r \in [0, \infty)$, we can write this map as $f(r, u) = \min(r, 1) \cdot u$. Since the exponential map is expanding, the map $f : M \to \overline{\mathbb{B}}^n$ is contracting and, in particular, is a Lipschitz map.

Recall that $\omega = dx_1 \wedge dx_2 \wedge \cdots \wedge dx_n$ is the volume form on $\overline{\mathbb{B}}^n$. It can also be written as $r^{n-1} dr \wedge d\sigma_0$, where $d\sigma_0$ is the volume form of the standard sphere $\mathbb{S}^{n-1}$. It follows that $f^*\omega = 0$ on the set $\{x \in M \mid d(o, x) > 1\}$ and $f^*\omega$ has compact support and is integrable. Denote by $U_1 = \{x \in M \mid d(o, x) < 1\}$ the Riemannian open unit ball in M. The restriction of f to U_1 is a diffeomorphism onto $\mathbb{B}^n$.

Therefore,

$$\int\limits_M f^*\omega = \int\limits_{U_1} f^*\omega = \int\limits_{\mathbb{B}^n} \omega = \mathrm{Vol}\,(\mathbb{B}^n) > 0.$$

The next lemma implies that $|\Lambda^k f| \in L^p(M)$ if

$$\frac{1+\varepsilon}{p} < \frac{k}{n-1}$$

and $|\Lambda^{n-k} f| \in L^{q'}(M)$ if

$$\frac{1+\varepsilon}{q'} < \frac{n-k}{n-1}.$$

Note that the inequality

$$\frac{1+\varepsilon}{q'} < \frac{n-k}{n-1}$$

is equivalent to the inequality

$$\frac{k-1}{n-1} + \varepsilon < \frac{1+\varepsilon}{q}$$

since $q' = q/(q-1)$. Likewise, $|\Lambda^{n-k} f| \in L^{p'}(M)$ if

$$\frac{k-1}{n-1} + \varepsilon < \frac{1+\varepsilon}{p}.$$

In conclusion, the map f satisfies all the assumptions of Theorem 2.1, as soon as assumption (A) or (B) of Theorem 1.3 is fulfilled. The proof of Theorem 1.3 is complete.

Lemma 3.1. *The map* $f : M \to \overline{\mathbb{B}}^n$ *satisfies* $|\Lambda^m f| \in L^s(M)$ *if*

$$\frac{1+\varepsilon}{s} < \frac{m}{n-1}.$$

Proof. By the Gauss lemma from Riemannian geometry, we know that, in the polar coordinates, $M \simeq [0, \infty) \times \mathbb{S}^{n-1}/(\{0\} \times \mathbb{S}^{n-1})$, the Riemannian metric can be written as

$$g = dr^2 + g_r,$$

where g_r is a Riemannian metric on the sphere $\mathbb{S}^{n-1}$. The Rauch comparison theorem tells us that if the sectional curvature of g satisfies $K \leqslant -1$, then

$$g_r \leqslant (\sinh(r))^2 \, g_0, \tag{3.1}$$

where g_0 is the standard metric on the sphere $\mathbb{S}^{n-1}$ (see any textbook on Riemannian geometry, for example, [12, Sect. 6.2, Corollary 2.4] or [9, Corollary 4.6.1]). Using the fact that the Euclidean metric on $\mathbb{R}^n = T_o M$ is written in the polar coordinates as $ds^2 = dr^2 + r^2 g_0$ and taking into account the inequality (3.1), we find

$$|f^*(\theta)| \leqslant \frac{r}{\sinh(r)} |\theta|$$

for any covector $\theta \in T^*_{(r,u)}M$, orthogonal to dr. Since $f^*(dr)$ has compact support, we conclude that

$$|f^*(\varphi)| \leqslant \text{const} \left(\frac{r}{\sinh(r)}\right)^m |\varphi|$$

for any m-form $\varphi \in \Lambda^m(T^*_{(r,u)}M)$. In other words, we have the pointwise estimate

$$|\Lambda^m f|_{(r,u)} \leqslant \text{const} \left(\frac{r}{\sinh(r)}\right)^m . \tag{3.2}$$

By the Ricci curvature comparison estimate, $\text{Ric} \geqslant -(1+\varepsilon)^2(n-1)$ implies that the volume form of (M, g) satisfies

$$\text{dvol} \leqslant \left(\frac{\sinh((1+\varepsilon)r)}{1+\varepsilon}\right)^{n-1} dr \wedge d\sigma_0, \tag{3.3}$$

where $d\sigma_0$ is the volume form of the standard sphere $\mathbb{S}^{n-1}$ (see, for example, [12, Sect. 9.1.1]). The above inequalities give us a control of the growth of $|\Lambda^m f|^s_{(r,u)}\text{dvol}$. To be precise, let us choose a number t such that

$$\frac{m(1+\varepsilon)}{n-1} < t < s.$$

Then (3.2) and (3.3) imply

$$|\Lambda^m f|^s_{(r,u)}\text{dvol} \leqslant \text{const } \mathrm{e}^{-ar} dr \wedge d\sigma_0$$

with $a = mt - (n-1)(1+\varepsilon) > 0$. The last inequality implies the integrability of $|\Lambda^m f|^s_{(r,u)}$:

$$\int_M |\Lambda^m f|^s_{(r,u)}\text{dvol} \leqslant \text{Vol}\,(\mathbb{S}^{n-1}) \int_0^\infty \mathrm{e}^{-ar} dr < \infty.$$

The lemma is proved. □

Acknowledgement. Research of the first author was supported in part by Israel Science Foundation.

References

1. Gol'dshtein, V.M., Kuz'minov, V.I., Shvedov, I.A.: Differential forms on Lipschitz Manifolds (Russian). Sib. Mat. Zh. **23**, no. 2, 16-30 (1982); English transl.: Sib. Math. J. **23**, 151-161 (1982)

2. Gol'dshtein, V.M., Kuz'minov, V.I., Shvedov, I.A.: Dual spaces of spaces of differential forms (Russian). Sib. Mat. Zh. **27**, no. 1, 45-56 (1986); English transl.: Sib. Math. J. **27**, 35–44 (1986)
3. Gol'dshtein, V.M., Kuz'minov, V.I., Shvedov, I.A.: Reduced L_p-cohomology of warped cylinders (Russian). Sib. Mat. Zh. **31**, no. 5, 10-23 (1990); English transl.: Sib. Math. J. **31**, no. 5, 716-727 (1990)
4. Gol'dshtein, V.M., Kuz'minov, V.I., Shvedov, I.A.: L_p-Cohomology of warped cylinders (Russian). Sib. Mat. Zh. **31**, no. 6, 55-63 (1990); English transl.: Sib. Math. J. **31**, no. 6, 919-925 (1990)
5. Gol'dshtein, V., Troyanov, M.: Sobolev inequality for differential forms and $L_{q,p}$-cohomology. J. Geom. Anal. **16**, no 4, 597-631 (2006)
6. Gol'dshtein, V., Troyanov, M.: On the naturality of the exterior differential. To appear in Mathematical Reports of the Canadian Academy of Sciences (also on arXiv:0801.4295v1)
7. Gol'dshtein, V., Troyanov, M.: Distortion of Mappings and $L_{q,p}$-Cohomology. Preprint. arXiv:0804.0025v1.
8. Gromov, M.: Asymptotic invariants of infinite groups. In: Geometric Group Theory. Vol. 2. Cambridge Univ. Press, Cambridge (1992)
9. Jost, J.: Riemannian Geometry and Geometric Analysis. Springer, Berlin (2005)
10. Maz'ya, V.G.: Sobolev Spaces. Springer, Berlin-Tokyo (1985)
11. Pansu, P.: Cohomologie L^p et pincement Comment. Math. Helv. **83**, 327-357 (2008)
12. Petersen, P.: Riemannian Geometry. Springer, New York (2006)

Volume Growth and Escape Rate of Brownian Motion on a Cartan–Hadamard Manifold

Alexander Grigor'yan and Elton Hsu

Abstract We prove an upper bound for the escape rate of Brownian motion on a Cartan–Hadamard manifold in terms of the volume growth function. One of the ingredients of the proof is the Sobolev inequality on such manifolds.

1 Introduction

Let M be a geodesically complete noncompact Riemannian manifold. We denote by $d(x, y)$ the geodesic distance between x and y and by μ the Riemannian volume measure. We use $\mathbb{P}_x$ to denote the diffusion measure generated by the Laplace–Beltrami operator Δ. Let $X = \{X_t,\ t \in \mathbb{R}_+\}$ be the coordinate process on the path space $W(M) = C(\mathbb{R}_+, M)$. By definition, $\mathbb{P}_x$ is a probability measure on $W(M)$ under which X is a Brownian motion starting from x.

Fix a reference point $z \in M$, and let $\rho(x) = d(x, z)$. We say that a function $R(t)$ is an *upper rate function* for Brownian motion on M if

$$\mathbb{P}_z\{\rho(X_t) \leqslant R(t) \text{ for all sufficiently large } t\} = 1.$$

The purpose of this paper is to study the rate of escape of Brownian motion on M in terms of the volume growth function. Let us first point out that the notion of an upper rate function makes sense only if the lifetime of Brownian motion is infinite. In this case, the manifold M is called stochastically

Alexander Grigor'yan
University of Bielefeld, 33501 Bielefeld, Germany,
e-mail: grigor\@math.uni-bielefeld.de

Elton Hsu
Northwestern University, Evanston, IL 60208, USA,
e-mail: ehsu\@math.northwestern.edu

V. Maz'ya (ed.), *Sobolev Spaces in Mathematics II*,
International Mathematical Series.
doi: 10.1007/978-0-387-85650-6,

complete. The stochastic completeness is equivalent to the identity

$$\int_M p(t,x,y)\,d\mu(y) = 1,$$

where $p(t,x,y)$ is the minimal heat kernel on M, which is also the transition density function of Brownian motion on M.

Let $B(z,R)$ be the geodesic ball of radius R centered at z. It was proved [3] that M is stochastically complete if

$$\int^{\infty} \frac{r\,dr}{\log \mu(B(z,R))} = \infty. \tag{1.1}$$

The integral in (1.1) will be used in this paper to construct an upper rate function. Before we state the result, let us briefly survey the existing estimates of escape rate.

- The classical Khinchin law of the iterated logarithm says that for a Brownian motion in $\mathbb{R}^n$ with probability 1

  $$\limsup_{t\to\infty} \frac{\rho(X_t)}{\sqrt{4t\log\log t}} = 1$$

  (the factor 4 instead of the classical 2 appears because, in our setting, a Brownian motion is generated by Δ rather than $\frac{1}{2}\Delta$). It follows that for any $\varepsilon > 0$

  $$R(t) = \sqrt{(4+\varepsilon)t\log\log t} \tag{1.2}$$

  is an upper rate function.
- If M has nonnegative Ricci curvature, then (1.2) is again an upper rate function on M (see [9, Theorem 1.3] and [7, Theorem 4.2]).
- If the volume growth function is at most polynomial, i.e.,

  $$\mu(B(z,r)) \leqslant Cr^D$$

  for large enough r and some positive constants C and D, then the function

  $$R(t) = \text{const}\ \sqrt{t\log t} \tag{1.3}$$

  is an upper rate function (see [15, Theorem 5.1], [7, Theorem 1.1], and [9, Theorem 1.1]). Note that the logarithm in (1.3) is single in contrast to (1.2) and, in general, cannot be replaced by the iterated logarithm (see [1, 10]).
- If the volume growth function admits a sub-Gaussian exponential estimate

  $$\mu(B(z,r)) \leqslant \exp(Cr^{\alpha})$$

where $0 < \alpha < 2$, then the function

$$R(t) = \text{const}\, t^{\frac{1}{2-\alpha}}$$

is an upper rate function (see [7, Theorem 4.1]).

Note that (1.1) is satisfied if the volume growth function admits the Gaussian exponential estimate

$$\mu(B(z,r)) \leqslant \exp(Cr^2) \tag{1.4}$$

(under the condition (1.4), the stochastic completeness was also proved by different methods in [15], [12], and [2]). However, none of the existing results provided any estimates of escape rate under the condition (1.4), let alone under the volume growth function $\exp(Cr^2 \log r)$ and the like.

We construct an upper rate function under the most general condition (1.1). However, we assume, in addition, that M is a Cartan–Hadamard manifold, i.e., a geodesically complete simply connected Riemannian manifold of nonnegative sectional curvature. The property of Cartan–Hadamard manifolds that we use is the Sobolev inequality: if $N = \dim M$, then for any function $f \in C_0^\infty(M)$

$$\left(\int_M |f|^{\frac{N}{N-1}} d\mu \right)^{\frac{N-1}{N}} \leqslant C_N \int_M |\nabla f| d\mu, \tag{1.5}$$

where C_N is a constant depending only on N (see [11]). The Sobolev inequality allows us to carry through the Moser iteration argument in [14] and prove a mean value estimate for solutions of the heat equation on M, which is one of the ingredients of our proof.

Now, we state our main result.

Theorem 1.1. *Let M be a Cartan–Hadamard manifold. Assume that the following volume estimate holds for a fixed point $z \in M$ and all sufficiently large R:*

$$\mu(B(z,R)) \leqslant \exp(f(R)), \tag{1.6}$$

where $f(R)$ is a positive, strictly increasing, and continuous function on $[0, +\infty)$ such that

$$\int^\infty \frac{r\,dr}{f(r)} = \infty. \tag{1.7}$$

Let $\varphi(t)$ be the function on $\mathbb{R}_+$ defined by

$$t = \int_0^{\varphi(t)} \frac{r\,dr}{f(r)}. \tag{1.8}$$

Then $R(t) = \varphi(Ct)$ is an upper rate function for Brownian motion on M for some absolute constant C (for example, for any $C > 128$).

If we set $f(R) = \log \mu(B(z,R))$ for large R, then the condition (1.7) becomes identical to (1.1). Under this condition, Theorem 1.1 guarantees the existence of an upper rate function $R(t)$. This, in particular, means that in a finite time Brownian motion stays with probability one in a bounded set, which implies that the life time of Brownian motion is infinite almost surely. Hence the manifold M is stochastically complete. This recovers the above cited result that (1.1) on geodesically complete manifolds implies the stochastic completeness, although under the additional assumption that M is Cartan–Hadamard.

Let us show some examples.

- If
$$\mu(B(z,R)) \leqslant CR^D \tag{1.9}$$
for some constants C and D, then (1.6) holds with
$$f(R) = D \log R + \text{const}$$
and (1.8) yields
$$t \simeq \frac{\varphi^2}{2D \log \varphi}.$$
It follows that $\log t \simeq \log \varphi^2$ and
$$\varphi(t) \simeq \sqrt{Dt \log t}.$$
Hence the function
$$R(t) = \sqrt{CDt \log t}$$
is an upper rate function which matches the above cited results of [7, 9, 15].
- If $\mu(B(z,R)) \leqslant \exp(Cr^\alpha)$ for some $0 < \alpha < 2$, then (1.6) holds with $f(R) = Cr^\alpha$ and (1.8) yields $t \simeq \varphi(t)^{2-\alpha}$. Hence we obtain the upper rate function
$$R(t) = Ct^{\frac{1}{2-\alpha}}$$
which matches the above cited result [7].
- If
$$\mu(B(z,R)) \leqslant \exp(CR^2),$$
then $f(R) = CR^2$. Then (1.8) yields $t \simeq \ln \varphi(t)$. Hence we obtain the upper rate function
$$R(t) = \exp(Ct).$$
This result is new. Similarly, if
$$\mu(B(z,R)) \leqslant \exp(CR^2 \log R),$$

then (1.8) yields $t \simeq \log\log\varphi$. Hence

$$R(t) = \exp(\exp Ct)).$$

The hypothesis that M is Cartan–Hadamard can be replaced by the requirement that the Sobolev inequality (1.5) holds on M. Furthermore, the method goes through also in the setting of weighted manifolds, when measure μ is not necessarily the Riemannian measure, but has a smooth positive density, say $\sigma(x)$, with respect to the Riemannian measure. Then, instead of the Laplace–Beltrami operator, one should consider the weighted Laplace operator

$$\Delta_\mu = \frac{1}{\sigma}\operatorname{div}(\sigma\nabla)$$

which is symmetric with respect to μ. Theorem 1.1 extends to the weighted manifolds that are geodesically complete and satisfy the Sobolev inequality (1.5).

This paper is organized as follows. Section 2 contains the proof of an upper bound for certain positive solutions of the heat equation. In Sect. 3, we prove the main result stated above. In Sect. 4, we compute the sharp upper rate function on model manifold and show that, for a certain range of volume growth functions, the upper rate function of Theorem 1.1 is sharp up to a constant factor in front of t.

2 Heat Equation Solution Estimates

In this section, we prove a pointwise upper bound of certain solutions of the heat equation on a Cartan–Hadamard manifold M (Theorem 2.3). It is an easy consequence of an L^2-bound for a general complete manifold and a mean value type inequality for a Cartan–Hadamard manifold. These two upper bounds are known, and we state them as lemmas.

For any set $A \subset M$ let A_r be the open r-neighborhood of A in M.

Lemma 2.1. *Let M be a geodesically complete Riemannian manifold. Suppose that $u(x,t)$ is a smooth subsolution to the heat equation in the cylinder $A_r \times [0,T]$, where $A \subset M$ is a compact set and $r, T > 0$ (see Fig. 1). Assume also that $0 \leqslant u(x,t) \leqslant 1$ and $u(x,0) = 0$ on A_r. Then for any $t \in (0,T]$*

$$\int_A u^2(x,t)d\mu(x) \leqslant \mu(A_r)\max\left(1, \frac{r^2}{2t}\right)\exp\left(-\frac{r^2}{2t}+1\right).$$

For the proof see [6, Theorem 3] (see also [7, Proposition 3.6]). Note that no geometric assumption about M is made except for the geodesic completeness.

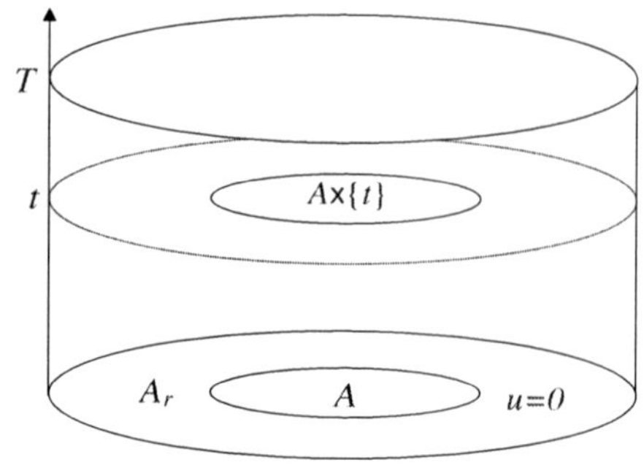

Fig. 1 Illustration to Lemma 2.1.

The proof exploits essentially the property of the geodesic distance function that $|\nabla d| \leqslant 1$.

Takeda [15] proved a similar estimate for $\int\limits_A u(x,t)d\mu(x)$ by using a different probabilistic argument. However, it is more convenient for us to work with the L^2 rather than with L^1 estimate in view of the following lemma.

Lemma 2.2. *Let M be a Cartan–Hadamard manifold of dimension N. Suppose that $u(x,t)$ is a smooth nonnegative subsolution to the heat equation in a cylinder $B(y,r) \times [0,T]$, where $r, T > 0$ (see Fig. 2). Then*

$$u(y,T)^2 \leqslant \frac{C_N}{\min(\sqrt{T},\, r)^{N+2}} \int\limits_0^T \int\limits_{B(y,r)} u^2(x,t)\, d\mu(x)dt, \tag{2.1}$$

where C_N is a constant depending only on N.

Proof. As was already mentioned above, a Cartan–Hadamard manifold admits the Sobolev inequality (1.5). By a standard argument, (1.5) implies the Sobolev–Moser inequality

$$\int\limits_M |f|^{2+\frac{4}{N}}\, d\mu \leqslant C_N \Big(\int\limits_M |f|^2\, d\mu \Big)^{2/N} \int\limits_M |\nabla f|^2\, d\mu,$$

which leads, by the Moser iteration argument [14], to the mean value inequality (2.1). Note that the value of C_N may be different in all the above inequalities.

An alternative proof of the implication (1.5)⇒(2.1) can be found in [4] (see also [5, Theorem 3.1 and formula (3.4)]). □

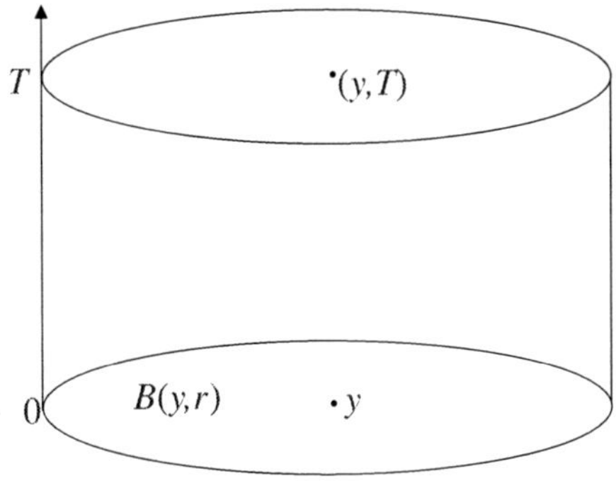

Fig. 2 Illustration to Lemma 2.2.

With these two preliminary results, we prove the main inequality in this section. This inequality gives an upper bound for probabilities of escaping times not exceeding a given upper bound.

Theorem 2.3. *Let M be a Cartan–Hadamard manifold of dimension N. Suppose that $u(x,t)$ is a smooth subsolution to the heat equation in a cylinder $B(y,2r)\times[0,T]$, where $r,T>0$. If $0\leqslant u\leqslant 1$ in this cylinder and $u(x,0)=0$ on $B(y,2r)$, then*

$$u(y,T)\leqslant C_N\sqrt{\mu(B(y,2r))}\frac{\max(\sqrt{T},r)}{\min(\sqrt{T},r)^{1+N/2}}\exp\left(-\frac{r^2}{4T}\right). \tag{2.2}$$

Proof. We use Lemma 2.1 with $A=\overline{B(y,r)}$. Then $A_r=B(y,2r)$ and for any $0<t\leqslant T$

$$\int\limits_{B(y,r)} u^2(x,t)d\mu(x)\leqslant \mu(B(y,2r))\max\left(1,\frac{r^2}{2t}\right)\exp\left(-\frac{r^2}{2t}+1\right)$$

(see Fig. 3). Hence

$$\int\limits_{T/2}^{T}\int\limits_{B(y,r)} u^2(x,t)d\mu(x)dt$$
$$\leqslant 2\mu(B(y,2r))T\max\left(1,\frac{r^2}{T}\right)\exp\left(-\frac{r^2}{2T}\right).$$

Applying Lemma 2.2 in the cylinder $B(y,r)\times[T/2,T]$, we obtain

$$u(y,T)^2 \leqslant C_N \mu(B(y,2r)) \frac{\max(T,r^2)}{\min(\sqrt{T},r)^{N+2}} \exp\left(-\frac{r^2}{2T}\right).$$

Hence (2.2) follows. □

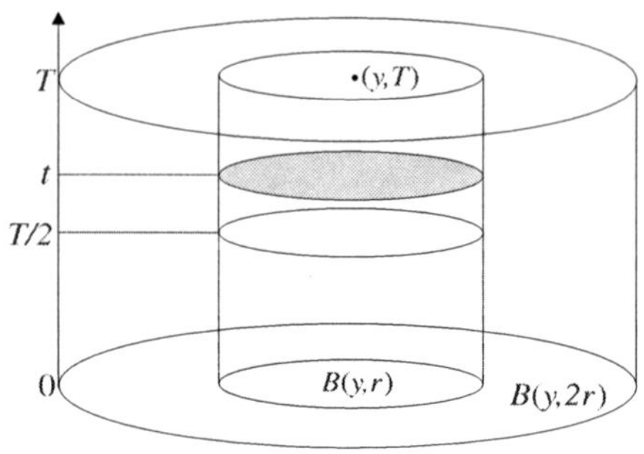

Fig. 3 Illustration to the proof of Theorem 2.3.

3 Escape Rate of Brownian Motion

We first explain the main idea of the proof. For any open set $\Omega \subset M$ we denote by τ_Ω the first exit time from Ω, i.e.,

$$\tau_\Omega = \inf\{t > 0 : X_t \notin \Omega\}.$$

Recall that $B(x,r)$ denotes the geodesic ball of radius r centered at x. Fix a reference point $z \in M$ and set $\rho(x) = d(x,z)$.

Let $\{R_n\}_{n=1}^{\infty}$ be a sequence of strictly increasing radii to be fixed later such that $\lim_{n\to\infty} R_n = \infty$. Consider the following sequence of stopping times:

$$\tau_n = \tau_{B(z,R_n)}.$$

Then $\tau_n - \tau_{n-1}$ is the amount of time the Brownian motion X_t takes to cross from $\partial B(z,R_{n-1})$ to $\partial B(z,R_n)$ for the first time (if $n = 0$, then we set $R_0 = 0$ and $\tau_0 = 0$). Let $\{c_n\}_{n=1}^{\infty}$ be a sequence of positive numbers to be fixed later. Suppose that we can show that

$$\sum_{n=1}^{\infty} \mathbb{P}_z\{\tau_n - \tau_{n-1} \leqslant c_n\} < \infty. \tag{3.1}$$

Then, by the Borel–Cantelli lemma, with $\mathbb{P}_z$-probability 1 we have

$$\tau_n - \tau_{n-1} > c_n \quad \text{for all large enough } n. \tag{3.2}$$

For any $n \geqslant 1$ we set

$$T_n = \sum_{k=1}^{n} c_k.$$

From (3.2) it follows that for all sufficiently large n

$$\tau_n > T_n - T_0,$$

where T_0 is a large enough (random) number. In other words, we have the implication

$$t \leqslant T_n - T_0 \;\Rightarrow\; \rho(X_t) \leqslant R_n \text{ if } \; n \text{ is large enough.} \tag{3.3}$$

Let ψ be an increasing bijection of $\mathbb{R}_+$ onto itself such that

$$T_{n-1} - \psi(R_n) \to +\infty \text{ as } n \to \infty. \tag{3.4}$$

We claim that ψ^{-1} is an upper rate function. Indeed, for large enough t we choose n such that

$$T_{n-1} - T_0 < t \leqslant T_n - T_0.$$

If t is large enough, then also n is large enough so that, by (3.3),

$$\rho(X_t) \leqslant R_n$$

and, by (3.4),

$$T_{n-1} - \psi(R_n) > T_0.$$

It follows that

$$t > T_{n-1} - T_0 > \psi(R_n).$$

Hence

$$\rho(X_t) \leqslant R_n < \psi^{-1}(t),$$

which proves that ψ^{-1} is an upper rate function.

Now, let us find c_n such that (3.1) is true. By the strong Markov property of Brownian motion, we have

$$\mathbb{P}_z\{\tau_n - \tau_{n-1} \leqslant c_n\} = \mathbb{E}_z \mathbb{P}_{X_{\tau_{n-1}}}\{\tau_n \leqslant c_n\}. \tag{3.5}$$

Note that $X_{\tau_{n-1}} \in \partial B(z, R_{n-1})$. If a Brownian motion starts from a point $y \in \partial B(z, R_{n-1})$, then it has to travel no less than the distance

$$r_n = R_n - R_{n-1}$$

before it reaches $\partial B(z, R_n)$ (see Fig. 4). Hence

$$\mathbb{P}_y\{\tau_n \leqslant c_n\} \leqslant \mathbb{P}_y\{\tau_{B(y,r_n)} \leqslant c_n\}, \quad y \in \partial B(z, R_{n-1}).$$

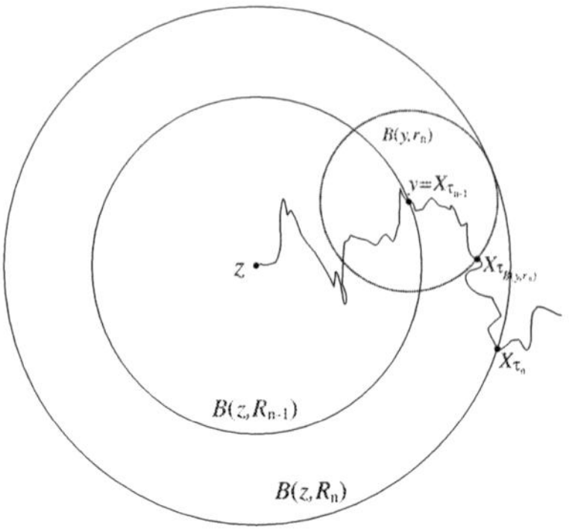

Fig. 4 Brownian motion X_t exits the ball $B(y, r_n)$ before $B(z, R_n)$.

From the above inequality and (3.5) we obtain

$$\mathbb{P}_z\{\tau_n - \tau_{n-1} \leqslant c_n\} \leqslant \sup_{y \in \partial B(z,R_{n-1})} \mathbb{P}_y\{\tau_{B(y,r_n)} \leqslant c_n\}. \tag{3.6}$$

For a fixed $y \in \partial B(z, R_{n-1})$ we consider the function

$$u(x,t) = \mathbb{P}_x\{\tau_{B(y,r)} \leqslant t\}.$$

Clearly, $u(x,t)$ is the solution of the heat equation in the cylinder $B(y,r) \times \mathbb{R}_+$. Furthermore, $0 \leqslant u \leqslant 1$ and

$$u(x,0) = 0 \quad \text{for } x \in B(y,r).$$

The probability we wanted to estimate is the value of the solution at the center of the ball:

$$\mathbb{P}_y\{\tau_{B(y,r_n)} \leqslant c_n\} = u(y, c_n).$$

Applying the estimate (2.2) of Theorem 2.3 in the cylinder $B(y, r_n) \times [0, c_n]$ and noting that $B(y, r_n) \subset B(z, R_n)$ so that

$$\mu(B(y, r_n)) \leqslant \exp(f(R_n)),$$

we obtain

$$u(y, c_n) \leqslant C_N \exp(f(R_n)/2) \frac{\max(\sqrt{c_n}, r_n)}{\min(\sqrt{c_n}, r_n)^{1+N/2}} \exp\left(-\frac{r_n^2}{16c_n}\right). \tag{3.7}$$

Now, we choose c_n to satisfy the identity

$$\frac{r_n^2}{16c_n} = f(R_n)$$

i.e.,

$$c_n = \frac{1}{16} \frac{r_n^2}{f(R_n)}.$$

Noticing that $c_n < r_n^2$ for large enough n, form (3.7) we obtain

$$\begin{aligned}\mathbb{P}_y\{\tau_{B(y,r_n)} \leqslant c_n\} &\leqslant C_N \frac{r_n}{\sqrt{c_n}^{1+N/2}} \exp(-f(R_n)/2) \\ &= C_N r_n^{-N/2} f(R_n)^{\frac{2+N}{4}} \exp(-f(R_n)/2) \\ &\leqslant C_N r_n^{-N/2}.\end{aligned}$$

Set now $R_n = 2^n$ so that $r_n = 2^{n-1}$. The above estimate together with (3.6) obviously implies

$$\sum_{n=1}^{\infty} \mathbb{P}_z\{\tau_n - \tau_{n-1} \leqslant c_n\} \leqslant C_N \sum_{n=1}^{\infty} r_n^{-N/2} < \infty,$$

i.e., (3.1).

Knowing the sequences $\{R_n\}$ and $\{c_n\}$, we can now determine a function ψ that satisfies (3.4). Indeed, we have

$$\begin{aligned}T_n = c_1 + \ldots + c_n &= \frac{1}{16} \sum_{k=1}^{n} \frac{r_k^2}{f(R_k)} = \frac{1}{128} \sum_{k=1}^{n} \frac{R_{k+1}(R_{k+1} - R_k)}{f(R_k)} \\ &\geqslant \frac{1}{128} \sum_{k=1}^{n} \int_{R_k}^{R_{k+1}} \frac{rdr}{f(r)} = \frac{1}{128} \int_{R_1}^{R_{n+1}} \frac{rdr}{f(r)}.\end{aligned}$$

Setting

$$\psi(r) = c \int_0^r \frac{rdr}{f(r)},$$

where $c < \frac{1}{128}$, and using (1.7), we obtain

$$T_n - \psi(R_{n+1}) \to \infty \text{ as } n \to \infty,$$

which is equivalent to (3.4). Therefore, ψ^{-1} is an upper rate function. Clearly, $\psi^{-1}(t) = \varphi(Ct)$, where φ is defined by (1.8) and $C = c^{-1}$, which completes the proof of our main result, Theorem 1.1.

4 Escape Rate on Model Manifolds

In this section, we compute sharp upper rate function on model manifolds and compare it to the one from Theorem 1.1. We first illustrate the method in a simple case when M is a hyperbolic space.

4.1 Constant curvature

Let M be the hyperbolic space $\mathbb{H}_K^N$ of dimension N and of the constant sectional curvature $-K$. Then $\mu(B(z,r)) \leqslant Ce^{(N-1)Kr}$ so that we can take $f(r) = (N-1)Kr$. Theorem 1.1 yields the following upper rate function:

$$R(t) = CK(N-1)t.$$

In this case, a sharp upper rate function can be computed as follows. The radial process $r_t = \rho(X_t)$ satisfies the identity (see [13])

$$r_t = \sqrt{2}\,W_t + (N-1)\int_0^t K \coth Kr_s\,ds,$$

where W_t is a one-dimensional Brownian motion. We have

$$r_t \to \infty \text{ and } \frac{W_t}{t} \to 0$$

as $t \to \infty$. Therefore,

$$\frac{r_t}{t} \to (N-1)K.$$

Hence a sharp upper rate function is

$$R(t) = (1+\varepsilon)K(N-1)t,$$

where $\varepsilon > 0$.

4.2 General model manifolds

Here, M is not necessarily Cartan–Hadamard, but we do assume that M has a pole z, i.e., the exponential map $\exp : T_zM \to M$ is a diffeomorphism. Then the polar coordinates (ρ, θ) are defined on $M \setminus \{z\}$. The manifold M is said to be a *model* if the Riemannian metric of M is spherically symmetric, i.e., has the form

$$ds^2 = dr^2 + h(r)^2 d\theta^2, \tag{4.1}$$

where $h(r)$ is a smooth positive function of $r > 0$ and $d\theta^2$ is the canonical metric on $\mathbb{S}^{N-1}$ (note that θ varies in $\mathbb{S}^{N-1}$). For example, $\mathbb{R}^N$ is a model with $h(r) = r$ and the hyperbolic space $\mathbb{H}^N_K$ is a model with $h(r) = K^{-1} \sinh Kr$. The volume growth function of the metric (4.1) is

$$V(r) := \mu(B(z,r)) = \omega_N \int_0^r h(s)^{N-1} ds,$$

where ω_N is the $(N-1)$-volume of the unit sphere $\mathbb{S}^{N-1}$. The Laplace operator of the metric (4.1) is represented in the polar coordinates as follows:

$$\Delta = \frac{\partial^2}{\partial r^2} + m(r)\frac{\partial}{\partial r} + \frac{1}{h^2(r)}\Delta_{\mathbb{S}^{n-1}},$$

where $\Delta_{\mathbb{S}^{n-1}}$ is the Laplacian in the variable θ with respect to the canonical metric of $\mathbb{S}^{N-1}$ and

$$m(r) := (N-1)\frac{h'}{h} = \frac{V''}{V'}.$$

The function $m(r)$ plays an important role in what follows. Clearly, m satisfies the identity

$$V'(r) = V'(r_0) \exp\left(\int_{r_0}^r m(s) ds \right) \tag{4.2}$$

for all $r > r_0 > 0$. We assume in the sequel that

$$m(r) > 0 \text{ and } m'(r) \geqslant 0 \text{ for large enough } r \tag{4.3}$$

and

$$\int^{\infty} \frac{dr}{m(r)} = \infty. \tag{4.4}$$

For example, we have $m(r) = \frac{N-1}{r}$ in $\mathbb{R}^N$ and $m(r) = (N-1)K \coth Kr$ in $\mathbb{H}^N_K$. In neither case is the hypothesis (4.3) satisfied. On the other hand, if $V'(r) = \exp(r^\alpha)$, then $m(r) = \alpha r^{\alpha-1}$, and both (4.3) and (4.4) are satisfied provided that $1 \leqslant \alpha \leqslant 2$. If $V'(r) = \exp(r^2 \log^\beta r)$, then (4.3) and (4.4) hold for all $0 \leqslant \beta \leqslant 1$.

We claim that, under the condition (4.3), the Brownian motion on M is transient and, under the conditions (4.3)–(4.4), M is stochastically complete. We use the following well-known results (see [8]) that for model manifolds the transience is equivalent to

$$\int^{\infty} \frac{dr}{V'(r)} = \infty \tag{4.5}$$

and the stochastic completeness is equivalent to

$$\int^{\infty} \frac{V(r)}{V'(r)} dr = \infty. \tag{4.6}$$

Clearly, (4.3) implies $m(r) \geqslant c$ for some positive constant c and all large enough r. From (4.2) it follows that $V'(r)$ grows at least exponentially as $r \to \infty$, which implies (4.5). To prove (4.6), we observe that for large enough $r > r_0$

$$\begin{aligned} V(r) - V(r_0) &= \int_{r_0}^{r} V'(s)ds = \int_{r_0}^{r} \frac{V''(s)}{m(s)} ds \\ &\geqslant \frac{1}{m(r)} \int_{r_0}^{r} V''(s)ds = \frac{1}{m(r)} (V'(r) - V(r_0)). \end{aligned}$$

Therefore,

$$\frac{1}{m(r)} \leqslant \frac{V(r) - V(r_0)}{V'(r) - V'(r_0)} \sim \frac{V(r)}{V'(r)} \text{ as } r \to \infty.$$

Hence (4.6) follows from (4.4).

Let us define the function $r(t)$ by the identity

$$t = \int_0^{r(t)} \frac{ds}{m(s)}. \tag{4.7}$$

Our main result in this section is as follows.

Theorem 4.1. *Under the above assumptions, the function $r((1+\varepsilon)t)$ is the upper rate function for Brownian motion on M for any $\varepsilon > 0$, and is not for any $\varepsilon < 0$.*

Let us compare the function $r(t)$ with the upper rate function $R(t)$ given by Theorem 1.1, which is defined by the identity

$$\int_0^{R(t)} \frac{rdr}{\log V(r)} = Ct.$$

For "nice" functions $V(r)$ we have

$$\frac{V''}{V'} \simeq \frac{V'}{V} = (\log V)' \simeq \frac{\log V(r)}{r}, \tag{4.8}$$

which means that the functions $r(t)$ and $R(t)$ are comparable up to a constant multiple in front of t. For example, (4.8) holds for functions like $V(r) = \exp(r^\alpha)$ and $V(t) = \exp(r^\alpha \log^\beta r)$, where $\alpha > 0$, etc. On the other hand, it is easy to construct an example of $V(r)$ where $r(t)$ may be significantly less than $R(t)$ because one can modify a "nice" function $V(r)$ to make the second derivative $V''(r)$ very small in some intervals without affecting too much the values of V' and V. Then the function $r(t)$ in (4.7) will drop significantly, while $R(t)$ will not change very much.

Proof of Theorem 4.1. By the Ito decomposition, the radial process $r_t = \rho(X_t)$ satisfies the identity

$$r_t = \sqrt{2}W_t + \int_0^t m(r_s)\, ds, \tag{4.9}$$

where W_t is a one-dimensional Brownian motion (see [13]). Since the process X_t is transient, $r_t \to \infty$ as $t \to \infty$ with probability 1. Hence $m(r_t) \geqslant c$ for large enough t so that the second term on the right-hand side of (4.9) grows at least linearly in t. Since $W_t = o(t)$ as $t \to \infty$, we have with probability 1

$$r_t \sim \int_0^t m(r_s)\, ds \text{ as } t \to \infty. \tag{4.10}$$

Consider the function

$$u(t) = \int_0^t m(r_s) ds$$

From (4.10) it follows that for any $C > 1$ and large enough t

$$r_t \leqslant Cu(t). \tag{4.11}$$

Hence, by the monotonicity of m,

$$m(r_t) \leqslant m(Cu(t)).$$

Since $\frac{du}{dt}(t) = m(r_t)$, we obtain the differential inequality for $u(t)$

$$\frac{du}{dt} \leqslant m(Cu(t)).$$

Solving it by separation of variables, for large enough t_0 and all $t > t_0$ we obtain

$$\int_{Cu(t_0)}^{Cu(t)} \frac{d\xi}{m(\xi)} \leqslant C(t-t_0).$$

Hence

$$\int_{0}^{Cu(t)} \frac{d\xi}{m(\xi)} \leqslant Ct + C_0, \tag{4.12}$$

where C_0 is a large enough (random) constant. Comparing (4.12) with (4.7) and using again (4.11), we obtain

$$r_t \leqslant Cu(t) \leqslant r(Ct + C_0) \leqslant r(C^2 t)$$

for large enough r with probability 1. Since $C > 1$ was arbitrary, this proves that $r((1+\varepsilon)t)$ is an upper rate function for any $\varepsilon > 0$. In the same way, one proves that $r_t \geqslant r(C^{-2}t)$ for large enough t so that $r((1-\varepsilon)t)$ is not an upper rate function. □

Acknowledgement. The research of the first author was supported by the German Research Council (grant no. SFB 701). The research of the second author was supported in part by the NSF (grant no. DMS-0407819). This work was done during a visit of the second author at the University of Bielefeld, which was supported by grant no. SFB 701.

References

1. Barlow, M.T., Perkins, E.A.: Symmetric Markov chains in $\mathbb{Z}^d$: how fast can they move? Probab. Th. Rel. Fields **82**, 95-108 (1989)
2. Davies. E.B.: Heat kernel bounds, conservation of probability and the Feller property. J. Anal. Math. **58**, 99-119, (1992)
3. Grigor'yan, A.: On stochastically complete manifolds (Russian). Dokl. Akad. Nauk SSSR **290**, no. 3, 534-537 (1986); English transl.: Sov. Math., Dokl. **34**, no. 2, 310-313 (1987)
4. Grigor'yan, A.: The heat equation on noncompact Riemannian manifolds (Russian). Mat. Sb. **182**, no. 1, 55-87 (1991); English transl.: Math. USSR–Sb. **72**, no. 1, 47-77 (1992)
5. Grigor'yan, A.: Heat kernel upper bounds on a complete noncompact manifold. Rev. Mat. Iberoam. **10**, no. 2, 395-452 (1994)
6. Grigor'yan, A.: Integral maximum principle and its applications. Proc. R. Soc. Edinb., Sect. A, Math. **124**, 353-362 (1994)
7. Grigor'yan, A.: Escape rate of Brownian motion on weighted manifolds. Appl. Anal. **71**, no. 1-4, 63-89 (1999)
8. Grigor'yan, A.: Analytic and geometric background of recurrence and non-explosion of the Brownian motion on Riemannian manifolds. Bull. Am. Math. Soc. **36**, 135-249 (1999)

9. Grigor'yan, A., Kelbert M.: Range of fluctuation of Brownian motion on a complete Riemannian manifold. Ann. Prob. **26**, 78-111 (1998)
10. Grigor'yan, A., Kelbert M.: On Hardy–Littlewood inequality for Brownian motion on Riemannian manifolds. J. London Math. Soc. (2) **62**, 625-639 (2000)
11. Hoffman, D., Spruck, J.: Sobolev and isoperimetric inequalities for Riemannian submanifolds. Commun. Pure Appl. Math. **27**, 715–727 (1974); A correction to: "Sobolev and isoperimetric inequalities for Riemannian submanifolds." Commun. Pure Appl. Math. **28**, no. 6, 765–766 (1975)
12. Hsu, E.P.: Heat semigroup on a complete Riemannian manifold. Ann. Probab. **17**, 1248-1254 (1989)
13. Hsu, E.P.: Stochastic Analysis on Manifolds. Am. Math. Soc., Providence, RI (2002)
14. Moser J.: A Harnack inequality for parabolic differential equations. Commun. Pure Appl. Math. **17**, 101-134 (1964); Correction to "A Harnack inequality for parabolic differential equations." Commun. Pure Appl. Math. **20**, 231-236 (1967)
15. Takeda, M.: On a martingale method for symmetric diffusion process and its applications. Osaka J. Math **26**, 605-623 (1989)

Sobolev Estimates for the Green Potential Associated with the Robin–Laplacian in Lipschitz Domains Satisfying a Uniform Exterior Ball Condition

Tünde Jakab, Irina Mitrea, and Marius Mitrea

Dedicated to the memory of S.L. Sobolev

Abstract We show that if $u = G_\lambda f$ is the solution operator for the Robin problem for the Laplacian, i.e., $\Delta u = f$ in Ω, $\partial_\nu u + \lambda u = 0$ on $\partial\Omega$ (with $0 \leqslant \lambda \leqslant \infty$), then $G_\lambda : L^p(\Omega) \to W^{2,p}(\Omega)$ is bounded if $1 < p \leqslant 2$ and $\Omega \subset \mathbb{R}^n$ is a bounded Lipschitz domain satisfying a uniform exterior ball condition. This extends the earlier results of V. Adolfsson, B. Dahlberg, S. Fromm, D. Jerison, G. Verchota, and T. Wolff, who have dealt with Dirichlet ($\lambda = \infty$) and Neumann ($\lambda = 0$) boundary conditions. Our treatment of the end-point case $p = 1$ works for arbitrary Lipschitz domains and is conceptually different from the proof given by the aforementioned authors.

1 Introduction

Let Ω be a bounded Lipschitz domain in $\mathbb{R}^n$. We consider the Poisson problem for the Laplacian with Dirichlet and Neumann boundary conditions, i.e.,

$$(D)\begin{cases} \Delta u = f & \text{in } \Omega, \\ \operatorname{Tr} u = 0 & \text{on } \partial\Omega, \end{cases} \tag{1.1}$$

Tünde Jakab
University of Virginia, Charlottesville, VA 22904, USA, e-mail: tj8y@virginia.edu

Irina Mitrea
University of Virginia, Charlottesville, VA 22904, USA, e-mail: im3p@virginia.edu

Marius Mitrea
University of Missouri, Columbia, MO 65211, USA, e-mail: marius@math.missouri.edu

V. Maz'ya (ed.), *Sobolev Spaces in Mathematics II*, *International Mathematical Series*.
doi: 10.1007/978-0-387-85650-6,

and

$$(N)\begin{cases} \Delta u = f & \text{in } \Omega, \\ \partial_\nu u = 0 & \text{on } \partial\Omega. \end{cases} \tag{1.2}$$

Above, Tr stands for the boundary trace operator and ∂_ν denotes the directional derivative along ν, the outward unit normal to $\partial\Omega$. A natural question arising in this setting is whether there are numbers $p \in (1,\infty)$ for which the following implication holds:

$$f \in L^p(\Omega) \implies u \in W^{2,p}(\Omega). \tag{1.3}$$

Here and elsewhere, $W^{s,p}(\Omega)$, $1 \leqslant p \leqslant \infty$, $s \in \mathbb{R}$, stands for the L^p-based Sobolev space of order s in Ω. It has long been understood that this regularity issue is intimately linked to the analytic and geometric properties of the underlying domain Ω. To illustrate this point, let us briefly consider the case where $\Omega \subset \mathbb{R}^2$ is a polygonal domain with at least one re-entrant corner. In this scenario, let $\omega_1, \dots, \omega_N$ be the internal angles of Ω satisfying $\pi < \omega_j < 2\pi$, $1 \leqslant j \leqslant N$. Denote by $P_1, \dots, P_N$ the corresponding vertices. Then the solution to the Poisson problem equipped with a homogeneous Dirichlet boundary condition

$$\Delta u = f \in L^2(\Omega), \quad u \in W_0^{1,2}(\Omega) := \text{closure of } C_0^\infty(\Omega) \text{ in } W^{1,2}(\Omega), \tag{1.4}$$

permits the representation

$$u = \sum_{j=1}^{N} \lambda_j v_j + w, \qquad \lambda_j \in \mathbb{R}, \tag{1.5}$$

where $w \in W^{2,2}(\Omega) \cap W_0^{1,2}(\Omega)$ and, for each j, v_j is a function exhibiting a singular behavior at P_j of the following nature. Given $j \in \{1, \dots, N\}$, choose polar coordinates (r_j, θ_j) taking P_j as the origin and so that the internal angle is spanned by the half-lines $\theta_j = 0$ and $\theta_j = \omega_j$. Then

$$v_j(r_j, \theta_j) = \varphi_j(r_j, \theta_j) r_j^{\pi/\omega_j} \sin(\pi\theta_j/\omega_j), \qquad 1 \leqslant j \leqslant N, \tag{1.6}$$

where φ_j is a C^∞-smooth cut-off function of small support, which is identically one near P_j. In this scenario, $v_j \in W^{s,2}(\Omega)$ for every $s < 1 + (\pi/\omega_j)$, though $v_j \notin W^{1+(\pi/\omega_j),2}(\Omega)$. This implies that the best regularity statement regarding the solution of (1.4) is

$$u \in W^{s,2}(\Omega) \text{ for every } s < 1 + \frac{\pi}{\max\{\omega_1, \dots, \omega_N\}} \tag{1.7}$$

and this fails for the critical value of s. In particular, this provides a quantifiable way of measuring the failure of the implication (1.3) when $p = 2$ for

Lipschitz, piecewise C^∞ domains exhibiting inwardly directed irregularities. The interested reader is referred to the monograph [10] for a first-rate survey of the state of the art of this area, as well as for pertinent references to the rather vast literature dealing with partial differential equations in domains with isolated singularities.

From a different perspective, the $W^{2,p}$-regularity result (1.3) holds (both for (1.1) and (1.2)) whenever Ω is sufficiently smooth. For example, it is not too difficult to show that for a given $p \in (1, \infty)$

$$\partial\Omega \in C^{1,r} \text{ for some } r > 1 - 1/p \Longrightarrow \text{(1.3) holds.} \tag{1.8}$$

For near optimal conditions of this nature, see the excellent monograph [17] by Maz'ya and Shaposhnikova.

In this paper, we, however, are interested in the case where Ω is an irregular domain. More specifically, we assume that $\Omega \subset \mathbb{R}^n$ is a bounded Lipschitz domain (i.e., a bounded open set lying on just one side of its topological boundary which, in turn, is a surface locally described as the graph of a real-valued Lipschitz function defined in $\mathbb{R}^{n-1}$). As apparent from (1.7), the implication (1.3) fails in non-convex polygons when $p = 2$ so, in this case, extra conditions on Ω need to be imposed. More specifically, besides being Lipschitz, we require that Ω satisfies a *uniform exterior ball condition.* Heuristically, this ensures that all the singularities of $\partial\Omega$ are directed outwardly. This class of domains contains all (geometrically) convex domains $\Omega \subset \mathbb{R}^n$.

In the class of convex domains, fairly simple counterexamples (see, for example, the discussion in [2]) show that (1.3) is actually false if $p > 2$. Hence $p = 2$ is shaping up as a critical exponent in this class. As regards the end-point case $p = 1$ a more sophisticated counterexample was produced, originally, by Dahlberg [4], and subsequently sharpened by Jerison and Kenig [8], to the effect that there exist a bounded C^1 domain in Ω along with some function $f \in C^\infty(\overline{\Omega})$, such that the Lax–Milgram solution of (1.1) satisfies

$$u \notin W^{2,1}(\Omega). \tag{1.9}$$

In particular, this shows that the case $p = 1$ and $r = 0$ of (1.8) is, generally speaking, false.

Turning to positive results for irregular domains, in the case where $1 < p \leqslant 2$ and Ω is convex the question was affirmatively answered by Dahlberg, Verchota and Wolff, and Adolfsson, Fromm (see the discussion in [1], [6]) for the Dirichlet problem (1.1) and Adolfsson and Jerison [2] for the Neumann problem (1.2).

The aim of this work is to generalize the results in [1, 2, 4, 6], by proving $W^{2,p}$ estimates for the solution of the Poisson problem for the Laplacian with Robin boundary conditions in bounded Lipschitz domains in $\mathbb{R}^n$ satisfying a uniform exterior ball condition. Our main result reads as follows:

Theorem 1.1. *Let Ω be a bounded Lipschitz domain in $\mathbb{R}^n$ satisfying a uniform exterior ball condition (see Definition 2.1 below). Also, assume that $1 < p \leqslant 2$ and $0 \leqslant \lambda \leqslant \infty$. Then the boundary value problem*

$$(R_\lambda) \begin{cases} \Delta u = f \in L^p(\Omega), \\ u \in W^{2,p}(\Omega), \\ \partial_\nu u + \lambda \operatorname{Tr} u = 0 \text{ on } \partial\Omega, \end{cases} \tag{1.10}$$

is well posed. In particular, there exists a finite constant $C = C(\Omega, \lambda, p) > 0$ such that

$$\|u\|_{W^{2,p}(\Omega)} \leqslant C\|f\|_{L^p(\Omega)}. \tag{1.11}$$

As a corollary, if $u = G_\lambda f$ denotes the solution operator for the Robin problem (1.10), *then*

$$G_\lambda : L^p(\Omega) \longrightarrow W^{2,p}(\Omega) \tag{1.12}$$

is well defined, linear, and bounded.

It should be mentioned that (1.1) and (1.2) are special cases of (1.10) corresponding to the cases $\lambda = \infty$ and $\lambda = 0$ respectively. Thus, from this perspective, the inhomogeneous Dirichlet and Neumann problems are particular instances (indeed, end-point cases) of the above Robin problem, which naturally bridges between the two as the parameter λ ranges in $[0, \infty]$. The results of Adolfsson, Dahlberg, Fromm, Jerison, Verchota and Wolff, quoted earlier, correspond to the statement that (1.12) holds if $\Omega \subset \mathbb{R}^n$ is a bounded convex domain, $1 < p \leqslant 2$, and $\lambda \in \{0, \infty\}$ (parenthetically, we wish to point out that, in contrast to Theorem 1.1, this is not a well-posedness statement, per se).

For the sake of this introduction, let us now heuristically describe the main steps that go into the proof of Theorem 1.1 in the special case where Ω is a convex domain of class C^2 (our constants will, however, not depend on the smoothness of $\partial\Omega$, but rather only on the diameter and Lipschitz character of Ω).

`Step I.` *The operator* (1.12) *is bounded when* $p = 2$.

In this scenario, we recall the following useful integral identity. If $u \in W^{2,2}(\Omega)$, then

$$\int_\Omega |\Delta u|^2\,dx = \sum_{j,k=1}^n \int_\Omega \left(\frac{\partial^2 u}{\partial x_j \partial x_k}\right)^2 dx - 2\int_{\partial\Omega} \nabla_{tan}(\partial_\nu u)\cdot\nabla_{tan} u\,d\sigma$$

$$- 2 \int_{\partial\Omega} \left\{ \mathcal{K}(\nabla_{tan} u, \nabla_{tan} u) + (\operatorname{tr} \mathcal{K})(\partial_\nu u)^2 \right\} d\sigma. \tag{1.13}$$

Above, $d\sigma$ denotes the surface measure on $\partial\Omega$, ∇_{tan} stands for the tangential gradient, $\mathcal{K}$ is the second fundamental quadratic form on $\partial\Omega$, regarded as C^2-submanifold of $\mathbb{R}^n$, and $\operatorname{tr} \mathcal{K}$ stands for trace of $\mathcal{K}$. More specifically, if $x_o \in \partial\Omega$, then $\mathcal{K}$ at x_o is the bilinear form

$$\mathcal{K}(\xi, \eta) := - \sum_{i,j=1}^{n-1} \Big\langle \frac{\partial \nu}{\partial s_i}, \Theta_j \Big\rangle \xi_i \, \eta_j = - \Big\langle \frac{\partial \nu}{\partial \xi}, \eta \Big\rangle, \tag{1.14}$$

where ξ, η are tangent vectors to $\partial\Omega$ at x_o with components $\{\xi_1, \cdots, \xi_{n-1}\}$, $\{\eta_1, \cdots, \eta_{n-1}\}$ in the basis $\{\Theta_1, \cdots, \Theta_{n-1}\}$ provided by the tangent vectors at x_o to $n-1$ curves $\Lambda_1, \cdots, \Lambda_{n-1}$ which pass through x_o and are orthogonal at x_o. Furthermore, $s_1, \cdots, s_{n-1}$ denote the arc lengths along $\Lambda_1, \cdots, \Lambda_{n-1}$ and $\langle \cdot, \cdot \rangle$ denotes the scalar product. The trace of this form is then defined as

$$\operatorname{tr} \mathcal{K} := - \sum_{i=1}^{n-1} \Big\langle \frac{\partial \nu}{\partial s_i}, \Theta_i \Big\rangle \tag{1.15}$$

For a proof of (1.13) we refer, for example, to [7].

The crux of the matter is that, if $\partial\Omega \in C^2$, then

$$\Omega \quad \text{convex} \implies \mathcal{K} \leqslant 0 \quad \text{and} \quad \operatorname{tr} \mathcal{K} \leqslant 0 \quad \text{pointwise on } \partial\Omega. \tag{1.16}$$

For the purposes we have in mind, we use this in conjunction with the observation that, if u satisfies the Robin boundary condition $\partial_\nu u + \lambda \, u = 0$ on $\partial\Omega$, then

$$\nabla_{tan}(\partial_\nu u) \cdot \nabla_{tan} u = -\lambda |\nabla_{tan} u|^2 \leqslant 0. \tag{1.17}$$

In concert, (1.16) and (1.17) allow one to conclude that if $u \in W^{2,2}(\Omega)$ satisfies the Robin boundary condition in (1.10) then

$$\|\nabla^2 u\|_{L^2(\Omega)} \leqslant \|\Delta u\|_{L^2(\Omega)}. \tag{1.18}$$

In turn, this entails

$$\nabla^2 G_\lambda : L^2(\Omega) \longrightarrow L^2(\Omega) \tag{1.19}$$

is bounded, which is the key ingredient in ensuring that the claim made in the formulation of Step I holds. For related results see [15, 16].

Moving on, one would naturally be tempted to try to show that the operator (1.12) is also bounded when $p = 1$, with a bound for operator norm *independent of smoothness*. However, the correct point of view is to actually treat the case where $p < 1$, in which scenario one has to replace the standard (Lebesgue-based) Sobolev space $W^{2,p}(\Omega)$ by the Hardy-based Sobolev-like space $h^{2,p}(\Omega)$. Specifically, we prove the following:

Step II. *For any bounded Lipschitz domain Ω there exists $\varepsilon > 0$ depending only on the Lipschitz character of Ω such that the operator*

$$G_\lambda : h_0^p(\Omega) \longrightarrow h^{2,p}(\Omega) \tag{1.20}$$

is well defined, linear, and bounded whenever $1 - \varepsilon < p < 1$.

Here, $h_0^p(\Omega)$ is the space of distributions in Ω consisting of the restrictions to Ω of those distributions in the Triebel–Lizorkin space $F_0^{p,2}(\mathbb{R}^n)$ which are supported in $\overline{\Omega}$. Also, $h^{2,p}(\Omega)$ is the Hardy-based Sobolev-like space of order 2 (see the body of the paper for more details).

This step is based on work developed in connection to the solution of Chang–Krantz–Stein conjecture from [13, 14]. It is important to point out that our result in the second step holding for $1 - \varepsilon < p \leqslant 1$ has been established for the entire class of bounded Lipschitz domains in $\mathbb{R}^n$. In particular, this result does not depend on the convexity, nor on the uniform exterior ball condition property of the domain Ω. Under the assumption of convexity or a uniform exterior ball condition property, it is conceivable that a better lower endpoint, in the class of Hardy spaces, can be established. This is a problem we plan to return to on a different occasion.

Step III. *The problem* (1.10) *is well posed whenever Ω is a bounded Lipschitz domain satisfying a uniform exterior ball condition and $1 < p \leqslant 2$.*

Existence and estimates are obtained by relying on the fact that the Green operator (1.12) is well defined, linear, and bounded if $1 < p \leqslant 2$. The latter claim is a consequence of Steps I-II and interpolation. Finally, uniqueness is proved by relying on the fundamental work of Dahlberg and Kenig [5].

2 Preliminaries

We call a bounded open set $\Omega \subset \mathbb{R}^n$ a *bounded Lipschitz domain* if there exists a finite open covering $\{\mathcal{O}_j\}_{1 \leqslant j \leqslant N}$ of $\partial\Omega$ with the property that, for every $j \in \{1, \ldots, N\}$, $\mathcal{O}_j \cap \Omega$ coincides with the portion of $\mathcal{O}_j$ lying in the upper graph of a Lipschitz function $\varphi_j : \mathbb{R}^{n-1} \to \mathbb{R}$ (where $\mathbb{R}^{n-1} \times \mathbb{R}$ is a new system of coordinates obtained from the original one via a rigid motion). As is well known, for a Lipschitz domain Ω the surface measure $d\sigma$ is well defined on $\partial\Omega$ and there exists an outward pointing normal vector $\nu = (\nu_1, \ldots, \nu_n)$ at almost all points on $\partial\Omega$.

Given $\alpha > 0$, we set

$$\Gamma_\alpha(x) := \{y \in \Omega : |x-y| \leqslant (1+\alpha)\,\mathrm{dist}\,(y, \partial\Omega)\} \tag{2.1}$$

and denote by $M = M_\alpha$ the nontangential maximal operator associated with Ω, i.e., for a function u defined in Ω we set

$$(Mu)(x) := \sup_{y \in \Gamma_\alpha(x)} |u(y)|, \qquad x \in \partial\Omega. \tag{2.2}$$

Also, define the nontangential pointwise trace by

$$u\big|_{\partial\Omega}(x) := \lim_{\substack{y\in\Gamma_\alpha(x)\\ y\to x}} u(y), \qquad x \in \partial\Omega. \tag{2.3}$$

For a Lipschitz domain $\Omega \subset \mathbb{R}^n$ we introduce the tangential gradient of a real-valued function f defined on $\partial\Omega$ by

$$\nabla_{tan} f := \Big(\sum_{k=1}^{n} \nu_k \partial_{\tau_{kj}} f\Big)_{1\leqslant j\leqslant n}, \tag{2.4}$$

where $\partial_{\tau_{jk}} := \nu_j \partial_k - \nu_k \partial_j$, $1 \leqslant j, k \leqslant n$, are tangential derivative operators on $\partial\Omega$.

Next, we discuss layer potential operators associated with a given Lipschitz domain $\Omega \subset \mathbb{R}^n$. To set the stage, we denote by Γ the canonical fundamental solution for the Laplacian $\Delta = \sum\limits_{j=1}^{n} \partial_j^2$ in $\mathbb{R}^n$, i.e.,

$$\Gamma(x) := \begin{cases} \dfrac{1}{\omega_{n-1}(2-n)} \dfrac{1}{|x|^{n-2}}, & \text{if } n \geqslant 3, \\[2ex] \dfrac{1}{2\pi} \log|x|, & \text{if } n = 2, \end{cases} \qquad x \in \mathbb{R}^n \setminus \{0\}, \tag{2.5}$$

where ω_n is the surface measure of the unit sphere S^{n-1} in $\mathbb{R}^n$. Next, we recall the harmonic single layer and its boundary version given respectively by the formulas

$$\mathcal{S}f(x) := \int_{\partial\Omega} \Gamma(x-y) f(y)\, d\sigma(y), \quad x \in \Omega, \tag{2.6}$$

$$Sf(x) := \int_{\partial\Omega} \Gamma(x-y) f(y)\, d\sigma(y), \quad x \in \partial\Omega. \tag{2.7}$$

We also recall here that

$$\mathcal{S}f\big|_{\partial\Omega} = Sf \qquad \text{on } \partial\Omega \tag{2.8}$$

and the following jump-formula for the normal derivative of the single layer potential operator holds:

$$\partial_\nu \mathcal{S} f = \left(-\tfrac{1}{2} I + K^*\right) f \qquad \text{on } \partial\Omega, \tag{2.9}$$

where, with p.v. denoting principal value, we have set

$$K^* f(x) := \frac{1}{\omega_{n-1}} \text{p.v.} \int_{\partial\Omega} \frac{\langle x - y, \nu(x)\rangle}{|x-y|^n} f(y)\, d\sigma(y), \quad x \in \partial\Omega. \tag{2.10}$$

Moving on, we briefly recall the Newtonian volume potential for the Laplacian. Specifically, for a given function $f \in L^1(\Omega)$ we set

$$\Pi f(x) := \int_{\Omega} \Gamma(x-y) f(y)\, dy, \qquad x \in \mathbb{R}^n, \tag{2.11}$$

and note that

$$\Delta \Gamma f = f \quad \text{in} \quad \Omega. \tag{2.12}$$

Finally, we record here the following definition.

Definition 2.1. An open set $\Omega \subset \mathbb{R}^n$ is said to satisfy a *uniform exterior ball condition* (abbreviated by UEBC) if there exists $r > 0$ with the following property: For each $x \in \partial\Omega$ there exists a point $y = y(x) \in \mathbb{R}^n$ such that

$$\overline{B(y,r)} \backslash \{x\} \subseteq \mathbb{R}^n \backslash \Omega \text{ and } x \in \partial B(y,r). \tag{2.13}$$

The largest radius r satisfying the above property is referred to as the UEBC *constant* of Ω.

3 Smoothness Spaces on Lipschitz Boundaries and Lipschitz Domains

We begin by briefly reviewing the Besov and Triebel–Lizorkin scales in $\mathbb{R}^n$. One convenient point of view is offered by the classical Littlewood–Paley theory (see, for example, [21, 23]). More specifically, let Ξ be the collection of all systems $\{\zeta_j\}_{j=0}^{\infty}$ of Schwartz functions with the following properties:

(i) there exist positive constants A, B, C such that

$$\begin{aligned} &\operatorname{supp}(\zeta_0) \subset \{x : |x| \leqslant A\}; \\ &\operatorname{supp}(\zeta_j) \subset \{x : B2^{j-1} \leqslant |x| \leqslant C2^{j+1}\} \quad \text{if } j \in \mathbb{N}; \end{aligned} \tag{3.1}$$

(ii) for every multiindex α there is a positive, finite constant C_α such that

$$\sup_{x \in \mathbb{R}^n} \sup_{j \in \mathbb{N}} 2^{j|\alpha|} |\partial^\alpha \zeta_j(x)| \leqslant C_\alpha; \tag{3.2}$$

(iii)

$$\sum_{j=0}^{\infty} \zeta_j(x) = 1 \text{ for every } x \in \mathbb{R}^n. \tag{3.3}$$

Let $s \in \mathbb{R}$, $0 < q \leqslant \infty$. Fix some family $\{\zeta_j\}_{j=0}^{\infty} \in \Xi$. Also, let $\mathcal{F}$ and $S'(\mathbb{R}^n)$ denote the Fourier transform and the class of tempered distributions in $\mathbb{R}^n$ respectively. Then the Triebel–Lizorkin space $F_s^{p,q}(\mathbb{R}^n)$ is defined for each $0 < p < \infty$ as

$$F_s^{p,q}(\mathbb{R}^n) := \Big\{ f \in S'(\mathbb{R}^n) : \|f\|_{F_s^{p,q}(\mathbb{R}^n)}$$

$$:= \Big\| \Big(\sum_{j=0}^{\infty} |2^{sj} \mathcal{F}^{-1}(\zeta_j \mathcal{F} f)|^q \Big)^{1/q} \Big\|_{L^p(\mathbb{R}^n)} < \infty \Big\}. \tag{3.4}$$

If $0 < p \leqslant \infty$ then the Besov space $B_s^{p,q}(\mathbb{R}^n)$ can be defined as

$$B_s^{p,q}(\mathbb{R}^n) := \Big\{ f \in S'(\mathbb{R}^n) : \|f\|_{B_s^{p,q}(\mathbb{R}^n)}$$

$$:= \Big(\sum_{j=0}^{\infty} \|2^{sj} \mathcal{F}^{-1}(\zeta_j \mathcal{F} f)\|_{L^p(\mathbb{R}^n)}^q \Big)^{1/q} < \infty \Big\}. \tag{3.5}$$

A different choice of the system $\{\zeta_j\}_{j=0}^{\infty} \in \Xi$ yields the same spaces (3.4)-(3.5), albeit equipped with equivalent norms. Furthermore, the class of Schwartz functions in $\mathbb{R}^n$ is dense in both $B_s^{p,q}(\mathbb{R}^n)$ and $F_s^{p,q}(\mathbb{R}^n)$ provided that $s \in \mathbb{R}$ and $0 < p, q < \infty$.

It has long been known that many classical smoothness spaces are encompassed by the Besov and Triebel–Lizorkin scales. For example,

$$C^s(\mathbb{R}^n) = B_s^{\infty,\infty}(\mathbb{R}^n), \quad 0 < s \notin \mathbb{Z}, \tag{3.6}$$

$$L^p(\mathbb{R}^n) = F_0^{p,2}(\mathbb{R}^n), \quad 1 < p < \infty, \tag{3.7}$$

$$L_s^p(\mathbb{R}^n) = F_s^{p,2}(\mathbb{R}^n), \quad 1 < p < \infty, \quad s \in \mathbb{R}, \tag{3.8}$$

$$W^{k,p}(\mathbb{R}^n) = F_k^{p,2}(\mathbb{R}^n), \quad 1 < p < \infty, \quad k \in \mathbb{N}, \tag{3.9}$$

$$h^p(\mathbb{R}^n) = F_0^{p,2}(\mathbb{R}^n), \quad 0 < p \leqslant 1, \tag{3.10}$$

$$h^{k,p}(\mathbb{R}^n) = F_k^{p,2}(\mathbb{R}^n), \quad 0 < p \leqslant 1, \quad k \in \mathbb{Z}, \tag{3.11}$$

$$\mathrm{bmo}(\mathbb{R}^n) = F_0^{\infty,2}(\mathbb{R}^n). \tag{3.12}$$

Above, given $1 < p < \infty$ and $s \in \mathbb{R}$, $L_s^p(\mathbb{R}^n)$ stands for the Bessel potential space defined by

$$L_s^p(\mathbb{R}^n) := \Big\{(I-\Delta)^{-s/2}g : g \in L^p(\mathbb{R}^n)\Big\}$$
$$= \Big\{\mathcal{F}^{-1}(1+|\xi|^2)^{-s/2}\mathcal{F}g : g \in L^p(\mathbb{R}^n)\Big\}, \qquad (3.13)$$

equipped with the norm

$$\|f\|_{L_s^p(\mathbb{R}^n)} := \|\mathcal{F}^{-1}(1+|\xi|^2)^{s/2}\mathcal{F}f\|_{L^p(\mathbb{R}^n)}. \qquad (3.14)$$

As is well known, when the smoothness index is a natural number, say $s = k \in \mathbb{N}$, this can be identified with the classical Sobolev space

$$W^{k,p}(\mathbb{R}^n) := \Big\{f \in L^p(\mathbb{R}^n) : \|f\|_{W^{k,p}(\mathbb{R}^n)} := \sum_{|\gamma|\leqslant k} \|\partial^\gamma f\|_{L^p(\mathbb{R}^n)} < \infty\Big\}, \qquad (3.15)$$

i.e.,

$$L_k^p(\mathbb{R}^n) = W^{k,p}(\mathbb{R}^n), \qquad k \in \mathbb{N}_o,\ 1 < p < \infty. \qquad (3.16)$$

Also, $C^s(\mathbb{R}^n)$, $h^p(\mathbb{R}^n)$, $h^{k,p}(\mathbb{R}^n)$ stand for the Hölder, (local) Hardy, and Hardy-based Sobolev-like spaces in $\mathbb{R}^n$.

We next wish to adapt some of these smoothness classes to the situation where the Euclidean space is replaced by the boundary of a Lipschitz domain Ω. To get started, for each $1 < p < \infty$, the space $L^p(\partial\Omega)$ is the space of p-integrable functions on $\partial\Omega$ with respect to the surface measure $d\sigma$. For $a \in \mathbb{R}$ we set $(a)_+ := \max\{a, 0\}$. Consider three parameters p, q, and s subject to

$$0 < p, q \leqslant \infty, \qquad (n-1)(1/p-1)_+ < s < 1 \qquad (3.17)$$

and assume that $\Omega \subset \mathbb{R}^n$ is the upper graph of a Lipschitz function $\varphi : \mathbb{R}^{n-1} \to \mathbb{R}$. We then define $B_s^{p,q}(\partial\Omega)$ as the space of locally integrable functions f on $\partial\Omega$ for which the assignment $\mathbb{R}^{n-1} \ni x \mapsto f(x, \varphi(x))$ belongs to $B_s^{p,q}(\mathbb{R}^{n-1})$, the classical Besov space in $\mathbb{R}^{n-1}$. We equip this space with the (quasi) norm

$$\|f\|_{B_s^{p,q}(\partial\Omega)} := \|f(\cdot, \varphi(\cdot))\|_{B_s^{p,q}(\mathbb{R}^{n-1})}. \qquad (3.18)$$

As far as Besov spaces with a negative amount of smoothness are concerned, in the same context as above we set

$$f \in B_{s-1}^{p,q}(\partial\Omega) \Longleftrightarrow f(\cdot, \varphi(\cdot))\sqrt{1+|\nabla\varphi(\cdot)|^2} \in B_{s-1}^{p,q}(\mathbb{R}^{n-1}), \qquad (3.19)$$

$$\|f\|_{B_{s-1}^{p,q}(\partial\Omega)} := \|f(\cdot, \varphi(\cdot))\sqrt{1+|\nabla\varphi(\cdot)|^2}\|_{B_{s-1}^{p,q}(\mathbb{R}^{n-1})}. \qquad (3.20)$$

As is well known, the case where $p = q = \infty$ corresponds to the usual (inhomogeneous) Hölder spaces $C^s(\partial\Omega)$, defined by the requirement that

$$\|f\|_{C^s(\partial\Omega)} := \|f\|_{L^\infty(\partial\Omega)} + \sup_{\substack{x\neq y\\ x,y\in\partial\Omega}} \frac{|f(x)-f(y)|}{|x-y|^s} < +\infty, \tag{3.21}$$

i.e.,

$$B_s^{\infty,\infty}(\partial\Omega) = C^s(\partial\Omega) \qquad \text{for } s\in(0,1). \tag{3.22}$$

All the above definitions then readily extend to the case of (bounded) Lipschitz domains in $\mathbb{R}^n$ via a standard partition of unity argument.

We now proceed to discuss Triebel–Lizorkin spaces defined on the boundary of a bounded Lipschitz domain $\Omega \subset \mathbb{R}^n$, denoted in the sequel by $F_s^{p,q}(\partial\Omega)$. Compared with the Besov scale, the most important novel aspect here is the possibility of allowing the endpoint case $s=1$ as part of the general discussion if $q=2$. To discuss this in more detail, assume that either

$$0<p<\infty, \quad 0<q\leqslant\infty, \quad (n-1)\Big(\frac{1}{\min\{p,q\}}-1\Big)_+ < s < 1, \tag{3.23}$$

or

$$\frac{n-1}{n} < p < \infty, \quad q=2, \quad s=1. \tag{3.24}$$

In this scenario, the Triebel–Lizorkin scale in $\mathbb{R}^{n-1}$ is invariant under pointwise multiplication by Lipschitz maps, as well as composition by Lipschitz diffeomorphisms. When s, p, and q are as above and Ω is a Lipschitz domain in $\mathbb{R}^n$ lying above the graph of a Lipschitz function $\varphi:\mathbb{R}^{n-1}\to\mathbb{R}$, we may therefore define the space $F_s^{p,q}(\partial\Omega)$ as the collection of all locally integrable functions f on $\partial\Omega$ such that

$$f(\cdot,\varphi(\cdot)) \in F_s^{p,q}(\mathbb{R}^{n-1}), \tag{3.25}$$

endowed with the norm

$$\|f\|_{F_s^{p,q}(\partial\Omega)} := \|f(\cdot,\varphi(\cdot))\|_{F_s^{p,q}(\mathbb{R}^{n-1})}. \tag{3.26}$$

Also, if $\mathrm{Lip}_0(\partial\Omega)$ stands for the collection of all compactly supported Lipschitz functions on $\partial\Omega$, the space $F_{s-1}^{p,q}(\partial\Omega)$ is defined as the collection of all functionals $f\in(\mathrm{Lip}_0(\partial\Omega))'$ such that

$$f(\cdot,\varphi(\cdot))\sqrt{1+|\nabla\varphi(\cdot)|^2} \in F_{s-1}^{p,q}(\mathbb{R}^{n-1}), \tag{3.27}$$

and we equip it with the quasinorm

$$\|f\|_{F_{s-1}^{p,q}(\partial\Omega)} := \|f(\cdot,\varphi(\cdot))\sqrt{1+|\nabla\varphi(\cdot)|^2}\|_{F_{s-1}^{p,q}(\mathbb{R}^{n-1})}. \tag{3.28}$$

Hereafter, we introduce

$$H^s(\partial\Omega) := F_s^{2,2}(\partial\Omega), \qquad -1 \leqslant s \leqslant 1. \tag{3.29}$$

In particular, the following duality result holds:

$$\big(H^s(\partial\Omega)\big)^* = H^{-s}(\partial\Omega), \qquad -1 \leqslant s \leqslant 1. \tag{3.30}$$

When $\Omega \subset \mathbb{R}^n$ is a *bounded Lipschitz domain* and (s,p,q) are as in (3.23)-(3.24), we define $F_s^{p,q}(\partial\Omega)$ and $F_{s-1}^{p,q}(\partial\Omega)$ via localization (using a smooth, finite partition of unity) and pull-back to $\mathbb{R}^{n-1}$ (in the manner described above, for graph-Lipschitz domains). When equipped with the natural quasi-norms, the Triebel–Lizorkin spaces just introduced are quasi-Banach, and different partitions of unity yield equivalent quasinorms.

Going further, we adapt the Besov and Triebel–Lizorkin spaces $B_s^{p,q}(\mathbb{R}^n)$, $F_s^{p,q}(\mathbb{R}^n)$, $0 < p,q \leqslant \infty$, $s \in \mathbb{R}$, originally defined in the entire Euclidean setting, to arbitrary open subsets of $\mathbb{R}^n$. Concretely for a given arbitrary open subset Ω of $\mathbb{R}^n$ we denote by $f|_\Omega$ the restriction of a distribution f in $\mathbb{R}^n$ to Ω. For $0 < p,q \leqslant \infty$ and $s \in \mathbb{R}$, both $B_s^{p,q}(\mathbb{R}^n)$ and $F_s^{p,q}(\mathbb{R}^n)$ are spaces of (tempered) distributions, hence it is meaningful to define

$$\begin{aligned} &A_s^{p,q}(\Omega) := \{f \text{ distribution in } \Omega : \exists\, g \in A_s^{p,q}(\mathbb{R}^n) \text{ such that } g|_\Omega = f\},\\ &\|f\|_{A_s^{p,q}(\Omega)} := \inf\{\|g\|_{A_s^{p,q}(\mathbb{R}^n)} : g \in A_s^{p,q}(\mathbb{R}^n),\ g|_\Omega = f\}, \qquad f \in A_s^{p,q}(\Omega), \end{aligned} \tag{3.31}$$

where $A = B$ or $A = F$. Throughout the paper, the subscript *loc* appended to one of the function spaces already introduced indicates the local version of that particular space.

The existence of an universal extension operator for Besov and Triebel–Lizorkin spaces in an arbitrary Lipschitz domain $\Omega \subset \mathbb{R}^n$ was established by Rychkov [22]. This allows one transferring a number of properties of the Besov-Triebel–Lizorkin spaces in the Euclidean space $\mathbb{R}^n$ to the setting of a bounded Lipschitz domain $\Omega \subset \mathbb{R}^n$. Here, we only wish to mention a few of these properties. First, if $0 < p \leqslant \infty$, $0 < q < \infty$, and $s \in \mathbb{R}$, then

$$B_s^{p,\min(p,q)}(\Omega) \hookrightarrow F_s^{p,q}(\Omega) \hookrightarrow B_s^{p,\max(p,q)}(\Omega) \tag{3.32}$$

and

$$B_{s_0}^{p_0,p}(\Omega) \hookrightarrow F_s^{p,q}(\Omega) \hookrightarrow B_{s_1}^{p_1,p}(\Omega) \tag{3.33}$$

if $0 < p_0 < p < p_1 \leqslant \infty$ and $\dfrac{1}{p_0} - \dfrac{s_0}{n} = \dfrac{1}{p} - \dfrac{s}{n} = \dfrac{1}{p_1} - \dfrac{s_1}{n}$. Second, if k is a nonnegative integer and $1 < p < \infty$, then

$$F_k^{p,2}(\Omega) = W^{k,p}(\Omega) := \{f \in L^p(\Omega) : \partial^\alpha f \in L^p(\Omega),\ |\alpha| \leqslant k\}, \tag{3.34}$$

the classical Sobolev spaces in Ω. Third, if $k \in \mathbb{N}_0$ and $0 < s < 1$, then

$$B^{\infty,\infty}_{k+s}(\Omega) = C^{k+s}(\Omega), \tag{3.35}$$

where

$$C^{k+s}(\Omega) := \Big\{ u \in C^k(\Omega) : \ \|u\|_{C^{k+s}(\Omega)} < \infty \text{ and } \|u\|_{C^{k+s}(\Omega)} := \sum_{j=0}^{k} \|\nabla^j u\|_{L^\infty(\Omega)} + \sum_{|\alpha|=k} \sup_{x\neq y\in\Omega} \frac{|\partial^\alpha u(x) - \partial^\alpha u(y)|}{|x-y|^s} \Big\}. \tag{3.36}$$

Going further, for $0 < p, q \leqslant \infty$, $s \in \mathbb{R}$, we set

$$\begin{aligned} A^{p,q}_{s,0}(\Omega) &:= \{f \in A^{p,q}_s(\mathbb{R}^n) : \operatorname{supp} f \subseteq \overline{\Omega}\}, \\ \|f\|_{A^{p,q}_{s,0}(\Omega)} &:= \|f\|_{A^{p,q}_s(\mathbb{R}^n)}, \quad f \in A^{p,q}_{s,0}(\Omega), \end{aligned} \tag{3.37}$$

where we use the convention that either $A = F$ and $p < \infty$ or $A = B$. Thus, $B^{p,q}_{s,0}(\Omega)$ and $F^{p,q}_{s,0}(\Omega)$ are closed subspaces of $B^{p,q}_s(\mathbb{R}^n)$ and $F^{p,q}_s(\mathbb{R}^n)$ respectively.

Next we record an assertion from [14] which extends work done in [8].

Theorem 3.1. *Let Ω be a bounded Lipschitz domain in $\mathbb{R}^n$. Assume that the indices p, s satisfy $\frac{n-1}{n} < p \leqslant \infty$ and $(n-1)(\frac{1}{p}-1)_+ < s < 1$. Then the following holds.*

(i) *The restriction to the boundary extends to a linear, bounded operator*

$$\mathrm{Tr} : B^{p,q}_{s+\frac{1}{p}}(\Omega) \longrightarrow B^{p,q}_s(\partial\Omega) \quad \text{for} \quad 0 < q \leqslant \infty. \tag{3.38}$$

Moreover, for this range of indices, Tr *is onto and has a bounded right inverse*

$$\mathrm{Ex} : B^{p,q}_s(\partial\Omega) \longrightarrow B^{p,q}_{s+\frac{1}{p}}(\Omega). \tag{3.39}$$

(ii) *Similar considerations hold for*

$$\mathrm{Tr} : F^{p,q}_{s+\frac{1}{p}}(\Omega) \longrightarrow B^{p,p}_s(\partial\Omega) \tag{3.40}$$

with the convention that $q = \infty$ if $p = \infty$. More specifically, Tr *in (3.40) is a linear, bounded, operator which has a linear, bounded right inverse*

$$\mathrm{Ex} : B^{p,p}_s(\partial\Omega) \longrightarrow F^{p,q}_{s+\frac{1}{p}}(\Omega). \tag{3.41}$$

Denote by $(\cdot,\cdot)_{\theta,q}$ and $[\cdot,\cdot]_\theta$ the real and complex method of interpolation respectively. A proof of the following result can be found in [9].

Theorem 3.2. *Suppose that Ω is a bounded Lipschitz domain in $\mathbb{R}^n$. Let $\alpha_0, \alpha_1 \in \mathbb{R}$, $\alpha_0 \neq \alpha_1$, $0 < q_0, q_1, q \leqslant \infty$, $0 < \theta < 1$, $\alpha = (1-\theta)\alpha_0 + \theta\alpha_1$. Then*

$$(F^{p,q_0}_{\alpha_0}(\Omega), F^{p,q_1}_{\alpha_1}(\Omega))_{\theta,q} = B^{p,q}_{\alpha}(\Omega), \quad 0 < p < \infty, \tag{3.42}$$

$$(B^{p,q_0}_{\alpha_0}(\Omega), B^{p,q_1}_{\alpha_1}(\Omega))_{\theta,q} = B^{p,q}_{\alpha}(\Omega), \quad 0 < p \leqslant \infty. \tag{3.43}$$

Furthermore, if $\alpha_0, \alpha_1 \in \mathbb{R}$, $0 < p_0, p_1 \leqslant \infty$ *and* $0 < q_0, q_1 \leqslant \infty$ *are such that*

$$\text{either} \quad \max\{p_0, q_0\} < \infty \quad \text{or} \quad \max\{p_1, q_1\} < \infty, \tag{3.44}$$

then

$$[F^{p_0,q_0}_{\alpha_0}(\Omega), F^{p_1,q_1}_{\alpha_1}(\Omega)]_\theta = F^{p,q}_{\alpha}(\Omega), \tag{3.45}$$

where $0 < \theta < 1$, $\alpha = (1-\theta)\alpha_0 + \theta\alpha_1$, $\dfrac{1}{p} = \dfrac{1-\theta}{p_0} + \dfrac{\theta}{p_1}$ *and* $\dfrac{1}{q} = \dfrac{1-\theta}{q_0} + \dfrac{\theta}{q_1}$.

On the other hand, if $\alpha_0, \alpha_1 \in \mathbb{R}$, $0 < p_0, p_1, q_0, q_1 \leqslant \infty$ *are such that*

$$\min\{q_0, q_1\} < \infty, \tag{3.46}$$

then also

$$[B^{p_0,q_0}_{\alpha_0}(\Omega), B^{p_1,q_1}_{\alpha_1}(\Omega)]_\theta = B^{p,q}_{\alpha}(\Omega), \tag{3.47}$$

where θ, α, p, q *are as above.*

Finally, the same interpolation results are valid if the spaces $B^{p,q}_s(\Omega)$, $F^{p,q}_s(\Omega)$ *are replaced by* $B^{p,q}_{s,0}(\Omega)$ *and* $F^{p,q}_{s,0}(\Omega)$ *respectively.*

The following is a collection of mapping and invertibility properties of the layer potential operators introduced at the end of the previous section which will be useful in the sequel. Specifically, we have the following assertion from [14].

Theorem 3.3. *Let* Ω *be a bounded Lipschitz domain in* $\mathbb{R}^n$. *Then for each* $0 < p < \infty$ *and* $(n-1)(1/p-1)_+ < s < 1$ *the following operators are well defined, linear, and bounded:*

$$K^* : B^{p,p}_{s-1}(\partial\Omega) \longrightarrow B^{p,p}_{s-1}(\partial\Omega), \tag{3.48}$$

$$S : B^{p,p}_{s-1}(\partial\Omega) \longrightarrow B^{p,p}_{s}(\partial\Omega), \tag{3.49}$$

$$\mathcal{S} : B^{p,p}_{s-1}(\partial\Omega) \longrightarrow F^{p,2}_{s+\frac{1}{p}}(\Omega). \tag{3.50}$$

Regarding the invertibility of boundary harmonic layer potentials, we record here the following version of Theorem 1.5 of [12] (see also [14]).

Theorem 3.4. *For each bounded Lipschitz domain* Ω *in* $\mathbb{R}^n$ *there exists* $\varepsilon = \varepsilon(\Omega) \in (0,1]$ *with the following significance. Assume that* $\dfrac{n-1}{n} < p \leqslant \infty$, $(n-1)(1/p-1)_+ < s < 1$, *are such that either one of the four conditions*

$$(I): \quad \tfrac{n-1}{n-1+\varepsilon} < p \leqslant 1 \quad \text{and} \quad (n-1)\left(\tfrac{1}{p} - 1\right) + 1 - \varepsilon < s < 1;$$

$$(II): \quad 1 \leqslant p \leqslant \tfrac{2}{1+\varepsilon} \quad \textit{and} \quad \tfrac{2}{p} - 1 - \varepsilon < s < 1; \tag{3.51}$$

$$(III): \quad \tfrac{2}{1+\varepsilon} \leqslant p \leqslant \tfrac{2}{1-\varepsilon} \quad \textit{and} \quad 0 < s < 1;$$

$$(IV): \quad \tfrac{2}{1-\varepsilon} \leqslant p \leqslant \infty \quad \textit{and} \; 0 < s < \tfrac{2}{p} + \varepsilon,$$

is satisfied if $n \geqslant 3$, and either one of the following three conditions

$$(I'): \quad \tfrac{2}{1+\varepsilon} \leqslant p \leqslant \tfrac{2}{1-\varepsilon} \quad \textit{and} \quad 0 < s < 1;$$

$$(II'): \quad \tfrac{2}{3+\varepsilon} < p < \tfrac{2}{1+\varepsilon} \quad \textit{and} \quad \tfrac{1}{p} - \tfrac{1+\varepsilon}{2} < s < 1; \tag{3.52}$$

$$(III'): \quad \tfrac{2}{1-\varepsilon} < p \leqslant \infty \quad \textit{and} \quad 0 < s < \tfrac{1}{p} + \tfrac{1+\varepsilon}{2},$$

is satisfied if $n = 2$. Then, with $0 < q \leqslant \infty$, the operators

$$\pm\tfrac{1}{2}I + K^* : B^{p,q}_{s-1}(\partial\Omega) \longrightarrow B^{p,q}_{s-1}(\partial\Omega) \quad \textit{are Fredholm with index zero.} \tag{3.53}$$

Let us also record here the fact (see [24] for a proof) that

$$S : L^2(\partial\Omega) \longrightarrow H^1(\partial\Omega) \quad \text{is an isomorphism.} \tag{3.54}$$

We conclude this section by discussing finite energy solutions for the inhomogeneous problem for the Laplacian with Robin boundary conditions in a bounded Lipschitz domain $\Omega \subset \mathbb{R}^n$. To set the stage, let ${}_{X^*}\langle \cdot, \cdot \rangle_X$ denote the duality pairing between a Banach space X and its dual X^*. Then for a given bounded Lipschitz domain with outward unit normal ν we consider the Neumann trace operator

$$\partial_\nu : \{u \in W^{1,2}(\Omega) : \Delta u \in L^2(\Omega)\} \longrightarrow H^{-1/2}(\partial\Omega) \tag{3.55}$$

given by

$${}_{(H^{1/2}(\partial\Omega))^*}\langle \partial_\nu u, \varphi \rangle_{H^{1/2}(\partial\Omega)} := \int_\Omega \nabla u \cdot \nabla \Phi \, dx + \int_\Omega (\Delta u)\Phi \, dx. \tag{3.56}$$

Above, $\varphi \in H^{1/2}(\partial\Omega)$ is arbitrary and $\Phi \in W^{1,2}(\Omega)$ is such that $\operatorname{Tr}\Phi = \varphi$ on $\partial\Omega$. Then (3.55)-(3.56) is a well-defined operator (whose definition is unaffected by the particular extension Φ of $\varphi \in H^{1/2}(\partial\Omega)$ to a function in $W^{1,2}(\Omega)$), which is linear and bounded.

Proposition 3.5. *Let Ω be a bounded Lipschitz domain, and let $\lambda > 0$. Then the boundary value problem*

$$\begin{cases} \Delta u = f \in L^2(\Omega) \quad in \quad \Omega, \\ \partial_\nu u + \lambda u = g \in H^{-1/2}(\partial\Omega) \quad on \quad \partial\Omega, \\ u \in W^{1,2}(\Omega), \end{cases} \tag{3.57}$$

is uniquely solvable. In addition, the solution satisfies

$$\|u\|_{W^{1,2}(\Omega)} \leqslant C\|f\|_{L^2(\Omega)} + C\|g\|_{H^{-1/2}(\partial\Omega)} \tag{3.58}$$

for some $C > 0$ which depends only on the Lipschitz character of Ω and λ, and

$$u \in W^{2,2}_{loc}(\Omega). \tag{3.59}$$

Proof. It is well known that the Neumann trace operator (3.55)-(3.56) is onto. Granted this, there is no loss of generality in assuming that $g = 0$. Assuming that this is the case, the idea is now to implement the standard Lax–Milgram lemma. Specifically, let

$$B : W^{1,2}(\Omega) \times W^{1,2}(\Omega) \longrightarrow \mathbb{R} \tag{3.60}$$

be given by

$$B(u,v) := \int_\Omega \nabla u \cdot \nabla v \, dx + \lambda \int_{\partial\Omega} \operatorname{Tr} u \operatorname{Tr} v \, d\sigma. \tag{3.61}$$

Then B is bilinear, bounded, symmetric and coercive, where this last property follows from the Poincaré inequality (see, for example, [3]). Then, according to the Lax–Milgram lemma, for each function $f \in L^2(\Omega) \hookrightarrow \big(W^{1,2}(\Omega)\big)^*$ there exists a unique $u \in W^{1,2}(\Omega)$ such that

$$\|u\|_{W^{1,2}(\Omega)} \leqslant C(\Omega,\lambda)\|f\|_{L^2(\Omega)}, \tag{3.62}$$

for some finite $C = C(\Omega, \lambda) > 0$ and

$$-B(u,v) = {}_{(W^{1,2}(\Omega))^*}\langle f, v\rangle_{W^{1,2}(\Omega)} \quad \forall\, v \in W^{1,2}(\Omega). \tag{3.63}$$

Thus,

$$\int_\Omega \nabla u \cdot \nabla v \, dx + \lambda \int_{\partial\Omega} \operatorname{Tr} u \operatorname{Tr} v \, d\sigma = -\int_\Omega f v \, dx \qquad \forall\, v \in W^{1,2}(\Omega). \tag{3.64}$$

Specializing this to the case $v \in C_0^\infty(\Omega)$, we prove that $\Delta u = f$ in Ω. With this in hand, then (3.55)-(3.56) show that also $\partial_\nu u + \lambda \operatorname{Tr} u = 0$ in $H^{-1/2}(\partial\Omega)$, from which the conclusion in the first part of the proposition follows.

As for the membership of u to $W^{2,2}_{loc}(\Omega)$, we pick an arbitrary function $\eta \in C_0^\infty(\Omega)$ and consider a domain $D \subset \Omega$ of class C^∞ such that $\operatorname{supp} \eta \subset D$.

Then $\Delta(\eta u) \in L^2(D)$ and $\eta u \in W_0^{1,2}(D)$. Since ∂D is of class C^∞, well-known elliptic estimates in smooth domains imply that $\eta u \in W^{2,2}(D)$. Thus, $u \in W_{loc}^{2,2}(\Omega)$, as required. □

4 The Case of C^2 Domains

Throughout this section, Ω is a bounded C^2 domain in $\mathbb{R}^n$ with a C^2 defining function ρ, i.e.,

$$\rho < 0 \text{ in } \Omega, \quad \rho > 0 \text{ in } \mathbb{R}^n \setminus \overline{\Omega}, \quad \nabla\rho \neq 0 \text{ on } \partial\Omega = \{\rho = 0\}. \tag{4.1}$$

In particular, $\nu = \dfrac{\nabla\rho}{|\nabla\rho|}$ is the outward unit normal vector to $\partial\Omega$. Define the Hessian matrix of ρ:

$$\text{Hess}\,(\rho) := \sum_{1\leqslant j,k\leqslant n} \frac{\partial^2\rho}{\partial x_j\partial x_k} e_j \otimes e_k \equiv \left(\frac{\partial^2\rho}{\partial x_j\partial x_k}\right)_{1\leqslant j,k\leqslant n}, \tag{4.2}$$

where $(e_j)_{1\leqslant j\leqslant n}$ is the standard orthonormal basis for $\mathbb{R}^n$.

Theorem 4.1. *Consider Ω and ρ as above and assume that $w \in W^{2,2}(\Omega)$. Then, with σ denoting the surface area on $\partial\Omega$, ν denoting the outward unit normal, ∂_ν the normal derivative and ∇_{tan} the tangential gradient on $\partial\Omega$, one has*

$$\begin{aligned}\int_\Omega |\Delta w|^2\,dx &= \sum_{j,k=1}^n \int_\Omega \left(\frac{\partial^2 w}{\partial x_j\partial x_k}\right)^2 dx \\ &\quad + \int_{\partial\Omega} [\text{Tr}\,(\text{Hess}\,(\rho)) - \langle \text{Hess}\,(\rho)\nu, \nu\rangle]\,(\partial_\nu w)^2\,d\sigma \\ &\quad + \int_{\partial\Omega} \langle \text{Hess}\,(\rho)\nabla_{tan}w, \nabla_{tan}w\rangle\,d\sigma \\ &\quad -2\,{}_{H^{1/2}(\partial\Omega)}\big\langle \nabla_{tan}w, \nabla_{tan}(\partial_\nu w)\big\rangle_{H^{-1/2}(\partial\Omega)},\end{aligned} \tag{4.3}$$

where ${}_{H^{1/2}(\partial\Omega)}\langle f, g\rangle_{H^{-1/2}(\partial\Omega)}$ is the duality pairing between $f \in H^{1/2}(\partial\Omega)$ and $g \in H^{-1/2}(\partial\Omega) = \big(H^{1/2}(\partial\Omega)\big)^*$ with respect to the pivotal space $L^2(\partial\Omega)$.

Proof. This follows by specializing results from [18] where a similar formula was established in a more general framework of differential forms. Below, we employ relatively standard notation and terminology inherent to the language of differential forms in $\mathbb{R}^n$. For example, d is the exterior derivative operator, $\wedge$ stands for the exterior product of forms, and δ is its formal adjoint. Also, $\vee$ denotes the interior product of forms (which happens to be the adjoint of "wedging"). Generally speaking, for a given differential form ω in $\overline{\Omega}$ we define its normal and tangential components respectively by

$$\omega_{nor} := \nu \wedge (\nu \vee \omega), \qquad \omega_{tan} := \nu \vee (\nu \wedge \omega), \tag{4.4}$$

where the outward unit normal $\nu = (\nu_1, \dots, \nu_n)$ is identified with the 1-form $\nu = \sum_{j=1}^{n} \nu_j dx_j$.

Following [18], we also introduce the boundary versions of the operators d, δ:

$$d_\partial := -\nu \wedge d(\nu \vee \cdot) \quad \text{and} \quad \delta_\partial := -\nu \vee \delta(\nu \wedge \cdot). \tag{4.5}$$

Then, according to Theorem 4.1 in [18], for each $\ell \in \{0, \dots, n\}$ and any two differential ℓ-forms $u = \sum_{|I|=\ell} u_I dx^I$, $v = \sum_{|I|=\ell} v_I dx^I$ (where $\sum_{|I|=\ell}$ indicates that the sum is performed over increasing multiindices I of length ℓ), with coefficients in $W^{1,2}(\Omega)$, we have

$$\begin{aligned}
&\int_\Omega \big[\langle du, dv\rangle + \langle \delta u, \delta v\rangle\big]\, dx \\
&= \sum_{|I|=\ell} \sum_{j=1}^{n} \int_\Omega \frac{\partial u_I}{\partial x_j} \frac{\partial v_I}{\partial x_j}\, dx + \int_{\partial\Omega} \sum_{j,k=1}^{n} \frac{\partial^2 \rho}{\partial x_j \partial x_k} \langle dx_j \wedge u_{nor}, dx_k \wedge v_{nor}\rangle\, d\sigma \\
&+ \int_{\partial\Omega} \sum_{j,k=1}^{n} \frac{\partial^2 \rho}{\partial x_j \partial x_k} \langle dx_j \vee u_{tan}, dx_k \vee v_{tan}\rangle\, d\sigma \\
&- {}_{H^{1/2}(\partial\Omega)}\langle \nu \vee u, \delta_\partial v_{tan}\rangle_{H^{-1/2}(\partial\Omega)} + {}_{H^{1/2}(\partial\Omega)}\langle \nu \wedge u, d_\partial v_{nor}\rangle_{H^{-1/2}(\partial\Omega)}.
\end{aligned} \tag{4.6}$$

We use (4.6) in a special case where $u = v := dw \in W^{1,2}(\Omega)$, regarded as 1-form. This allows us to make the natural identifications

$$(dw)_{nor} \equiv (\partial_\nu w)\nu, \quad (dw)_{tan} \equiv \nabla_{tan} w. \tag{4.7}$$

Also, it is straightforward to check that $d_\partial (dw)_{nor} = -\nu \wedge d(\partial_\nu w)$ so that

$$
\begin{aligned}
{}_{H^{1/2}(\partial\Omega)}\langle \nu \wedge dw, d_{\partial}(dw)_{nor}\rangle_{H^{-1/2}(\partial\Omega)} \\
&= -{}_{H^{1/2}(\partial\Omega)}\langle dw, \nu \vee (\nu \wedge d(\partial_\nu w))\rangle_{H^{-1/2}(\partial\Omega)} \\
&= -{}_{H^{1/2}(\partial\Omega)}\langle \nabla w, \nabla_{tan}(\partial_\nu w)\rangle_{H^{-1/2}(\partial\Omega)} \\
&= -{}_{H^{1/2}(\partial\Omega)}\langle \nabla_{tan} w, \nabla_{tan}(\partial_\nu w)\rangle_{H^{-1/2}(\partial\Omega)}. \qquad (4.8)
\end{aligned}
$$

Also, with Div denoting the surface divergence operator on $\partial\Omega$ (i.e., the formal adjoint of the tangential gradient introduced in (2.4)), we may write

$$
\begin{aligned}
{}_{H^{1/2}(\partial\Omega)}\langle \nu \vee dw, \delta_{\partial}(dw)_{tan}\rangle_{H^{-1/2}(\partial\Omega)} \\
&= -{}_{H^{1/2}(\partial\Omega)}\langle \partial_\nu w, \mathrm{Div}(\nabla_{tan} w)\rangle_{H^{-1/2}(\partial\Omega)} \\
&= -{}_{H^{1/2}(\partial\Omega)}\langle \nabla_{tan} w, \nabla_{tan}(\partial_\nu w)\rangle_{H^{-1/2}(\partial\Omega)}. \qquad (4.9)
\end{aligned}
$$

Finally, using that $d^2 = 0$, $\delta(dw) = \mathrm{div}(\nabla w) = \Delta w$, and the fact that $u_{nor} = v_{nor} = (\partial_\nu w)\nu$, identity (4.3) follows from (4.6) and the above identifications in a straightforward fashion, once we note that

$$
\begin{aligned}
\langle dx_j \wedge u_{nor}, dx_k \wedge v_{nor}\rangle &= (\partial_\nu w)^2 \langle dx_j \wedge \nu\,,\, dx_k \wedge \nu\rangle \\
&= (\partial_\nu w)^2 (\delta_{jk} - \nu_j \nu_k) \qquad (4.10)
\end{aligned}
$$

and

$$
\langle dx_j \vee \nabla_{tan} w\,,\, dx_k \vee \nabla_{tan} w\rangle = (\nabla_{tan} w)_j (\nabla_{tan} w)_k \qquad (4.11)
$$

for every $j, k \in \{1, \ldots, n\}$. □

Proposition 4.2. *Assume that the domain $\Omega \subset \mathbb{R}^n$ is given as the upper graph of a C^2 function $\varphi : \mathbb{R}^{n-1} \to \mathbb{R}$, and satisfies an uniform exterior ball condition. Then there exists $\rho \in C^2$ defining function for Ω (in the sense of (4.1)) such that* Hess (ρ) *is bounded from below by a negative constant. More specifically, there exists $C_0 > 0$, depending only on the Lipschitz and UEBC constants of Ω, such that*

$$
\langle \mathrm{Hess}\,(\rho)\xi, \xi\rangle \geqslant -C_0|\xi|^2 \quad \forall \xi \in \mathbb{R}^n. \qquad (4.12)
$$

Proof. Define

$$
\rho(x) := \varphi(x') - x_n, \quad x = (x', x_n) \in \mathbb{R}^{n-1} \times \mathbb{R} = \mathbb{R}^n, \qquad (4.13)
$$

so that

$$\frac{\partial^2 \rho}{\partial x_j \partial x_k}(x) = \begin{cases} \dfrac{\partial^2 \varphi}{\partial x_j \partial x_k}(x'), & 1 \leqslant j,k \leqslant n-1, \\ 0, & 1 \leqslant j \leqslant n-1 \text{ and } k = n, \\ 0, & j = k = n. \end{cases} \tag{4.14}$$

Thus, based on (4.13) and (4.14), we have

$$\begin{aligned} \langle \mathrm{Hess}\,(\rho)(x)\xi, \xi\rangle &= \big\langle \big(\mathrm{Hess}\,(\varphi)(x')\xi', 0\big), \xi\big\rangle \\ &= \langle \mathrm{Hess}\,(\varphi)(x')\xi', \xi'\rangle \geqslant -C_0|\xi'|^2 \geqslant -C_0|\xi|^2, \end{aligned} \tag{4.15}$$

where the next-to-last inequality follows from Lemma 6.3 in [18]. □

Lemma 4.3. *If H is an $n\times n$ symmetric matrix which is bounded from below by some possibly negative constant $-C_0$ in the sense of* (4.12), *then*

$$\mathrm{tr}H - \langle H\nu, \nu\rangle \geqslant -C_0(n-1) \tag{4.16}$$

for any unit vector $\nu \in \mathbb{R}^n$.

Proof. Diagonalizing H, we obtain $H = U^{-1}DU$, where D is an $n\times n$ diagonal matrix with entries $d_{ij} := \delta_{ij} d_j$, $i, j \in \{1, \dots, n\}$, and U is a unitary matrix (i.e., $U^{-1} = U^*$). Since, by assumption, H is bounded from below by $-C_0$, it readily follows that

$$d_j \geqslant -C_0 \quad \forall\, j \in \{1, \dots, n\}. \tag{4.17}$$

Let ν be a unit vector in $\mathbb{R}^n$. Introducing $\xi := U\nu$, we have

$$\begin{aligned} \mathrm{tr}H - \langle H\nu, \nu\rangle &= \sum_{j=1}^n d_j - \langle D\xi, \xi\rangle \\ &= \sum_{j=1}^n d_j(1-\xi_j^2) \geqslant -C_0 \sum_{j=1}^n (1-\xi_j^2) = -C_0(n-1), \end{aligned} \tag{4.18}$$

where the inequality above follows from (4.17) and the fact that for each $j \in \{1, \dots, n\}$ we have $1 - \xi_j^2 \geqslant 0$ as $|\xi| = |U\nu| = |\nu| = 1$. The last equality in (4.18) also uses the fact that $|\xi| = 1$. □

5 Approximation Scheme

In principle, the approximation scheme discussed in this section is known (see, for example, [18, 19, 7] for the case of convex domains). Nonetheless, since this plays an important role in our analysis, for the benefit of the reader we describe it here in some detail.

Proposition 5.1. *Given $\Omega \subset \mathbb{R}^n$ a Lipschitz domain satisfying a uniform exterior ball condition, there exist smooth domains Ω_ε, $\varepsilon > 0$, with Lipschitz character bounded in ε such that $\Omega_\varepsilon \nearrow \Omega$ and each Ω_ε has a defining function ρ_ε satisfying*

$$\operatorname{Hess}(\rho_\varepsilon) \geqslant -C_0\, I_{n\times n} \quad \textit{for} \quad C_0 > 0 \quad \textit{independent of } \varepsilon, \tag{5.1}$$

where $I_{n\times n}$ is the $n \times n$ identity matrix. In addition, for each $\xi \in C_0^\infty(\mathbb{R}^n)$

$$\int\limits_{\partial\Omega_\varepsilon} \xi \, d\sigma_\varepsilon \to \int\limits_{\partial\Omega} \xi \, d\sigma \qquad \textit{as} \quad \varepsilon \to 0, \tag{5.2}$$

where $d\sigma$ and $d\sigma_\varepsilon$ stand for the surface measures on $\partial\Omega$ and $\partial\Omega_\varepsilon$ respectively.

Property (5.2) is typically not explicitly mentioned, but it follows from an inspection of the specific way Ω_ε's are constructed. To shed more light on this, below we briefly review the construction and check (5.2) in the technically somewhat simpler case where Ω is a graph Lipschitz domain satisfying a uniform exterior ball condition (the case of a bounded domain is handled similarly by patching together local results). With this goal in mind, let Ω be a graph Lipschitz domain which satisfies a uniform exterior ball condition in the sense of Definition 2.1. In particular

$$\Omega := \{(x', x_n) \in \mathbb{R}^n : x_n > \varphi(x')\}, \tag{5.3}$$

where $\varphi : \mathbb{R}^{n-1} \to \mathbb{R}$ is a Lipschitz function.

Pick $\Phi \in C_0^\infty(\mathbb{R}^{n-1})$, $0 \leqslant \Phi \leqslant 1$, $\Phi \equiv 0$ for $|x'| > 1$ and $\displaystyle\int\limits_{\mathbb{R}^{n-1}} \Phi(x')\, dx' = 1$ and for each $\varepsilon > 0$ let $\Phi_\varepsilon(x') := \varepsilon^{-(n-1)}\Phi(x'/\varepsilon)$. Going further, for each $\varepsilon > 0$ we consider

$$\varphi^\varepsilon(x') := C\varepsilon + \int\limits_{\mathbb{R}^{n-1}} \Phi_\varepsilon(y')\varphi(x' - y')\, dy', \quad x' \in \mathbb{R}^{n-1}, \tag{5.4}$$

where C is a sufficiently large constant, to be specified momentarily. Note that $\varphi^\varepsilon \in C^\infty(\mathbb{R}^{n-1})$ and $\|\nabla\varphi^\varepsilon\|_{L^\infty(\mathbb{R}^{n-1})} \leqslant \|\nabla\varphi\|_{L^\infty(\mathbb{R}^{n-1})}$ for all $\varepsilon > 0$. A direct calculation shows that if we take $C > \|\nabla\varphi\|_{L^\infty(\mathbb{R}^{n-1})}$, then

$$\frac{d}{d\varepsilon}[\varphi^\varepsilon(x')] > 0 \qquad \forall\, x' \in \mathbb{R}^{n-1}. \tag{5.5}$$

Hence for each $x' \in \mathbb{R}^{n-1}$ we have $\varphi^\varepsilon(x') \searrow \varphi(x')$ as $\varepsilon \searrow 0$. Setting

$$\Omega_\varepsilon := \{(x', x_n) : x_n > \varphi^\varepsilon(x')\}, \tag{5.6}$$

by Lemma 6.2 in [18], we conclude that $\Omega_\varepsilon \in C^\infty$ is Lipschitz uniformly in ε, satisfies a uniform exterior ball condition uniformly in ε, and $\Omega_\varepsilon \nearrow \Omega$ as $\varepsilon \searrow 0$. Note that $\rho_\varepsilon(x) := \varphi^\varepsilon(x') - x_n$ is the defining function for Ω_ε in the sense of (4.1). Denote by σ_ε the surface measure on $\partial\Omega_\varepsilon$.

Lemma 5.2. *With the notation introduced above, for each $\xi \in C_0^\infty(\mathbb{R}^n)$ we have*

$$\int\limits_{\partial\Omega_\varepsilon} \xi\, d\sigma_\varepsilon \to \int\limits_{\partial\Omega} \xi\, d\sigma, \qquad as \quad \varepsilon \to 0. \tag{5.7}$$

Proof. Indeed,

$$\int\limits_{\partial\Omega_\varepsilon} \xi\, d\sigma_\varepsilon = \int\limits_{\mathbb{R}^{n-1}} \xi(x', \varphi^\varepsilon(x'))\sqrt{1+|\nabla\varphi^\varepsilon(x')|^2}\, dx' \tag{5.8}$$

$$\overset{\varepsilon\to 0}{\longrightarrow} \int\limits_{\mathbb{R}^{n-1}} \xi(x', \varphi(x'))\sqrt{1+|\nabla\varphi(x')|^2}\, dx' = \int\limits_{\partial\Omega} \xi\, d\sigma,$$

where the convergence above follows from invoking the Lebesgue dominated convergence theorem (note that $\varphi^\varepsilon \to \varphi$, $\nabla\varphi^\varepsilon \to \nabla\varphi$ pointwise almost everywhere, and $\|\xi\|_{L^\infty(\mathbb{R}^n)}\sqrt{1+|\nabla\varphi|^2_{L^\infty(\mathbb{R}^{n-1})}}\chi_{\operatorname{supp}\xi}$ provides uniform domination). □

Finally, we present a corollary of Lemma 5.2 which will be useful in the sequel.

Corollary 5.3. *Let $\Omega_\varepsilon \nearrow \Omega$ as $\varepsilon \searrow 0$ be as constructed above. Denote by Tr_ε the trace operator corresponding to $\partial\Omega_\varepsilon$ and by $d\sigma_\varepsilon$ the surface measure on $\partial\Omega_\varepsilon$. Then for all $u \in W^{1,2}(\Omega)$ and $\xi \in C^\infty(\overline{\Omega})$*

$$\int\limits_{\partial\Omega_\varepsilon} \mathrm{Tr}_\varepsilon(u)\xi\Big|_{\partial\Omega_\varepsilon} d\sigma_\varepsilon \longrightarrow \int\limits_{\partial\Omega} \mathrm{Tr}(u)\xi\, d\sigma \quad as \quad \varepsilon \to 0. \tag{5.9}$$

Proof. Let $u \in W^{1,2}(\Omega)$ be arbitrary. Fix $\psi \in C^\infty(\overline{\Omega})$. We write $\mathrm{Tr}_\varepsilon(u) = \mathrm{Tr}_\varepsilon(u-\psi) + \mathrm{Tr}_\varepsilon(\psi)$. Further,

$$\int\limits_{\partial\Omega_\varepsilon} \mathrm{Tr}_\varepsilon(u)\xi\Big|_{\partial\Omega_\varepsilon} d\sigma_\varepsilon = I + II, \tag{5.10}$$

where

$$I := \int\limits_{\partial\Omega_\varepsilon} \mathrm{Tr}_\varepsilon(u-\psi)\xi\Big|_{\partial\Omega_\varepsilon} d\sigma_\varepsilon \tag{5.11}$$

and

$$II := \int_{\partial\Omega_\varepsilon} \mathrm{Tr}_\varepsilon(\psi)\xi\Big|_{\partial\Omega_\varepsilon} d\sigma_\varepsilon. \tag{5.12}$$

Next, $H^{1/2}(\partial\Omega_\varepsilon) \hookrightarrow L^2(\partial\Omega_\varepsilon)$ with accompanying bounds for the inclusion maps independent of ε. Also, denoting by R_ε the restriction operator from Ω to Ω_ε, we find that

$$\mathrm{Tr}_\varepsilon : W^{1,2}(\Omega_\varepsilon) \longrightarrow H^{1/2}(\partial\Omega_\varepsilon) \text{ and } R_\varepsilon : W^{1,2}(\Omega) \longrightarrow W^{1,2}(\Omega_\varepsilon) \tag{5.13}$$

are bounded operators with bounds controlled uniformly in $\varepsilon > 0$. Thus, there exists $C > 0$ independent of ε such that

$$|I| \leqslant \|\mathrm{Tr}_\varepsilon(u-\psi)\|_{L^2(\Omega_\varepsilon)}\|\xi|_{\partial\Omega_\varepsilon}\|_{L^2(\partial\Omega_\varepsilon)} \leqslant C\|u-\psi\|_{W^{1,2}(\Omega)}. \tag{5.14}$$

Recall now that $C^\infty(\overline{\Omega})$ is densely embedded into $W^{1,2}(\Omega)$ and note that

$$\int_{\partial\Omega} \mathrm{Tr}(\psi)\xi\, d\sigma \longrightarrow \int_{\partial\Omega} \mathrm{Tr}(u)\xi\, d\sigma \quad \text{as} \quad \psi \to u \quad \text{in} \quad W^{1,2}(\Omega). \tag{5.15}$$

As for II in (5.12), we invoke Lemma 5.2 to conclude that

$$\lim_{\varepsilon\to 0} II = \int_{\partial\Omega} \mathrm{Tr}(\psi)\xi\, d\sigma. \tag{5.16}$$

Finally, the conclusion of the Corollary follows from (5.10), (5.14), and (5.15) by letting $\psi \to u$ in $W^{1,2}(\Omega)$. □

6 Proof of Step I

Let Ω be a bounded Lipschitz domain in $\mathbb{R}^n$ satisfying an exterior ball condition in the sense of Definition 2.1. Fix $\lambda > 0$, $f \in L^2(\Omega)$, and assume that u solves

$$\begin{cases} \Delta u = f \in L^2(\Omega) & \text{in} \quad \Omega, \\ \partial_\nu u + \lambda u = 0 & \text{on} \quad \partial\Omega, \\ u \in W^{1,2}(\Omega), \end{cases} \tag{6.1}$$

and satisfies

$$\|u\|_{W^{1,2}(\Omega)} \leqslant C\|f\|_{L^2(\Omega)}, \tag{6.2}$$

for some $C > 0$ which depends only on the Lipschitz character of Ω and λ (see Proposition 3.5). Next, for each $\varepsilon > 0$ let Ω_ε be the domain constructed as in (5.6) with outward unit normal vector ν_ε, and surface measure $d\sigma_\varepsilon$. Also, we denote by Tr_ε the trace operator on $\partial\Omega_\varepsilon$. By once again relying on Proposition 3.5, we then solve

$$\begin{cases} \Delta u_\varepsilon = 0 & \text{in } \Omega_\varepsilon, \\ \partial_{\nu_\varepsilon} u_\varepsilon + \lambda \,\mathrm{Tr}_\varepsilon\, u_\varepsilon = \partial_{\nu_\varepsilon} u + \lambda \,\mathrm{Tr}_\varepsilon\, u \in H^{-1/2}(\partial\Omega_\varepsilon) & \text{on } \partial\Omega_\varepsilon, \\ u_\varepsilon \in W^{1,2}(\Omega_\varepsilon). \end{cases} \tag{6.3}$$

From (3.62) we know that $\|u_\varepsilon\|_{W^{1,2}(\Omega_\varepsilon)}$ is controlled uniformly in $\varepsilon > 0$. Since the Robin data of u_ε in (6.3) is the Robin data of the function $u \in W^{2,2}(\Omega_\varepsilon)$ (this, however, without control of the norm in ε) and $\partial\Omega_\varepsilon$ is of class C^∞, standard elliptic regularity results give that $u_\varepsilon \in W^{2,2}(\Omega_\varepsilon)$ (without control of the norm with respect to ε). Consider

$$w_\varepsilon := u\big|_{\Omega_\varepsilon} - u_\varepsilon, \qquad \varepsilon > 0, \tag{6.4}$$

where u_ε is the same as in (6.3). Then

$$\begin{cases} w_\varepsilon \in W^{2,2}(\Omega_\varepsilon) \quad \text{(without control in } \varepsilon), \\ w_\varepsilon \in W^{1,2}(\Omega_\varepsilon) \quad \text{(with control in } \varepsilon), \\ \Delta w_\varepsilon = f|_{\Omega_\varepsilon} \quad \text{in} \quad \Omega_\varepsilon, \\ \partial_{\nu_\varepsilon} w_\varepsilon + \lambda \,\mathrm{Tr}_\varepsilon\, w_\varepsilon = 0 \quad \text{on} \quad \partial\Omega_\varepsilon. \end{cases} \tag{6.5}$$

Next, denoting by ρ_ε the defining function of Ω_ε, using (5.1) and Lemma 4.3, we conclude that there exists C_0 finite positive constant which is independent of ε and

$$\mathrm{tr}(\mathrm{Hess}\,(\rho_\varepsilon)) - \langle \mathrm{Hess}\,(\rho_\varepsilon)\nu, \nu\rangle \geqslant -C_0 \quad \forall\, \varepsilon > 0. \tag{6.6}$$

Applying Theorem 4.1 to the domain Ω_ε, with defining function ρ_ε, and to the function w_ε, using (4.3) and the boundary condition in (6.5), we find

$$\begin{aligned} \sum_{j,k=1}^n \int_{\Omega_\varepsilon} \Big(\frac{\partial^2 w_\varepsilon}{\partial x_j \partial x_k}\Big)^2 dx &\leqslant \int_{\Omega_\varepsilon} |\Delta w_\varepsilon|^2\, dx + C_0 \int_{\partial\Omega_\varepsilon} |\nabla w_\varepsilon|^2\, d\sigma_\varepsilon \\ &= \int_{\Omega_\varepsilon} |f|^2\, dx + C_0 \int_{\partial\Omega_\varepsilon} |\nabla w_\varepsilon|^2\, d\sigma_\varepsilon. \end{aligned} \tag{6.7}$$

On the other hand, Proposition 3.5 gives that

$$\int_{\Omega_\varepsilon} \left[|\nabla w_\varepsilon|^2 + |w_\varepsilon|^2\right] dx \leqslant C \int_{\Omega} |f|^2 \, dx \tag{6.8}$$

for some $C > 0$ independent of $\varepsilon > 0$. Combining (6.7) and (6.8), we may then conclude that

$$\|w_\varepsilon\|^2_{W^{2,2}(\Omega_\varepsilon)} \leqslant C \int_{\Omega_\varepsilon} |f|^2 \, dx + C \int_{\partial\Omega_\varepsilon} |\nabla w_\varepsilon|^2 \, d\sigma_\varepsilon \tag{6.9}$$

for some $C > 0$ independent of $\varepsilon > 0$.

At this point, we bring in the following boundary trace inequality. For each $\theta > 0$ there exists C_θ depending on θ (and the Lipschitz character of Ω_ε which is controlled uniformly in $\varepsilon > 0$) such that

$$\int_{\partial\Omega_\varepsilon} |\nabla w_\varepsilon|^2 \, d\sigma_\varepsilon \leqslant \theta \|w_\varepsilon\|^2_{W^{2,2}(\Omega_\varepsilon)} + C_\theta \|w_\varepsilon\|^2_{W^{1,2}(\Omega_\varepsilon)}. \tag{6.10}$$

To prove (6.10), we fix $j \in \{1, \dots, n\}$ and set $v := \partial_j w_\varepsilon \in W^{1,2}(\Omega_\varepsilon)$. Also, fix $h \in [C_0^\infty(\mathbb{R}^n)]^n$, a transversal vector field to $\partial\Omega_\varepsilon$, uniformly in ε, i.e., there exists $\varkappa > 0$ such that

$$h \cdot \nu_\varepsilon \geqslant \varkappa \quad \text{a.e. on } \partial\Omega_\varepsilon \qquad \forall\, \varepsilon > 0. \tag{6.11}$$

Then for a given $\theta > 0$ the divergence theorem yields

$$\begin{aligned}
\int_{\partial\Omega_\varepsilon} |v|^2 \, d\sigma_\varepsilon &\leqslant \varkappa^{-1} \int_{\partial\Omega_\varepsilon} \langle v^2 h, \nu_\varepsilon \rangle \, d\sigma_\varepsilon = \varkappa^{-1} \int_{\Omega_\varepsilon} \operatorname{div}(v^2 h) \, dx \\
&= \varkappa^{-1} \int_{\Omega_\varepsilon} [v^2 (\operatorname{div} h) + 2v(\nabla v \cdot h)] \, dx \\
&\leqslant \theta \|\nabla v\|^2_{L^2(\Omega_\varepsilon)} + C_\theta \|v\|^2_{L^2(\Omega_\varepsilon)}.
\end{aligned} \tag{6.12}$$

From this, (6.10) readily follows. Thus, making use of (6.5) and (3.58), we obtain

$$\int_{\partial\Omega_\varepsilon} |\nabla w_\varepsilon|^2 \, d\sigma_\varepsilon \leqslant \theta \|w_\varepsilon\|^2_{W^{2,2}(\Omega_\varepsilon)} + C_\theta \|f\|^2_{L^2(\Omega)}. \tag{6.13}$$

Employing (6.13) for, say $\theta = \frac{1}{2C}$, together with (6.9), allows us to obtain

$$\|w_\varepsilon\|_{W^{2,2}(\Omega_\varepsilon)} \leqslant C \|f\|_{L^2(\Omega)}, \tag{6.14}$$

where $C > 0$ is independent of $\varepsilon > 0$. Consequently,

$$\begin{aligned} &\exists\, w \in L^2_{loc}(\Omega) \quad \text{such that} \quad \text{for every } \alpha \in \mathbb{N}_0^n \text{ with } |\alpha| \leqslant 2 \\ &\partial^\alpha w_\varepsilon \longrightarrow \partial^\alpha w \quad \text{as} \quad \varepsilon \to 0 \quad L^2\text{-weakly on compact sets of } \Omega. \end{aligned} \tag{6.15}$$

Moving on, let $K \subset \Omega$ be an arbitrary fixed compact set. Using (6.15), we further infer that for some finite $C > 0$ independent of K

$$\int_K |\partial^\alpha w|^2\, dx \leqslant \limsup_{\varepsilon \to 0} \int_K |\partial^\alpha w_\varepsilon|^2\, dx \leqslant C \int_\Omega |f|^2\, dx. \tag{6.16}$$

This follows from the general simple fact to the effect that if $(X, \|\cdot\|_X)$ is a Banach space and X^* is its dual, then $v_\varepsilon \in X^*$ converges weekly to $v \in X^*$ implies $\|v\|_{X^*} \leqslant \limsup\limits_{\varepsilon \to 0} \|v_\varepsilon\|_{X^*}$. Indeed, let $x \in X$ with $\|x\|_X = 1$. Then

$$|{}_{X^*}\langle v, x\rangle_X| = \lim_{\varepsilon \to 0} |{}_{X^*}\langle v_\varepsilon, x\rangle_X| \leqslant \limsup_{\varepsilon \to 0} \|v_\varepsilon\|_{X^*}$$

which implies the desired conclusion.

Now, using (6.16), we let $K \nearrow \Omega$ and conclude that $w \in W^{2,2}(\Omega)$ and

$$\|w\|^2_{W^{2,2}(\Omega)} = \sum_{|\alpha| \leqslant 2} \int_\Omega |\partial^\alpha w|^2\, dx \leqslant C \int_\Omega |f|^2\, dx. \tag{6.17}$$

Our goal is to show that $w = u$ which, in light of the above estimate, completes the proof of Step I. To this end, thanks to (6.4), it suffices to show that

$$u_\varepsilon \longrightarrow 0 \quad \text{as} \quad \varepsilon \to 0 \quad L^2\text{-weakly on compact subsets of } \Omega. \tag{6.18}$$

Since $\|u_\varepsilon\|_{W^{1,2}(\Omega_\varepsilon)}$ is controlled independently in $\varepsilon > 0$ it follows that there exists $\omega \in W^{1,2}(\Omega)$ such that for each compact subset K of Ω we have

$$\partial^\alpha u_\varepsilon \longrightarrow \partial^\alpha \omega \quad \text{weakly in } L^2(K) \qquad \forall\, \alpha \in \mathbb{N}_0^n,\ |\alpha| \leqslant 1. \tag{6.19}$$

Note that

$$\Delta\omega = 0 \quad \text{in} \quad \Omega \tag{6.20}$$

in the sense of distributions. Indeed, let $\varphi \in C_0^\infty(\Omega)$. Select a compact subset K of Ω such that $\operatorname{supp}\varphi$ is a subset of the interior of K. Using (6.19), we are then able to write

$$\begin{aligned} \langle \omega, \Delta\varphi\rangle_{L^2(\Omega)} &= \langle \omega, \Delta\varphi\rangle_{L^2(K)} = \lim_{\varepsilon \to 0} \langle u_\varepsilon, \Delta\varphi\rangle_{L^2(K)} \\ &= \lim_{\varepsilon \to 0} \langle \Delta u_\varepsilon, \varphi\rangle_{L^2(K)} = 0. \end{aligned} \tag{6.21}$$

If we show that

$$\partial_\nu \omega + \lambda \operatorname{Tr} \omega = 0 \quad \text{in } H^{-1/2}(\partial\Omega), \tag{6.22}$$

then, by Proposition 3.5, we have $\omega \equiv 0$ in Ω. Then the claim (6.18) follows from (6.19).

At this point, we are left with showing that (6.22) holds and start by considering a function $\varphi \in H^{1/2}(\partial\Omega)$ along with some $\Phi \in W^{1,2}(\Omega)$ such that $\operatorname{Tr} \Phi = \varphi$. From (3.56) we have

$${}_{(H^{1/2}(\partial\Omega))^*}\langle \partial_\nu \omega, \varphi \rangle_{H^{1/2}(\partial\Omega)} = \int_\Omega \langle \nabla\omega, \nabla\Phi \rangle \, dx = \lim_{\varepsilon \to 0} \int_{\Omega_\varepsilon} \langle \nabla\omega, \nabla\Phi \rangle \, dx. \tag{6.23}$$

Fix for the moment $\delta > 0$. Let K_δ be a compact subset of Ω for which $\|\nabla\Phi\|_{L^2(\Omega\setminus K_\delta)} \leqslant \delta$. Using (6.23), we obtain

$$\begin{aligned} &{}_{(H^{1/2}(\partial\Omega))^*}\langle \partial_\nu \omega, \varphi \rangle_{H^{1/2}(\partial\Omega)} \\ &= \lim_{\varepsilon \to 0} \Big[\int_{\Omega_\varepsilon \setminus K_\delta} \langle \nabla\omega, \nabla\Phi \rangle \, dx + \int_{K_\delta} \langle \nabla\omega, \nabla\Phi \rangle \, dx \Big] =: \lim_{\varepsilon \to 0} [I + II]. \end{aligned} \tag{6.24}$$

Due to the choice of K_δ,

$$\limsup_{\varepsilon \to 0} |I| \leqslant \|\nabla\Phi\|_{L^2(\Omega\setminus K_\delta)} \|\nabla\omega\|_{L^2(\Omega)} \leqslant \delta \, \|\nabla\omega\|_{L^2(\Omega)}. \tag{6.25}$$

Also, using (6.19), we may write

$$\begin{aligned} II &= \lim_{\varepsilon \to 0} \int_{K_\delta} \langle \nabla u_\varepsilon, \nabla\Phi \rangle \, dx \\ &= \lim_{\varepsilon \to 0} \Big[\int_{\Omega_\varepsilon} \langle \nabla u_\varepsilon, \nabla\Phi \rangle \, dx - \int_{\Omega_\varepsilon \setminus K_\delta} \langle \nabla u_\varepsilon, \nabla\Phi \rangle \, dx \Big] \\ &=: \lim_{\varepsilon \to 0} [III - IV]. \end{aligned} \tag{6.26}$$

Since $\|u_\varepsilon\|_{W^{1,2}(\Omega)}$ is controlled uniformly in $\varepsilon > 0$, we conclude that there exists a constant $C > 0$ independent of ε such that

$$\limsup_{\varepsilon \to 0} |IV| \leqslant \|\nabla u_\varepsilon\|_{L^2(\Omega)} \|\nabla\Phi\|_{L^2(\Omega\setminus K_\delta)} \leqslant C\delta. \tag{6.27}$$

Thus, using (6.23)-(6.27), we obtain (employing the usual "big O" notation)

$${}_{(H^{1/2}(\partial\Omega))^*}\langle \partial_\nu \omega, \varphi \rangle_{H^{1/2}(\partial\Omega)} = \lim_{\varepsilon \to 0} \int_{\Omega_\varepsilon} \langle \nabla u_\varepsilon, \nabla\Phi \rangle \, dx + O(\delta). \tag{6.28}$$

Thus, by letting $\delta \to 0$, we conclude

$$ {}_{(H^{1/2}(\partial\Omega))^*}\langle \partial_\nu \omega, \varphi\rangle_{H^{1/2}(\partial\Omega)} = \lim_{\varepsilon\to 0}\int_{\Omega_\varepsilon} \langle \nabla u_\varepsilon, \nabla \Phi\rangle\, dx. \tag{6.29} $$

Going further, we use the integration by parts formula (3.56) in order to write

$$ \begin{aligned} \int_{\Omega_\varepsilon} \langle \nabla u_\varepsilon, \nabla \Phi\rangle\, dx &= {}_{(H^{1/2}(\partial\Omega_\varepsilon))^*}\langle \partial_{\nu_\varepsilon} u_\varepsilon, \mathrm{Tr}_\varepsilon \Phi\rangle_{H^{1/2}(\partial\Omega_\varepsilon)} \\ &= \lambda \int_{\partial\Omega_\varepsilon} (\mathrm{Tr}_\varepsilon\, u - \mathrm{Tr}_\varepsilon\, u_\varepsilon)\, \mathrm{Tr}_\varepsilon\, \Phi\, d\sigma_\varepsilon \\ &\quad + {}_{(H^{1/2}(\partial\Omega_\varepsilon))^*}\langle \partial_{\nu_\varepsilon} u, \mathrm{Tr}_\varepsilon\, \Phi\rangle_{H^{1/2}(\partial\Omega_\varepsilon)}, \end{aligned} \tag{6.30} $$

where the last equality follows from the boundary condition in (6.3). We claim next that

$$ \int_{\partial\Omega_\varepsilon} \mathrm{Tr}_\varepsilon\, u_\varepsilon\, \mathrm{Tr}_\varepsilon\, \Phi\, d\sigma_\varepsilon \longrightarrow \int_{\partial\Omega} \mathrm{Tr}\, \omega\, \varphi\, d\sigma \quad \text{as} \quad \varepsilon \to 0, \tag{6.31} $$

$$ \int_{\partial\Omega_\varepsilon} \mathrm{Tr}_\varepsilon\, u\, \mathrm{Tr}_\varepsilon\, \Phi\, d\sigma_\varepsilon \longrightarrow \int_{\partial\Omega} \mathrm{Tr}\, u\, \varphi\, d\sigma \quad \text{as} \quad \varepsilon \to 0, \tag{6.32} $$

and

$$ {}_{(H^{1/2}(\partial\Omega_\varepsilon))^*}\langle \partial_{\nu_\varepsilon} u, \mathrm{Tr}_\varepsilon\, \Phi\rangle_{H^{1/2}(\partial\Omega_\varepsilon)} \longrightarrow {}_{(H^{1/2}(\partial\Omega))^*}\langle \partial_\nu u, \varphi\rangle_{H^{1/2}(\partial\Omega)} \tag{6.33} $$

as $\varepsilon \to 0$.

Turning our attention to proving (6.31), for each $\varepsilon > 0$ we consider the extension operator (i.e., a right-inverse for the restriction operator mapping distributions in Ω to distributions in Ω_ε)

$$ \mathcal{E}_\varepsilon : W^{1,2}(\Omega_\varepsilon) \longrightarrow W^{1,2}(\Omega). \tag{6.34} $$

Matters can be arranged so that the operator norm of (6.34) is bounded uniformly in $\varepsilon > 0$ (see [22]). Relying on this and the fact that $\|u_\varepsilon\|_{W^{1,2}(\Omega)}$ is controlled uniformly in $\varepsilon > 0$, we may deduce that the sequence $\{\mathcal{E}_\varepsilon u_\varepsilon\}_{\varepsilon>0}$ is bounded in $W^{1,2}(\Omega)$. Thus, by the Rellich selection lemma, we may further conclude that there exists a function $v \in W^{1,2}(\Omega)$ with the property that for each $r \in (1/2, 1)$

$$ \mathcal{E}_\varepsilon u_\varepsilon \longrightarrow v \quad \text{as} \quad \varepsilon \to 0 \quad \text{with convergence in} \quad W^{1-r,2}(\Omega). \tag{6.35} $$

Now, from (6.35) and (6.19) we deduce that $v = \omega$. Next, for each function $\xi \in C^\infty(\overline{\Omega})$ we write

$$\int_{\partial\Omega_\varepsilon} \mathrm{Tr}_\varepsilon\, u_\varepsilon\, \xi\, d\sigma_\varepsilon = \int_{\partial\Omega_\varepsilon} \mathrm{Tr}_\varepsilon(\mathcal{E}_\varepsilon u_\varepsilon)\, \xi\, d\sigma_\varepsilon \tag{6.36}$$

$$= \int_{\partial\Omega_\varepsilon} \mathrm{Tr}_\varepsilon(\mathcal{E}_\varepsilon u_\varepsilon - \omega)\, \xi\, d\sigma_\varepsilon + \int_{\partial\Omega_\varepsilon} \mathrm{Tr}_\varepsilon\, \omega\, \xi\, d\sigma_\varepsilon.$$

Analogously to the argument leading to (5.14) in Corollary 5.3, we infer that

$$\left| \int_{\partial\Omega_\varepsilon} \mathrm{Tr}_\varepsilon(\mathcal{E}_\varepsilon u_\varepsilon - \omega)\xi\, d\sigma_\varepsilon \right| \leqslant C_\xi \,\|\mathrm{Tr}_\varepsilon(\mathcal{E}_\varepsilon u_\varepsilon - \omega)\|_{L^2(\partial\Omega_\varepsilon)} \tag{6.37}$$

$$\leqslant C_\xi \,\|\mathcal{E}_\varepsilon u_\varepsilon - \omega\|_{W^{1-r,2}(\Omega)} \longrightarrow 0 \quad \text{as} \quad \varepsilon \to 0.$$

For the second term in the second line of (6.36) we employ Corollary 5.3 in order to conclude that

$$\int_{\partial\Omega_\varepsilon} \mathrm{Tr}_\varepsilon\, \omega\, \xi\, d\sigma_\varepsilon \longrightarrow \int_{\partial\Omega} \mathrm{Tr}\, \omega\, \xi\, d\sigma. \tag{6.38}$$

With this in hand, (6.31) follows from (6.36)-(6.38) and a simple density argument (involving approximating $\Phi \in W^{1,2}(\Omega)$ with functions $\xi \in C^\infty(\overline{\Omega})$ in $W^{1,2}(\Omega)$). Furthermore, (6.32) follows from Corollary 5.3 and the same type of density argument as above. Finally, as regards (6.33), we write

$$\begin{aligned} {}_{(H^{1/2}(\partial\Omega_\varepsilon))^*}\langle \partial_{\nu_\varepsilon} u, \mathrm{Tr}_\varepsilon\, \Phi\rangle_{H^{1/2}(\partial\Omega_\varepsilon)} &= \int_{\Omega_\varepsilon} \nabla u \cdot \nabla\Phi\, dx + \int_{\Omega_\varepsilon} f\Phi\, dx \\ &\overset{\varepsilon\to 0}{\longrightarrow} \int_{\Omega} \nabla u \cdot \nabla\Phi\, dx + \int_{\Omega} f\,\Phi\, dx \\ &= {}_{(H^{1/2}(\partial\Omega))^*}\langle \partial_\nu u, \varphi\rangle_{H^{1/2}(\partial\Omega)}. \end{aligned} \tag{6.39}$$

At this stage, using (6.31)-(6.33) and (6.30) as passing to the limit in (6.29), we conclude that

$$\begin{aligned} {}_{(H^{1/2}(\partial\Omega))^*}\langle \partial_\nu \omega, \varphi\rangle_{H^{1/2}(\partial\Omega)} &= \lambda \int_{\partial\Omega} \mathrm{Tr}\, u\, \varphi\, d\sigma - \lambda \int_{\partial\Omega} \mathrm{Tr}\, \omega\, \varphi\, d\sigma \\ &\quad + {}_{(H^{1/2}(\partial\Omega))^*}\langle \partial_\nu u, \varphi\rangle_{H^{1/2}(\partial\Omega)} \\ &= -\lambda \int_{\partial\Omega} \mathrm{Tr}\, \omega\, \varphi\, d\sigma \end{aligned} \tag{6.40}$$

since $\partial_\nu u + \lambda \operatorname{Tr} u = 0$ on $\partial\Omega$. Hence (6.22) immediately follows. As already remarked, this finishes the proof of Step I. Here, we only wish to point out that, as a corollary,

$$G_\lambda : L^2(\Omega) \longrightarrow W^{2,2}(\Omega) \tag{6.41}$$

is well defined and bounded.

7 Proof of Step II

For this segment in our analysis we assume that $\Omega \subset \mathbb{R}^n$ is a bounded Lipschitz domain. We seek a solution for the inhomogeneous Robin boundary value problem

$$\begin{cases} \Delta u = f & \text{in } \Omega, \\ \partial_\nu u + \lambda u = 0 & \text{on } \partial\Omega, \end{cases} \tag{7.1}$$

in the form $u = \Pi f - w$, where Π is as in (2.11) and w solves

$$\begin{cases} \Delta w = 0 & \text{in } \Omega, \\ \partial_\nu w + \lambda w = \partial_\nu \Pi f + \lambda \operatorname{Tr} \Pi f & \text{on } \partial\Omega. \end{cases} \tag{7.2}$$

Formally, a solution for (7.2) is given by

$$w := \mathcal{S}(-\tfrac{1}{2}I + K^* + \lambda S)^{-1}(\partial_\nu \Pi f + \lambda \operatorname{Tr}(\Pi f)). \tag{7.3}$$

Our goal is to show that if $f \in h^p_0(\Omega) := F^{p,2}_{0,0}(\Omega)$ for $1 - \varepsilon < p < 1$ (for some small $\varepsilon = \varepsilon(\Omega) > 0$), then w above is well defined and satisfies $\nabla^2 w \in h^p(\Omega)$ plus a natural estimate. To this end, we first claim that

$$-\tfrac{1}{2}I + K^* + \lambda S \ \text{ is one-to-one on } \ L^2(\partial\Omega). \tag{7.4}$$

In order to show (7.4), let $g \in L^2(\partial\Omega)$ be such that $(-\frac{1}{2}I + K^* + \lambda S)g = 0$, and let $v := \mathcal{S}g$ in Ω. Then from the standard Calderón–Zygmund theory we have

$$\begin{cases} \Delta v = 0 \ \text{ in } \ \Omega, \\ M(\nabla v) \in L^2(\partial\Omega), \\ \partial_\nu v + \lambda v = 0 \text{ on } \partial\Omega, \end{cases} \tag{7.5}$$

where the nontangential maximal operator M was introduced in (2.2). As a consequence of this and Green's formula (which holds in this degree of generality), we find

$$0 \leqslant \int_{\Omega} |\nabla v|^2\, dx = \int_{\partial\Omega} \partial_\nu v \cdot v\, d\sigma = -\lambda \int_{\partial\Omega} |v|^2 d\sigma \leqslant 0. \tag{7.6}$$

This forces $v \equiv c$ in Ω and $v|_{\partial\Omega} \equiv 0$, which further implies that $v \equiv 0$ in Ω. Thus, $\mathcal{S}g \equiv 0$ in Ω and, consequently, $Sg = 0$ on $\partial\Omega$ in view of (2.8). With this in hand, a reference to (3.54) then yields $g = 0$, finishing the proof of (7.4).

Taking into account that $-\frac{1}{2}I + K^*$ is Fredholm with index zero on $L^2(\partial\Omega)$ and S is compact on $L^2(\partial\Omega)$, we may conclude that the operator $-\frac{1}{2}I + K^* + \lambda S$ is Fredholm with index zero on $L^2(\partial\Omega)$. In concert with (7.4), this further implies that

$$-\tfrac{1}{2}I + K^* + \lambda\, S : L^2(\partial\Omega) \longrightarrow L^2(\partial\Omega) \ \text{ is an isomorphism.} \tag{7.7}$$

Recalling (3.53) and observing that $S : B^{p,p}_{s-1}(\partial\Omega) \to B^{p,p}_{s-1}(\partial\Omega)$ is compact for any $p \in (\frac{n-1}{n}, \infty)$ and $s \in (0,1)$, we conclude that

$$\begin{aligned} &-\tfrac{1}{2}I + K^* + \lambda\, S \quad \text{is Fredholm with index zero on } B^{p,p}_{s-1}(\partial\Omega) \\ &\text{for all pairs of indices } s, p \text{ as in the statement of Theorem 3.4.} \end{aligned} \tag{7.8}$$

With (7.8) in hand, the global stability result presented in Theorem 2.10 of [11], together with (7.7), gives

$$\begin{aligned} &-\tfrac{1}{2}I + K^* + \lambda\, S \quad \text{is invertible on } B^{p,p}_{s-1}(\partial\Omega) \\ &\text{for } s, p \text{ as in the statement of Theorem 3.4.} \end{aligned} \tag{7.9}$$

In order to proceed from here, we recall from [14] that the Newtonian potential has the property that its normal derivative

$$\partial_\nu \varPi : F^{p,q}_{s+\frac{1}{p}-2,0}(\Omega) \longrightarrow B^{p,p}_{s-1}(\partial\Omega) \tag{7.10}$$

is a bounded operator whenever $\frac{n-1}{n} < p \leqslant \infty$, $(n-1)(1/p-1)_+ < s < 1$ and $0 < q \leqslant \infty$. Based on this, Theorem 3.1, and the fact that the Newtonian potential is smoothing of order two on the Triebel–Lizorkin scale, we then conclude that, whenever s and p are the same as in Theorem 3.4 and $f \in F^{p,2}_{s+1/p-2,0}(\Omega)$, the function

$$h := \partial_\nu \varPi f + \lambda\, \mathrm{Tr}\,(\varPi f) \in B^{p,p}_{s-1}(\partial\Omega) \tag{7.11}$$

is meaningfully defined and satisfies a natural estimate. Consequently, whenever s and p are the same as in Theorem 3.4, from (7.9) it follows that $g := (-\frac{1}{2}I + K^* + \lambda S)^{-1} h \in B^{p,p}_{s-1}(\partial\Omega)$ is meaningfully defined and satisfies a natural estimate. If we now recall the mapping properties of the single layer potential operator from Theorem 3.3, we may conclude that

$$w := \mathcal{S}g \in F^{p,2}_{s+\frac{1}{p}}(\Omega) \tag{7.12}$$

satisfies a natural estimate.

When $f \in h^p_0(\Omega) = F^{p,2}_{0,0}(\Omega)$, the above considerations hold if $1-\varepsilon < p < 1$ since for such p's the pair s, p with $s := 2 - \frac{1}{p}$ satisfies the conditions in Theorem 3.4. Thus, in this particular scenario,

$$w \in h^{2,p}(\Omega) = F^{p,2}_2(\Omega) \quad \text{and} \quad \|w\|_{h^{2,p}(\Omega)} \leqslant C\|f\|_{h^p_0(\Omega)}. \tag{7.13}$$

Since we also have

$$\Pi f \in h^{2,p}(\Omega) \quad \text{and} \quad \|\Pi f\|_{h^{2,p}(\Omega)} \leqslant C\|f\|_{h^p_0(\Omega)}, \tag{7.14}$$

we may finally conclude that $u = \Pi f - w \in h^{2,p}(\Omega)$ and

$$\|u\|_{h^{2,p}(\Omega)} \leqslant C\|f\|_{h^p_0(\Omega)} \tag{7.15}$$

whenever u solves (7.1) with $f \in h^p_0(\Omega)$ for some $1-\varepsilon < p < 1$ (where $\varepsilon > 0$ is as in Theorem 3.4). Hence, as a corollary, the Green operator

$$G_\lambda : F^{p,2}_0(\Omega) \longrightarrow F^{p,2}_2(\Omega) \tag{7.16}$$

is well defined and bounded whenever $1 - \varepsilon < p < 1$. This completes the proof of Step II. □

8 Proof of Step III

Granted (6.41) and (7.16), the interpolation results in the last part of Theorem 3.2 imply that

$$G_\lambda : F^{p,2}_{0,0}(\Omega) \to F^{p,2}_2(\Omega) \text{ is well defined and bounded if } 1-\varepsilon < p \leqslant 2. \tag{8.1}$$

In light of the identifications in (3.6)-(3.12), when further restricted to the range $1 < p \leqslant 2$, this shows that the Green operator G_λ in (1.12) is well defined and bounded. Granted this, a routine argument shows that the inhomogeneous Robin boundary value problem (1.10) is solvable, for a solution which satisfies (1.11).

Thus, at this point, it remains to establish the uniqueness part for the boundary value problem (1.10). To this end, we find it useful to recall the following general regularity result from [20].

Theorem 8.1. *Let Ω be a bounded Lipschitz domain in $\mathbb{R}^n$, $n \geqslant 2$. Assume that L is a homogeneous, constant (real) coefficient, symmetric, strongly elliptic differential operator of order $2m$, $m \in \mathbb{N}$. Then if $w \in F^{p,q}_{m-1+1/p}(\Omega)$*

for some $\frac{n-1}{n} < p \leqslant 2$, $0 < q < \infty$, *and* $Lw = 0$ *in* Ω, *it follows that* $M(\nabla^{m-1}w) \in L^p(\partial\Omega)$ *and a natural estimate holds.*

Based on this theorem (applied with $w = \nabla u$, $q = 2$, $1 < p \leqslant 2$, $m = 1$, and $L = \Delta$; note that $W^{1,p}(\Omega) \hookrightarrow F^{p,2}_{1/p}(\Omega)$), we may then deduce that if u is a null-solution of (1.10) for some $1 < p \leqslant 2$, then $M(\nabla u) \in L^p(\partial\Omega)$. With this in hand, the results in [5] then give that, necessarily,

$$u = \mathcal{S}g \quad \text{in} \quad \Omega \qquad \text{for some } g \in L^p(\partial\Omega). \tag{8.2}$$

Hence, on the boundary,

$$0 = \partial_\nu u + \lambda\, u = (-\tfrac{1}{2}I + K^* + \lambda S)g. \tag{8.3}$$

Thus, we may conclude that $g = 0$ and, further, $u = 0$, as soon as we show that

$$-\tfrac{1}{2}I + K^* + \lambda S : L^p(\partial\Omega) \to L^p(\partial\Omega) \text{ is an isomorphism, if } 1 < p \leqslant 2. \tag{8.4}$$

This, however, follows by observing that (8.4) holds when $\lambda = 0$ (see [5]) and then proceeding as in the case of (7.7).

Acknowledgment. The research of the authors was supported in part by the NSF.

References

1. Adolfsson, V.: L^p-integrability of the second order derivatives of Green potentials in convex domains. Pacific J. Math. **159**, no. 2, 201–225 (1993)
2. Adolfsson, V., Jerison, D.: L^p-integrability of the second order derivatives for the Neumann problem in convex domains. Indiana Univ. Math. J. **43**, no. 4, 1123–1138 (1994)
3. Brown, R., Mitrea, I.: The Mixed Problem for the Lamé System in a Class of Lipschitz Domains. Preprint (2007)
4. Dahlberg, B.E.J.: L^q-estimates for Green potentials in Lipschitz domains. Math. Scand. **44**, no. 1, 149–170 (1979)
5. Dahlberg, B., Kenig, C.: Hardy spaces and the L^p–Neumann problem for Laplace's equation in a Lipschitz domain. Ann. Math. **125**, 437–465 (1987)
6. Fromm, S.J.: Potential space estimates for Green potentials in convex domains. Proc. Am. Math. Soc. **119**, no. 1, 225–233 (1993)
7. Grisvard, P.: Elliptic Problems in Nonsmooth Domains. Pitman, Boston, MA (1985)
8. Jerison, D., Kenig, C.: The inhomogeneous Dirichlet problem in Lipschitz domains. J. Funct. Anal. **130**, no. 1, 161–219 (1995)
9. Kalton, N., Mayboroda, S., Mitrea, M.: Interpolation of Hardy–Sobolev–Besov–Triebel–Lizorkin spaces and applications to problems in partial differential equations. Contemp. Math. **445**, 121–177 (2007)

10. Kozlov, V.A., Maz'ya, V.G., Rossmann, J.: Elliptic Boundary Value Problems in Domains with Point Singularities. Am. Math. Soc., Providence, RI (1997)
11. Kalton, N.J., Mitrea, M.: Stability results on interpolation scales of quasi-Banach spaces and applications. Trans. Am. Math. Soc. **350**, no. 10, 3903–3922 (1998)
12. Mayboroda, S., Mitrea, M.: Layer potentials and boundary value problems for Laplacian in Lipschitz domains with data in quasi-Banach Besov spaces. Ann. Mat. Pura Appl. (4) **185**, no. 2, 155–187 (2006)
13. Mayboroda, S., Mitrea, M.: Sharp estimates for Green potentials on non-smooth domains. Math. Res. Lett. **11**, 481–492 (2004)
14. Mayboroda, S., Mitrea, M.: The solution of the Chang–Krein–Stein conjecture. In: Proc. Conf. Harmonic Analysis and its Applications (March 24–26, 2007), pp. 61–154. Tokyo Woman's Cristian University, Tokyo (2007)
15. Maz'ya, V.G.: Solvability in $\dot{W}_2^2$ of the Dirichlet problem in a region with a smooth irregular boundary (Russian). Vestn. Leningr. Univ. **22**, no. 7, 87–95 (1967)
16. Maz'ya, V.G.: The coercivity of the Dirichlet problem in a domain with irregular boundary (Russian). Izv. VUZ, Ser. Mat. no. 4, 64–76 (1973)
17. Maz'ya, V.G., Shaposhnikova, T.O.: Theory of Multipliers in Spaces of Differentiable Functions. Pitman, Boston, MA (1985)
18. Mitrea, M.: Dirichlet integrals and Gaffney–Friedrichs inequalities in convex domains. Forum Math. **13**, no. 4, 531–567 (2001)
19. Mitrea, M., Taylor, M., Vasy, A.: Lipschitz domains, domains with corners, and the Hodge Laplacian. Commun. Partial Differ. Equ. **30**, no. 10–12, 1445–1462 (2005)
20. Mitrea, M., Wright, M.: Layer Potentials and Boundary Value Problems for the Stokes system in Arbitrary Lipschitz Domains. Preprint (2008)
21. Runst, T., Sickel, W.: Sobolev Spaces of Fractional Order, Nemytskij Operators, and Nonlinear Partial Differential Operators. de Gruyter, Berlin–New York (1996)
22. Rychkov, V.: On restrictions and extensions of the Besov and Triebel–Lizorkin spaces with respect to Lipschitz domains. J. London Math. Soc. (2) **60**, no. 1, 237–257 (1999)
23. Triebel, H.: The Structure of Functions. Basel, Birkhäuser (2001)
24. Verchota, G.: Layer potentials and boundary value problems for Laplace's equation in Lipschitz domains. J. Funct. Anal. **59**, 572–611 (1984)

Properties of Spectra of Boundary Value Problems in Cylindrical and Quasicylindrical Domains

Sergey Nazarov

To the memory of S.L. Sobolev

Abstract General formally self-adjoint boundary value problems with spectral parameter are investigated in domains with cylindrical and quasicylindrical (periodic) outlets to infinity. The structure of the spectra is studied for operators generated by the corresponding sesquilinear forms. In addition to general results, approaches and methods are discussed to get a piece of information on continuous, point, and discrete spectra, in particular, for specific problems in the mathematical physics.

1 Statement of Problems and Preliminary Description of Results

1.1 Introduction

We consider one particular question in the theory of waves; namely, cylindrical and periodic (quasicylindrical) waveguides, acoustic, elastic, and piezoelectric. The formulation of the corresponding problems in the mathematical physics usually, but not always (see Subsect. 3.8 below) leads to a formally self-adjoint boundary value problem for a system of partial differential equations with spectral parameter λ proportional to the squared oscillation frequency. The subject of the investigation becomes a structure of the spectrum of the operator of the boundary value problem since a wave can propagate

Sergey Nazarov
Institute of Problems in Mechanical Engineering, Russian Acad. Sci., V.O., Bolshoi pr. 61, St. Petersburg 199178, Russia, e-mail: serna@snark.ipme.ru

V. Maz'ya (ed.), *Sobolev Spaces in Mathematics II*, *International Mathematical Series*.
doi: 10.1007/978-0-387-85650-6, © Springer Science + Business Media, LLC 2009

only in the case where the point λ belongs to the continuous spectrum. We also pay attention to particular features of the spectra such as the point spectrum and opening gaps in the continuous spectrum which have clear physical interpretation.

There are two methods for investigating boundary value problems in cylindrical and quasicylindrical domains. The first one was formed within the framework of the theory of elliptic boundary value problems in domains with piecewise smooth boundaries (see the key works [25, 36, 37, 44]), and the second one (see reviews [29, 31]) is related with the operator theory in Hilbert spaces. These methods usually provide results of equal strength. However, certain specific questions may become prerogative for one in the couple. For example, the lost of the Fredholm property by the problem operator in a proper Sobolev space is equivalent to the fact that the point λ belongs to the essential spectrum of a self-adjoint unbounded operator in the Lebesgue space, but asymptotic expansions of solutions at infinity ought to be concluded by means of the first method and the segmental structure of the spectrum by means of the second one.

In the paper, we employ both methods and also many of both, widely used and little known approaches providing an additional information on spectra.

1.2 Spectral boundary value problem

Let $n \geqslant 2$, and let Σ be a domain in the Euclidian space $\mathbb{R}^n$ included into the layer $\{x = (x', x_n) \in \mathbb{R}^n : x_n \in (0,1)\}$. We introduce the periodic set Π (the quasicylinder, see Fig. 1) as the interior of the union of the closures of periodicity cells

$$\bigcup_{j \in \mathbb{Z}} \overline{\Sigma_j}.$$

Here, $\mathbb{Z} = \{0, \pm 1, \dots\}$ and $\Sigma_j = \{x : (x', x_n - j) \in \Sigma\}$. We assume that Π is a domain with Lipschitz boundary $\partial\Pi$, in particular, Π is a connected set. Let also Ω be a subdomain of $\mathbb{R}^n$ with Lipschitz boundary $\partial\Omega$ which, outside the ball $\mathbb{B}^n_{R^0} = \{x : |x| < R^0\}$ of a large radius $R^0 > 1$, coincides with the half-cylinder $\Pi_+ = \{x \in \Pi : x_n > 0\}$ (see Fig. 2).

We introduce an $(N \times k)$-matrix $\mathcal{D}(\nabla_x)$ of homogeneous differential first order operators with constant, in general complex, coefficients where $N, k \in \mathbb{N} := \{1, 2, \dots\}$ and $N \geqslant k$. We assume that this matrix is algebraically complete [54], i.e., there exists a number $\tau_{\mathcal{D}} \in \mathbb{N}$ such that, for any $\tau \geqslant \tau_{\mathcal{D}}$ and any row $P(\xi) = (P_1(\xi), \dots, P_k(\xi))$ of homogeneous polynomials of degree τ, one may find a polynomial string $Q(\xi) = (Q_1(\xi), \dots, Q_N(\xi))$ satisfying the relation

$$P(\xi) = Q(\xi)\mathcal{D}(\xi), \quad \xi = (\xi_1, \dots, \xi_n) \in \mathbb{R}^n. \tag{1.1}$$

Let $\mathcal{A}$ and $\mathcal{B}$ be Hermitian matrix-valued functions of size $N \times N$ and $k \times k$ respectively. They have measurable entries and satisfy the following positivity and "stabilization" conditions:

$$c_{\mathcal{A}}|a|^2 \leqslant \overline{a}^\top \mathcal{A}(x)a \leqslant C_{\mathcal{A}}|a|^2, \quad c_{\mathcal{B}}|b|^2 \leqslant \overline{b}^\top \mathcal{B}(x)b \leqslant C_{\mathcal{B}}|b|^2 \\ \text{for } a \in \mathbb{C}^N, b \in \mathbb{C}^k \text{ and a.a. } x \in \Omega, \tag{1.2}$$

$$|\mathcal{A}_{jl}(x) - \mathcal{A}^0_{jl}(x)| + |\mathcal{B}_{pq}(x) - \mathcal{B}^0_{pq}(x)| \leqslant c_0 \exp(-\delta_0 x_n) \\ \text{for a.a. } x \in \Omega \setminus \mathbb{B}^n_R. \tag{1.3}$$

Here, $\top$ stands for transposition and the bar for complex conjugation; furthermore, $c_{\mathcal{A}}$, $C_{\mathcal{A}}$, $c_{\mathcal{B}}$, $C_{\mathcal{B}}$ and c_0, δ_0 are positive constants. The matrix-valued functions $\mathcal{A}^0 = (\mathcal{A}^0_{jl})$ and $\mathcal{B}^0 = (\mathcal{B}^0_{pq})$ are Hermitian, measurable, bounded, and positive definite as well. Moreover, they are periodic[1] in x_n and the inequalities (1.2) are valid with positive constants $c_{\mathcal{A}^0}$, $C_{\mathcal{A}^0}$ and $c_{\mathcal{B}^0}$, $C_{\mathcal{B}^0}$. We determine the following $(k \times k)$-matrix $\mathcal{L}$ of second order differential operators

$$\mathcal{L}(x, \nabla_x) = \mathcal{D}(\nabla_x)^* \mathcal{A}(x) \mathcal{D}(\nabla_x), \tag{1.4}$$

where $\mathcal{D}(\nabla_x)^* = \overline{\mathcal{D}(-\nabla_x)}^\top$ is the formal adjoint differential matrix operator for $\mathcal{D}(\nabla_x)$. By the requirements made on the matrices $\mathcal{D}$ and $\mathcal{A}$, the operator (1.4) is formally positive [54] and possesses the polynomial property [41, 44], hence it is elliptic and formally self-adjoint. We emphasize that the radius vector x and the gradient operator ∇_x are interpreted as the columns $(x_1, \ldots, x_n)^\top$ and $(\partial/\partial x_1, \ldots, \partial/\partial x_n)^\top$ respectively.

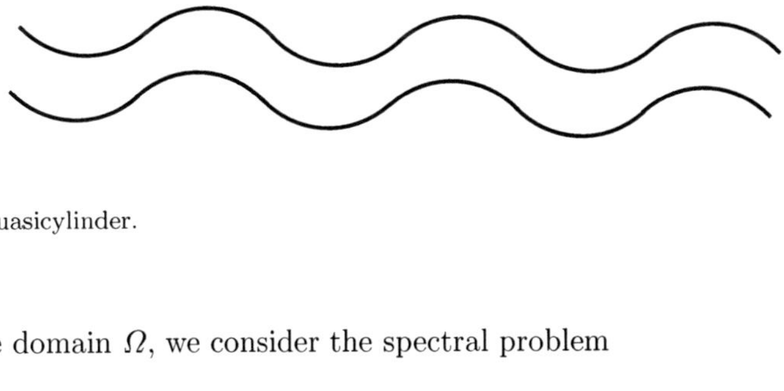

Fig. 1 Quasicylinder.

In the domain Ω, we consider the spectral problem

$$\mathcal{L}(x, \nabla_x)u(x) = \lambda \mathcal{B}(x)u(x), \quad x \in \Omega, \tag{1.5}$$

$$\mathcal{T}(x, \nabla_x)u(x) = 0, \quad x \in \partial\Omega. \tag{1.6}$$

Here, $\lambda \in \mathbb{C}$ is the spectral parameter and $\mathcal{T}$ is a matrix operator of the boundary conditions. The classical formulation of the problem expects smooth data and in the sequel we give its variational formulation. We now

[1] The period is equal to one if another period is not announced in advance.

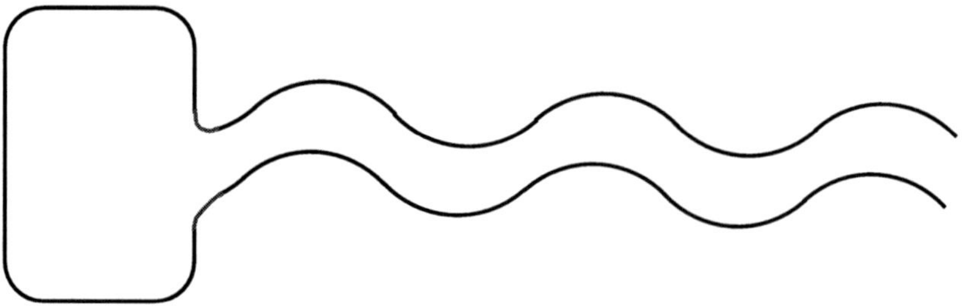

Fig. 2 Periodic waveguide.

describe a procedure [35] to construct the matrix operator $\mathcal{T}$, which provides a symmetric operator of the boundary value problem in the Sobolev space $H^1(\Omega)^k$. Let $\mathcal{S}$ be a unitary matrix-valued function with measurable entries on the boundary $\partial\Omega$ while

$$\mathcal{S}(x) = \mathcal{S}^0(x) \text{ for a.a. } x \in \partial\Omega \setminus \mathbb{B}_R, \tag{1.7}$$

and $\mathcal{S}^0$ depends periodically on x_n. We determine the rows $\mathcal{T}_{(q)}$ and $\mathcal{Q}_{(q)}$ of the matrices $\mathcal{T}$ and $\mathcal{Q}$ by one of the following couples of formulas:

$$\begin{aligned} &\mathcal{T}_{(q)}(x, \nabla_x) = \mathcal{S}_{(q)}(x), \ \ \mathcal{Q}_{(q)}(x, \nabla_x) = -\mathcal{S}_{(q)}(x)\overline{\mathcal{D}(\nu(x))}^\top \mathcal{A}(x)\mathcal{D}(\nabla_x) \\ &\text{or} \\ &\mathcal{T}_{(q)}(x, \nabla_x) = \mathcal{S}_{(q)}(x)\overline{\mathcal{D}(\nu(x))}^\top \mathcal{A}(x)\mathcal{D}(\nabla_x), \quad \mathcal{Q}_{(q)}(x, \nabla_x) = \mathcal{S}_{(q)}(x). \end{aligned} \tag{1.8}$$

Here, ν stands for the unit vector (column) of the outer normal on the boundary of the domain Ω which is properly defined for almost all $x \in \partial\Omega$. Again, under additional smoothness of the data, the symmetric Green formula

$$(\mathcal{L}u, v)_\Omega + (\mathcal{T}u, \mathcal{Q}v)_{\partial\Omega}(u, \mathcal{L}v)_\Omega + (\mathcal{Q}u, \mathcal{T}v)_{\partial\Omega} \tag{1.9}$$

is valid, where $(\, , \,)_\Xi$ is the natural inner product in the Lebesgue space $L_2(\Xi)$, either scalar or vector, and the test functions u and v belong to the linear space $C_c^\infty(\overline{\Omega})^k$, i.e., they are smooth and have compact support. Furthermore, the problem (1.5), (1.6) can be reformulated as the integral identity [33]

$$a(u, v; \Omega) := (\mathcal{A}\mathcal{D}(\nabla_x)u, \mathcal{D}(\nabla_x)v)_\Omega = \lambda(\mathcal{B}u, v)_\Omega, \quad v \in \mathcal{H}(\Omega), \tag{1.10}$$

where $\mathcal{H}(\Omega)$ is the subspace of vector-valued functions $v = (v_1, \dots, v_k)^\top \in H^1(\Omega)^k$ satisfying the stable boundary conditions (see [35] for terminology), i.e., $\mathcal{T}_{(q)}(x)v(x) = \mathcal{S}_{(q)}(x)v(x) = 0$, $x \in \partial\Omega$, for $q = 1, \dots, k$ such that the first couple in the definition (1.8) is chosen. We emphasize that vector functions $u \in H^2(\Omega)^k$ and $v \in H^1(\Omega)^k$ satisfy the following "halved" variant of the

Green formula:

$$(\mathcal{L}u, v)_{\Omega} + (\mathcal{N}u, v)_{\partial\Omega} a(u, v; \Omega). \tag{1.11}$$

By $\mathcal{N}(x, \nabla_x)$ we understand the operator $\overline{\mathcal{D}(\nu(x))}^{\top} \mathcal{A}(x)\mathcal{D}(\nabla_x)$ of the Neumann boundary conditions. If the inclusions $u, v \in \mathcal{H}(\Omega)$ are valid in addition, then

$$(\mathcal{N}u, v)_{\partial\Omega} = \sum_{q=1}^{k} (\mathcal{S}_{(q)}\mathcal{N}_{(q)}u, \mathcal{S}_{(q)}v)_{\partial\Omega} = 0 \tag{1.12}$$

since for all $q = 1, \ldots, k$ and any choice of operators in (1.8) one of the multipliers in each inner product vanishes. Formulas (1.11) and (1.12) provide the variational formulation (1.10) of the problem.

Apart from the spectral problem (1.10), the corresponding inhomogeneous problem with the fixed parameter λ

$$a(u, v; \Omega) - \lambda(\mathcal{B}u, v)_{\Omega} = f(v), \quad v \in \mathcal{H}(\Omega) \tag{1.13}$$

is of further use as well. Here, $f \in \mathcal{H}(\Omega)^*$ is a linear functional in the space $\mathcal{H}(\Omega)$.

Remark 1.1. It is allowed to set different couples of boundary conditions in (1.8) on disjoint parts of the boundary $\partial\Omega$. However, outside the ball with a certain large radius, the collision surfaces for different boundary conditions must follow the periodic structure of the quasicylinder $\partial\Pi$. To simplify the notation, the possible simple generalization is ignored in Sects. 1 and 2; however, in Sect. 3, we employ it in the simplest situation; namely, the Dirichlet or Neumann conditions at the end of a straight cylinder are accompanied with boundary conditions of other type on the lateral side. Moreover, in Subsects. 3.5 and 3.6, we permit a cylindrical collision submanifold of dimension $n - 2$. □

Remark 1.2. Further results keep the validity in the case where the set $\Omega \backslash \mathbb{B}_{R^0}$ is an exponentially decaying perturbation of the set $\Pi_+ \setminus \mathbb{B}_{R^0}$. The same is true for the matrix $\mathcal{S}$ (the condition (1.7) can be weakened) and the collision surfaces mentioned in Remark 1.1. To simplify the presentation, we do not pay attention to this possible generalization. □

1.3 Polynomial property and the Korn inequality

We formulate the above-mentioned polynomial property [41, 44]. It means that for any domain $\Xi \subset \mathbb{R}^n$ the assertion

$$a(u, u; \Xi) = 0, u \in C^{\infty}(\overline{\Xi})^k \quad \Leftrightarrow \quad u \in \mathcal{P} \tag{1.14}$$

is valid, where $\mathcal{P}$ is a finite dimensional subspace of vector polynomials $p(x) = (p_1(x), \ldots, p_k(x))^\top$. It is clear that

$$\mathcal{P} = \{p : \mathcal{D}(\nabla_x)p(x) = 0\} \tag{1.15}$$

due to the algebraic completeness of the matrix $\mathcal{D}$ and the positive definiteness of the matrix $\mathcal{A}$. Furthermore, according to (1.1), the degree of a scalar polynomial p_j in the column $p \in \mathcal{P}$ cannot exceed $\tau_{\mathcal{D}}$ (see [41] and [44] for details). The subspace $\mathcal{P}$ figures in the description of attributes and main properties of an operator of an elliptic problem $\{\mathcal{L}, \mathcal{T}\}$ in either bounded domains with smooth or piecewise smooth boundaries or unbounded domains with conical, cylindrical or periodic outlets to infinity (see [44, 53]).

The following proposition based on the condition (1.1) is proved in [54].

Proposition 1.1. *For any domain $\Xi \subset \mathbb{R}^n$ with a Lipschitz boundary and a compact closure the Korn inequality*

$$\|\nabla_x u; L_2(\Xi)\| \leqslant C(\Xi)\left(\|\mathcal{D}(\nabla_x)u; L_2(\Xi)\| + \|u; L_2(\Xi)\|\right), \tag{1.16}$$

is valid while the factor $C(\Xi)$ does not depend on the vector-valued function $u \in H^1(\Xi)^k$.

The review [44] contains plenty of examples of specific problems in the mathematical physics which fulfills the above requirements, but in the introduction we mention only two simplest ones. In Subsect. 3.8, we also discuss the piezoelectricity model in mechanics of continuous media.

Example 1.1. Let $k = 1$ and $N = n$. Then $\mathcal{D}(\nabla_x) = \nabla_x$ and $\mathcal{L}(x, \nabla_x) = -\nabla_x^\top \mathcal{A}(x)\nabla_x$ is a scalar second order differential operator in the divergence form. The linear space (1.15) consists of constants. If $\mathcal{A}$ is the unit matrix and $\mathcal{B} = \text{const}$, then the system (1.5) turns into the Helmgoltz equation. □

Example 1.2. The operator $\mathcal{L}$ of the three-dimensional elasticity system, describing deformation of an anisotropic and inhomogeneous solid with the stiffness matrix $\mathcal{A}(x)$ gets the numbers $k = 3$, $N = 6$ and the matrix

$$\mathcal{D}(\xi)^\top = \begin{pmatrix} \xi_1 & 0 & 0 & 0 & 2^{-1/2}\xi_3 & 2^{1/2}\xi_2 \\ 0 & \xi_2 & 0 & 2^{-1/2}\xi_3 & 0 & 2^{-1/2}\xi_1 \\ 0 & 0 & \xi_3 & 2^{-1/2}\xi_2 & 2^{-1/2}\xi_1 & 0 \end{pmatrix}. \tag{1.17}$$

It is straightforward to verify the algebraic completeness (1.1) with $\tau_{\mathcal{D}} = 2$ of the (6×3)-matrix $\mathcal{D}$. By the factors $2^{-1/2}$, the strain column

$$\mathcal{D}(\nabla_x)u = \left(\varepsilon_{11}(u), \varepsilon_{22}(u), \varepsilon_{33}(u), 2^{1/2}\varepsilon_{23}(u), 2^{1/2}\varepsilon_{31}(u), 2^{1/2}\varepsilon_{12}(u)\right)^\top \tag{1.18}$$

(see monographs [34, 45]) has the same natural norm as the strain tensor $(\varepsilon_{jk}(u))_{j,k=1}^3$ with the Cartesian components

$$\varepsilon_{jk}(u) = \frac{1}{2}\left(\frac{\partial u_j}{\partial x_k} + \frac{\partial u_k}{\partial x_j}\right)$$

and, therefore, any orthogonal transform of the Cartesian coordinates leads to an orthogonal transform of column (1.18) (see [45]).

Similarly to (1.18), the column

$$\mathcal{A}\mathcal{D}(\nabla_x)u = \left(\sigma_{11}(u), \sigma_{22}(u), \sigma_{33}(u), 2^{1/2}\sigma_{23}(u), 2^{1/2}\sigma_{31}(u), 2^{1/2}\sigma_{12}(u)\right)^\top \tag{1.19}$$

contains components of the strain tensor so that the rule (1.8) provides all physically meaningful boundary conditions. The inequality (1.16) implies the Korn inequality [27] which plays an important role in elasticity problems (see, for example, [54, 26, 7, 45]). Finally, the linear space $\mathcal{P}$ in formula (1.15) consists but of rigid motions $a + x \times b$ and gets dimension six; here, a and b are columns in $\mathbb{R}^3$ and the cross stands for the vector product in $\mathbb{R}^3$. The term a gives rise to a translation while $x \times b$ is a rotation. □

1.4 Formulation of the problem in the operator form

Applying the inequality (1.16) on the cells Σ_j, $j = j_0, j_0 + 1, \ldots$, and the set $\Omega \cap \mathbb{B}_{R^0}$, which all together cover the domain Ω, yields the following assertion.

Lemma 1.1. *For any vector-valued function $u \in H^1(\Omega)^k$ the Korn inequality (1.16) is valid in the infinite domain $\Xi = \Omega$ whilst $C(\Omega) = \max\{C(\Omega \cap \mathbb{B}_{R^0}), C(\Sigma)\}$.*

Taking Lemma 1.1 and the inequality (1.2) into account, we endow the Hilbert space $\mathcal{H}(\Omega)$ with the specific inner product

$$\langle u, v\rangle = a(u, v; \Omega) + b(u, v; \Omega). \tag{1.20}$$

We also introduce a positive, continuous and symmetric, therefore self-adjoint operator $\mathcal{K}$ by the formula

$$\langle \mathcal{K}u, v\rangle = b(u, v; \Omega) := (\mathcal{B}u, v)_\Omega, \quad u, v \in \mathcal{H}(\Omega). \tag{1.21}$$

The operator $\mathcal{K}$ cannot be compact and its spectrum cannot consist of the only point $\mu = 0$ because the set Ω is not bounded.

The variational formulation (1.10) is equivalent to the abstract spectral equation

$$\mathcal{K}u = \mu u \tag{1.22}$$

with the new spectral parameter

$$\mu = (1 + \lambda)^{-1}. \tag{1.23}$$

By the definitions (1.20) and (1.21), the norm of the operator $\mathcal{K}$ does not exceed one, i.e., $\mathbb{C} \setminus \{\mu : \operatorname{Im} \mu = 0, \operatorname{Re} \mu \in [0,1]\}$ is certainly belongs to the resolvent field of the operator. Owing to the relation (1.23) the λ-spectrum of the problem (1.10) (or the problem (1.5), (1.6) if the data are smooth) inherits all the properties of the μ-spectrum of the operator $\mathcal{K}$, except for those attributed to the point $\mu = 0$. Thus, in the sequel, we mainly study the spectrum of the operator $\mathcal{K}$ on the segment $(0,1]$ of the real axis and only reformulate the obtained results for the problem (1.10) itself.

1.5 Contents of the paper

In many situations, the process of wave propagation in cylindrical and quasicylindrical (periodic) waveguides admits a stationary formulation which requires to investigate the spectrum of a formally self-adjoint boundary value problem for an elliptic system of partial differential equations. Since the geometrical domain is unbounded, the boundary value problem gets a continuous spectrum whose points give rise to *propagating* waves which drive energy to/from infinity. The discrete and point spectra bring about exponentially decaying waves, i.e., the so-called *trapped modes*. At bottom of fact, the identification of the above-mentioned spectra becomes the subject of investigation in many applicable disciplines.

One may allot the following two approaches to study solutions of elliptic boundary value problems in domains with cylindrical and quasicylindrical outlets to infinity. Within the framework of elliptic boundary value problems, in domains with piecewise smooth boundaries (see the key works [25, 36, 37, 44] and, for example, monographs [53, 28]), the first approach delivers results on the Fredholm property of the problem operator in standard and weighted Sobolev spaces, asymptotic expansions for solutions at infinity, and a correct formulation of problems with asymptotic conditions at infinity in function spaces with detached asymptotics. Moreover, the model problem (2.8) in the periodicity cell Σ is interpreted as the polynomial pencil $\eta \mapsto \mathfrak{A}(\eta; \lambda)$ in the dual variable arising from either the Gel'fand transform (2.1) or the Fourier transform (2.37) while the parameter λ has to be fixed. Usually, the theory under discussion deals with boundary value problems in the classical formulation (see the differential expressions (1.5), (1.6)), i.e., their data, the boundary and coefficients of differential operators, are supposed smooth. However, to employ the variational formulation of the problems (see the integral identities (1.10) and (1.13)) does not causes a serious impediment due to the Parseval theorem for both transforms (see formula (2.3) for the Gel'fand transform). At the same time, the author does not know a publication where variational problems are consistently studied within this framework, and in Sect. 2 we shortly present necessary calculations and arguments following the standard scheme (see again [25, 36, 37, 44] and [53, 28]).

The second approach appeals to the theory of operators in Hilbert space. This approach is expounded, for example, in [29, 59, 30] and requires to fix the dual variable η and to take λ as a spectral parameter. The model problem (2.8) in the periodicity cell Σ gives rise to an unbounded self-adjoint operator in the space $L_2(\Sigma)^k$ generated by the Hermitian form (2.6) and the discrete spectrum of this operator provides an inference on the structure of the spectrum in problem Ω with an outlet to infinity.

In the present paper, it is convenient to employ both approaches. Since the right-hand side of the system (1.5) includes the variable matrix $\mathcal{B}(x)$, we are forced to deal with the specific inner product (1.20) in the Hilbert space $\mathcal{H}(\Omega)$ and the self-adjoint, positive and continuous operator $\mathcal{K} : \mathcal{H}(\Omega) \to \mathcal{H}(\Omega)$. The continuity of the operator does not bring any profit, but encumbers the application of the maxmin principle (see, for example, [1, Theorem 10.2.2]) which becomes one of the main tools in Sect. 3. We stress that the above-mentioned innovation is accepted with the only purpose; namely, to consider bodies and media with inhomogeneous and anisotropic densities $\mathcal{B}(x)$ which are quite often in applications, for example, in mechanics of composite materials.

In addition to the main theorem (Theorem 2.1) on the Fredholm property of the operator of the problem (1.13) and the essential spectrum of the operator $\mathcal{K}$, in Sect. 2 we establish the exponential decay for vector eigenfunctions (Corollary 2.1) and the finite dimension of the kernel in the problem (1.13) which occurs independently of the spectrum of the problem in the periodicity cell (Proposition 2.3). The latter fact, in particular, shows that all points $\mu \in (0,1]$ in the essential spectrum of the operator $\mathcal{K}$ actually lie in the continuous spectrum. Finally, in Proposition 2.4 (see also Remark 2.2), we discuss the particular case of the straight cylinder

$$\Pi = \omega \times \mathbb{R}, \tag{1.24}$$

the cross-section ω of which is a domain in the space $\mathbb{R}^{n-1}$ with Lipschitz boundary $\partial\omega$ and compact closure $\overline{\omega} = \omega \cup \partial\omega$. We prove that the essential spectrum of the operator $\mathcal{K}$ coincides with the segment $[0, \mu^\dagger]$ and, thus, cannot have a gap. The cut-off $\mu^\dagger$ belongs to the segment $(0, 1]$.

In Sect. 3, we describe certain approaches and tricks which give a piece of information about the spectrum structure of the problem (1.5), (1.6) in sufficiently general, however still particular situations. Let us list these approaches and tricks while indicating the corresponding primary source and the section of the paper where it is described.

i. The absence of the point spectrum ([56] and Subsect. 3.3).

ii. The comparison principles ([18] and Subsect. 3.3).

iii. Artificial boundary conditions and the point spectrum in the continuous spectrum ([8] and Subsect. 3.4).

iv. Concentration of the discrete and point spectra ([48] and Subsects. 3.2 and 3.4).

v. The variational approach for searching eigenvalues below the cut-off ([20, 21] and Subsect. 3.5).

vi. Opening gaps in the continuous spectrum ([50] and Subsect. 3.6).

If an approach or a trick works in the general situation, we treat the boundary value problem (1.5), (1.6). However, in the impediment case, we turn to particular problems in the mathematical physics in Examples 1.1 and 1.2. In the last two subsections, we consider two issues in mechanics of continuous media; namely, cracks and piezoelectricity.

2 The Model Problem and the Operator Pencil

2.1 Model problem in the quasicylinder

The Gel'fand transform (a discrete Fourier transform; see [12] and, for example, [29, 53, 30]) is defined by the formula

$$v(x) \mapsto \widehat{v}(x', z, \eta) = (2\pi)^{-1/2} \sum_{j \in \mathbb{Z}} \exp(-i\eta(j+z)) v(x', j+z), \tag{2.1}$$

where $x \in \Pi$ and $(x', z) \in \Sigma$, $\eta \in [0, 2\pi)$ and i is the imaginary unit. This transform establishes the isometric isomorphism

$$L_2(\Pi) \cong L_2(0, 2\pi; L_2(\Sigma)).$$

Here, $L_2(0, 2\pi; \mathfrak{B})$ is a space of abstract functions with the norm

$$\|v; L_2(0, 2\pi; \mathfrak{B})\| = \Big(\int_0^{2\pi} \|v(\eta); \mathfrak{B}\|^2 \, d\eta \Big)^{1/2}.$$

and $\mathfrak{B}$ is a Banach space. The inverse transform is given by the formula

$$\widehat{v}(x', z, \eta) \mapsto v(x', z) = (2\pi)^{-1/2} \int_0^{2\pi} \exp(i\eta z) \widehat{v}(x', z - [z], \eta) \, d\eta, \tag{2.2}$$

where $[z] = \max\{q \in \mathbb{Z} : q \leqslant z\}$ is the integral part of a number $z \in \mathbb{R}$. Since $\widehat{v}(x', 0, \eta) = \widehat{v}(x', 1, \eta)$ for $v \in C_c^\infty(\overline{\Pi})$ and

$$\widehat{P(\nabla_x) v}(x', z, \eta) = P(\nabla_{x'}, \partial_z + i\eta) \widehat{v}(x', z, \eta), \quad \partial_z = \partial/\partial z,$$

the transform (2.1) gives rise to the isomorphism

$$H^1(\Pi) \approx L_2(0, 2\pi; H^1_{per}(\Sigma)),$$

where $H^1_{per}(\Sigma)$ is the subspace of functions in the Sobolev space which are periodic in the variable z on the cell Σ.

The Parseval formula

$$(u,v)_\Pi = \int_0^{2\pi} (\widehat{u}(\cdot,\eta), \widehat{v}(\cdot,\eta))_\Sigma \, d\eta \tag{2.3}$$

is valid and, thus, the Gel'fand transform establishes the isomorphism for spaces of functionals

$$H^1(\Pi)^* \approx L_2(0,2\pi; H^1_{per}(\Sigma)^*).$$

We "freeze" coefficients of the differential operators $\mathcal{L}$ and $\mathcal{T}$ at infinity, i.e., perform the changes $\mathcal{A} \mapsto \mathcal{A}^0$, $\mathcal{B} \mapsto \mathcal{B}^0$ and $\mathcal{S} \mapsto \mathcal{S}^0$. According to the stabilization conditions (1.3) and (1.7), from (1.13) we derive the model problem in the quasicylinder

$$a^0(0,u,v;\Pi) - \lambda b^0(u,v;\Pi) = f(v), \quad v \in \mathcal{H}(\Pi). \tag{2.4}$$

Applying the transform (2.1) and formula (2.3), we obtain the family of model problems in the periodicity cell

$$a^0(\eta,U,V;\Sigma) - \lambda b^0(U,V;\Sigma) = F(\eta;V), \quad V \in \mathcal{H}_{per}(\Sigma), \quad \text{for a.a. } \eta \in (0,2\pi). \tag{2.5}$$

Here, $\mathcal{H}(\Pi)$ is the subspace of vector-valued functions in $H^1(\Pi)^k$ which are subject to the condition $\mathcal{S}^0_{(q)}v = 0$ on the surface $\partial\Pi$ for $q = 1,\dots,k$ such that the first couple is chosen in (1.8). In addition to the condition $\mathcal{S}^0_{(q)}V = 0$ on $\partial\Sigma \cap \partial\Pi$, elements in the subspace $\mathcal{H}_{per}(\Sigma) \subset H^1(\Sigma)^k$ satisfy the periodicity conditions on the bases $\partial\Sigma \cap \Pi$ of the cell. Finally, f and $F = \widehat{f}$ are linear functionals on the above-mentioned subspaces and the forms on the left-hand side of the integral identities are defined as follows:

$$a^0(\eta,U,V;\Sigma) = (\mathcal{A}^0\mathcal{D}(\nabla_{x'},\partial_z + i\eta)U, \mathcal{D}(\nabla_{x'},\partial_z + i\overline{\eta})V)_\Sigma, \tag{2.6}$$

$$b^0(u,v;\Pi) = (\mathcal{B}^0 u, v)_\Pi. \tag{2.7}$$

The parameter λ is fixed, and the complex conjugate $\overline{\eta}$ will be used below.

In the case $F = 0$, from (2.5) we obtain the spectral problem

$$a^0(\eta,U,V;\Sigma) - \lambda b^0(U,V;\Sigma) = 0, \quad V \in \mathcal{H}_{per}(\Sigma), \quad \text{for a.a. } \eta \in (0,2\pi), \tag{2.8}$$

which involves the squared spectral parameter η. According to the Riesz representation theorem, the problem (2.8) generates the quadratic pencil

$$\mathbb{C} \ni \eta \mapsto \mathfrak{A}(\eta;\lambda) : \mathcal{H}_{per}(\Sigma) \to \mathcal{H}_{per}(\Sigma). \tag{2.9}$$

The operator $\mathfrak{A}(\eta_1;\lambda) - \mathfrak{A}(\eta_2;\lambda)$ is compact in the space $\mathcal{H}_{per}(\Sigma)$ for any $\eta_1, \eta_2 \in \mathbb{C}$ because the difference $a^0(\eta_1, U, V; \Sigma) - a^0(\eta_2, U, V; \Sigma)$ gets rid of first order derivatives of entries in the vector functions U and V while the embedding $H^1(\Sigma) \subset L_2(\Sigma)$ is compact. Hence, in view of the general results in [13, Theorem 1.5.1], the eigenvalues of the pencil (2.9) are normal and have the only accumulation point at infinity. If η is an eigenvalue and $U \in \mathcal{H}_{per}(\Sigma)$ is the corresponding eigenvector, then the number $\eta \pm 2\pi$ and the vector-valued function $(x', z) \mapsto \exp(\pm 2\pi i z) U(x', z)$ compose a spectral pair as well. The same is true for associated vectors (see [38] and [53, Subsect. 3.4] for details). Thus, the pencil spectrum is invariant with respect to shifts for $\pm 2\pi$ along the real axis.

2.2 The Fredholm property of the problem operator

If the segment $[0, 2\pi) \subset \mathbb{R} \subset \mathbb{C}$ is free of the spectrum of the pencil (2.9), then for almost all $\eta \in (0, 2\pi)$ problem (2.5) with the right-hand side $F \in L_2(0, 2\pi; \mathcal{H}_{per}(\Sigma)^*)$ admits a solution $U \in \mathcal{H}_{per}(\Sigma)$ and, for the solution family, the estimate

$$\|U; L_2(0, 2\pi; \mathcal{H}_{per}(\Sigma))\| \leqslant c \|F \in L_2(0, 2\pi; \mathcal{H}_{per}(\Sigma)^*)\|$$

is valid. Properties of the Gel'fand transform ensure that the inverse transform $U = \widehat{u} \mapsto u$ (see formula (2.2)) gives a solution $u \in \mathcal{H}(\Pi)$ of the problem (2.4) together with the inequality

$$\|u; \mathcal{H}(\Pi)\| \leqslant c \|f; \mathcal{H}(\Pi)^*\|. \tag{2.10}$$

Theorem 2.1. *A point $\mu \in \mathbb{C}$ belongs to either the resolvent field or the discrete spectrum of the operator $\mathcal{K}$ if and only if for $\lambda = \mu^{-1} - 1$ the segment $[0, 2\pi)$ is free of the spectrum of the pencil $\eta \mapsto \mathfrak{A}(\eta; \lambda)$.*

Proof. 1. *Necessity.* Assume that for a certain λ the segment $[0, 2\pi)$ does not contain eigenvalues of the pencil. By the Korn inequality (1.16), the auxiliary problem

$$a(u^R, v; \Omega_R) - (\lambda - \lambda_R) b(u^R, v; \Omega_R) = f^R(v), \quad v \in \mathcal{H}(\Omega_R), \tag{2.11}$$

has a unique solution $u^R \in \mathcal{H}^0(\Omega_R)$ for any functional $f^R \in \mathcal{H}^0(\Omega_R)^*$ in the case $\lambda_R \geqslant 1 + |\lambda| + C(\Omega_R)$. Here, $\Omega_R = \{x \in \Omega : x_n < 2R\}$ and $\mathcal{H}^0(\Omega_R)$ is the subspace of elements in $\mathcal{H}(\Omega)$ vanishing for $x_n \geqslant 2R$. Moreover, the estimate

$$\|u^R; \mathcal{H}^0(\Omega_R)\| \leqslant C \|f^R; \mathcal{H}^0(\Omega_R)^*\| \tag{2.12}$$

is valid with the factor C independent of R.

Let us construct the right parametrix for the operator of the problem (1.13). Given the right-hand side $\mathbf{f} \in \mathcal{H}(\Omega)^*$, we search for a solution of the problem in the form

$$\mathbf{u} = \chi_R^{+1} u^R + (1 - \chi_R^{-1}) u^\Pi, \tag{2.13}$$

where $\chi_R^p \in C^\infty(\mathbb{R})$ is a cut-off function which is equal to one for $x_n \leqslant R(1 + (1+p)/3)$ and zero for $x_n \geqslant R(1 + (2+p)/3)$; in the sequel, we need the indexes $p = 0, \pm 1$. We choose the components on the right-hand side of (2.13) as the solutions $u^R \in \mathcal{H}(\Omega_R)$ and $u^\Pi \in \mathcal{H}(\Pi)$ of the problems (2.11) and (2.4) with the right-hand sides $f^R(v) = \mathbf{f}(\chi_R^0 v)$ and $f^\Pi(v) = \mathbf{f}((1-\chi_R^0)v)$ respectively. We emphasize that $R \geqslant R^0$ and, therefore, in view of the requirement (1.7), the vector-valued function $(1-\chi_R^{-1})u^\Pi$ satisfies the necessary boundary conditions on the boundary $\partial\Omega$ and falls into the space $\mathcal{H}(\Pi)$. According to the estimates (2.12) and (2.10), we obtain the relation

$$\begin{aligned} \|u; \mathcal{H}(\Omega)\| &\leqslant c(\|u^R; \mathcal{H}^0(\Omega_R)\| + \|u^\Pi; \mathcal{H}(\Pi)\|) \\ &\leqslant c(\|f^R; \mathcal{H}^0(\Omega_R)^*\| + \|f^\Pi; \mathcal{H}(\Pi)^*\|) \\ &\leqslant cR\|\mathbf{f}; \mathcal{H}(\Omega)^*\|. \end{aligned} \tag{2.14}$$

All the factors c in formula (2.14) and further do not depend on R while the additional multiplier R occurs on the last norm due to the estimate $|\partial_z \chi_R^p(z)| \leqslant c_p R^{-1}$.

We fix a vector-valued function $v \in \mathcal{H}(\Omega)$ and choose $v^R = \chi_R^{+1} v \in \mathcal{H}^0(\Omega_R)$ and $v^\Pi = (1 - \chi_R^{-1})v \in \mathcal{H}(\Pi)$ as test functions in the integral identities (1.13) and (2.10) respectively. Note that

$$\begin{aligned} &f^R(v^R) + f^\Pi(v^\Pi) \\ &= \mathbf{f}(\chi_R^{+1}\chi_R^0 v) + \mathbf{f}((1-\chi_R^{-1})(1-\chi_R^0)v)\mathbf{f}(\chi_R^0 v) + \mathbf{f}((1-\chi_R^0)v) \\ &= \mathbf{f}(v), \\ &b(u^R, v^R; \Omega_R) + b^0(u^\Pi, v^\Pi; \Pi) = b(u, v; \Omega) - \widetilde{b}((1-\chi_R^{-1})u^\Pi, v), \\ &\left|\widetilde{b}((1-\chi_R^{-1})u^\Pi, v)\right| = \left|((\mathcal{B} - \mathcal{B}^0)(1-\chi_R^{-1})u^\Pi, v)_{\{x \in \Pi : x_n > R\}}\right| \\ &\qquad\qquad \leqslant c\exp(-\delta_0 R)\|u^\Pi; \mathcal{H}(\Pi)\|\,\|v; \mathcal{H}(\Omega)\|. \end{aligned} \tag{2.15}$$

Furthermore,

$$\begin{aligned} &a(u^R, v^R; \Omega_R) a(\chi_R^{+1} u^R, v; \Omega) + a_R(u^R, v), \\ &\left|a_R(u^R, v)\right| \\ &\leqslant c\left|(\mathcal{A}\mathcal{D}(\nabla_x)u^R, \mathcal{D}(\nabla_x \chi_R^{+1})v; \Omega) - (\mathcal{A}\mathcal{D}(\nabla_x \chi_R^{+1})u^R, \mathcal{D}(\nabla_x)v; \Omega)\right| \\ &\leqslant c(\|u^R; \mathcal{H}^0(\Omega_R)\|\,\|v; L_2(\Omega_R)\| + \|u^R; L_2(\Omega_R)\|\,\|v; \mathcal{H}(\Omega)\|). \end{aligned} \tag{2.16}$$

We emphasize that $\mathcal{D}(\nabla_x \chi_R^{+1})$ denotes a matrix-valued function with the matrix norm bounded from above by cR^{-1}. According to the last relation in (2.16), the estimate (2.12), and the compact embedding $\mathcal{H}^0(\Omega_R) \subset L_2(\Omega_R)$, the mapping $f \mapsto a_R(u^R, \cdot)$ in the space $\mathcal{H}(\Omega)^*$ is also compact.

Finally,

$$a^0(u^\Pi, v^\Pi; \Pi) a((1-\chi_R^{-1})u^\Pi, v; \Omega) + a_\Pi(u^\Pi, v) + \widetilde{a}(u^\Pi, v),$$

$$\begin{aligned} |\widetilde{a}(u^\Pi, v)| &= \big|((\mathcal{A}-\mathcal{A}^0)\mathcal{D}(\nabla_x)u^\Pi, \mathcal{D}(\nabla_x)(1-\chi_R^{-1})v)_{\{x\in\Pi : x_n > R\}}\big| \\ &\leqslant c\exp(-\delta_0 R)\|u^\Pi; \mathcal{H}(\Pi)\|\,\|v; \mathcal{H}(\Omega)\|, \end{aligned} \tag{2.17}$$

$$\begin{aligned} &|a_\Pi(u^\Pi, v)| \\ &\leqslant c\big|-(\mathcal{A}\mathcal{D}(\nabla_x)u^\Pi, \mathcal{D}(\nabla_x\chi_R^{-1})v; \Pi) + (\mathcal{A}\mathcal{D}(\nabla_x\chi_R^{-1})u^R, \mathcal{D}(\nabla_x)v; \Pi)\big| \\ &\leqslant c(\|u^\Pi; \mathcal{H}(\Pi)\|\,\|v; L_2(\Omega_R)\| + \|u^R; L_2(\Pi\cap\Omega_R)\|\,\|v; \mathcal{H}(\Omega)\|). \end{aligned}$$

The forms $\widetilde{a}$ and a_Π give rise to the small and compact operators in $\mathcal{H}(\Omega)$ respectively.

Thus, the sum (2.13) satisfies the integral identity

$$a(\mathbf{u}, v; \Omega) - \lambda b(\mathbf{u}, v; \Omega) = \mathbf{f}(v) + \mathcal{F}^s(\mathbf{f}; v) + \mathcal{F}^c(\mathbf{f}; v), \quad v \in \mathcal{H}(\Omega). \tag{2.18}$$

In the space $\mathcal{H}(\Omega)^*$, the mapping $\mathbf{f} \mapsto \mathcal{F}^c(\mathbf{f}; \cdot)$ is compact and the norm of the mapping $\mathbf{f} \mapsto \mathcal{F}^s(\mathbf{f}; \cdot)$ does not exceed $cR\exp(-\delta_0 R)$ (see formulas (2.14) and (2.15), (2.17)) while $\delta_0 > 0$ is the exponent in the stabilization conditions (1.3). Hence for a large $R \geqslant R^0$ the abstract equation

$$f(\cdot) = \mathbf{f}(\cdot) + \mathcal{F}^s(\mathbf{f}; \cdot) + \mathcal{F}^c(\mathbf{f}; \cdot)$$

enjoys the Fredholm alternative, i.e., under a finite number of orthogonality conditions on f, there exists a functional $\mathbf{f}$ such that formula (2.18) turns into the integral identity (1.13) which has the solution $u = \mathbf{u} \in \mathcal{H}(\Omega)$. In other words, the right parametrix has constructed.

We check up in Proposition 2.34 that the kernel of the operator of the problem (1.13)

$$\mathcal{H}(\Omega) \ni u \mapsto f \in \mathcal{H}(\Omega)^* \tag{2.19}$$

is of finite dimension even in the case where the segment $[0, 2\pi)$ contains eigenvalues of the pencil (2.9). Together with the above conclusion, this fact implies that the problem (1.13) operator is Fredholm. In other words, the point λ (the point $\mu = (1+\lambda)^{-1}$) does not belong to the essential spectrum of the problem (1.10) (the operator $\mathcal{K}$).

2. *Sufficiency.* Let the segment $[0, 2\pi)$ include the eigenvalue η of the pencil $\mathfrak{A}(\cdot; \lambda)$. The construction of the singular Weyl sequence for the operator $\mathcal{K}$ at the point (1.23) is standard (see, for example, [25] and [53, Subsect. 3.1]) so that we do not comment on the construction. Let U be an eigenvector corresponding to the eigenvalue η. We set

$$\mathbf{u}^q(x) = 2^{-q}\chi(x_n - R^0 - 2^q)\chi(R^0 + 2^{q+1} - x_n)\exp(i\eta x_n)U(x', x_n), \quad (2.20)$$

where $\chi \in C^\infty(\mathbb{R})$ and $\chi(t) = 1$ for $t \geqslant 1$, $\chi(t) = 0$ for $t \leqslant 0$. Since the product of the cut-off functions on the left-hand side of (2.20) is equal to one in the interval $(R^0 + 2^q, R^0 + 2^{q+1}) \ni x_n$ and the vector-valued function U is periodic in x_n, we derive the inequality

$$b(\mathbf{u}^q, \mathbf{u}^q; \Omega) \geqslant c2^{-q}(2^{q+1} - 2^q - 2),$$

while the factor c depends on $\mathcal{B}$, χ and η, U, but is independent of $q \in \mathbb{N}$. It is clear that

$$a(\mathbf{u}^q, \mathbf{u}^q; \Omega) + b(\mathbf{u}^q, \mathbf{u}^q; \Omega) \leqslant c2^{-q}2^{q+1}.$$

Thus, the relation

$$0 < c \leqslant \|\mathbf{u}^q; \mathcal{H}\| \leqslant C,$$

holds and the sequence $\{\mathbf{u}^q\}$ converges to null weakly in the space $\mathcal{H}$ because the supports of the vector-valued functions $\mathbf{u}^q$ and $\mathbf{u}^p$ do not intersect each other as $q \neq p$. It suffices to verify the third property of the Weyl sequence (see [1, Subsect. 9.1]), namely,

$$\|\mathcal{K}\mathbf{u}^q - \mu\mathbf{u}^q; \mathcal{H}\| \to 0, \quad q \to \infty. \quad (2.21)$$

We have

$$\begin{aligned}\|\mathcal{K}\mathbf{u}^q - \mu\mathbf{u}^q; \mathcal{H}\| &= \sup |\langle \mathcal{K}\mathbf{u}^q - \mu\mathbf{u}^q, \mathbf{v}\rangle| \\ &= (1+\lambda)^{-1} \sup |a(\mathbf{u}^q, \mathbf{v}; \Omega) - \lambda b(\mathbf{u}^q, \mathbf{v}; \Omega)|. \quad (2.22)\end{aligned}$$

Here, the supremum is taken over all $\mathbf{v} \in \mathcal{H}$ such that $\|\mathbf{v}; \mathcal{H}\| = 1$. Note that

$$a(\mathbf{u}^q, \mathbf{v}; \Omega) - \lambda b(\mathbf{u}^q, \mathbf{v}; \Omega) = a^0(\mathbf{u}^q, \mathbf{v}; \Pi) - \lambda b^0(\mathbf{u}^q, \mathbf{v}; \Pi) + \widetilde{a}(\mathbf{u}^q, v) - \lambda \widetilde{b}(\mathbf{u}^q, \mathbf{v}),$$

$$|\widetilde{a}(\mathbf{u}^q, v)| + |\widetilde{b}(\mathbf{u}^q, \mathbf{v})| \leqslant c\exp(-\delta_0(R^0 + 2^q))$$

and for any vector-valued function $v \in \mathcal{H}$ with support in the set $\Pi^q = \{x \in \Pi : x_n \in (R^0 + 2^q + 2, R^0 + 2^{q+1} - 2\}$ the equality

$$a^0(\mathbf{u}^q, v; \Pi) - \lambda b^0(\mathbf{u}^q, v; \Pi) = a^0(\eta, U, \widehat{v}(\cdot, \eta); \Sigma) - \lambda b^0(U, \widehat{v}(\cdot, \eta); \Sigma) = 0$$

is valid by virtue of formula (2.1) for the Gel'fand image $\widehat{v}(\cdot, \eta) \in \mathcal{H}_{per}(\Sigma)$ and the identity (2.8), where $V = \widehat{v}(\cdot, \eta)$ (recall that $\{U, \eta\}$ is an eigenpair of the pencil (2.9)). It is clear that the expression (2.22) does not exceed the quantity

$$c2^{-q}\sup\{\|\mathbf{v}; H^1(\Omega \setminus \Pi^q)\|\} \leqslant C2^{-q}.$$

Thus, the convergence (2.21) was verified. □

Remark 2.1. While proving the first part of Theorem 2.1 we used the following fact (see, for example, [60]). Let $\mathfrak{H}$ be a Hilbert space which is compactly embedded in a Banach space $\mathfrak{L}$. Let also $\mathfrak{k}(\mathfrak{u}, \mathfrak{v})$ be a form subject to the estimate

$$|\mathfrak{k}(\mathfrak{u}, \mathfrak{v})| \leqslant c(\|\mathfrak{u}; \mathfrak{H}\| \, \|\mathfrak{v}; \mathfrak{L}\| + \|\mathfrak{u}; \mathfrak{L}\| \, \|\mathfrak{v}; \mathfrak{H}\|). \tag{2.23}$$

Then the operator $\mathfrak{K}$ in the space $\mathfrak{H}$ defined by the formula

$$(\mathfrak{K}\mathfrak{u}, \mathfrak{v})_{\mathfrak{H}} = \mathfrak{k}(\mathfrak{u}, \mathfrak{v}), \quad \mathfrak{u}, \mathfrak{v} \in \mathfrak{H},$$

is compact. Indeed, if $\{\mathfrak{u}_p\}$ is a sequence bounded in the space $\mathfrak{H}$, then

$$\mathfrak{u}_{p_q} \to \mathfrak{u}_\infty \quad \text{weakly in } \mathfrak{H} \quad \text{and strongly in } \mathfrak{L} \tag{2.24}$$

along a certain infinitely large subsequence $\{p_q\} \subset \mathbb{N}$. Since the operator $\mathfrak{K}$ is continuous, we derive that

$$\mathfrak{K}\mathfrak{u}_{p_q} \to \mathfrak{K}\mathfrak{u}_\infty \quad \text{weakly in } \mathfrak{H} \quad \text{and strongly in } \mathfrak{L}. \tag{2.25}$$

The last two terms in the relation

$$\begin{aligned} &\|\mathfrak{K}\mathfrak{u}_{p_q}; \mathfrak{H}\|^2 - \|\mathfrak{K}\mathfrak{u}_\infty; \mathfrak{H}\|^2 \\ &= \mathfrak{k}(\mathfrak{u}_{p_q} - \mathfrak{u}_\infty, \mathfrak{K}(\mathfrak{u}_{p_q} - \mathfrak{u}_\infty)) + \mathfrak{k}(\mathfrak{u}_{p_q} - \mathfrak{u}_\infty, \mathfrak{K}\mathfrak{u}_\infty) + \mathfrak{k}(\mathfrak{u}_\infty, \mathfrak{K}(\mathfrak{u}_{p_q} - \mathfrak{u}_\infty)) \end{aligned}$$

decay as $p_q \to +\infty$ owing to the weak continuity of the form in both arguments. The first term on the right-hand side has the null limit by the inequality (2.23) and formulas (2.24), (2.25). Hence $\|\mathfrak{K}\mathfrak{u}_{p_q}; \mathfrak{H}\| \to \|\mathfrak{K}\mathfrak{u}_\infty; \mathfrak{H}\|$ and, in view of the weak convergence (2.25), the strong convergence $\mathfrak{K}\mathfrak{u}_{p_q} \to \mathfrak{K}\mathfrak{u}_\infty$ occurs in $\mathfrak{H}$. The desired assertion follows from the last fact. □

2.3 Exponential decay and finite dimension of the kernel

Assume that for a certain λ the segment $[0, 2\pi) \subset \mathbb{C}$ and, therefore, the whole real axis, is free of the spectrum of the pencil $\eta \mapsto \mathfrak{A}(\eta, \lambda)$. Then, according to [13, Chapter 1], there exists $\beta_0 > 0$ such that the rectangle $Q = \{\eta : \operatorname{Re} \eta \in [0, 2\pi), |\operatorname{Im} \eta| < \beta_0\}$ does not contain an eigenvalue as well. Let $\beta \in (-\beta_0, \beta_0)$. In the problem (2.4), we make the formal changes $u \mapsto u^\beta = \exp(-\beta x_n) u$ and $v \mapsto v^{-\beta} = \exp(\beta x_n) v$. The inclusions $u^\beta \in \mathcal{W}_\beta(\varPi)$ and $v^{-\beta} \in \mathcal{W}_{-\beta}(\varPi)$ are valid, where $\mathcal{W}_\gamma(\varPi)$ is the completion of the linear space $C_c^\infty(\overline{\varPi})^k \cap \mathcal{H}(\varPi)$ (infinitely differentiable vector-valued functions having compact support and satisfying the stable boundary conditions on $\partial\varPi$) with respect to the norm $\|w; \mathcal{W}_\gamma\| = \|\exp(\gamma x_n) w; \mathcal{H}(\varPi)\|$. By these changes, we generate the problem in the quasicylinder

$$a^0(i\beta, u^\beta, v^{-\beta}; \Pi) - \lambda b^0(u^\beta, v^{-\beta}; \Pi) = f^\beta(v^{-\beta}), \quad v^{-\beta} \in \mathcal{H}(\Pi), \qquad (2.26)$$

and the corresponding model problem in the periodicity cell

$$a^0(\eta + i\beta, U^\beta, V^{-\beta}; \Sigma) - \lambda b^0(U^\beta, V^{-\beta}; \Sigma) = F^\beta(\eta; V^{-\beta}),$$
$$V^{-\beta} \in \mathcal{H}_{per}(\Sigma) \quad \text{for a.a.}\, \eta \in (0, 2\pi).$$

Here, $f^\beta(v^{-\beta}) = f(\exp(-\beta x_n)v^{-\beta})$ and F^β is the Gel'fand image of f^β.

In the case $f^\beta \in \mathcal{H}(\Pi)^*$, by the same scheme as above; namely, applying the direct transform, solving the problem on the periodicity cell with the parameter η, and applying the inverse transform, we determine the solution $u^\beta \in \mathcal{H}(\Pi)$ of the problem (2.26). Moreover, if $v \mapsto f(\exp(\gamma x_n)v)$ is an element of the space $\mathcal{H}(\Pi)^*$ with $\gamma = 0$ and $\gamma = \beta > 0$, therefore, for any $\gamma \in [0, \beta]$, then the above-mentioned procedure gives the family of solutions $u^{\beta 0} = \exp(\beta x_n)u^\beta \in \mathcal{W}_\beta(\Pi)$ of the integral identity (2.4) with test functions $v \in C_c^\infty(\overline{\Pi})^k \cap \mathcal{H}(\Pi)$. The following assertion ensures that $u^{\beta 0} = u$, i.e., the solution of the problem (2.4), which was constructed in the previous subsection, decays exponentially as $x_n \to \pm\infty$.

Proposition 2.1. *Let the segment $[0, 2\pi) \subset \mathbb{C}$ be free of the spectrum of the pencil $\eta \mapsto \mathfrak{A}(\eta, \lambda)$, and let $f \in \mathcal{W}_\gamma(\Pi)^*$ with $\gamma = 0$ and $\gamma = \beta \in (-\beta_0, \beta_0)$. Then the solution $u \in \mathcal{H}(\Pi) = \mathcal{W}_0(\Pi)$ of the problem (2.4) falls into the space $\mathcal{W}_\beta(\Pi)$ and satisfies the estimates*

$$\|u; \mathcal{W}_\gamma(\Pi)\| \leqslant c_\gamma(\|f; \mathcal{W}_\gamma(\Pi)^*\|, \qquad (2.27)$$

where $\gamma = 0$ and $\gamma = \beta$.

Proof. We can assume that $\beta > 0$; otherwise, we make the change $x_n \mapsto -x_n$. We extend the Gel'fand transform (2.1) to complex values of the parameter η. The following formula holds;

$$u^{\beta 0}(x', z) = (2\pi)^{-1/2} \int_0^{2\pi} \exp(i(\eta - i\beta)z)\mathfrak{A}(\eta - i\beta, \lambda)^{-1}\widehat{f}(\eta - i\beta)\, d\eta. \quad (2.28)$$

By the conditions on f, the abstract function $\eta \mapsto \widehat{f}(\cdot, \eta) \in \mathcal{H}(\Pi)^*$ is holomorphic in the rectangle $Q_\beta = \{\eta : \operatorname{Re}\eta \in [0, 2\pi), \operatorname{Im}\eta \in (-\beta, 0)\}$ and continuous in the closed rectangle. According to [13], the operator-valued function $\eta \mapsto \mathfrak{A}(\eta, \lambda)^{-1} : \mathcal{H}(\Pi)^* \to \mathcal{H}(\Pi)$ is holomorphic in a neighborhood of the set $\overline{Q}_\beta$. By (2.28), the difference $u - u^{\beta 0}$ is represented as a path integral along the boundary of the rectangle; indeed, the integrals over the lateral sides cancel each because the integrand is 2π-periodic in $\operatorname{Re}\eta$ and the paths are directed oppositely. Thus, the equality $u = u^{\beta 0}$ and, finally, the assertion of the proposition follows from the abstract variant of the Cauchy theorem (see [14]). □

Corollary 2.1. *Let the assumptions of Proposition* 2.1 *be valid, and let* $f \in \mathcal{W}_\beta(\Omega)^*$ *for a certain exponent* $\beta \in (0, \beta_0)$ *subject additionally to the inequality* $\beta \leqslant \delta_0$ (*see* (1.3)). *Then the solution* $u \in \mathcal{H}(\Omega) = \mathcal{W}_0(\Omega)$ *of the problem* (1.13) *belongs to the space* $\mathcal{W}_\beta(\Omega)$ *and admits the estimate* $\|u; \mathcal{W}_\beta(\Omega)\| \leqslant c_\beta \|f; \mathcal{W}_\beta(\Omega)^*\|$.

Proof. For $\beta > 0$ the norm $\|u; \mathcal{W}_{-\beta}(\Omega)\| = \|\exp(-\beta x_n)u; \mathcal{H}(\Omega)\|$ contains a weight function which decays exponentially due to the inequality $x_n > -R^0$ (see the condition on the domain Ω in Subsect. 1.2). Hence the inclusion $f \in \mathcal{W}_\beta(\Omega)^*$ provides $f \in \mathcal{H}(\Omega)^*$. Aiming to apply Proposition 2.1, we multiply the solution with an appropriate cut-off function, for example, $1 - \chi_R^0$, and note that, according to the restriction (1.3), the terms $((\mathcal{A} - \mathcal{A}^0)\mathcal{D}(\nabla_x)u, \mathcal{D}(\nabla_x)((1 - \chi_R^0)v))_\Pi$ and $\lambda((\mathcal{B} - \mathcal{B}^0)u, (1 - \chi_R^0)v)_\Pi$ give rise to continuous forms on the space $\mathcal{H}(\Omega) \times \mathcal{W}_{-\beta}(\Omega)$. □

If $\beta \to +0$, the operator of the problem (2.26) is a small perturbation of the operator of the problem (2.4) and, hence, the assertion, checked up above, can be established with the help of the classical perturbation theory (see [22, Chapters 7, 8]). However, the above approach leads to an assertion which follows directly from formula (2.28) and is used in the sequel.

Proposition 2.2. *Let the segment* $[0, 2\pi) \subset \mathbb{C}$ *contain an eigenvalue of the pencil* $\eta \mapsto \mathfrak{A}(\eta, \lambda)$. *There exists a number* $\beta_0 > 0$ *such that for* $\beta \in (-\beta_0, 0) \cup (0, \beta_0)$, *the problem* (2.4) *admits a unique solution* $u^\beta \in \mathcal{W}_\beta(\Pi)$ *for any functional* $f^\beta \in \mathcal{W}_{-\beta}(\Pi)^*$ *and the estimate* (2.27) *is valid with exponent* $\gamma = \beta$ *whilst the factor* c_β *grows indefinitely as* $\beta \to 0$.

The following assertion was used in the first part of the proof of Theorem 2.1.

Proposition 2.3. *The operator* (2.19) *of the problem* (1.13) *has the kernel of finite dimension for any* λ.

Proof. If there are no eigenvalues of the pencil $\eta \mapsto \mathfrak{A}(\eta, \lambda)$ on the segment $[0, 2\pi)$, we set $\beta = 0$. Otherwise, $\beta \in (-\beta_0, 0)$ (see Proposition 2.1). Since the norm in the space $\mathcal{W}_\beta(\Omega)$ includes an exponentially decaying weight in the case $\beta < 0$, the kernel of the operator (2.19) belongs to the kernel $\mathcal{N}_\beta$ of the operator of the problem (1.13) realized as the mapping

$$\mathcal{W}_\beta(\Omega) \ni u \mapsto f \in \mathcal{W}_{-\beta}(\Omega)^*$$

(see the mapping (2.19) in the case $\beta = 0$).

We are going to check up that $\dim \ker \mathcal{N}_\beta < \infty$. Let $u \in \mathcal{N}_\beta$, i.e., the vector-valued function $u \in \mathcal{W}_\beta(\Omega)$ satisfies the integral identity (1.10) with any test function $v \in \mathcal{W}_{-\beta}(\Omega)$. The product $u^R = \chi_R^0 u \in \mathcal{H}(\Omega_R)$ (here cut-off functions are the same as in formula (2.13)) satisfies the uniquely solvable problem (2.11) with the right-hand side

$$
\begin{aligned}
f^R(v) =& \lambda_R(\mathcal{B}u, \chi_R^0 v)_{\Omega_R} - (\mathcal{A}\mathcal{D}(\nabla_x)u, \mathcal{D}(\nabla_x \chi_R^0)v)_{\Omega_R} \\
&+ (\mathcal{A}\mathcal{D}(\nabla_x \chi_R^0)u, \mathcal{D}(\nabla_x)v)_{\Omega_R}.
\end{aligned}
$$

Since the expression does not include products of the first order derivatives of components of the vector-valued functions u^R and v, according to Remark 2.1 the mapping $\mathcal{W}_\beta(\Omega) \ni u \mapsto F^R \in \mathcal{H}(\Omega_R)^*$ is compact and, therefore, the operator $\mathcal{N}_\beta \ni u \mapsto \chi_R^0 u \in \mathcal{W}_\beta(\Omega)$ is compact as well.

We consider the product $u^\Pi = (1 - \chi_R^0)u \in \mathcal{W}_\beta(\Pi)$ which satisfies the integral identity (2.4) with a test function $v \in \mathcal{W}_{-\beta}(\Pi)$ and the right-hand side

$$
\begin{aligned}
f^\Pi(v) &= ((\mathcal{A} - \mathcal{A}^0)\mathcal{D}(\nabla_x)u, \mathcal{D}(\nabla_x)((1 - \chi_R^0)v))_\Pi \\
&- \lambda((\mathcal{B} - \mathcal{B}^0)u, (1 - \chi_R^0)v)_\Pi + (\mathcal{A}\mathcal{D}(\nabla_x)u, \mathcal{D}(\nabla_x \chi_R^0)v)_\Pi \\
&- (\mathcal{A}\mathcal{D}(\nabla_x \chi_R^0)u, \mathcal{D}(\nabla_x)v)_\Pi. \qquad (2.29)
\end{aligned}
$$

By Theorem 2.1 with $\beta = 0$ and Proposition 2.2 with $\beta \in (-\beta_0, 0)$, the obtained model problem is uniquely solvable in the class $\mathcal{W}_\beta(\Pi)$. By the condition (1.3), the first two terms on the right-hand side of (2.29) generate an operator, the norm $O(R\exp(-\delta_0 R))$ of which can be made arbitrarily small by choosing an appropriate R. According to Remark 2.1, the other two terms yield compact operators. Hence the operator $\mathcal{N}_\beta \ni u \mapsto (1 - \chi_R^0)u \in \mathcal{W}_\beta(\Omega)$ is also compact.

We have proved that the embedding of subspace $\mathcal{N}_\beta$ into $\mathcal{W}_\beta(\Omega)$ is compact and, therefore, the subspace has a finite dimension. □

2.4 Continuous spectrum

In the case of the real parameter η, we endow the space $\mathcal{H}_{per}(\Sigma)$ with the specific inner product

$$
\langle U, V\rangle_\eta = a^0(\eta, U, V; \Sigma) + b^0(U, V; \Sigma) \qquad (2.30)
$$

generated by the Hermitian forms (2.6) and (2.7). The necessary properties of the form (2.30) follow from the relations (1.2) and (1.16). Let also $\mathcal{K}_\eta$ be an operator in $\mathcal{H}_{per}(\Sigma)$ defined analogously to (1.21); namely,

$$
\langle \mathcal{K}_\eta U, V\rangle_\eta = b^0(U, V; \Sigma), \quad U, V \in \mathcal{H}_{per}(\Sigma). \qquad (2.31)
$$

Clearly, this operator is continuous, positive, self-adjoint, and compact. Moreover, its norm does not exceed one. Thus, by general results of the operator theory in Hilbert spaces (see, for example, [1, Theorem 9.2.1]), the spectrum of the operator $\mathcal{K}_\eta$ consists of the point $M^{(\infty)} = 0$ in the essential spectrum and the infinitesimal positive sequence of eigenvalues

$$1 \geqslant M^{(1)}(\eta) \geqslant M^{(2)}(\eta) \geqslant \cdots \geqslant M^{(p)}(\eta) \geqslant \to +0,$$

where they are listed according to multiplicity.

For a fixed $\eta \in \mathbb{R}$, we treat (2.8) as a spectral problem with parameter $\varLambda = \lambda$. Owing to the definitions (2.30) and (2.31), this problem has the eigenvalues $\varLambda^{(p)}(\eta) = M^{(p)}(\eta)^{-1} - 1$ forming the infinitely large positive sequence

$$\varLambda^{(1)}(\eta) \leqslant \varLambda^{(2)}(\eta) \leqslant \cdots \leqslant \varLambda^{(p)}(\eta) \leqslant \ldots \to +\infty. \tag{2.32}$$

If $\eta \in \mathbb{R}$ is an eigenvalue of the pencil $\eta \mapsto \mathfrak{A}(\eta;\lambda)$ in (2.9), then $\lambda = \varLambda^{(p)}(\eta)$ for some $p \in \mathbb{N}$. The contrary assertion is true as well; namely, $\eta \in \mathbb{R}$ is an eigenvalue for the pencil in the case $\lambda = \varLambda^{(p)}(\eta)$. Since the η-spectrum of the pencil $\mathfrak{A}$ is invariant with respect to shifts $\pm 2\pi$ along the real axis, the functions $\mathbb{R} \ni \eta \mapsto \varLambda^{(p)}(\eta)$ are 2π-periodic. They are continuous by a general result in [22, Chapter 8].

The above observations, together with Theorem 2.1, demonstrate that the operator (2.19) of the problem (1.13) is not Fredholm if and only if

$$\lambda \in \bigcup_{p \in \mathbb{N}} \varUpsilon^{(p)}, \tag{2.33}$$

where

$$\varUpsilon^{(p)} = [\varLambda_-^{(p)}, \varLambda_+^{(p)}], \quad \pm\varLambda_\pm^{(p)} = \max\{\pm\varLambda^{(p)}(\eta) \mid \eta \in [0, 2\pi)\}. \tag{2.34}$$

Let us formulate the derived result for the operator $\mathcal{K}$ introduced in formula (1.21) and related to the spectral parameter (1.23).

Theorem 2.2. *The essential spectrum of the operator $\mathcal{K}$ consists of the point $\mu = 0$ and the union of closed segments $[\mu_+^{(p)}, \mu_-^{(p)}]$, where $\mu_\pm^{(p)} = (1 + \varLambda_\pm^{(p)})^{-1}$ and the numbers $\varLambda_\pm^{(p)}$ are defined by (2.34). Each of the segments contains only points of the continuous spectrum.*

Proof. By Theorem 2.1, it suffices to verify that the segments $[\mu_+^{(p)}, \mu_-^{(p)}]$ lie in the continuous spectrum. We assume that the point $\mu \in [\mu_+^{(p)}, \mu_-^{(p)}]$ does not belong to the continuous spectrum. Hence the operator $\mathcal{K}$ at the point μ has the singular Weyl sequence $\{u^p\}$ with the following three properties:

$$\inf_{p \in \mathbb{N}} \|u^p; \mathcal{H}\| > 0, \quad u^p \to 0 \quad \text{weakly in} \quad \mathcal{H}, \quad \|\mathcal{K}u^p - \mu u^p; \mathcal{H}\| \to 0$$

(see the second part of the proof of Theorem 2.1). We are going to show that the resolvent $(\mathcal{K} - \mu)^{-1}$ cannot satisfy the estimate

$$\|(\mathcal{K} - \mu)^{-1} u; \mathcal{H}\| \leqslant c\|u; \mathcal{H}(\varOmega)\|, \quad u \in \mathcal{H}(\varOmega) \ominus \mathcal{N}(\mu), \tag{2.35}$$

where $\mathcal{N}(\mu)$ denotes the kernel of the operator $\mathcal{K} - \mu$ which has a finite dimension by Proposition 2.3. The violation of the estimate (2.35) just means that μ is a point in the continuous spectrum, i.e., contradicts the above assumption. Let $v^p = P(\mu)u^p$, where $P(\mu)$ is the orthogonal projector into the subspace $\mathcal{H}(\Omega) \ominus \mathcal{N}(\mu)$. By the third property of the Weyl sequence, we obtain

$$\|\mathcal{K}v^p - \mu v^p; \mathcal{H}(\Omega)\| = \|\mathcal{K}u^p - \mu u^p; \mathcal{H}(\Omega)\| \to 0.$$

Since $v^p = (\mathcal{K} - \mu)^{-1}(\mathcal{K}v^p - \mu v^p)$, the estimate (2.35) surely does not hold under the condition that a certain subsequence $\{v^{p_q}\}_{q=1}^{\infty}$ satisfies

$$\inf_q \|v^{p_q}; \mathcal{H}(\Omega)\| > 0.$$

Assume that this infimum vanishes for any subsequence and consider some subsequence. We have $\|v^{p_q}; \mathcal{H}(\Omega)\| \to 0$ and, hence, $v^{p_q} \to 0$ strongly in $\mathcal{H}(\Omega)$. The difference $u^{p_q} - v^{p_q} \in \mathcal{N}(\mu)$ converges to null weakly due to the second property of the Weyl sequence. This convergence is strong because the dimension of the kernel is finite. This conclusion contradicts the first property of the Weyl sequence. Thus, the estimate (2.35) cannot be valid and the point μ belongs to the continuous spectrum. □

In the sequel, the number $\Lambda_-^{(1)}$ (the number $(1 + \Lambda_-^{(1)})^{-1}$) is denoted by $\lambda^\dagger$ (by $\mu^\dagger$) and is called the minimal (maximal) cut-off in the spectrum of the problem (1.10) (of the operator $\mathcal{K}$).

The structure (2.33) of the continuous spectrum in the problem (1.10) permits for opening gaps, i.e., intervals which have ends in the continuous spectrum, but can contain only points of the discrete spectrum (see the example constructed in Subsect. 3.5). Such a gap occurs in the case where for a certain $q \geqslant 2$ the inequality

$$\max\left\{\Lambda_+^{(1)}, \dots, \Lambda_+^{(q-1)}\right\} < \Lambda_-^{(q)} \tag{2.36}$$

is valid, i.e., the segments $\Upsilon^{(1)}, \dots, \Upsilon^{(q-1)}$ do not intersect the segment $\Upsilon^{(q)}$. We mention that, by formulas (2.32) and (2.34), the relations $\Lambda_-^{(p)} \geqslant \Lambda^{(q)}$ are valid for $p \geqslant q$. Thus, the inequality (2.36) and formula (2.33) really provide the absence of the continuous spectrum on the nonempty interval

$$\left(\max\left\{\Lambda_+^{(1)}, \dots, \Lambda_+^{(q-1)}\right\}, \Lambda_-^{(q)}\right).$$

The following assertion demonstrates that gaps do not open in the case of the straight cylinder (1.24) and the matrices $\mathcal{A}^0$, $\mathcal{B}^0$ independent of the longitudinal variable $x_n = z$.

Proposition 2.4. *Under the above assumptions, the maximal cut-off* $\mu^\dagger \in (0, 1]$ *is the only cut-off, in other words, the closed segment* $(0, \mu^\dagger]$ *is covered with the continuous spectrum of the operator* $\mathcal{K}$ *and the segment* $(\mu^\dagger, 1]$ *can contain points of the discrete spectrum only.*

Proof. Since the model problem is invariant with respect to any shift along the cylinder axis, it is worth to use the partial Fourier transform

$$v(x) \mapsto \widehat{v}(x', \eta) = (2\pi)^{-1/2} \int\limits_{\mathbb{R}} \exp(-i\eta z) v(x', z)\, dz \tag{2.37}$$

instead of the Gel'fand transform (2.1). As a result, we obtain the following $(n-1)$-dimensional model problem in the cross-section ω:

$$\begin{gathered} \mathcal{L}^0(x', \nabla_{x'}, i\eta)\widehat{u}(x', \eta) = \lambda \mathcal{B}^0(x')\widehat{u}(x', \eta), \quad x' \in \omega, \\ \mathcal{T}^0(x', \nabla_{x'}, i\eta)\widehat{u}(x', \eta) = 0, \quad x' \in \partial\omega. \end{gathered} \tag{2.38}$$

The variational formulation of the problem (2.38) reads:

$$(\mathcal{A}^0 \mathcal{D}(\nabla_{x'}, i\eta)\widehat{u}, \mathcal{D}(\nabla_{x'}, i\eta)\widehat{v})_\omega = \lambda(\mathcal{B}\widehat{u}, \widehat{v})_\omega, \quad v \in \mathcal{H}(\omega). \tag{2.39}$$

Here, $\mathcal{H}(\omega)$ is a subspace of vector-valued functions $v \in H^1(\omega)^k$ subject to the stable boundary conditions, i.e., $\mathcal{T}^0_{(q)}(x')v(x') = \mathcal{S}^0_{(q)}(x')v(x') = 0$, $x' \in \partial\omega$, for $q = 1, \dots, k$ such that the first couple of formulas in (1.8) is chosen. For $q \in \{2, 3, \dots\}$, we insert into the Korn inequality

$$\|\nabla_x u; L_2(\varPi)\|^2 \leqslant c(\varPi)(\|\mathcal{D}(\nabla_x)u; L_2(\varPi)\|^2 + \|u; L_2(\varPi)\|^2) \tag{2.40}$$

(see Lemma 1.1) the vector-valued function

$$\varPi \ni x \mapsto u(x) = \chi(x_n - 2^q)\chi(2^{q+1}) \exp(i\eta x_n)\widehat{u}(x')$$

(see the definition (2.20)) and, after calculating integrals in $x_n \in (2^q + 1, 2^{q+1} - 1)$, we derive the relation

$$\begin{aligned} &(2^{q+1} - 2^q - 2)\left(\|\nabla_{x'}\widehat{u}; L_2(\omega)\|^2 + \eta^2\|\widehat{u}; L_2(\omega)\|^2\right) \\ &\leqslant (2^{q+1} - 2^q - 2)\Big(c(\varPi)\left(\|\mathcal{D}(\nabla_{x'}, i\eta)\widehat{u}; L_2(\omega)\|^2 + \|\nabla_{x'}\widehat{u}; L_2(\omega)\|^2\right) \\ &+ C(2^{q+1} - 2^q - 2)^{-1}\|\widehat{u}; H^1(\omega)\|^2\Big). \end{aligned}$$

The last term is caused by cut-off functions in the formula which manifest themselves only inside the finite cylinders $\omega \times (2^q, 2^q + 1)$ and $\omega \times (2^{q+1} - 1, 2^{q+1})$ of the unit length. Reducing the multiplier $2^{q+1} - 2^q - 2 > 0$ in the inequality (2.40) and sending q to infinity, we arrive at the estimate

$$\begin{aligned} &\|\nabla_{x'}\widehat{u}; L_2(\omega)\|^2 + \eta^2\|\widehat{u}; L_2(\omega)\|^2 \\ &\leqslant c(\varPi)\left(\|\mathcal{D}(\nabla_{x'}, i\eta)\widehat{u}; L_2(\omega)\|^2 + \|\widehat{u}; L_2(\omega)\|^2\right). \end{aligned}$$

Inserting the eigenvector U of the operator of the problem (2.8) corresponding to the eigenvalue $\lambda^{(p)}(\eta)$ and taking the inequalities (1.2) into account, we obtain

$$\lambda^{(p)}(\eta)\|U;L_2(\omega)\|^2 \geqslant C_{\mathcal{B}}^{-1}b^0(U,U;\Sigma) = c_{\mathcal{B}}^{-1}a(\eta,U,U;\Sigma)$$
$$\geqslant C_{\mathcal{B}}^{-1}c_{\mathcal{A}}\|\mathcal{D}(\nabla_{x'},i\eta)U;L_2(\omega)\|^2 \geqslant C_{\mathcal{B}}^{-1}c_{\mathcal{A}}(c(\Pi)^{-1}\eta^2-1)\|U;L_2(\omega)\|^2.$$

Thus, we find constants c_0 and $c_1>0$ such that $\lambda^{(1)}(\eta)\geqslant c_1\eta^2-c_0$. Hence the range of the continuous function $\mathbb{R}\ni\eta\mapsto\Lambda^{(1)}(\eta)$ is the ray $[\lambda^\dagger,+\infty)$, where $\lambda^\dagger=\min\{\Lambda^{(1)}(\eta)\,|\,\eta\in\mathbb{R}\}\geqslant 0$.

Let $\eta\in[2\pi q,2\pi(q+1))$ with $q\in\mathbb{Z}$. Then the vector-valued function $\Sigma\ni(x',x_n)\mapsto\exp(i\eta x_n)U(x')$ satisfies the problem (2.8) with parameters $\eta-2\pi q\in[0,2\pi)$, $\lambda=\lambda^{(1)}(\eta)$ and, therefore, $\lambda^{(1)}(\eta)=\Lambda^{(p)}(\eta-2\pi q)$ for a certain $p\in\mathbb{N}$.

The above-mentioned facts, together with formula (2.33) for the continuous spectrum, complete the proof of Proposition 2.4. □

Remark 2.2. The cylindrical waveguide (1.24) can be regarded as a periodic one with the cylindrical periodicity cell $\Sigma=\omega\times(0,1)$. At the end of the proof of Proposition 2.4, we observed why the union of segments $\Upsilon^{(p)}$ on the right-hand side of (2.33) implies the ray $[\lambda^\dagger,+\infty)$ in the case of a straight cylinder. Indeed, the first eigenvalue $\lambda^{(1)}(\eta)\geqslant c_1\eta^2-c_0$ of the problem (2.39) gives rise to the infinite family $\{\Lambda^{(m_q)}(\eta)\}_{q\in\mathbb{N}}$ of eigenvalues in the problem (2.8) while the corresponding segments $\Upsilon^{(m_q)}$ cover the ray in the way indicated in Fig. 3 for the right branch of the eigenvalue $\lambda^{(1)}(\eta)$. □

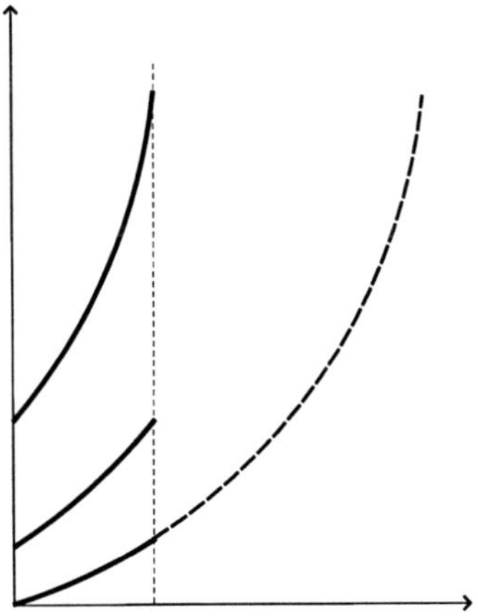

Fig. 3 The formation of the segmentary continuous spectrum in the cylindrical waveguide.

2.5 On the positive threshold

The polynomial property (1.14) mentioned in Subsect. 1.3 allows us to formulate a simple criterion for the positive threshold of the continuous spectrum, i.e., the nonempty segment $[0, \lambda^\dagger)$ of the closed positive real semiaxis can include only points of the discrete spectrum in the problem (1.10).

Proposition 2.5. *The continuous spectrum* (2.33) *of the problem* (1.10) *lies inside the ray* $[\lambda^\dagger, +\infty)$ *with endpoint* $\lambda^\dagger > 0$ *if and only if the subspace*

$$\{u \in \mathcal{P} : \mathcal{T}^0(x, \nabla_x)u(x) = 0, \quad x \in \partial\Sigma \cap \partial\Pi\} \tag{2.41}$$

of vector polynomials in the polynomial property (1.14) *of the Hermitian form* $a^0(\cdot, \cdot; \Xi)$ *(see the definition* (1.15)*) is trivial. If the dimension of the subspace* (2.41) *is positive, the point* $\lambda = 0$ *belongs to the continuous spectrum of the problem* (1.10).

Proof. If the linear subspace (2.41) contains the vector polynomial

$$x_n^q p^0(x') + x_n^{q-1} p^1(x') + \cdots + p^q(x'), \quad p^0 \neq 0,$$

then, by the changes of coordinates $x \mapsto (x', x_n - j)$ with $j \in \mathbb{Z}$, we observe that the polynomial p^0 also belongs to this subspace. Hence the pencil $\mathfrak{A}(\eta; 0)$ at $\lambda = 0$ has the eigenvalue $\eta = 0$ and the eigenvector p^0.

Let $\mathfrak{A}(\eta; 0)$ have the eigenvalue $\eta_0 \in [0, 2\pi)$ and the eigenvector $U \in \mathcal{H}_{per}(\Sigma)$. Inserting the vector-valued function

$$\Sigma \ni x_n \mapsto u(x) = \exp(-i\eta_0 x_n)U(x) \tag{2.42}$$

in the Green formula (1.11) and taking the integral identity (2.8) into account, we obtain the relation

$$0 = a^0(\eta_0, U, U; \Sigma) = a^0(0, u, u; \Sigma),$$

which means that $u \in \mathcal{P}$ due to formula (1.14). The vector-valued function (2.42) can imply a polynomial only in the case $\eta_0 = 0$ while, being a solution of the problem (2.8) with $\lambda = 0$, it belongs to the subspace (2.41). Indeed, the stable boundary conditions are included into the definition of the space $\mathcal{H}_{per}(\Sigma)$ and the natural boundary conditions ought to be derived by integrating by parts (see [33]). □

3 Specific Properties of Spectra in Particular Situations

3.1 The absence of the point spectrum

Let Π be the straight cylinder (1.24), and let

$$\Omega = \{x : x' \in \omega, x_n > H(x')\}. \tag{3.1}$$

Here, H is a function in the Hölder space $C^{2,\alpha}(\omega)$ and $\omega \subset \mathbb{R}^{n-1}$ is a domain with compact closure and boundary of class $C^{2,\alpha}$. We also assume that the matrix-valued functions $\mathcal{A}$ and $\mathcal{B}$ do not depend on the variable x_n and their entries belong to the same Hölder space. Finally, the operator $\mathcal{T}$ in the boundary conditions (1.6) on the lateral side $\partial\Omega \setminus \overline{\Upsilon}$ is chosen according to the rule (1.8) with the unitary matrix $\mathcal{S} \in C^{2,\alpha}(\partial\omega)^{k\times k}$; however, $\mathcal{T}u = u$ on the waveguide end $\Upsilon = \{x : x' \in \omega, x_n = H(x')\}$ (see Remark 1.1).

The following assertion is confirmed with the help of an approach developed in [56] for the Dirichlet problem with the Helmgoltz operator. We denote by $\nu(x') = (\nu_1(x'), \dots, \nu_n(x'))^\top$ the unit vector of the outward normal at the point $(x', H(x')) \in \Upsilon$. It is clear that

$$\nu_n(x') < 0, \quad x' \in \omega. \tag{3.2}$$

Let $u \in H^1(\Omega)^k$ be a solution of the problem (1.10) with a certain number $\lambda \in \mathbb{C}$. We recall the properties of this solution assured by the theory of elliptic boundary value problems in domains with piecewise smooth boundaries (see the key works [25, 36, 37, 44] and monographs [53, 28]). First, under the above-mentioned smoothness conditions, the weak solution $u \in H^1(\Omega)^k$ falls into the space $H^2_{loc}(\overline{\Omega} \setminus \vartheta)^k$, i.e., the solution cannot be classical only at the edge $\vartheta = \{x : x' \in \partial\omega, x_n = H(x')\}$ (see, for example, [53, Subsects. 8.4 and 10.1]). At the same time, owing to the Dirichlet conditions on the end Υ, we find a positive number δ_ϑ depending on $\mathcal{A}$, $\mathcal{T}$ and ω, H (see [53, Theorems 8.4.4 and 8.4.9] and [44, Theorem 2.11]) such that

$$r^{p-1-\delta_\vartheta} \nabla^p_x u \in L_2(\Omega), \tag{3.3}$$

where r is a positive functions in $\overline{\Omega} \setminus \vartheta$ which is equal to one for $x_n > R$ and is equivalent to the distance to the edge in a neighborhood of ϑ. Finally, according to Proposition 2.1, the solution lies in the space $W^2_\beta(\Omega \setminus \mathbb{B}_R)^k$ with exponent $\beta > 0$, i.e., it decays exponentially at infinity.

Proposition 3.1. *Let*

$$\delta_\vartheta > 1/2. \tag{3.4}$$

Then any vector-valued function $u \in H^1(\Omega)^k$ satisfying the problem (1.10) *with a certain $\lambda \in \mathbb{C}$ vanishes in Ω everywhere.*

Proof. The assertion, of course, must be proved for $\lambda \in \overline{\mathbb{R}_+}$ only (see Subsect. 1.4). We insert the vector-valued functions u and $\mathcal{U} = \partial u/\partial x_n$ into the Green formula (1.9) in the domain $\Omega_d = \{x \in \Omega : r > d\}$ with a small $d > 0$. Since the coefficients of differential operators do not depend on the variable x_n, the derivative $\mathcal{U}$ satisfies the system of differential equations in the set Ω_d and the boundary conditions on the surface $\partial\Omega \setminus (\Upsilon \cup \overline{\Omega_d})$ in the problem (1.5), (1.6). We derive that

$$\begin{aligned}
&\mathrm{Re} \int\limits_{\partial\Omega_d \cap \Omega} \left(\overline{\mathcal{D}(\nu(x))\mathcal{U}(x)}^\top \mathcal{A}(x')\mathcal{D}(\nabla_x)u(x) \right) ds_x \\
&= \mathrm{Re} \int\limits_{\Omega_d} \left(\overline{\mathcal{D}(\nabla_x)\mathcal{U}(x)}^\top \mathcal{A}(x')\mathcal{D}(\nabla_x)u(x) - \overline{\mathcal{U}(x)}^\top \mathcal{B}(x')u(x) \right) dx \\
&= \frac{1}{2} \int\limits_{\Omega_d} \frac{\partial}{\partial x_n} \left(\overline{\mathcal{D}(\nabla_x)u(x)}^\top \mathcal{A}(x')\mathcal{D}(\nabla_x)u(x) - \overline{u(x)}^\top \mathcal{B}(x')u(x) \right) dx \\
&= \frac{1}{2} \int\limits_{\partial\Omega_d \cap \Omega} \nu_n(x) \left(\overline{\mathcal{D}(\nabla_x)u(x)}^\top \mathcal{A}(x')\mathcal{D}(\nabla_x)u(x) - \overline{u(x)}^\top \mathcal{B}(x')u(x) \right) ds_x \\
&+ \frac{1}{2} \int\limits_{\Upsilon \setminus \Omega_d} \nu_n(x) \left(\overline{\mathcal{D}(\nabla_x)u(x)}^\top \mathcal{A}(x')\mathcal{D}(\nabla_x)u(x) - \overline{u(x)}^\top \mathcal{B}(x')u(x) \right) ds_x.
\end{aligned} \tag{3.5}$$

A well-known result on traces of functions in weighted spaces (see [25] and, for example, [28]) demonstrates that formula (3.3) guarantees the inclusions

$$r^{-\delta_\vartheta + p - 1/2} \nabla_x^p u \in L_2(\Upsilon), \quad p = 0, 1,$$

which, in view of the condition (3.4), provides the convergence of the last integral at $d = 0$. Furthermore, the norms

$$\|r^{-\delta_\vartheta + p - 1/2} \nabla_x^p u; L_2(\partial\Omega_d \cap \Omega)\|, \quad p = 0, 1,$$

are uniformly bounded with respect to the parameter d and, therefore, the integrals over the small surfaces $\partial\Omega_d \cap \Omega$ in (3.5) are equal to $O(d^{2\delta_\vartheta - 1})$ and vanish after the limit passage $d \to +0$.

Thus, the equality (3.5) is left with the only integral which, thanks to the Dirichlet conditions on the surface Υ, takes the form

$$\frac{1}{2} \int\limits_{\Upsilon} \nu_n(x) \overline{\mathcal{D}(\nabla_x)u(x)}^\top \mathcal{A}(x')\mathcal{D}(\nabla_x)u(x)\, ds_x. \tag{3.6}$$

By virtue of formula (3.2) and the positive definiteness of the matrix $\mathcal{A}$, the fact that the integral (3.6) becomes null due to the limit passage in (3.5) leads to the equality $\mathcal{D}(\nabla_x)u(x) = 0$, $x \in \Upsilon$. Hence the solution u itself and its gradient $\nabla_x u$ vanish on the end Υ (see Lemma 3.1 below). That is why the vector-valued function u extended by zero from Ω onto Π belongs to the space $W^1_\beta(\Pi)^k$ with any negative exponent β and satisfies the homogeneous model problem (2.4). As a result, we conclude that $u = 0$ by virtue of Proposition 2.2. □

Lemma 3.1. *If $u = 0$ and $\mathcal{D}(\nabla_x)u = 0$ on the surface Υ, then $\nabla_x u(x) = 0$, $x \in \Upsilon$.*

Proof. We fix a point $x' \in \omega$. Without loosing generality, we can assume that $\nu(x') = -e_n$ (otherwise, we turn the Cartesian coordinate system while preserving the algebraic completeness of the matrix $\mathcal{D}$). According to the Dirichlet conditions $u(x) = 0$, $x \in \Upsilon$, and the sufficient smoothness of the surface Υ, we have $\nabla_{x'} u(x', z)\big|_{z=H(x')} = 0$. Hence the equality $\mathcal{D}(\nabla_x)u = 0$ at the point $(x', H(x'))$ turns into the following one:

$$\mathcal{D}(e_n)\frac{\partial u}{\partial x_n}(x', H(x')) = 0, \quad e_n = (0, \dots, 0, 1)^\top. \tag{3.7}$$

The $(N \times k)$-matrix $\mathcal{D}(e_n)$ has rank k. Indeed, if a column $b \in \mathbb{C}^k$ satisfies $\mathcal{D}(e_n)b = 0 \in \mathbb{C}^N$, then the linear space (1.15) of vector polynomials contains the vector-valued function $b\varphi(x_n)$ which surely is not a polynomial for $\varphi \in C^\infty_c(\mathbb{R})$, $\varphi \neq 0$. Now, from formula (3.7) we derive that $\dfrac{\partial u}{\partial x_n}(x', H(x')) = 0$ and thus complete the proof. □

3.2 Concentration of the discrete spectrum

Let $\mathcal{A}$ and $\mathcal{B}$ be number matrices, and let the domain Ω be still determined by formula (3.1), but, on the end Υ and the lateral side $\partial\Omega \setminus \overline{\Upsilon}$, we impose the Neumann and Dirichlet boundary conditions respectively (see (1.11)). According to Proposition 2.5, the spectrum of the problem (1.5), (1.6) has the positive cut-off $\lambda^\dagger$ and the interval $(0, \lambda^\dagger)$ can contain the discrete spectrum only. We are going to show that, in contrast to Proposition 3.1, for any given numbers $l > 0$ and $N > 0$, real and integer, one may find a profile function $H \in C^\infty(\overline{\omega})$ such that the interval $(0, l)$ includes at least N eigenvalues of the problem (1.10). This fact implies the concentration of the discrete spectrum.

Let the ε-neighborhood $\mathbb{B}^{n-1}_\varepsilon$ of the point $x' = 0$ belong to the cross-section ω and let $h \in C^\infty[0, \varepsilon]$ be a strict monotone increasing function on the segment $[1/3, 2/3]$ while $h(r) = 0$ for $r \in [0, 1/3]$ and $h(r) = 1$ for $r \in [2/3, 1]$. We set

$$H(x') = \begin{cases} 0, & x' \in \omega \setminus \mathbb{B}^{n-1}_\varepsilon, \\ -\tau h(\varepsilon^{-1}|x'|), & x' \in \mathbb{B}^{n-1}_\varepsilon, \end{cases}$$

where $\tau > 0$ and $\varepsilon > 0$ are large and small parameters. We apply the maxmin principle [1, Theorem 10.2.2] for the operator $-\mathcal{K}$ (with minus) defined by formula (1.21):

$$-\mu^{(p)} = \max_{\mathcal{H}_p \subset \mathcal{H}} \inf_{v \in \mathcal{H}_p \setminus \{0\}} \frac{\langle -\mathcal{K}v, v\rangle}{\langle v, v\rangle}. \tag{3.8}$$

Here, $\mathcal{H}_p$ is an arbitrary subspace in $\mathcal{H}$ of co-dimension $p-1$ (i.e., $\dim(\mathcal{H} \ominus \mathcal{H}_p) = p-1$ and, in particular, $\mathcal{H}_1 = \mathcal{H}$) while $-\mu_1 \leqslant \cdots \leqslant -\mu_P$ imply the list (maybe, empty or infinite) of eigenvalues of the operator $-\mathcal{K}$ in the segment $[-1, -\mu^\dagger)$ containing only the discrete spectrum by Theorem 2.1.

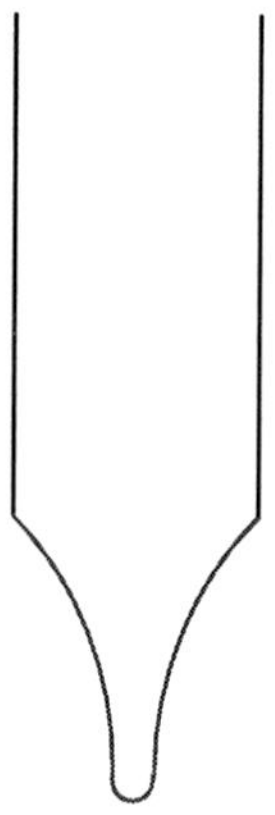

Fig. 4 The waveguide with the needle.

We take a test vector-valued function v which is equal to $V(-\tau^{-1}x_n)$ for $x_n < 0$ and zero for $x_n > 0$, where V is a smooth vector function on the segment $[0,1]$ and $V(0) = 0$. We have

$$\begin{aligned}\langle \mathcal{K}v, v\rangle &= \varepsilon^{n-1} \int_{-\tau}^{0} m(\tau^{-1}x_n)\, \overline{V(\tau^{-1}x_n)}^\top \mathcal{B}V(\tau^{-1}x_n)\, dx_n \\ &= \varepsilon^{n-1}\tau \int_0^1 m(z)\, \overline{V(z)}^\top \mathcal{B}V(z)\, dz = \varepsilon^{n-1}\tau I_0(V,V),\end{aligned}$$

$$\begin{aligned}
&\langle v, v\rangle \\
&= \varepsilon^{n-1} \int\limits_{-\tau}^{0} m(\tau^{-1}x_n) \big(\overline{\mathcal{D}(\nabla_x)V(\tau^{-1}x_n)}^{\top} \mathcal{A}\mathcal{D}(\nabla_x)V(\tau^{-1}x_n) \\
&+ \overline{V(\tau^{-1}x_n)}^{\top} \mathcal{B}V(\tau^{-1}x_n)\big)\, dx_n \\
&= \varepsilon^{n-1}\tau \int\limits_{0}^{1} m(z) \left(\tau^{-2}\overline{\mathcal{D}(e_n)\partial_z V(z)}^{\top} \mathcal{A}\mathcal{D}(e_n)\partial_z V(z) + \overline{V(z)}^{\top} \mathcal{B}V(z)\right) dz \\
&= \varepsilon^{n-1}\tau \left(\tau^{-2} I_1(V,V) + I_0(V,V)\right),
\end{aligned}$$

where $m(z) = \text{meas}\,_{n-1}\left(\mathbb{B}^{n-1}_{g(z)}\right)$, $e_n = (0,\dots,0,1)^{\top}$ is the unit vector of the axis x_n, and $[0,1] \ni z \mapsto g(z)$ is the inverse function for the profile function $[1/3, 2/3] \ni r \mapsto f(r)$. Hence

$$\frac{\langle -\mathcal{K}v, v\rangle}{\langle v, v\rangle} = \frac{I_0(V,V)}{\tau^{-2}I_1(V,V) + I_0(V,V)}. \tag{3.9}$$

Since no restrictive condition is imposed on the vector-valued function V, we can choose the set $\{V^1,\dots,V^N\}$ such that

$$I_0(V^j, V^k) = I_1(V^j, V^k) = 0, \quad j \neq k, \quad I_0(V^j, V^j) = 1.$$

Any subspace $\mathcal{H}_N$ includes nontrivial linear combination v of the vector-valued functions v^q,constructed from V^q according to the above rule, namely:

$$v = a_1 v^1 + \dots + a_N v^N, \quad |a_1|^2 + \dots + |a_N|^2 = 1.$$

As a result, we find that the right-hand side of (3.9) does not exceed the quantity

$$-(1+\tau^{-2}I_1)^{-1} \leqslant -1 + \tau^{-2} I_1.$$

It suffices to fix a large parameter τ such that $-1+\tau^{-2}I_1 \leqslant -(1+l)^{-1} < -(1+\lambda^{\dagger})^{-1} = -\mu^{\dagger}$ and to conclude, by [1, Theorem 10.2.2], that at least N eigenvalues of the operator $-\mathcal{K}$ fall into the segment $[-1, -(1+l)^{-1})$. By the relationship (1.23), we find out on the segment $[0,l) \subset [0,\lambda^{\dagger})$ the same number of eigenvalues of the problem (1.10) (or (1.5), (1.6)).

We emphasize that the discovered concentration of eigenvalues in the vicinity of the point $\lambda = 0$ is stimulated by the needle at the end of the waveguide (see Fig. 4); this needle is long (the large parameter τ) and thin (the small parameter ε). It is known (see, for example, [39, 44] and [45, Subsect. 5.2]) that for many elliptic systems, in particular, the elasticity system, a family of test functions with necessary properties can be constructed with the help of only one parameter ε (a thin needle of the unit length). To construct such test functions, it is sufficient that the linear space $\mathcal{P}$ includes a nontrivial polynomial of the variable x_n (see [51]).

3.3 Comparison principles

The consideration in the previous subsection is based on a simple and elegant idea [18], which is explained here for the Dirichlet problem (1.5), (1.6), i.e., $\mathcal{T}u = u$. If the spectral problem in the truncated domain (see the proof of Proposition 3.1)

$$\begin{aligned} &\mathcal{L}(x, \nabla_x)u^{R^0}(x) = \lambda_{R^0}\mathcal{B}(x)u^{R^0}(x), \quad x \in \Omega_{R^0}, \\ &u^{R^0}(x) = 0, \quad x \in \partial\Omega_{R^0}, \end{aligned} \tag{3.10}$$

has an eigenvalue λ_{R^0} in the interval $(0, \lambda^\dagger)$, then the same interval gets a point of the discrete spectrum of the Dirichlet problem (1.5), (1.6) in the unbounded domain. The verification of this fact uses the maxmin principle with the test function obtained by the null extension of the eigenfunction u^{R^0} from Ω_{R^0} onto Ω.

Clearly, the number of eigenvalues of the problem (3.10) in the interval $(0, \lambda^\dagger)$ estimates from below the total multiplicity $\varkappa^\dagger$ of the discrete spectrum of the operator $\mathcal{K}$ in the segment $(\mu^\dagger, 1]$. The estimate of $\varkappa^\dagger$ from above is given by the number of eigenvalues in the segment $[0, \lambda^\dagger)$ for the mixed boundary value problem

$$\begin{aligned} &\mathcal{L}(x, \nabla_x)v^{R^0}(x) = \lambda_{R^0}\mathcal{B}(x)v^{R^0}(x), \quad x \in \Omega_{R^0}, \\ &v^{R^0}(x) = 0, \quad x \in \partial\Omega_{R^0} \setminus \partial\Omega, \\ &\mathcal{N}(x, \nabla_x)v^{R^0}(x) = 0, \quad x \in \partial\Omega_{R^0} \cap \partial\Omega. \end{aligned} \tag{3.11}$$

The reason is the same as above; namely, the maxmin principle for the problem (3.11) operator with the test functions $v^{R^0,j}$ obtained by the restriction on Ω_{R^0} of the eigenfunctions u^j of the Dirichlet problem (1.5), (1.6). Note that, in the first case, the orthogonality conditions for eigenfunctions of the problem (3.10) are directly passed over to test functions in the domain Ω while, in the second case, it is necessary to verify the linear independence of the restrictions $v^{R^0,j}$; however, it usually becomes a simple task.

The two above-described tricks are often referred to as the comparison principles (see [3, 2]). They require the positive cut-off $\lambda^\dagger$ and do not work in the case $\lambda^\dagger = 0$.

General boundary conditions do not need a new argument. Note that the problems (3.10) and (3.11) must get on the artificial surface $\partial\mathbb{B}_{R^0} \cap \Omega$ the Dirichlet and Neumann boundary conditions respectively. The shape of the truncation surface may be arbitrary.

3.4 Artificial boundary conditions

To detect the point spectrum in the continuous one (the case $\lambda^\dagger = 0$), it is helpful to impose the artificial boundary conditions on the planes of geometrical symmetry and apply the comparison principles. This approach was proposed in [8].

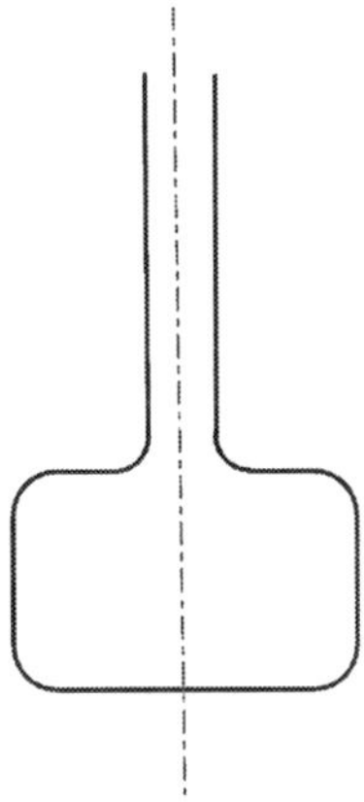

Fig. 5 The symmetric waveguide.

Let $\Pi = \omega \times \mathbb{R}$, and let a domain Ω be symmetric with respect to the plane $\{x : x_1 = 0\}$. We consider the Neumann problem for the Helmgoltz equation

$$\begin{aligned} &-\nabla_x^\top \nabla_x u(x) = \lambda u(x), \quad x \in \Omega, \\ &\mathcal{N}(x, \nabla_x) u(x) := \nu(x)^\top \nabla_x u(x) = 0, \quad x \in \partial\Omega, \end{aligned} \tag{3.12}$$

and the mixed boundary value problem, posed on the half $\Omega_+ = \{x \in \Omega : x_1 > 0\}$ of the domain and supplied with the Dirichlet conditions on the artificial boundary $\Theta = \{x \in \Omega : x_1 = 0\}$ (see Fig. 5),

$$\begin{aligned} &-\nabla_x^\top \nabla_x u_+(x) = \lambda_+ u_+(x), \quad x \in \Omega_+, \\ &\nu(x)^\top \nabla_x u_+(x) = 0, \quad x \in \partial\Omega_+ \setminus \overline{\Theta}, \\ &u_+(x) = 0, \quad x \in \Theta. \end{aligned} \tag{3.13}$$

According to Propositions 2.4 and 2.5, the continuous spectrum of the problem (3.12) coincides with the real semiaxis $[0, \infty)$, but the continuous spectrum of the problem (3.13) implies the ray $[\lambda_+^\dagger, \infty)$ while $\lambda_+^\dagger > 0$ is the first eigenvalue of the model problem on the cross-section

$$
\begin{aligned}
&-\nabla_{x'}^{\top}\nabla_{x'}U_+(x') = \Lambda_+U_+(x'), \quad x' \in \omega_+,\\
&\nu(x')^{\top}\nabla_{x'}U_+(x') = 0, \quad x' \in \partial\omega_+,\, x_1 > 0,\\
&U_+(x') = 0,\, x' \in \omega,\, x_1 = 0.
\end{aligned}
$$

Assume that the interval $(0, \lambda_+^{\dagger})$ contains a point of the discrete spectrum of the problem (3.13). We extend the corresponding eigenfunction from ω_+ onto ω with the property of being odd and obtain an eigenfunction of the problem (3.12) with the same eigenvalue. In other words, we detect a point of the point spectrum in the continuous one. The comparison principle permits for estimating the number of eigenvalues in the interval $(0, \lambda_+^{\dagger})$ (see [23]). As was mentioned at the beginning of the subsection, the desired eigenvalue can be found, for example, in the case where the first eigenvalue of the problem

$$
\begin{aligned}
&-\nabla_{x}^{\top}\nabla_{x}u_0(x) = \lambda_0u_0(x), \quad x \in \Omega_+^-,\\
&\nu(x)^{\top}\nabla_{x}u_0(x) = 0, \quad x \in \partial\Omega_+^- \cap \partial\Omega,\\
&u_0(x) = 0,\, x \in \partial\Omega_+^- \setminus \partial\Omega,
\end{aligned}
$$

is less than $\lambda_+^{\dagger}$. Here, $\Omega_+^- = \{x \in \Omega : x_1 > 0, x_n < 0\}$ is the overshadowed subdomian in Fig. 6.

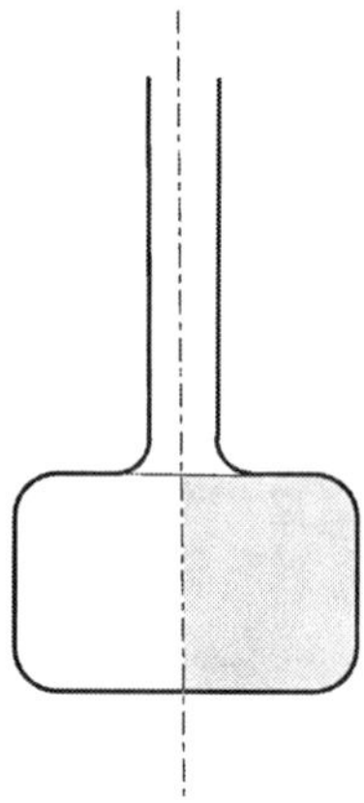

Fig. 6 The artificial boundary in the symmetric waveguide.

Clearly, the same trick is fit for quasicylinders in $\mathbb{R}^n$ (see Fig. 7) and scalar differential operators with periodic coefficients under an evident restriction. It is not easy to apply the approach [8] of artificial boundary conditions to systems of differential equations because the odd extension of a vector-valued function from Ω_+ onto Ω usually violates the system of differential equations on the artificial surface Θ. This violation, for example, happens in the two-dimensional elasticity system for the isotropic half-strip $\Pi = (-l, l) \times (0, +\infty)$. That is why, in addition to the artificial boundary conditions

$$u_1(0, x_2) = 0, \quad \sigma_{12}(u; 0, x_2) = 0, \quad x_2 > 0, \tag{3.14}$$

the paper [58] employs new arguments to verify that a trapped mode exists in the isotropic half-strip $\Pi_+ = (-1/2, 1/2) \times (0, +\infty)$ with the null Poisson ratio and the Neumann (traction-free) boundary conditions on the boundary. The artificial boundary conditions (3.14) annuls on the ray $\Theta = \{x = (x_1, x_2) : x_1 = 0, x_2 > 0\}$ the *normal* displacement u_1 and the *tangent component* $\sigma_{12}(u)$ of traction (see Example 1.2 and, in particular, the definition (1.19) of the stress column). It is remarkable that the odd extension of u_1 and the even extension of u_2 from the thin half-strip $\Pi_+^+ = (0, 1/2) \times (0, +\infty)$ onto the whole half-strip Π_+ keep the Lamé system valid in Π_+ due to the complete symmetry of elastic properties. At the same time, the boundary conditions (3.14) make the subspace (2.41) one-dimensional, but nontrivial (the constant vector $(0, c)$ is annulled by the Neumann conditions as well) and, therefore, the positive cut-off $\lambda^\dagger$ does not occur.

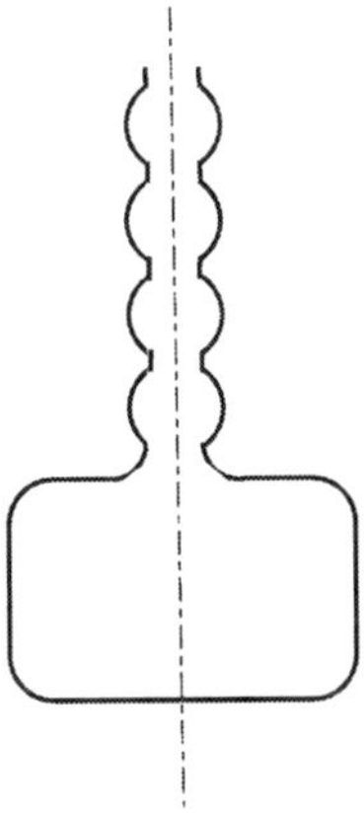

Fig. 7 The symmetric waveguide with the periodic cross-section.

For three-dimensional elastic solids the paper [48] introduces another artificial boundary conditions. For the sake of simplicity, we assume that the straight cylinder Π has circle cross-section $\omega = \mathbb{B}^2_{r0}$ and is filled with the homogeneous and isotropic material. We create the artificial surfaces $\Theta_0 = \{x : r > 0, \varphi = 0\}$ and $\Theta_\Phi = \{x : r > 0, \varphi = \Phi\}$, where (r, φ, z) are cylindrical coordinates and $\Phi \in (0, \pi]$. We impose the following artificial boundary conditions:

$$u_\varphi = 0, \ \sigma_{r\varphi} = \sigma_{z\varphi} = 0, \quad \text{on} \quad \Theta_0, \tag{3.15}$$

$$u_r = u_z = 0, \quad \sigma_{\varphi\varphi} = 0, \quad \text{on} \quad \Theta_\Phi. \tag{3.16}$$

Here, u_r, u_z, and u_φ stand for the projections of the displacement vector on the axes r, z, and φ respectively and in the analogy the tangential $\sigma_{r\varphi}$, $\sigma_{z\varphi}$ and normal $\sigma_{\varphi\varphi}$ projections of traction are determined. The even extension for the displacement components u_r, u_z and the odd extension for u_φ through the half-plane Θ_0 keep the validity of the Lamé system. The same is true for the odd extensions of u_r, u_z and the even extension of u_φ through the half-plane Θ_Φ. Hence, in the case $\Phi = \pi/k$ with $k \in \mathbb{N}$, a solution of the problem with the boundary conditions (3.15) on $\Pi_+ \cap \Theta_0$ and (3.16) on $\Pi_+ \cap \Theta_\Phi$ is extended smoothly to the whole cylinder Π_+.

It is straightforward to confirm that for $\Phi \in (0, \pi/2)$, therefore, for $\Phi = \pi/k$ and $k \geqslant 3$ the rigid motion $a + x \times b$ subject to the artificial boundary conditions (3.15) and (3.16) vanishes (see Example 1.2). In other words, the condition in Proposition 2.5 is fulfilled and the spectrum of the elasticity problem in the sectorial part $\{x \in \Pi_+ : \varphi \in (0, \pi/k)\}$ of the cylinder acquires the positive cut-off $\lambda^\dagger$.

The trick described above provides in [48] the following fact. For any given $l > 0$ and $N \in \mathbb{N}$ the interval $(0, l)$ in the continuous spectrum of the cylindrical elastic waveguide with damping gasket, which is hard and heavy and has a sharp edge (see Fig. 8) can be filled with at least N eigenvalues by a proper choice of the physical properties of the gasket. In the analogy with Subsect. 3.2, we can speak about the concentration of the point spectrum in the continuous spectrum.

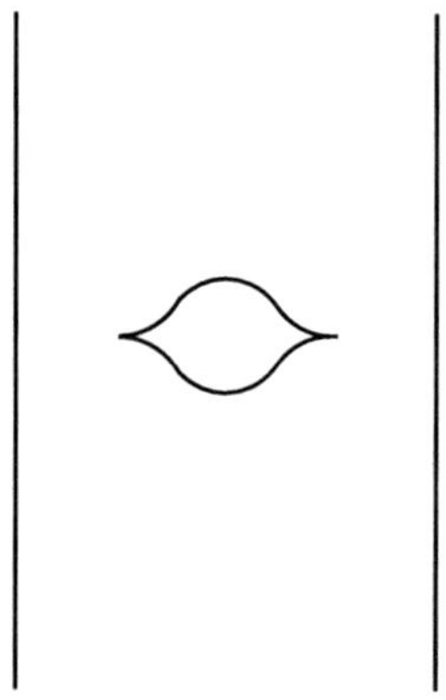

Fig. 8 The gasket with sharp edge in the elastic waveguide.

We emphasize that the artificial boundary conditions (3.15), (3.16) are fit for a waveguide with three planes of elastic and geometrical symmetry, i.e., the solid Ω obtained by perturbation of the cylinder Π_+ inside of the ball $\mathbb{B}_{R^0}$ (see Subsect. 1.2) can be anisotropic and the cross-section ω can differ from a circle. Sadly enough, the described trick cannot be used in the two-dimensional case.

3.5 Opening gaps in the continuous spectrum

It is known (see, for example, [9, 15, 17, 11, 62, 10]) that a gap opens in the continuous spectrum of the second order operator $\mathcal{L}(x, \nabla_x) = -\nabla_x^\top \mathcal{A}(x)\nabla_x$ and the Maxwell system with periodic coefficients in the whole space $\mathbb{R}^n$. Examples of such gaps are constructed by varying the coefficients of differential operators. In [50], an elastic waveguide is constructed (see Example 1.2) which has a gap in the continuous spectrum as well. We comment on the method while dealing with the Helmgoltz equation (see Example 1.1).

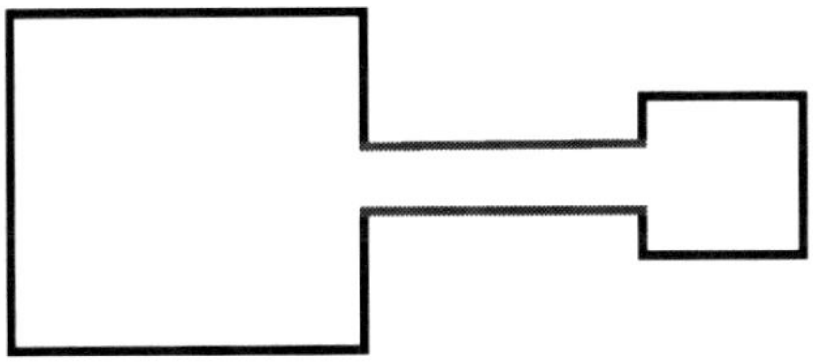

Fig. 9 The periodicity cell for the Helmgoltz equation.

Let $n = 2$, and let the periodicity cell Σ be the union of the squares $\Sigma_- = (-1, 0) \times (0, 1)$, $\Sigma_+ = (3/4, 5/4) \times (1/4, 3/4)$ and the thin ($h > 0$ is a small parameter) rectangle $\Sigma_h = \{x : x_1 \in (-1, 5/4), |x_2 - 1/2| < h/2\}$ (see Fig. 9). We impose the Dirichlet conditions on the left side $\sigma = \{-1\} \times (0, 1)$ of the bigger square, the periodicity conditions on its bases $(-1, 0) \times \{0\}$ and $(-1, 0) \times \{1\}$, and the Neumann conditions on the remaining part of the boundary of the junction.

Lemma 3.2. *If a function $u \in H^1(\Sigma)$ vanishes on σ, then the inequality*

$$\|u; L_2(\Sigma_- \cup \Sigma_h)\|^2 + \|u_\perp; L_2(\Sigma_+)\|^2 + h\,|\overline{u}|^2 \leqslant c\|\nabla_x u; L_2(\Sigma)\|^2 \qquad (3.17)$$

is valid with a constant c independent of the parameter $h \in (0, 1/4)$ and the function u. Here, $u_\perp$ and $\overline{u}$ are components in the decomposition

$$u(x) = u_\perp(x) + \overline{u}, \quad \overline{u} = 4 \int\limits_{\Sigma_+} u(x)\, dx, \quad \int\limits_{\Sigma_+} u_\perp(x)\, dx = 0. \tag{3.18}$$

Proof. By the condition $u(x_1, -1) = 0$, the Friedrichs inequalities

$$\|u; L_2(\Sigma_-)\|^2 \leqslant \frac{\pi^2}{4} \|\nabla_x u; L_2(\Sigma_-)\|^2,$$

$$\|u; L_2(\Sigma_h)\|^2 \leqslant \frac{81}{16} \pi^2 \|\nabla_x u; L_2(\Sigma_h)\|^2$$

are valid. By the Poincaré inequality,

$$\|u_\perp; L_2(\Sigma_+)\|^2 \leqslant 8\pi^2 \|\nabla_x u_\perp; L_2(\Sigma_+)\|^2 = 8\pi^2 \|\nabla_x u; L_2(\Sigma_+)\|^2.$$

It suffices to estimate the mean value $\overline{u}$ which, according to the decomposition (3.18), is calculated as follows:

$$\frac{h}{2}\, \overline{u} = \int\limits_{\Sigma_h \cap \Sigma_+} (u(x) - u_\perp(x))\, dx.$$

Namely, the Schwarz inequality gives the relation

$$|\overline{u}|^2 \leqslant \frac{4}{h^2} \frac{h}{2} 2 \int\limits_{\Sigma_h \cap \Sigma_+} (|u(x)|^2 + |u_\perp(x)|^2)\, dx \leqslant \frac{c}{h} \|\nabla_x u; L_2(\Sigma)\|^2,$$

which completes the proof of the lemma. □

Let $\overset{\circ}{H}{}^1_\perp(\Sigma; \sigma)$ be the space of functions $u_\perp \in H^1(\Sigma)$ vanishing on σ and satisfying the orthogonality condition in (3.18). By the equality $\overline{u_\perp} = 0$ and the estimate (3.17), we obtain the relation

$$\|u; L_2(\Sigma)\|^2 \leqslant c(\Sigma, \sigma) \|\nabla_x u; L_2(\Sigma)\|^2, \quad u \in \overset{\circ}{H}{}^1_\perp(\Sigma; \sigma). \tag{3.19}$$

We apply the maxmin principle (see [1, Theorem 10.2.2]) for the mixed boundary value problem with the formal self-adjoint differential operator $-\partial_1^2 - (\partial_2 + i\eta)^2$, more precisely, for the unbounded self-adjoint operator in the Lebesgue space $L_2(\Sigma)$ generated [1, Subsect. 10.1] by the corresponding Hermitian form:

$$\Lambda^{(p)}(\eta) = \sup_{\mathcal{H}_p \subset \overset{\circ}{H}_{per}(\Sigma;\sigma)} \ \inf_{U \in \mathcal{H}_p \setminus \{0\}} \frac{\|\partial_1 U; L_2(\Sigma)\|^2 + \|(\partial_2 + i\eta) U; L_2(\Sigma)\|^2}{\|U; L_2(\Sigma)\|^2}. \tag{3.20}$$

As in formula (3.8), $p \in \mathbb{N}$ and $\mathcal{H}_p$ is a subspace in $\overset{\circ}{H}_{per}(\Sigma; \sigma)$ with codimension $p - 1$.

We set $p = 2$ and $\mathcal{H}_2 = \{U \in \mathring{H}^1_{per}(\Sigma;\sigma) : u = \exp(-i\eta x_2)U \in \mathring{H}^1_\perp(\Sigma;\sigma)\}$ while diminishing the right-hand side of (3.20). By the inequality (3.19), we find

$$\Lambda^{(2)}(\eta) \geqslant \inf_{u \in \mathring{H}^1_\perp(\Sigma;\sigma)} \frac{\|\nabla_x u; L_2(\Sigma)\|^2}{\|u; L_2(\Sigma)\|^2} \geqslant c(\Sigma,\sigma)^{-1}.$$

Considering formula (3.20) with $p = 1$ and $\mathcal{H}_1 = \mathring{H}_{per}(\Sigma;\sigma)$, we take the test function $U = \exp(i\eta x_2)v$, where $v(x) = 1$ for $x_1 > 1/2$ and $v(x) = 0$ for $x_1 < 0$. Note that $v = 1$ and $\nabla_x v = 0$ in Σ_+. Thus, increasing the right-hand side of (3.20), we derive that

$$\Lambda^{(1)}(\eta) \leqslant \frac{ch}{3h+1} \leqslant Ch.$$

Let us fix a small parameter $h > 0$ such that $c(\Sigma,\sigma)^{-1} > Ch$. The obtained relation

$$\max\{\Lambda^{(1)}(\eta) \,|\, \eta \in [0, 2\pi)\} < \min\{\Lambda^{(2)}(\eta) \,|\, \eta \in [0, 2\pi)\}$$

takes the form of the relation (2.36) with $q = 1$ and, therefore, implies opening a gap in the continuous spectrum of the mixed boundary value problem for the Helmgoltz equation in the quasicylinder with the periodicity cell in Fig. 9. Namely, the segment Υ_1 cannot intersect the segment Υ_2 (see Subsect. 2.4).

The described constructions in a quasicylinder Ω with the dumbbells-shaped cell Σ consisting of the two massive bodies $\Sigma_\pm$ jointed by the thin ligament Σ_h of diameter $O(h)$, can be employed for the problem (1.5), (1.6) in a general formulation under the following three conditions. First, a part σ of the lateral side of the body Σ_- must be supplied with the Dirichlet boundary conditions providing the positive cut-off $\lambda^\dagger$ and the remaining part of the boundary with the Neumann conditions. Second, one needs the Korn inequality analogous to (1.16) and (3.19) for vector-valued functions $u \in \mathring{H}^1(\Sigma;\sigma)^k$ subject to the orthogonality conditions

$$\int_{\Sigma_+} p(x)^\top u(x)\,dx = 0, \quad p \in \mathcal{P},$$

where $\mathcal{P}$ is the linear space of vector polynomials (1.15) figuring in the polynomial property (1.14). Third, one has to detect test functions which coincide with polynomials from $\mathcal{P}$ in the body Σ_+, vanish in the body Σ_-, and generate the small energy $a(v, v; \Sigma_h)$ of the ligament. A procedure to create such test functions is known (see [39, § 3]), and the only obstacle to prove the existence of gaps for general matrix differential operator (1.4) becomes the absence of a general Korn inequality on the ligament Σ_h which is asymptotically sharp with respect to the small parameter. Such an inequality is known for the elasticity problem (see [40, 46, 45, 49]). In [50], the above-described method establishes opening a gap in the continuous spectrum of an elastic spatial waveguide with the periodicity cell in Fig. 10. We emphasize that the gap is guaranteed between the fifth and sixth segments Υ^p, although the linear space $\mathcal{P}$ in elasticity is of dimension six (see Example 1.2). The reason for the removal of one rigid motion, namely the longitudinal translation in the ligament, appears from the anisotropic structure of the weighted Korn inequality on a thin rod (see [49, Sect. 2] for details). In other words, some polynomials in $\mathcal{P}$ do not possess the necessary properties that assist crucially in the complication of a generalization.

The paper [52] gives other periodic elastic inhomogeneous and anisotropic waveguides with opened gaps in the continuous spectrum (see Figs. 11 and 12 with a chain of massive bodies connected either with thin and short ligaments or soft washers). These waveguides get a gap between the first six segments and the seventh one while, in the case of rotational symmetry, it is possible to prove that there is no gap below l_{67} (see Fig. 13).

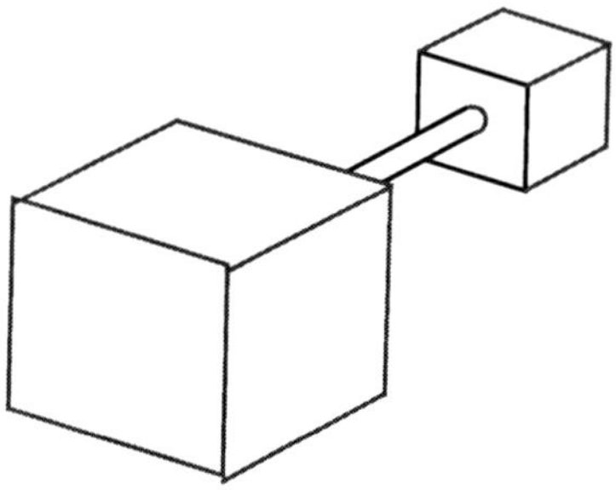

Fig. 10 The periodicity cell opening a gap in the continuous spectrum.

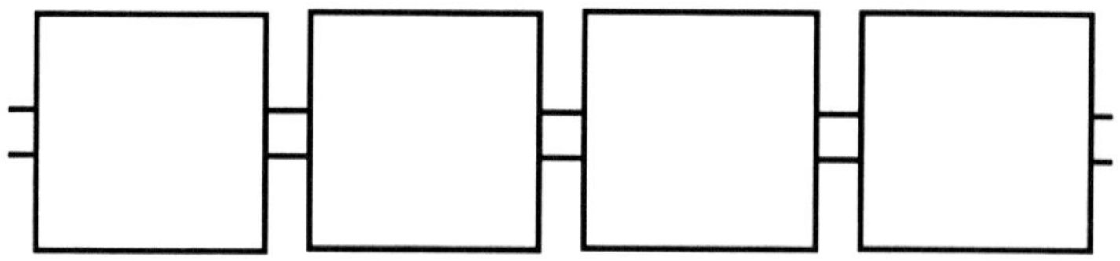

Fig. 11 Massive bodies connected with thin and short ligaments.

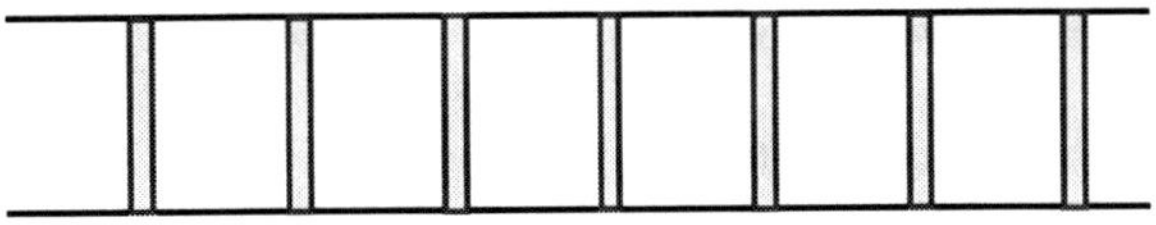

Fig. 12 Massive bodies connected with soft washers.

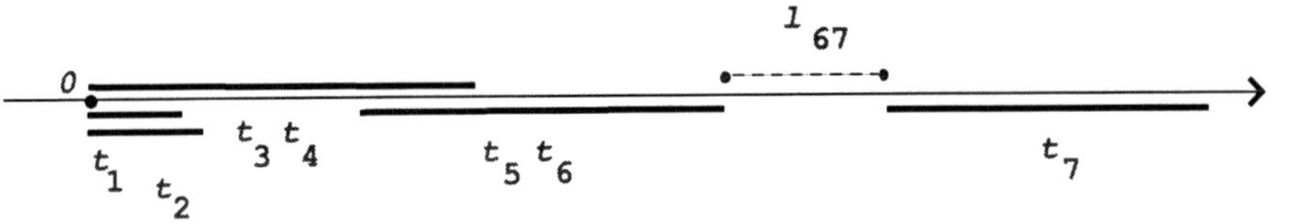

Fig. 13 The gap between the sixth and seventh segments of the continuous spectrum.

3.6 Variational methods for searching trapped modes below the cut-off

We assume that Π is a straight cylinder, the matrices $\mathcal{A}$, $\mathcal{B}$, and $\mathcal{S}^0$ do not depend on the variable x_n, and the inclusions

$$\Pi_+ \subset \Omega \subset \Pi \tag{3.21}$$

are valid. We also choose the boundary conditions (1.6) on the surface $\partial\Pi \cap \Omega$ according to the rule (1.8) and such that the conditions in Proposition 2.5 are satisfied, i.e., the positive cut-off $\lambda^\dagger$ occurs. Furthermore, if $\mathcal{T}_{(q)}(x, \nabla_x) = \mathcal{S}^0_{(q)}(x')$ for $x' \in \gamma_q \subset \partial\omega$, $x_n > 0$ and a certain $q = 1, \dots, k$, then $\mathcal{T}_{(q)}(x, \nabla_x) = \mathcal{S}^0_{(q)}(x')$ without any restriction on x_n (see Remark 1.2).

Let $\eta_0 \in \mathbb{R}$ denote an eigenvalue of the variational problem (2.39) (or the boundary value problem (2.38) under the smoothness assumption on the data), where $\lambda = \lambda^\dagger$. Note that the existence of a real eigenvalue follows from Theorem 2.1. With the corresponding eigenvector U^0, we set

$$u^\varepsilon(x) = \exp(i\eta_0 x_n) U^0(x'). \tag{3.22}$$

If $\varepsilon > 0$, then the inclusion $u^\varepsilon \in \mathcal{H}(\Omega)$ holds, in particular, due to the requirement on the domain and the structure of boundary conditions.

Suppose that the segment $[0, \lambda^\dagger)$ (the segment $(\mu^\dagger, 1])$ is free of the spectrum in the problem (1.10) (in Eq. (1.22)). Then, according to [1, Theorem 10.2.1], the relation

$$\frac{1}{1+\lambda^\dagger} = \mu^\dagger \leqslant \frac{\langle \mathcal{K}u, u\rangle}{\langle u, u\rangle} \tag{3.23}$$

is valid for any test function $u \in \mathcal{H}(\Omega)$, for example, $u = u^\varepsilon$. From the inequality (3.23) and definitions (1.20), (1.21) we derive that

$$0 \geqslant (\mathcal{A}\mathcal{D}(\nabla_x)u^\varepsilon, \mathcal{D}(\nabla_x)u^\varepsilon)_\Omega - \lambda^\dagger(\mathcal{B}u^\varepsilon, u^\varepsilon)_\Omega. \tag{3.24}$$

We compute the integrals on the right-hand side with error $O(\varepsilon)$ while dividing the integration set Ω into the parts Π_+ and $\Omega_- = \Omega\setminus\overline{\Pi_+}$ (see formula (3.21)). Taking into account that $\exp(-\varepsilon x_n) - 1 = O(\varepsilon)$ in the bounded set $\Omega_- \subset \mathbb{B}_{R^0}$ and $\displaystyle\int_0^\infty \exp(-2\varepsilon x_n)\,dx_n = \frac{1}{2\varepsilon}$, we conclude that

$$\begin{aligned}
&(\mathcal{A}\mathcal{D}(\nabla_x)u^\varepsilon, \mathcal{D}(\nabla_x)u^\varepsilon)_{\Omega_-} - \lambda^\dagger(\mathcal{B}u^\varepsilon, u^\varepsilon)_{\Omega_-}\\
&= (\mathcal{A}\mathcal{D}(\nabla_{x'}, i\eta_0)U^0, \mathcal{D}(\nabla_{x'}, i\eta_0)U^0)_{\Omega_-} - \lambda^\dagger(\mathcal{B}U^0, U^0)_{\Omega_-} + O(\varepsilon),\\
&(\mathcal{A}\mathcal{D}(\nabla_x)u^\varepsilon, \mathcal{D}(\nabla_x)u^\varepsilon)_{\Pi_+} - \lambda^\dagger(\mathcal{B}u^\varepsilon, u^\varepsilon)_{\Pi_+}\\
&= \frac{1}{2}\,\varepsilon^{-1}\left((\mathcal{A}\mathcal{D}(\nabla_{x'}, i\eta_0)U^0, \mathcal{D}(\nabla_{x'}, i\eta_0)U^0)_\omega - \lambda^\dagger(\mathcal{B}U^0, U^0)_\omega\right)\\
&- \frac{1}{2}((\mathcal{A}\mathcal{D}(\nabla_{x'}, i\eta_0)U^0, \mathcal{D}(e_n)U^0)_\omega\\
&+ (\mathcal{A}\mathcal{D}(e_n)U^0, \mathcal{D}(\nabla_{x'}, i\eta_0)U^0)_\omega) + O(\varepsilon).
\end{aligned} \tag{3.25}$$

The coefficient on ε^{-1} vanishes by the integral identity (2.39), where $\widehat{u} = \widehat{v} = U^0$, $\eta = \eta_0$ and $\lambda = \lambda^\dagger$. Hence the inequalities (3.24) and (3.23) are surely denied by the test function (3.22) with small $\varepsilon > 0$ in the case

$$(\mathcal{A}\mathcal{D}(\nabla_{x'}, i\eta_0)U^0, \mathcal{D}(\nabla_{x'}, i\eta_0)U^0)_{\Omega_-} - \lambda^\dagger(\mathcal{B}U^0, U^0)_{\Omega_-} > I_\omega, \tag{3.26}$$

where

$$I_\omega = \operatorname{Re}(\mathcal{A}\mathcal{D}(\nabla_{x'}, i\eta_0)U^0, \mathcal{D}(e_n)U^0)_\omega. \tag{3.27}$$

Thus, the inequality (3.26) guarantees the existence of eigenvalues of the problem (1.10) and the operator $\mathcal{K}$ in the segments $[0, \lambda^\dagger)$ and $(\mu^\dagger, 1]$ respectively

In certain situations, where a direct calculation is available, it is easy to confirm the inequality (3.26) for a specific shape of the domain Ω_-. Quite often the quantity (3.27) vanishes (see [19, 20, 21]). If additionally, the domain Ω is given by formula (3.1) with $H \leqslant 0$ (this fulfil the requirement (3.21)), then the left-hand side of (3.26) takes the form

$$J_\omega(H) = (H\mathcal{A}\mathcal{D}(\nabla_{x'}, i\eta_0)U^0, \mathcal{D}(\nabla_{x'}, i\eta_0)U^0)_\omega - \lambda^\dagger(H\mathcal{B}U^0, U^0)_\omega.$$

Applying the integral identity (2.39) with $\widehat{u} = U^0$, $\widehat{v} = HU^0$, $\eta = \eta_0$, and $\lambda = \lambda^\dagger$, we arrive at the equality

$$\begin{aligned} 0 &= (\mathcal{A}\mathcal{D}(\nabla_{x'}, i\eta_0)U^0, \mathcal{D}(\nabla_{x'}, i\eta_0)(HU^0))_\omega - \lambda^\dagger(\mathcal{B}U^0, HU^0)_\omega \\ &= J_\omega(H) + (\mathcal{A}\mathcal{D}(\nabla_{x'}, i\eta_0)U^0, \mathcal{D}(\nabla_{x'}H, i\eta_0)U^0)_\omega. \end{aligned}$$

Thus, to detect an eigenvalue of the problem (1.10) in the segment $[0, \lambda^\dagger)$, it remains to choose a profile function $H \in C^{2,\alpha}(\omega)$ such that $H \geqslant 0$ and

$$(\mathcal{A}\mathcal{D}(\nabla_{x'}, i\eta_0)U^0, \mathcal{D}(\nabla_{x'}H, i\eta_0)U^0)_\omega < 0. \tag{3.28}$$

It could be easy. Indeed, let $\mathcal{L}(\nabla_x) = -\nabla_x^\top \nabla_x$ be the Laplace operator, and let $\mathcal{B} = 1$ (see Example 1.1). We impose the Dirichlet and Neumann conditions on the surfaces $\partial\Omega \cap \partial\Pi$ and Υ respectively. Then $\eta_0 = 0$ whilst $\lambda^\dagger$ and U^0 are the first eigenvalue and the corresponding eigenfunction of the Dirichlet problem in the cross-section

$$-\nabla_{x'}^\top \nabla_{x'} U(x') = \lambda U(x'),\ x' \in \omega, \quad U(x') = 0,\ x' \in \partial\omega.$$

Note that U^0 can be fixed positive in Ω. Clearly, $I_\omega = 0$. Moreover, the inequality (3.28) takes the form

$$0 > (\nabla_{x'}U^0, U^0\nabla_{x'}H)_\omega = -\frac{1}{2}\,(|U^0|^2, \nabla_{x'}^\top \nabla_{x'}H)_\omega. \tag{3.29}$$

The relation (3.29) cannot hold with arbitrary H, however, using the fact that $U^0(x')$ vanishes on the boundary $\partial\omega$, it is straightforward to choose a profile H such that the requirement (3.29) is met (see Fig. 14 where the dotted line shows the rotation axis). As in [20], we take $\omega = (-1/2, 1/2)$ so that $\lambda^\dagger = \pi^2$ and $U^0(x_1) = \cos(\pi x_1)$. After the double integration by parts, the right-hand side of (3.29) converts into

$$\pi^2 \int\limits_{-1/2}^{1/2} H(x_1)\cos(2\pi x_1)\,dx_1.$$

In other words, the condition (3.29) is satisfied in the case where one of the Fourier coefficients of the profile function H stays positive.

If the quantity (3.27) vanishes and $H = 0$ in (3.1), we can continue the analysis. Namely, we insert into the relation (3.23) the test function $u^\varepsilon \pm \delta v$, where $u^\varepsilon(x)$ denotes the expression (3.22), $v \in C_c^\infty(\overline{\Pi_+})^k \cap \mathcal{H}(\Omega)$ and $\delta > 0$ is a new small parameter. Repeating the calculations (3.24)–(3.26) and taking

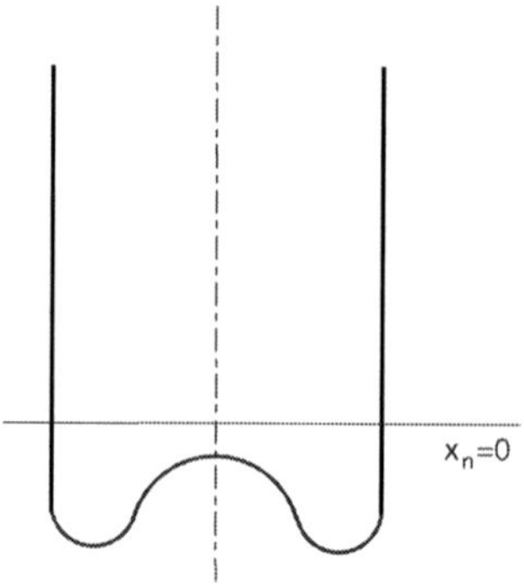

Fig. 14 The waveguide with a trapped mode below the cut-off.

$I_\omega = 0$ and $\Omega_- = \varnothing$ into account, we observe that

$$\begin{aligned}
&(\mathcal{A}\mathcal{D}(\nabla_x)(u^\varepsilon \pm \delta v), \mathcal{D}(\nabla_x)(u^\varepsilon \pm \delta v))_{\Pi_+} - \lambda^\dagger(\mathcal{B}(u^\varepsilon \pm \delta v), u^\varepsilon \pm \delta v)_{\Pi_+} \\
&= (\mathcal{A}\mathcal{D}(\nabla_x)u^\varepsilon, \mathcal{D}(\nabla_x)u^\varepsilon)_{\Pi_+} - \lambda^\dagger(\mathcal{B}u^\varepsilon, u^\varepsilon)_{\Pi_+} \\
&\pm 2\delta \mathrm{Re}\left((\mathcal{A}\mathcal{D}(\nabla_x)u^\varepsilon, \mathcal{D}(\nabla_x)v)_{\Pi_+} - \lambda^\dagger(\mathcal{B}u^\varepsilon, v)_{\Pi_+}\right) + O(\delta^2) \\
&= \pm 2\delta \mathrm{Re}\left((\mathcal{A}\mathcal{D}(\nabla_x)u^0, \mathcal{D}(\nabla_x)v)_{\Pi_+} - \lambda^\dagger(\mathcal{B}u^0, v)_{\Pi_+}\right) \\
&\qquad + O(\varepsilon(1+\delta) + \delta^2) \\
&= \pm 2\delta \mathrm{Re}\left(\int\limits_\omega \overline{v(x',0)}^\top \overline{\mathcal{D}(e_n)}^\top \mathcal{A}(x')\mathcal{D}(\nabla_{x'}, i\eta_0)U^0(x')\,dx'\right) \\
&\qquad + O(\varepsilon(1+\delta) + \delta^2). \qquad (3.30)
\end{aligned}$$

We emphasize that $I_\omega = 0$ by our assumption and, therefore, the second formula in (3.25) yields the relation

$$(\mathcal{A}\mathcal{D}(\nabla_x)u^\varepsilon, \mathcal{D}(\nabla_x)u^\varepsilon)_{\Pi_+} - \lambda^\dagger(\mathcal{B}u^\varepsilon, u^\varepsilon)_{\Pi_+} = O(\varepsilon),$$

which was applied in (3.30), together with the Green formula.

Thus, the inequality (3.23) is violated for $u = u^\varepsilon \pm \delta v$ provided that one finds a point $x' \in \omega$ such that

$$\overline{\mathcal{D}(e_n)}^\top \mathcal{A}(x')\mathcal{D}(\nabla_{x'}, i\eta_0)U^0(x') \neq 0. \qquad (3.31)$$

Indeed, to ratify the inequality

$$\frac{1}{1+\lambda^\dagger} = \mu^\dagger > \frac{\langle \mathcal{K}u, u\rangle}{\langle u, u\rangle}$$

which contradicts (3.23), it suffices to localize the vector-valued function v in a vicinity of the point $(x',0)$ and then to fix small parameters $\varepsilon > 0$ and $\delta > 0$ (the sign $\pm$ on v is to be adjusted with the right-hand side of (3.31)). As a result, under the conditions $I_\omega = 0$ and (3.31), the previous argument ensures the existence of an eigenvalue below the cut-off $\lambda^\dagger$. The condition (3.31) can be weakened by the creation cracks in the domain $\Omega = \Pi_+$ (see Remark 3.1 below).

Finally, if it happens (see [21]) that, at each point x' of the cross-section ω, the equality

$$\overline{\mathcal{D}(\nabla_{x'}, i\eta_0)U^0(x')}^\top \mathcal{A}(x')\mathcal{D}(\nabla_{x'}, i\eta_0)U^0(x') = \lambda^\dagger \overline{U^0(x')}^\top \mathcal{B}(x')U^0(x')$$

is valid, then the calculation (3.30) holds for any domain Ω subject to the condition (3.21) while the requirement (3.31) is replaced with the following:

$$\overline{\mathcal{D}(x)}^\top \mathcal{A}(x')\mathcal{D}(\nabla_{x'}, i\eta_0)U^0(x') \neq 0 \text{ for a certain } x \in \partial\Omega \setminus \partial\Pi. \qquad (3.32)$$

3.7 Remarks on cracks and edges

The example in Subsect. 3.2 shows that the trick in [56] does not work in the case of the Neumann boundary conditions at the end Υ of the cylinder (see also [61, 20] where an analogous mixed boundary problem for the Helmgoltz equation is considered). At first sight, it looks that the trick provides the result in the domain $\Omega^\sharp = \Omega \setminus \Gamma$, where $\Gamma = \gamma \times (\tau_0, \tau_0 + \tau)$ is a cylindrical crack, i.e., γ is a simple smooth closed contour inside the cross-section ω while τ_0 and τ are positive numbers. Indeed, in the case of constant coefficients, the differentiation of an eigenfunction in the variable x_n keeps both Dirichlet and Neumann boundary conditions. However, the inference on the absence of the point spectrum in the domain $\Omega^\sharp$ becomes false. The latter can be confirmed, for example, by the fact that the Neumann boundary conditions on the crack surfaces and the Dirichlet conditions on $\partial\Omega$ lead to nonempty discrete spectrum due to the construction in Subsect. 3.2.

The impossibility to prove that the integral (3.6) vanishes originates in the square-root singularities $O(r^{-1/2})$ of derivatives of the solution at the fronts $\gamma_0 = \gamma \times \{\tau_0\}$ and $\gamma_1 = \gamma \times \{\tau_0 + \tau\}$ of the crack (see, for example, [42, 6, 5, 43]). Indeed, the integration by parts in (3.5) requires to cut out the d-neighborhoods of the edges ϑ and γ_q, $q = 0, 1$, from the domain $\Omega^\sharp$. However, the integrals $J_q(d)$ over boundaries of the neighborhoods mentioned above are not infinitesimal as $d \to +0$ because $\delta_\gamma = 1/2$ and $2\delta_\gamma - 1 = 0$. Moreover, the differentiation along the crack gives rise to the Eshelby–Cherepanov–Rice invariant integrals [4, 57, 24] (see explanations in [42, 43]) which just compute the limit of $J_q(d)$ for $d \to +0$; namely, the quantity $J_q(0)$

is the squared $L_2(\gamma_q)$-norm of the coefficient on the singularity $O(r^{-1/2})$ on the front γ_q of the crack Γ.

We emphasize that the above observations confirm the necessity of the condition (3.4) in Proposition 3.1.

Remark 3.1. If Ω is a half-cylinder Π_+ with interior crack Γ, then, even under the violation of the requirement (3.31) that entails the equality $I_\omega = 0$ (see (3.27)), the problem (1.10) has an eigenvalue below the cut-off $\lambda^\dagger > 0$ provided that the requirement (3.32) is satisfied at any point x on the crack surfaces $\Gamma^\pm$. The argument remains the same as above. □

3.8 Piezoelectric bodies

The wave propagation in piezoelectric media is described (see, for example, [55, 16]) by the boundary value problem (1.5), (1.6), where $k = 4$, $N = 9$ and

$$\mathcal{D}(\nabla_x)^\top = \begin{pmatrix} \mathcal{D}^{\mathsf{M}}(\nabla_x)^\top & \mathbb{O}_{3\times 3} \\ \mathbb{O}_{1\times 6} & \nabla_x^\top \end{pmatrix}, \quad \mathcal{A}(x) = \begin{pmatrix} \mathcal{A}^{\mathsf{MM}}(x) & -\mathcal{A}^{\mathsf{ME}}(x) \\ \mathcal{A}^{\mathsf{EM}}(x) & \mathcal{A}^{\mathsf{EE}}(x) \end{pmatrix}, \quad (3.33)$$

$$\mathcal{B}(x)^\top = \operatorname{diag}\{\varrho(x), \varrho(x), \varrho(x), 0\}. \quad (3.34)$$

Here, $u^\top = ((u^{\mathsf{M}})^\top, u^{\mathsf{E}})$, $u^{\mathsf{M}} = (u_1^{\mathsf{M}}, u_2^{\mathsf{M}}, u_3^{\mathsf{M}})^\top$ is the displacement column and u^{E} the electric potential. Moreover, $\mathcal{D}^{\mathsf{M}}(\xi)^\top$ denotes the elasticity matrix (1.17), $\mathbb{O}_{p\times q}$ is the null $(p\times q)$-matrix, $\mathcal{A}^{\mathsf{MM}}(x)$ and $\mathcal{A}^{\mathsf{EE}}(x)$ are real symmetric and positive definite matrices of elastic modules and dielectric permeability coefficients of size 6×6 and 3×3 respectively. No restriction is imposed on piezoelectric modules composing the real (6×3)-matrix $\mathcal{A}^{\mathsf{ME}}(x) = \mathcal{A}^{\mathsf{EM}}(x)^\top$. The matrix $\mathcal{D}(\xi)$ keeps the algebraic completeness (1.1); however, the matrix $\mathcal{A}(x)$ is not symmetric due to the "wrong" sign on $\mathcal{A}^{\mathsf{ME}}(x)$ while the diagonal matrix (3.34) with the mechanical material density is not positive definite. Nevertheless, the results presented in Sect. 2 remain valid for the piezoelectricity problem and the approaches displayed in this section for investigating the spectrum of the problem. This optimistic announcement is based on the trick proposed in [47] (see also [32]). Using the fact that the spectral parameter λ is absent in the forth equation of the system (1.5) with matrices (3.33) and (3.34), the trick excludes the electric potential from the spectral problem. Let us explain the latter.

Consider the integral identity

$$(\mathcal{A}^{\mathsf{EE}}\nabla_x u^{\mathsf{E}}, \nabla_x w^{\mathsf{E}})_\Omega = F^{\mathsf{E}}(w^{\mathsf{E}}) := -(\mathcal{A}^{\mathsf{EM}}\mathcal{D}^{\mathsf{M}}(\nabla_x)u^{\mathsf{M}}, \nabla_x w^{\mathsf{E}})_\Omega \quad (3.35)$$

for $w^{\mathsf{E}} \in \mathcal{H}^{\mathsf{E}}(\Omega)$, where $\mathcal{H}^{\mathsf{E}}(\Omega)$ is the subspace of $H^1(\Omega)$ selected by the stable boundary conditions[2] for the electric potential (see (1.10)). It is known (see, for example, the introductory chapter in [53]) that, irrespectively of the boundary conditions, the problem (3.35) with any displacement vector $u^{\mathsf{M}} \in \mathcal{H}^{\mathsf{M}}(\Omega)$ (the "elasticity component" of the space $\mathcal{H}(\Omega)$) admits a solution u^{E} such that $\nabla_x u^{\mathsf{E}} \in L_2(\Omega)$ and

$$\|\nabla_x u^{\mathsf{E}}; \mathcal{H}^{\mathsf{E}}(\Omega)\| \leqslant c\|\mathcal{D}^{\mathsf{M}}(\nabla_x)u^{\mathsf{M}}, \mathcal{H}^{\mathsf{M}}(\Omega)\|. \tag{3.36}$$

Note that, in the case $\mathcal{H}^{\mathsf{E}}(\Omega) = H^1(\Omega)$, i.e., the "electricity component" of the boundary conditions (1.6) is of the Neumann type, the solution u^{E} is defined up to an additive constant which is eliminated by differentiation.

We introduce the continuous mapping

$$\mathcal{H}^{\mathsf{M}}(\Omega) \ni u^{\mathsf{M}} \quad \mapsto \quad \mathcal{G}u^{\mathsf{M}} = \nabla_x u^{\mathsf{E}} \in L_2(\Omega),$$

where u^{E} is a solution of the problem (3.36). The operator $\mathcal{J}$ in the space $\mathcal{H}^{\mathsf{M}}(\Omega)$ with the inner product

$$\langle u^{\mathsf{M}}, v^{\mathsf{M}}\rangle = (\mathcal{A}^{\mathsf{MM}}\mathcal{D}^{\mathsf{M}}(\nabla_x)u^{\mathsf{M}}, \mathcal{D}^{\mathsf{M}}(\nabla_x)v^{\mathsf{M}})_\Omega + (\varrho u^{\mathsf{M}}, v^{\mathsf{M}})_\Omega \tag{3.37}$$

for $u^{\mathsf{M}}, v^{\mathsf{M}} \in \mathcal{H}^{\mathsf{M}}(\Omega)$ is determined by the formula

$$\langle \mathcal{J}u^{\mathsf{M}}, v^{\mathsf{M}}\rangle = -(\mathcal{A}^{\mathsf{ME}}\mathcal{G}u^{\mathsf{M}}, \mathcal{D}^{\mathsf{M}}(\nabla_x)v^{\mathsf{M}})_\Omega, \quad u^{\mathsf{M}}, v^{\mathsf{M}} \in \mathcal{H}^{\mathsf{M}}(\Omega). \tag{3.38}$$

By the relations (3.36) and (3.37), the operator $\mathcal{J}$ is continuous.

Lemma 3.3. *The operator $\mathcal{J}$ is self-adjoint and positive.*

Proof. Let v^{E} be a solution of the problem (3.35) constructed from the test function $u^{\mathsf{M}} \in \mathcal{H}^{\mathsf{M}}(\Omega)$; namely,

$$(\mathcal{A}^{\mathsf{EE}}\nabla_x v^{\mathsf{E}}, \nabla_x w^{\mathsf{E}})_\Omega = -(\mathcal{A}^{\mathsf{EM}}\mathcal{D}^{\mathsf{M}}(\nabla_x)v^{\mathsf{M}}, \nabla_x w^{\mathsf{E}})_\Omega, \quad w^{\mathsf{E}} \in \mathcal{H}^{\mathsf{E}}(\Omega). \tag{3.39}$$

We set $w^{\mathsf{E}} = u^{\mathsf{E}} = \mathcal{G}u^{\mathsf{M}}$. Owing to the definition (3.38), formula (3.39) takes the form

$$\begin{aligned}(\nabla_x u^{\mathsf{E}}, \mathcal{A}^{\mathsf{EE}}\nabla_x v^{\mathsf{E}})_\Omega &= -(\nabla_x u^{\mathsf{E}}, \mathcal{A}^{\mathsf{EM}}\mathcal{D}^{\mathsf{M}}(\nabla_x)v^{\mathsf{M}})_\Omega \\ &= -(\mathcal{A}^{\mathsf{ME}}\nabla_x u^{\mathsf{E}}, \mathcal{D}^{\mathsf{M}}(\nabla_x)v^{\mathsf{M}})_\Omega \\ &= -(\mathcal{A}^{\mathsf{ME}}\mathcal{G}u^{\mathsf{M}}, \mathcal{D}^{\mathsf{M}}(\nabla_x)v^{\mathsf{M}})_\Omega = \langle \mathcal{J}u^{\mathsf{M}}, v^{\mathsf{M}}\rangle.\end{aligned} \tag{3.40}$$

The desired properties of the operator have become evident due to the symmetricity and positive definiteness of the matrix $\mathcal{A}^{\mathsf{EE}}$ on the left-hand side of (3.40). □

[2] The Dirichlet (stable) boundary conditions mean that the waveguide is in contact with an electroconductive medium while the Neumann (natural) boundary condition corresponds to an insulator encircling the body

Thus, the integral identity

$$(\mathcal{A}\mathcal{D}(\nabla_x)u, \mathcal{D}(\nabla_x)v)_\Omega = \lambda(\varrho u, v)_\Omega, \quad v \in \mathcal{H}(\Omega)$$

serving for the piezoelectricity problem (1.5), (1.6) with matrices (3.34) and (3.35) is not a variational problem, but reduces to the new integral identity

$$(\mathcal{A}^{\mathsf{MM}}\mathcal{D}(\nabla_x)u^{\mathsf{M}}, \mathcal{D}(\nabla_x)v^{\mathsf{M}})_\Omega + \langle \mathcal{J}u^{\mathsf{M}}, v^{\mathsf{M}}\rangle = \lambda(\varrho u^{\mathsf{M}}, v^{\mathsf{M}})_\Omega \tag{3.41}$$

for $v^{\mathsf{M}} \in \mathcal{H}^{\mathsf{M}}(\Omega)$. The quadratic form on the left-hand side of (3.41) is symmetric and positive definite according to Lemma 3.3. Indeed, by the positiveness of the operator $\mathcal{J}$, the term $\langle \mathcal{J}u^{\mathsf{M}}, v^{\mathsf{M}}\rangle$ with $v^{\mathsf{M}} = u^{\mathsf{M}}$ does not hinder to use the elasticity variant of the Korn inequality (1.16) (see references at the end of Example 1.2). Thus, nothing prevents from the reduction of the variational problem (3.41) to the abstract equation (1.22). The reduction permits us to employ methods of the operator theory in Hilbert space (see Subsect. 1.4). The theory of elliptic boundary value problems in domains with piecewise smooth boundaries, which does not require for the formal self-adjointness, directly applies to the piezoelectricity problem (see Subsect. 1.3). We only note that the operator $\mathcal{G}$ is not local and the transition to the model problems in the quasicylinder Π and the periodicity cell Σ needs to deal with the operator $\mathcal{J}^0$ in the space $\mathcal{H}(\Pi)$ by formula (3.38), where the matrix $\mathcal{A}^{\mathsf{ME}}$ is "frozen" at infinity and the integration domain is replaced by the set Π.

Acknowledgement. The work was supported by the Netherlands Organization for Scientific Research (NWO) and the Russian Foundation for Basic Research (grant no. 047.017.020).

References

1. Birman, M.S., Solomyak, M.Z.: Spectral Theory of Self-Adjoint operators in Hilbert Space. Reidel Publ. Company, Dordrecht (1986)
2. Bonnet-Bendhia, A.S., Duterte, J., Joly, P.: Mathematical analysis of elastic surface waves in topographic waveguides. Math. Meth. Appl. Sci. **9**, no. 5, 755-798 (1999)
3. Bonnet-Bendhia, A.S., Starling, F.: Guided waves by electromagnetic gratings and non-uniqueness examples for the diffraction problem. Math. Meth. Appl. Sci. **77**, 305-338 (1994)
4. Cherepanov, G.P.: Crack propagation in continuous media (Russian). Prikl. Mat. Mekh. **31**, 476-488 (1967); English transl.: J. Appl. Math. Mech. **31**, 503-512 (1967)
5. Costabel, M., Dauge, M.: Crack singularities for general elliptic systems. Math. Nachr. **235**, 29-49 (2002)
6. Duduchava, R., Wendland, W.L.: The Wiener–Hopf method for systems of pseudodifferential equations with an application to crack problems. Integral Eq. Operator Theory **23**, no. 3, 294-335 (1995)
7. Duvaut, G., Lions. J.L.: Les inéquations en mćanique et en physique. Dunod, Paris (1972)

8. Evans, D.V., Levitin, M., Vasil'ev, D.: Existence theorems for trapped modes. J. Fluid Mech. **261**, 21-31 (1994)
9. Figotin, A., Kuchment, P.: Band-gap structure of spectra of periodic dielectric and acoustic media. I. Scalar model. SIAM J. Appl. Math. **56**, 68-88 (1996); II. Two-dimensional photonic crystals. ibid. **56**, 1561-1620 (1996)
10. Filonov, N.: Gaps in the spectrum of the Maxwell operator with periodic coefficients. Commun. Math. Phys. **240**, no. 1-2, 161-170 (2003)
11. Friedlander, L.: On the density of states of periodic media in the large coupling limit. Commun. Partial Differ. Equ. **27**, 355-380 (2002)
12. Gel'fand, I.M.: Expansion in characteristic functions of an equation with periodic coefficients (Russian). Dokl. Akad. Nauk SSSR **73**, 1117-1120 (1950)
13. Gohberg, I.C., Krein, M.G.: Introduction to the Theory of Linear Nonselfadjoint Operators. Am. Math. Soc., Providence, RI (1969)
14. Gohberg, I.C., Sigal, E.I.: An operator generalization of the logarithmic residue theorem and the theorem of Rouché (Russian). Mat. Sb. **84**, 607-629. (1971); English transl.: Math. USSR–Sb. **13**, 603-625 (1971)
15. Green, E.L.: Spectral theory of Laplace–Beltrami operators with periodic metrics. J. Differ. Equations **133**, 15-29 (1997)
16. Grinchenko, V.T., Ulitko, A.F., Shul'ga, N.A.: Mechanics of Coupled Fields in Structural Elements (Russian). Naukova Dumka, Kiev (1989)
17. Hempel, R., Lineau, K.: Spectral properties of the periodic media in large coupling limit. Commun. Partial Differ. Equ. **25**, 1445-1470 (2000)
18. Jones, D.S.: The eigenvalues of $\nabla^2 u + \lambda u = 0$ when the boundary conditions are given on semi-infinite domains. Proc. Camb. Phil. Soc. **49**, 668-684 (1953)
19. Kamotskii, I.V., Nazarov, S.A.: Elastic waves localized near periodic sets of flaws (Russian). Dokl. Ross. Akad. Nauk. **368**, no. 6, 771-773 (1999); English transl.: Dokl. Physics **44**, no. 10, 715-717 (1999)
20. Kamotskii, I.V., Nazarov, S.A.: On eigenfunctions localized in a neighborhood of the lateral surface of a thin domain (Russian). Probl. Mat. Anal. **19**, 105-148 (1999); English transl.: J. Math. Sci. **101**, no. 2, 2941-2974 (2000)
21. Kamotskii, I.V., Nazarov, S.A.: Exponentially decreasing solutions of diffraction problems on a rigid periodic boundary (Russian). Mat. Zametki **73**, no. 1, 138-140 (2003); English transl.: Math. Notes. **73**, no. 1, 129-131 (2003)
22. Kato, T.: Perturbation Theory for Linear Operators. Springer (1966)
23. Khallaf, N.S.A., Parnovski, L., Vassiliev, D.: Trapped modes in a waveguide with a long obstacle. J. Fluid Mech. **403**, 251-261 (2000)
24. Knowles, J.K., Sternberg, E.: On a class of conservation laws in linearizad and finite elastostatics. Arch. Ration. Mech. Anal. **44**, no. 3, 187-211 (1972)
25. Kondratiev, V.A.: Boundary problems for elliptic equations in domains with conical or angular points (Russian). Tr. Mosk. Mat. Obshch. **16**, 209-292 (1967); English transl.: Trans. Mosc. Math. Soc. **16**, 227-313 (1967)
26. Kondratiev, V.A., Oleinik, O.A.: Boundary value problems for the system of elasticity theory in unbounded domains. Korn's inequalities (Russian). Uspekhi Mat. Nauk **43**, no. 5, 55-98 (1988); English transl.: Russian Math. Surveys **43**, no. 5, 65-119 (1988)
27. Korn, A.: Solution générale du probléme d'équilibre dans la théorie l'élasticité dans le cas où les efforts sont donnés à la surface. Ann. Univ. Toulouse, 165-269 (1908)
28. Kozlov, V.A., Maz'ya, V.G., Rossmann, J.: Elliptic Boundary Value Problems in Domains with Point Singularities. Am. Math. Soc., Providence, RI (1997)
29. Kuchment, P.A.: Floquet theory for partial differential equations (Russian). Uspekhi Mat. Nauk **37**, no. 4, 3-52 (1982); English transl.: Russian Math. Surveys **37**, no. 4, 1-60 (1982)
30. Kuchment, P.: Floquet Theory for Partial Differential Equations. Birchäuser, Basel (1993)

31. Kuchment, P.: The mathematics of photonic crystals. In: Mathematical Modeling in Optical Science, pp. 207-272. SIAM (2001)
32. Kulikov, A.A., Nazarov, S.A.: Cracks in piezoelectric and electro-conductive bodies (Russian). Sib. Zh. Ind. Mat. **8**, no. 1, 70-87 (2005)
33. Ladyzhenskaya, O.A.: Boundary Value Problems of Mathematical Physics. Springer-Verlag, New-York (1985)
34. Lekhnitskij, S.G.: Theory of Elasticity of an Anisotropic Body (Russian). Nauka, Moscow (1977); English transl.: Mir Publishers, Moscow, 1981.
35. Lions. J.L., Magenes. E.: *Non-Homogeneous Boundary Value Problems and Applications* (French). Dunod, Paris (1968); English transl.: Springer-Verlag, Berlin-Heidelberg-New York (1972)
36. Maz'ja, V.G., Plamenevskii, B.A.: On the coefficients in the asymptotics of solutions of elliptic boundary value problems in domains with conical points. Math. Nachr. **76**, 29-60 (1977); English transl.: Am. Math. Soc. Transl. (Ser. 2) **123**, 57-89 (1984)
37. Maz'ja, V.G., Plamenevskii, B.A.: Estimates in L_p and Hölder classes and the Miranda–Agmon maximum principle for solutions of elliptic boundary value problems in domains with singular points on the boundary. Math. Nachr. **81**, 25-82 (1978); English transl.: Am. Math. Soc. Transl. (Ser. 2) **123**, 1-56 (1984)
38. Nazarov, S.A.: Elliptic boundary value problems with periodic coefficients in a cylinder (Russian). Izv. Akad. Nauk SSSR. Ser. Mat. **45**, no. 1, 101-112 (1981); English transl.: Math. USSR.–Izv. **18**, no. 1, 89-98 (1982)
39. Nazarov, S.A.: A general scheme for averaging self-adjoint elliptic systems in multidimensional domains, including thin domains (Russian). Algebra Anal. **7**, no. 5, 1-92 (1995); English transl.: St. Petersbg. Math. J. **7**, no. 5, 681-748 (1996)
40. Nazarov, S.A.: Korn's inequalities for junctions of spatial bodies and thin rods. Math. Methods Appl. Sci. **20**, no. 3, 219-243 (1997)
41. Nazarov, S.A.: Self-adjoint elliptic boundary-value problems. The polynomial property and formally positive operators (Russian). Probl. Mat. Anal. **16**, 167-192 (1997); English transl.: J. Math. Sci. **92**, no. 6, 4338-4353 (1998)
42. Nazarov, S.A.: Weight functions and invariant integrals (Russian). Vychisl. Mekh. Deform. Tverd. Tela. **1**, 17-31 (1990)
43. Nazarov, S.A.: The interface crack in anisotropic bodies. Stress singularities and invariant integrals (Russian). Prikl. Mat. Mekh. **62**, no. 3, 489-502 (1998); English transl.: J. Appl. Math. Mech. **62**, no. 3, 453-464 (1998)
44. Nazarov, S.A.: The polynomial property of self-adjoint elliptic boundary-value problems and the algebraic description of their attributes (Russian). Uspekhi Mat. Nauk **54**, no. 5, 77-142 (1999); English transl.: Russian Math. Surveys **54**, no. 5, 947-1014 (1999)
45. Nazarov, S.A.: Asymptotic Theory of Thin Plates and Rods. Vol.1. Dimension Reduction and Integral Estimates (Russian). Nauchnaya Kniga IDMI, Novosibirsk (2002)
46. Nazarov, S.A.: Korn's inequality for an elastic junction of a body with a rod (Russian). In: Problems of Mechanics of Solids, pp. 234-240. St.-Petersburg State Univ. Press, St.-Petersburg (2002)
47. Nazarov, S.A.: Uniform estimates of remainders in asymptotic expansions of solutions to the problem on eigen-oscillations of a piezoelectric plate (Russian). Probl. Mat. Anal. **25**, 99-188 (2003); English transl.: J. Math. Sci. **114**, no. 5, 1657-1725 (2003)
48. Nazarov, S.A.: Trapped modes in a cylindrical elastic waveguide with a damping gasket (Russian). Zh. Vychisl. Mat. Mat. Fiz. **48**, no. 5, 863-881 (2008); English transl.: Comput. Math. Math. Phys. **48**, no. 5, 816-833 (2008)
49. Nazarov, S.A.: Korn inequalities for elastic junctions of massive bodies, thin plates and rods. inequalities (Russian). Uspekhi Mat. Nauk **63**, no. 5, 37-110 (2008); English transl.: Russian Math. Surveys **63**, no. 5 (2008)

50. Nazarov, S.A.: The Rayleigh waves in an elastic semi-layer with periodic boundary. Dokl. Ross. Akad. Nauk [Submitted]
51. Nazarov, S.A.: On the essential spectrum of boundary value problems for systems of differential equations in a bounded domain with a peak. Funkt. Anal. Pril. [To appear]
52. Nazarov, S.A.: Opening gaps in the continuous spectrum of periodic elastic waveguide with traction-free boundary. Zh. Vychisl. Mat. Mat. Fiz. [Submitted]
53. Nazarov, S.A., Plamenevskii, B.A.: Elliptic Problems in Domains with Piecewise Smooth Boundaries. Walter de Gruyter, Berlin, New York (1994)
54. Nečas. J.: Les méthodes in théorie des équations elliptiques. Masson-Academia, Paris-Prague (1967)
55. Parton, V.Z., Kudryavtsev, B.A.: Electromagnetoelasticity of Piezoelectronics and Conductive Solids (Russian). Nauka, Moscow (1988)
56. Rellich, F.: Über das asymptotische Verhalten der Lösungen von $\Delta u + \lambda u = 0$ in unendlichen Gebieten. J. Dtsch. Math.-Ver. **53**, no. 1, 57-65 (1943)
57. Rice, J.R.: A path independent integral and the approximate analysis of strain concentration by notches and cracks. J. Appl. Mech. **35**, no. 2, 379-386 (1968)
58. Roitberg, I., Vassiliev, D., Weidl, T.: Edge resonance in an elastic semi-strip. Quart. J. Appl. Math. **51**, no. 1, 1-13 (1998)
59. Skriganov, M.M.: Geometric and arithmetical methods in the spectral theory of multidimensional periodic operators (Russian). Tr. Mat. Inst. Steklova **171** (1985)
60. Sobolev, S.L.: Some Applications of Functional Analysis in Mathematical Physics (Russian). 3rd Ed. Nauka, Moscow (1988); English transl.: Am. Math. Soc., Providence, RI (1991)
61. Sukhinin, S.V.: Waveguide, anomalous, and whispering properties of a periodic chain of obstacles (Russian). Sib. Zh. Ind. Mat. **1**, no. 2, 175-198 (1998)
62. Zhikov, V.V.: Gaps in the spectrum of some elliptic operators in divergent form with periodic coefficients (Russian). Algebra Anal. **16**, no. 5, 34-58 (2004); English transl.: St. Petersbg. Math. J. **16**, no. 5, 773-790 (2005)

Estimates for Completely Integrable Systems of Differential Operators and Applications

Yuri Reshetnyak

Abstract A criterion for the complete integrability of some class of systems of differential equations is established. In the proof, the corresponding system for a matrix-valued function Z of class $W^{1,p}(\Omega)$ is used. Applications to differential geometry (in particular, the stability in the Bonnet theorem) are discussed.

1 Introduction

Let Ω be a bounded domain in the space $\mathbb{R}^n$. We consider vector-valued functions of the Sobolev space $W^{1,1}(\Omega)$, defined in Ω with the values in $\mathbb{R}^m$. On the set of such vector-valued functions, we introduce the system of differential operators

$$\Lambda_j z = \frac{\partial z}{\partial x_j} - A_j z, \quad j = 1, 2, \dots, n, \tag{1.1}$$

where A_j are $m \times m$-matrices whose entries are measurable p-integrable functions on Ω, $p > n$. We assume that the system Λ_j satisfies the following condition: the system of differential equations

$$\Lambda_j z = 0, \ j = 1, 2, \dots, n, \tag{1.2}$$

is completely integrable in the sense that for any point $x_0 \in \Omega$ and vector $\xi \in \mathbb{R}^m$ there exists a solution to the system (1.2) satisfying the Cauchy condition $z(x_0) = \xi$. As we will show below, any solution to the system (1.2)

Yuri Reshetnyak
Sobolev Institute of Mathematics SB RAS, 4 Pr. Koptyga, 630090 Novosibirsk, Russia; Novosibirsk State University, Prigova Str. 2, 630090 Novosibirsk, Russia, e-mail: Reshetnyak@math.nsc.ru

V. Maz'ya (ed.), *Sobolev Spaces in Mathematics II*, *International Mathematical Series*.
doi: 10.1007/978-0-387-85650-6, © Springer Science + Business Media, LLC 2009

is continuous. Hence we can ignor the fact that a function of class $W^{1,1}(\Omega)$ is defined only outside some "rare" set. In this paper, we obtain a criterion for the complete integrability of the system (1.2). The proof is based on the simultaneous consideration of the system (1.2) together with a system of equations for a matrix-valued function Z of class $W^{1,p}(\Omega)$.

More general differential equations were considered in [1], where a criterion for complete integrability was obtained. This criterion generalized the Frobenius criterion, known in the regular case.

The main result of this paper is an estimate for the $W^{1,p}$-norm of z in terms of the L_p-norm of Λ_j (see Theorem 4.2). The obtained results are applied to the study of stability in the Bonnet theorem. Note that this problem was considered by Ciarlet [2, 4, 3] in a different setting.

This paper continues the work of the author [5]. Owing to the above-mentioned criterion for complete integrability, the arguments of [5] become much simpler. In particular, it is not necessary to use the results of too complicated work [1]. Also, we correct the proof of Theorem 2 in [5] (which corresponds to Theorem 5.1 below). I thank N.N. Romanovskii who indicated me a mistake in the proof.

2 Notation and Preliminaries

Introduce the notation: Ω is a domain, i.e., a connected open set in the space $\mathbb{R}^n$, $n \geqslant 2$; $B(a,r)$ is an open ball in $\mathbb{R}^n$ with center a and radius $r > 0$; and $\overline{B}(a,r)$ is the closed ball with center a and radius r. The inner product of vectors $\mathbf{u}$ and $\mathbf{v}$ in $\mathbb{R}^N$ is denoted by $\langle \mathbf{u}, \mathbf{v}\rangle$. We say that a domain Ω in $\mathbb{R}^n$ is *star-shaped* relative to the ball $B(a,r) \subset \Omega$ if for any points $x \in \Omega$ and $y \in B(a,r)$ the segment joining x and y (i.e., the set of all points $u = \lambda x + (1-\lambda)y$, where $0 \leqslant \lambda \leqslant 1$) is contained in Ω. The spaces $L_p(\Omega)$ with $p \geqslant 1$ are introduced in a standard way.

For a function $u : \Omega \to \mathbb{R}^N$ we denote by $\|u\|_{L_\infty}(\Omega)$ the least number $h \leqslant \infty$ such that $|u(x)| \leqslant h$ for almost all $x \in \Omega$. The set of all functions u such that $\|u\|_{L_\infty}(\Omega) < \infty$ form a Banach space with respect to the norm $\|u\|_{L_\infty}(\Omega)$.

Consider the set $W^{l,p}(\Omega)$ of all functions defined on Ω that possess distributional derivatives of order l and are p-integrable with $p \geqslant 1$. The set $W^{l,p}$ is a Banach space relative to the norm defined as follows. Let

$$\|f\|_{L^{l,p}} = \left\{ \int\limits_\Omega \Big[\sum_{|\alpha|=l} \frac{l!}{\alpha!} \Big(\frac{\partial^\alpha z}{\partial x^\alpha}(x) \Big)^2 \Big]^{\frac{p}{2}} dx \right\}^{\frac{1}{p}}, \tag{2.1}$$

and let Π be a projection from $W^l_p(\Omega)$ to the set $\mathcal{P}_{l-1}$ of all polynomials of degree at most $l-1$. The set $\mathcal{P}_{l-1}$ is a finite-dimensional vector space, and any two norms in this space are equivalent. We set

$$\|f\|_{W^{l,p}} = \|\Pi f\|_{\mathcal{P}_{l-1}} + \|f\|_{L^{l,p}}.$$

In the case $l = 1$, elements of $\mathcal{P}_{l-1}$ are constants. Then for Π we take

$$\Pi u = \int\limits_{\Omega} u(x)\varphi(x)\,dx, \tag{2.2}$$

where $\varphi \in \mathcal{C}^\infty$ is a nonnegative function with support in the closed ball $\overline{B}(a,r) \subset \Omega$ such that

$$\int\limits_{\Omega} \varphi(x)\,dx = 1.$$

By [6], the norms in $W^{1,p}(\Omega)$ obtained for different operators Π of the form (2.2) are equivalent.

We need vector-valued and matrix-valued functions defined in a domain Ω. We say that a function

$$z : x \in \Omega \mapsto (z_1(x), z_2(x), \ldots, z_m(x))$$

belongs to the space $L_p(\Omega)$ (respectively, $W^{l,p}(\Omega)$, $\mathcal{C}^r$) if each of the real-valued functions z_i, $i = 1, 2, \ldots, m$, belongs to $L_p(\Omega)$ (respectively, $W^{l,p}(\Omega)$, $\mathcal{C}^r$). Similarly, a matrix-valued function $Z(x) = (z_{ij}(x))_{i,j=1,2,\ldots,m}$ defined in Ω belongs to the class $L_p(\Omega)$ (respectively, $W^{r,p}(\Omega)$) if each of the real-valued $z_{i,j}$, $i,j = 1,2,\ldots,m$, belongs to the class $L_p(\Omega)$ (respectively, $W^{r,p}(\Omega)$). For an $m \times m$-matrix Z we denote by $|Z|$ the operator norm of Z, i.e.,

$$|Z| = \sup_{|\xi| \leqslant 1} |Z\xi|.$$

The $L_p(\Omega)$-norm of a vector-valued function $z = (z_1(x), z_2(x), \ldots, z_m(x))$ of class $L_p(\Omega)$ (respectively, $W^{l,p}(\Omega)$) is defined as the sum of the L_p-norms (respectively, the $W^{l,p}$-norms) of components of the vector-valued function z.

We will use the Sobolev embedding theorem for functions of class $W^{1,p}$. Recall that these theorems are valid under some conditions on the boundary of a domain. We say that Ω is a Sobolev type domain (or a domain of class $\mathcal{S}$) if the Sobolev embedding theorems for $W^{1,p}$ are valid in this domain. Any domain in $\mathbb{R}^n$ whose boundary is an $(n-1)$-dimensional manifold of class $\mathcal{C}^1$ is a domain of class $\mathcal{S}$.

We also consider vector-valued functions z of class $W^{1,1}(\Omega)$ that are defined in a domain Ω, take the values in $\mathbb{R}^m$, and satisfy the system of differential equations

$$\frac{\partial z}{\partial x_i} = A_i(x) z(x), \quad i = 1, 2, \ldots, n, \tag{2.3}$$

where $A_i(x)$ for every $x \in \Omega$ is an $m \times m$-matrix; moreover, entries of the matrix-valued functions $A_i(x)$ are measurable functions in Ω.

Theorem 2.1 (see [5]). *Let Ω be a bounded domain of class S in $\mathbb{R}^n$, and let a function $z : \Omega \to \mathbb{R}^m$ of class $W^{1,1}(\Omega)$ be such that for every $i = 1, 2, \dots, n$ and almost all $x \in \Omega$*

$$\frac{\partial z}{\partial x_i} = A_i(x, z), \tag{2.4}$$

where the vector-valued functions $A_i(x, z)$ are defined on $\Omega \times \mathbb{R}^m$ and satisfy the condition

$$\forall i \in \{1, 2, \dots, n\} A_i(x, 0) \in L_p(\Omega), \quad |A_i(x, z) - A_i(x, \overline{z})| \leqslant M(x)|z - \overline{z}|, \tag{2.5}$$

where M is of class $L_p(\Omega)$, $p > n$. Then the vector-valued function $z : \Omega \to \mathbb{R}^m$ z belongs to the class $W^{1,p}$. In particular, the function $z(x)$ is continuous in the domain Ω.

The proof in [5] is based on an iterative procedure. By the assumptions of Theorem 2.1, $|A_i(x, z)| \leqslant |A_i(x, 0)| + M(x)|z|$. The function $x \mapsto A_i(x, 0)$ belongs to the class $L_p(\Omega)$. By assumption, $z \in W^{1,1}(\Omega)$. By the Sobolev embedding theorem, $z \in L_s$, where $s = \dfrac{n}{n-1}$. Hence the function $M|z|$ is q_1-integrable, $q_1 > s$. (Here, we essentially used the condition $p > n$.) Hence $z \in W^{1,r_1}$, where $r_1 > r_0 = 1$, which, in turn, means that $z \in L_{q_2}$, where $q_2 > q_1$. Therefore, $z \in W^{1.r_2}$, where $r_2 > r_1$. Repeating this argument, in a finite number of steps, we find that $z \in W^{1,r}$ for some $r > n$. Hence the function z is continuous on the set Ω and, consequently, is bounded there. Hence the right-hand side of (2.4) is a function of class $W^{1,p}(\Omega)$, which is required to prove. The common number of steps neccesary to reach $r > n$ is equal to $\dfrac{pn - p}{p - n} + \eta$, where $|\eta| \leqslant 1$ and, consequently, tends to ∞ as $p \to n$. We refer the reader to [5] for details.

The proof of the following assertion is obvious.

Corollary 2.1. *Suppose that Ω is a domain of class S and the matrix-valued functions A_i in (2.3) belong to the space $L_p(\Omega)$ with some $p > n$. If a vector-valued function $z : \Omega \to \mathbb{R}^m$ of class $W^{1,1}$ satisfies the system (2.3), then z belongs to the class $W^{1,p}$. In particular, the function $z(x)$ is continuous in the domain Ω.*

3 Remarks on Completely Integrable Linear Systems of Differential Equations

If P is an $(m \times m)$-matrix, then denote by P^* the transposed matrix of P. For any $(m \times m)$-matrices P and Q we have $(PQ)^* = Q^* P^*$. The symbol I_m denotes the identity $(m \times m)$-matrix.

In a domain Ω of the space $\mathbb{R}^n$, we consider the system of differential equations

$$\frac{\partial z}{\partial x_j} = A_j(x)z(x), \quad j = 1, 2, \ldots, n, \tag{3.1}$$

where $A_j(x)$ for every $x \in \Omega$ is an $(m \times m)$-matrix and the matrix-valued functions $A_j(x)$ belong to the class $L_p(\Omega)$, where $p > n$. By Theorem 2.1, any solution to this system is a function of class $W^{1,p}(\Omega)$ and, consequently, is a bounded continuous function in the domain Ω.

We say that the system of equations (3.1) is *completely integrable* if for any point $x_0 \in \Omega$ and vector $\xi \in \mathbb{R}^n$ there exists a solution $z(x)$ to the system (3.1) defined in the domain Ω and satisfying the Cauchy condition $z(x_0) = \xi$.

In addition to the system (3.1), we also consider the following system of equations for the matrix-valued functions:

$$\frac{\partial Z}{\partial x_i}(x) = A_i(x)Z(x), \quad j = 1, 2, \ldots, n, \tag{3.2}$$

where $X(x)$ is an $(m \times m)$-matrix.

If the system (3.1) is completely integrable, then the system (3.2) is also completely integrable in the sense that for any point $x_0 \in \Omega$ and an $(m \times m)$-matrix Ξ there exists a solution Z to the system (3.2) satisfying the Cauchy condition $Z(x_0) = \Xi$.

Indeed, let ξ_i be the ith column of the matrix Ξ, and let $\mathbf{z}^{(i)}$ be a solution to the system (4.1) satisfying the Cauchy condition $z(x_0) = \xi_i$. Then the matrix-valued function $Z(x)$ with the ith column $\mathbf{z}^{(i)}$, $i = 1, 2, \ldots, m$, is a solution to the system (3.2) such that $Z(x_0) = \Xi$.

If the system (3.2) for the matrix-valued function X is completely integrable, then the system (3.1) is also completely integrable. Indeed, assume that the system (3.2) is completely integrable. Choose arbitrarily a point $x_0 \in \Omega$ and a vector $\xi \in \mathbb{R}^m$. Let Ξ denote an $(m \times m)$-matrix with the first column ξ, and let $Z(x)$ be a solution to the Cauchy problem $Z(x_0) = \Xi$ for the system (4.2).

Let $z(x)$ be the first column of the matrix $Z(x)$. The vector-valued function $z(x)$ is a solution to the Cauchy problem $z(x_0) = \xi$ for Eq. (3.1), which is required.

If the system (3.2) has a solution $X = Z$ such that $\det Z(x) \neq 0$ for all $x \in \Omega$, this system is completely integrable. Indeed, let us assume that $Z(x)$ is such a solution of the system (3.2). We choose arbitrarily a point $x_0 \in \Omega$ and an $(m \times m)$-matrix Ξ. We set $X(x) = Z(x)H$, where $H = [Z(x_0)]^{-1}\Xi$. Since H is a matrix with constant entries, it is obvious that the matrix-valued function $X(x)$ satisfies the system of equations (3.2); moreover, $X(x_0) = \Xi$. Since the matrix Ξ and the point $x_0 \in \Omega$ are taken in an arbitrary way, thereby the complete integrability of the system (3.2) is proved.

Let $Z(x)$ be a solution of the system (3.2) such that $\det Z(x) \neq 0$ for all $x \in \Omega$. Then for all $x \in \Omega$ we can define the matrix $T(x) = [Z(x)]^{-1}$. The matrix-valued function $T(x)$ is continuous and belongs to the class $W^1_{p,\mathrm{loc}}(\Omega)$. This is a consequence of the fact that each entry of the matrix $T(x)$ is represented in the form of a fraction $\dfrac{P_{ij}[Z(x)]}{\det Z(x)}$, where P_{ij} is a polynomial of degree $m-1$. The numerator and denominator are functions of class $W^{1,p}(\Omega)$. Since we cannot exclude the case $\inf|\det Z(x)| = 0$, this means that $T \in W^1_{p,\mathrm{loc}}$. Differentiating both sides of the equality $T(x)Z(x) = I_m$ with respect to x_j, we find

$$\begin{aligned} 0 &= \frac{\partial T}{\partial x_j}(x)Z(x) + T(x)\frac{\partial Z}{\partial x_j}(x) \\ &= \frac{\partial T}{\partial x_j}(x)Z(x) + T(x)A_j(x)Z(x) \\ &= \left[\frac{\partial T}{\partial x_j}(x) + T(x)A_j(x)\right] Z(x). \end{aligned}$$

Since $\det Z(x) \neq 0$ for all $x \in \Omega$, we have $\dfrac{\partial T}{\partial x_j}(x) + T(x)A_j(x) = 0$ for $x \in \Omega$ at which the left-hand side of the equality is defined, i.e.,

$$\frac{\partial T}{\partial x_j}(x) + T(x)A_j(x) = 0 \tag{3.3}$$

for almost all $x \in \Omega$ for any $j = 1, 2, \ldots, n$.

Applying the transposition operation to both sides of (3.3), we obtain the system

$$\frac{\partial T^*}{\partial x_j}(x) = -[A_j(x)]^* T(x) = 0 \quad j = 1, 2, \ldots, n, \tag{3.4}$$

of the same form as the system (3.2). Since $\det T^*(x) = \det T(x) \neq 0$, the system of equation (3.4) is completely integrable.

The system of equations (3.4) is said to be *conjugate* to the system (3.2). Let $Z(x)$ be an arbitrary solution to the system (3.2), and let $T(x)$ be a solution to the system (3.3). Then the matrix-valued function $T(x)Z(x)$ is constant in the domain Ω. Indeed,

$$\begin{aligned} \frac{\partial}{\partial x_j}[T(x)Z(x)] &= \frac{\partial T}{\partial x_j}(x)Z(x) + T(x)\frac{\partial Z}{\partial x_j}(x) \\ &= -T(x)A_j(x)Z(x) + T(x)A_j(x)Z(x) = 0. \end{aligned}$$

Thus, the derivatives of the matrix-valued function $T(x)Z(x)$ vanish in the domain Ω, which means that it is constant in Ω.

Theorem 3.1. *In a domain Ω of the space $\mathbb{R}^n$, consider the system of differential equations (3.2) for the matrix-valued function $Z(x)$ such that the coefficients $A_j(x)$ are matrix-valued functions of class $L_{p,\mathrm{loc}}(\Omega)$. Assume that* *the system* (3.2) *is completely integrable. Let* $Z(x)$ *be a solution to the system* (3.2) *such that for some* $y \in \Omega$ *the determinant of the matrix* $Z(y)$ *does not vanish. Then* $\det Z(x) \neq 0$ *for all* $x \in \Omega$.

Proof. Let Σ be the set of all points $x \in \Omega$ where $\det Z(x) \neq 0$. By the continuity of $\det Z(x)$, the set Σ is open. We show that it is closed relative to Ω.

Let $t \in \Omega$ be a contact point of the set Σ. Since the system (3.2) is completely integrable, there exists a solution $Z_1(x)$ to this system satisfying the Cauchy condition $Z_1(t) = I_m$. By the continuity of $\det Z_1(x)$, there is $\delta > 0$ such that $\Omega \supset B(t, \delta)$ and $\det Z_1(x) > 0$ for all $x \in B(t, \delta)$. The matrix-valued function $T_1(x) = [Z_1(x)]^{-1}$, defined in the ball $B(t, \delta)$, satisfies there the system (3.3). In the ball $B(t, \delta)$, the function $Z(x)$ satisfies the system (3.2). Hence the product $T_1(x)Z(x)$ is constant in the set $B(t, \delta)$. Since t is a contact point of the set Σ, the neighborhood $B(t, \delta)$ of the point t contains points of the set Σ.

Let $t_0 \in \Sigma \cap B(t, \delta)$. We have $\det T_1(t_0) > 0$ and $\det Z(t_0) \neq 0$. Hence $\det\{T_1(t_0)Z(t_0)\} \neq 0$. Since the matrix-valued function $T_1(x)Z(x)$ is constant in the ball $B(t, \delta)$, we have

$$\det\{T_1(x)Z(x)\} = \det\{T_1(t_0)Z(t_0)\} \neq 0 \quad \text{for all } x \in B(t, \delta).$$

In particular, $\det\{T_1(t)Z(t)\} \neq 0$. It remains to note that $T_1(t) = I_m$ and, consequently, $\det Z(t) \neq 0$. Thus, $t \in \Sigma$.

The above arguments show that the set Σ is simultaneously open and closed relative to Ω. Since Σ is not empty and Ω is connected, we conclude that $\Sigma = \Omega$, i.e., $\det Z(x) \neq 0$ for all $x \in \Omega$. □

Corollary 3.1. *If the system of equations* (3.2) *is completely integrable, then the Cauchy problem* $Z(x_0) = \Xi$ *for the system* (3.2) *has a unique solution.*

Proof. It suffices to prove that for a solution $Z(x)$ to the system (3.2) such that $Z(x_0) = 0$ we have $Z(x) \equiv 0$. Let $Z_0(x)$ is a solution to the system (4.2) such that $\det Z_0(x) \neq 0$ for all $x \in \Omega$. We set $T(x) = [Z_0(x)]^{-1}$. The function $T(x)$ is a solution to the system (3.3).

By the above argument, the matrix-valued function $H = TZ$ is constant in the domain Ω. It is obvious that $H(x_0) = 0$. Hence $H(x) = 0$ for all $x \in \Omega$ and, consequently, $Z(x) = 0$ for all $x \in \Omega$, which is required. □

Let $Z(x)$ be a solution to the system (3.2) satisfying the condition $\det Z(x) \neq 0$ for all $x \in \Omega$. For arbitrary $x, y \in \Omega$ we set $\Theta(x, y) = Z(x)[Z(y)]^{-1}$. Then the matrix-valued function $X(x) = \Theta(x, x_0)\Xi$ is a solution to Eq. (3.2) satisfying the Cauchy condition $X(x_0) = \Xi$. The matrix-valued function $\Theta(x, y)$ is called the *characteristic matrix* for each of the systems (3.1) and (3.2).

For fixed y the matrix $\Theta(x, y)$ regarded as a function of the variable x satisfies the equations (3.2). For any vector $\xi \in \mathbb{R}^m$ the vector-valued function $z(x) = \Theta(x, y)\xi$ is a solution to (3.1) satisfying the Cauchy condition $z(y) = \Theta(y, y)\xi = \xi$.

Let Λ_i be a differential operator defined by the equality

$$(\Lambda_i z)(x) = \frac{\partial z}{\partial x_i}(x) - A_i(x) z(x),$$

where $A_i(x)$ are the matrices in the system (3.1). We say that $\Lambda_1, \Lambda_2, \dots, \Lambda_n$ is a system of differential operators subject to the *complete integrability* condition if the system (3.1) is completely integrable. Let $\Theta(x, y) = Z(x)[Z(y)]^{-1}$, be the characteristic matrix of the system (3.1). We have

$$\Lambda_i = Z \circ \frac{\partial}{\partial x_i} \circ Z^{-1} \tag{3.5}$$

or

$$\Lambda_i z(x) = Z(x) \frac{\partial}{\partial x_i} \{[Z(x)]^{-1} z(x)\}. \tag{3.6}$$

4 Estimates for Operators Satisfying the Complete Integrability Condition

The main results will be formulated for star-shaped domains relative to a ball. The proof of the general case differs by only techniques.

Consider the linear differential operators

$$\Lambda_i = \frac{\partial}{\partial x_i} - A_i(x), \quad i = 1, 3, \dots, n, \tag{4.1}$$

in a domain Ω such that the system of equations

$$\Lambda_i z = 0, \quad i = 1, 2, \dots, n, \tag{4.2}$$

is completely integrable in the sense of the definition of Sect. 3.

Assume that Ω is a star-shaped domain in $\mathbb{R}^n$ relative to the ball $B(a, r)$. Then any function $z \in W^{1,p}$, $p \geqslant 1$, admits the integral representation

$$z(x) = \int\limits_{\Omega} z(y)\varphi(y)\, dy + \int\limits_{\Omega} \sum_{k=1}^{n} K(x, y)(x_k - y_k) \frac{\partial z}{\partial x_i}(y) dy, \tag{4.3}$$

where φ is a nonnegative function of class C^∞ with support in the ball $B(a, r)$.

Setting $z \equiv 1$ in (4.3), we see that the integral of φ over the domain Ω is equal to 1. The function $K(x, y)$ admits the representation

$$K(x,y) = u(x, x-y) - v(x,y),$$

where

$$u(x,z) = \int_0^\infty \varphi(x - zt)t^{n-1}\,dt,$$
$$v(x,y) = \int_0^1 \varphi[x + (y-x)t]t^{n-1}\,dt. \tag{4.4}$$

The function $\lambda(x, z)$ is positively homogeneous of degree $-n$ and belongs to the class $\mathcal{C}^\infty$ on the set $\mathbb{R}^n \times (\mathbb{R}^n \setminus \{0\})$, whereas the function $\mu(x, y)$ belongs to the class $\mathcal{C}^\infty$ on the set $\mathbb{R}^n \times \mathbb{R}^n$. Using the equality (3.6), from (4.4) we obtain an integral representation of z in terms of the values of the differential operators Λ_i, $i = 1, 2, \dots, n$, satisfying the complete integrability condition.

Let the matrix-valued function be a solution of the system (3.2) such that $\det Z(x) \neq 0$ for all $x \in \Omega$. Replacing the function z with $[Z(x)]^{-1}z(x)$ in (4.4), multiplying both sides of the obtained equality from the left by the matrix $Z(x)$, and making obvious transformations, from the equalities (3.6) and (4.4) we obtain the integral representation

$$z(x) = \int_\Omega \Theta(x,y)z(y)\varphi(y)\,dy$$
$$+ \int_\Omega \Theta(x,y)\sum_{k=1}^n K(x,y)(x_k - y_k)L_k z(y)\,dy. \tag{4.5}$$

We set

$$(\Pi z)(x) = \int_\Omega \Theta(x,y)z(y)\varphi(y)\,dy, \tag{4.6}$$

where φ is the function appearing in the equalities (4.3) and (4.5).

Substituting an arbitrary solution $z(x)$ to the system (3.2) into (4.5), we see that $z(x) \equiv (\Pi z)(x)$ for any such a solution.

For an arbitrary function $z \in W^{1,p}(\Omega)$ we set

$$\|\Lambda z\|_{L_p(\Omega)} = \sum_{i=1}^n \|\Lambda_i z\|_{L_p(\Omega)}.$$

The following assertion is known.

Theorem 4.1 (about integrals with weak singularity). *Let Ω be a bounded domain in $\mathbb{R}^n$, and let $\omega(x,z)$ be a function of variables $x \in \Omega$ and $z \in (\mathbb{R}^n \setminus 0)$, homogeneous of zero degree with respect to the variable z. Suppose* *that the function ω has all partial derivatives of order at most l on the set $\Omega \times (\mathbb{R}^n \setminus 0)$; moreover, the derivatives are bounded on the set $\Omega \times S(0,1)$. For $p > 1$ and $f \in L_p(\Omega)$ we set*

$$Hf(x) = \int_{\Omega} \frac{\omega(x, x-y)}{|x-y|^{n-l}} \, dy; \tag{4.7}$$

$Hf(x)$ is defined and is finite for almost all $x \in \Omega$; moreover, it belongs to the class $W^{l,p}(\Omega)$. Then there exists $C < \infty$ independent of the choice of f such that

$$\|Hf\|_{W^{l,p}(\Omega)} \leqslant C\|f\|_{L_p(\Omega)}. \tag{4.8}$$

Theorem 4.2. *Let Λ_i, $i = 1, 2, \dots, n$ be a collection of differential operators of the form (4.1) that are defined in the domain Ω and such that the matrix-valued functions A_i belong to $L_p(\Omega)$, where $p > n$, and the system of differential equations (3.2) is completely integrable. Then there exists a constant $C < \infty$ such that for any function $z : \Omega \to \mathbb{R}^m$ of class $W^{1,p}(\Omega)$*

$$\|z - \Pi z\|_{W^{1,p}(\Omega)} \leqslant C\|\Lambda z\|_{L_p(\Omega)}, \tag{4.9}$$

where Π is the integral operator defined by (4.6).

Proof. Let $z : \Omega \to \mathbb{R}^m$ be an arbitrary function of class $W^{1,p}(\Omega)$ with $p > n$. From the integral representation (4.5) it follows that

$$z(x) - (\Pi z)(x) = \int_{\Omega} \Theta(x,y) \sum_{k=1}^{n} K(x,y)(x_k - y_k)\Lambda_k z(y) \, dy. \tag{4.10}$$

The matrix $\Theta(x,y)$ is represented in the form $\Theta(x,y) = Z(x)[Z(y)]^{-1}$. In addition, the matrix-valued function $Z(x)$ is a solution to the system (3.2) such that $\det Z(x) \neq 0$ for all $x \in \Omega$; moreover, $Z(y) = I_m$ for some point $y \in \Omega$.

A point y can be chosen in an arbitrary way. Assume that $Z(x_0) = I$. We set $\Lambda_i u = v_i$. The matrix-valued functions Z and Z^{-1} belong to the class $W^{1,p}(\Omega)$ and, consequently, are bounded and continuous in the set Ω. Let $M < \infty$ be such that for all $x \in \Omega$ the inequalities $|Z(x)| \leqslant M$ and $|[Z(x)]^{-1}| \leqslant M$ hold.

Consider separately the terms on the right-hand side of (4.10). We set

$$R_k v_i(x) = Z(x) \int_{\Omega} (x_k - y_k) K(x,y) [Z(y)]^{-1} \Lambda_i z(y) \, dy.$$

Then $R_k v_i(x) = S_k v_i(x) - T_k v_i(x)$, where

$$S_k v_i(x) = \int\limits_{\Omega} (x_k - y_k)\lambda(x, x-y)v_i(y)\,dy,$$

$$T_k v_i(x) = \int\limits_{\Omega} (x_k - y_k)\mu(x, y)v_i(y)\,dy$$

the functions $\lambda(x,z)$ and $\mu(x,y)$ are determined by the equalities (4.4). Setting $\omega(x,z) = z_k|z|^{n-1}\lambda(x,z)$ in Theorem 4.1, we find

$$\|S_k v_i\|_{W^{1,p}(\Omega)} \leqslant C\|v_i\|_{L_p(\Omega)} \leqslant MC\|u_i\|_{L_p(\Omega)}.$$

The function $\mu(x,y)v_i(y)$ has continuous derivatives of any order with respect to each of the variables x and y everywhere in $\mathbb{R}^n$. In the domain Ω, these derivatives are bounded. Hence the functions $T_k v_i(x)$ have continuous derivatives and can be found by using the known rules of differentiation of integrals with parameter. Hence the derivatives of the function $T_k v_i(x)$ on the set Ω are bounded and continuous and, consequently, $T_k v_i(x)$ belongs to the class $W^{1,p}(\Omega)$.

From the integral formula for the derivatives of $T_k v_i(x)$ we have

$$\|T_k v_i\|_{W^{1,p}(\Omega)} \leqslant C\|v_i\|_{L_p(\Omega)} \leqslant MC\|u_i\|_{L_p(\Omega)}.$$

The above argument shows that the functions $R_k v_i = S_k v_i - T_k v_i$ belong to the class $W^{1,p}(\Omega)$; moreover, the following *estimate* holds:

$$R_k v_{iW^{1,p}(\Omega)} \leqslant C\|u_i\|_{L_p(\Omega)}.$$

Let $u : \Omega \to \mathbb{R}^m$ be a vector-valued function of class $W^{1,p}(\Omega)$, $p > n$, and let $P(x)$ be the above matrix-valued function . Then $Pu \in W^{1,p}(\Omega)$ and for every $i = 1, 2, \ldots, n$

$$\frac{\partial(Pu)}{\partial x_i} = \frac{\partial P}{\partial x_i}u + P\frac{\partial u}{\partial x_i}.$$

This assertion can be easily proved by approximating the functions $P(x)$ and $u(x)$ by Sobolev averages. Then

$$\left\|\frac{\partial Zu}{\partial x_i}\right\|_{L_p(\Omega)} \leqslant \sup_{x\in\Omega}|u(x)|\left\|\frac{\partial P}{\partial x_i}\right\|_{L_p(\Omega)} + \sup_{x\in\Omega}|Z(x)|\left\|\frac{\partial u}{\partial x_i}\right\|_{L_p(\Omega)}. \tag{4.11}$$

Since $p > n$, the expression $\sup\limits_{x\in\Omega}|u(x)|$ is finite. Furthermore,

$$\sup_{x\in\Omega}|u(x)| \leqslant C\|u\|_{W^{1,p}(\Omega)}.$$

We find a constant $M < \infty$ such that $\max\limits_{x\in\Omega}|Z(x)| \leqslant M$ and

$$\left\| \frac{\partial Z}{\partial x_i} \right\|_{L_p(\Omega)} \leqslant M$$

for every $i = 1, 2, \ldots, n$. Then from the inequality (4.11) it follows that

$$\|Pu\|_{W^{1,p}(\Omega)} \leqslant C\|u\|W^{1,p}(\Omega),$$

where $C < \infty$.

The obtained estimates obviously lead to the required assertion. □

5 Stability of Solutions of Completely Linear Integrable Systems

Further, Ω is a star-shaped domain in $\mathbb{R}^n$ relative to a ball $B(x_0, r) \subset \Omega$. Consider completely integrable systems of differential equations in Ω:

$$\frac{\partial z}{\partial x_i}(x) - A_i(x)z(x) = 0, \tag{5.1}$$

$$\frac{\partial z}{\partial x_i}(x) - B_i(x)z(x) = 0, \tag{5.2}$$

where $A_i(x)$ and $B_i(x)$ are $m \times m$-matrices, $i = 1, 2, \ldots, n$. Assume that the matrix-valued functions A_i and B_i belong to the class $L_p(\Omega)$ for some $p > n$. We set

$$\delta_p(A, B) = \left\{ \sum_{i=1}^{n} \left\| |A_i - B_i| \right\|_{L_p}^{p} \right\}^{1/p}.$$

Theorem 5.1. *Let $z : \Omega \to \mathbb{R}^m$ and $\overline{z} : \Omega \to \mathbb{R}^m$ be solutions to* (5.1) *and* (5.2) *respectively. Suppose that these systems satisfy the above conditions and, in addition, $z(x_0) = \xi$, $\overline{z}(x_0) = \overline{\xi}$, $\xi, \overline{\xi} \in \mathbb{R}^m$. Then there exist constants $\varepsilon_0 > 0$, $C < \infty$, and $M < \infty$ such that $\delta_p(A, B) < \varepsilon_0$ implies the inequality*

$$\|\overline{z} - z\|_{W^{1,p}(\Omega)} \leqslant C\delta_p(A, B)|\overline{\xi}| + M|\xi - \overline{\xi}|.$$

Proof. Let z_0 be a solution to the system (5.1) satisfying the Cauchy condition $z_0(x_0) = \overline{\xi}$. We set $\zeta = \overline{z} - z_0$. From the equality (5.2) it follows that

$$\frac{\partial \overline{z}}{\partial x_i}(x) - A_i(x)\overline{z}(x) + [A_i(x) - B_i(x)]\overline{z}, \quad i = 1, 2, \ldots, n. \tag{5.3}$$

Substituting z_0 for z in (5.1) and subtracting term-by-term the equality obtained from (5.3), we find

$$\frac{\partial(\overline{z}-z_0)}{\partial x_i}(x) - A_i(x)[\overline{z}(x) - z_0(x)] = -[A_i(x) - B_i(x)]\overline{z}. \qquad (5.4)$$

Hence

$$\frac{\partial \zeta}{\partial x_i}(x) - A_i(x)\zeta(x) = -[A_i(x) - B_i(x)]z_0(x) - [A_i(x) - B_i(x)]\zeta(x). \quad (5.5)$$

Theorem 4.2 allows us to estimate $\|\zeta - \Pi\zeta\|_{W^{1,p}(\Omega)}$, where Π is the integral operator defined by the equality (4.6).

It is obvious that $\|\zeta\|_{W^{1,p}(\Omega)} \leqslant \|\zeta - \Pi\zeta\|_{W^1_p(\Omega)} + \|\Pi\zeta\|_{W^{1,p}(\Omega)}$. Since the kernel $\Theta(x,y)$ in the equality (4.6) admits the representation

$$\Theta(x,y) = P(x)[P(y)]^{-1},$$

the function $\Pi\zeta(x)$ has the form $\Pi\zeta(x) = P(x)\xi_0$, where

$$\xi_0 = \int\limits_{\Omega} [P(y)]^{-1}\zeta(y)\varphi(y)\, dy.$$

Thus, we can conclude that

$$\|\Pi\zeta\|_{W^1_p(\Omega)} = \|P\|_{W^1_p(\Omega)}|\xi_0| = M|\xi_0|,$$

where $M = \|P\|_{W^1_p(\Omega)}$. We set $N = \sup\limits_{y\in B(x_0,r)} |[P(y)]^{-1}|$. By Theorem 2.1, we can conclude that M and N are finite. Suppose that the function φ in the equality (4.6) and the subsequent formulas is nonnegative and vanishes outside the ball $B(x_0, r_1)$, $0 \leqslant r_1 \leqslant r$. The value of r_1 will be fixed later. By the Sobolev embedding theorem, for any $x \in B(x_0, r)$ we have

$$|\zeta(x)| = |\zeta(x) - \zeta(x_0)| \leqslant C\|\zeta\|_{W^{1,p}(\Omega)}|x - x_0|^\alpha,$$

where $\alpha = 1 - \dfrac{n}{p}$. Hence

$$\|\Pi\zeta\|_{W^1_p(\Omega)} \leqslant CMN \int\limits_{|x-x_0|\leqslant r_1} |x - x_0|^\alpha \varphi(x)\, dx \leqslant CMNr_a^\alpha.$$

Let r_1 be the least number among r and $r' = \dfrac{1}{[2CMN]^{1/\alpha}}$. It is obvious that

$$CNMr_1^\alpha \leqslant \frac{1}{2}.$$

The number r_1 depends only on parameters related to the system (5.5). Under this choice of r_1, we have

$$\|\zeta\|_{W_p^1(\Omega)} \leqslant \|\zeta - \Pi\zeta\|_{W_p^1(\Omega)} + \frac{1}{2}\|\zeta\|_{W_p^1(\Omega)},$$

which implies the estimate

$$\|\zeta\|_{W_p^1(\Omega)} \leqslant 2\|\zeta - \Pi\zeta\|_{W_p^1(\Omega)}.$$

Let $\Theta(x, y)$ be the fundamental matrix of the system of equations (5.1). Let $P(x) = \Theta(x, x_0)$, and let $Q(x) = [P(y)]^{-1}$ belong to the class $W^{1,p}$. By the Sobolev embedding theorems, we have

$$\|P\|_{L_\infty(\Omega)} = \sup_{x\in\Omega} |P(x)| = K < \infty.$$

Thus, $z_0(x) = P(x)\overline{\xi}$. Hence $|z_0(x)| \leqslant K|\overline{\xi}|$ for all $x \in \Omega$.

By Theorem 2.1, the function ζ belongs to the class $W^{1,p}(\Omega)$. Then, by the Sobolev embedding theorem, we have

$$\|\zeta\|_{L_\infty(\Omega)} \leqslant C\|\zeta\|_{W^{1,p}(\Omega)}.$$

The equalities (5.5) and Theorem 2.1 imply

$$\|\zeta\|_{W^{1,p}(\Omega)} \leqslant C\delta_p(A, B)|z|_{L_\infty(\Omega)} + C\delta_p(A, B)|\zeta|_{L_\infty(\Omega)}.$$

Hence

$$\|\zeta\|_{W^{1,p}(\Omega)} \leqslant C_1\delta_p(A, B)|\overline{\xi}| + C_2\delta_p(A, B)|\zeta|_{W^{1,p}(\Omega)},$$

where C_1 and C_2 are constants.

We determine $\varepsilon_0 > 0$ from the condition $C_2\varepsilon_0 \leqslant 1/2$. Then for $\delta_p(A, B) \leqslant \varepsilon_0$ we obtain the inequality

$$\|\zeta\|_{W^{1,p}(\Omega)} \leqslant C\delta_p(A, B)|\overline{\xi}|, \tag{5.6}$$

where $C = 2C_1$.

Since $z(x) = P(x)\xi$ and $z_0(x) = P(x)\overline{\xi}$, we have

$$\|z - z_0\|_{W^{1,p}(\Omega)} \leqslant \|P\|_{W^{1,p}(\Omega)}|\xi - \overline{\xi}| = M|\xi - \overline{\xi}|. \tag{5.7}$$

From (5.6) and (5.7) it follows that

$$\|\overline{z} - z\|_{W^{1,p}(\Omega)} \leqslant C\delta_p(A, B)|\overline{\xi}| + M|\xi - \overline{\xi}|.$$

The theorem is proved. □

6 Applications to Differential Geometry

Let Ω be a bounded domain of class S in the space $\mathbb{R}^n$, and let $\mathbf{r} : \Omega \to \mathbb{R}^{n+1}$ be a vector-valued function of class $W^{2,p}(\Omega)$. We assume that $p > n$. The partial derivatives of the vector-valued function $\mathbf{r}$ belong to the class $W^{1,p}$ and, consequently, are continuous. We set $\mathbf{r}_i = \dfrac{\partial \mathbf{r}}{\partial x^i}$.

We say that a mapping $\mathbf{r}$ is a hypersurface of class $W^{2,p}$ in $\mathbb{R}^{n+1}$ if the vectors $\mathbf{r}_1(x), \mathbf{r}_2(x), \ldots, \mathbf{r}_n(x)$ are linearly independent for any $x \in \Omega$.

Let $\mathbf{r} : \Omega \to \mathbb{R}^{n+1}$ be a hypersurface of class $W^{2,p}(\Omega)$ in $\mathbb{R}^n$. We introduce the first quadratic form of the surface $ds^2 = g_{ij}(x)dx_i dx_j$ in an ordinary way (we adopt the rule of summation over repeated indices from 1 to n), where $g_{ij} = \langle \mathbf{r}_i, \mathbf{r}_j \rangle$. The quadratic form ds^2 is positive definite since the vectors $\mathbf{r}_i(x)$, $i = 1, 2, \ldots, n$, are linearly independent. Since $\mathbf{r} \in W^{2,p}$ and $p > n$, the functions g_{ij} are continuous and belong to the class $W^{1,p}$.

For any point $x \in \Omega$ denote by $\mathbf{n}(x)$ the unit normal vector to the hyperspace $\mathbf{r}$ at the point x, i.e., $\mathbf{n}(x)$ satisfies the following conditions:

1) $\langle \mathbf{n}(x), \mathbf{r}_j(x) \rangle = 0$ for any $j = 1, 2, \ldots, n$.

2) $|\mathbf{n}(x)| = 1$.

3) The orientation of $\{\mathbf{r}_1(x), \ldots, \mathbf{r}_n(x), \mathbf{n}(x)\}$ is the same as that of the basis frame $\{\mathbf{e}_1, \ldots, \mathbf{e}_n, \mathbf{e}_{n+1}\}$ for the space $\mathbb{R}^{n+1}$, where ($\mathbf{e}_j$ is the vector in $\mathbb{R}^{n+1}$ whose jth coordinate is equal to 1, whereas the remaining coordinates vanish.

A collection of vectors $\mathbf{r}_1(x), \mathbf{r}_2(x), \ldots, \mathbf{r}_n(x), \mathbf{n}(x)$ is called the *accompanying frame* of the parametrization $\mathbf{r}$ of the surface V at the point x.

Let $R(x)$ be an $(n+1) \times (n+1)$-matrix such that for $1 \leqslant i \leqslant n$ the ith row of $R(x)$ is the vector $\mathbf{r}_i$ and the $(n+1)$th row is the vector $\mathbf{n}$. We say that $R(x)$ is the matrix representation of the accompanying frame at the point x of the hypersurface $\mathbf{r}$.

Condition 3) in the definition of $\mathbf{n}(x)$ means that $\det R(x) > 0$ for any $x \in \Omega$.

We set

$$b_{ij} = \left\langle \frac{\partial^2 \mathbf{r}}{\partial x^i \partial x^j}, \mathbf{n} \right\rangle.$$

The differential quadratic form $II = b_{ij} dx_i dx_j$ is called the *second quadratic form* of the surface $\mathbf{r}$.

We introduce matrix-valued functions $A_j(x)$ by setting

$$A_j(x) = \frac{\partial R}{\partial x_j}(x)[R(x)]^{-1}, \quad j = 1, 2, \ldots, n. \tag{6.1}$$

The derivatives $\dfrac{\partial R}{\partial x_j}$ belong to the class $L_p(\Omega)$. The matrix-valued function R is continuous and $\det R(x) > 0$ for all $x \in \Omega$. Hence the matrix-valued function A_j belongs to the class $L_{p,\mathrm{loc}}(\Omega)$. Note that $\inf \det R(x) > 0$ and $\sup |A_i(x)| < \infty$ imply $A_j \in L_p(\Omega)$.

By the definition of $A_j(x)$, for every $j = 1, 2, \ldots, n$ we have

$$\frac{\partial R}{\partial x_j}(x) = A_j(x)R(x), j = 1, 2, \ldots, n. \tag{6.2}$$

Equating the ith row of the matrix $\dfrac{\partial R}{\partial x_j}(x)$ with the ith row of the product $A_j(x)R(x)$ for every $i = 1, 2, \ldots, n+1$, we obtain the representation of the derivatives of vectors of the accompanying frame of the hypersurface $\mathbf{r}$ in the form of a linear combination of vectors of the accompanying frame. Such representations are referred to as the *derivation formulas* of the theory of surfaces. As is known, the elements of the matrix-valued functions A_j are expressed in terms of the coefficients of the first quadratic form of the surface, their first derivatives, and coefficients of the second quadratic form.

Let the matrix-valued functions $A_i(x)$ are defined by the equalities (6.1). Consider the system of differential equations

$$\frac{\partial \mathbf{z}}{\partial x_j}(x) = A_j(x)\mathbf{z}(x), \tag{6.3}$$

where $\mathbf{z}$ is a vector-valued function of class $W^{1,1}$ defined in the domain Ω and taking the valued in the space $\mathbb{R}^{n+1}$. The corresponding matrix system of equations

$$\frac{\partial Z}{\partial x_j}(x) = A_j(x)Z(x)$$

has a solution $Z(x)$ such that $\det Z(x) \neq 0$ for all $x \in \Omega$. Namely, $Z(x) \equiv R(x)$ is such a solution.

Assume that the hypersurfaces $\mathbf{r} : \Omega \to \mathbb{R}^{n+1}$ and $\overline{\mathbf{r}} : \Omega \to \mathbb{R}^{n+1}$ of class $W^2_p, p > n$, are given in $\mathbb{R}^{n+1}$, and $g_{ij}dx_idx_j, \overline{g}_{ij}dx_idx_j$ are the first quadratic forms of these surfaces. Assume that there is a constant $\gamma > 0$ such that the discriminant of each of these forms is bounded from below by the number γ^2 at every point of the domain Ω. (Recall that the discriminant of a quadratic form is the determinant of the matrix of coefficients.) Let b_{ij} and $\overline{b}_{ij}$ be the second quadratic forms of the hypersurfaces. We set

$$\Delta_p(\mathbf{r}, \overline{\mathbf{r}}) = \sum_{1 \leqslant i,j \leqslant n} \|\overline{g}_{ij} - g_{ij}\|_{W^{1,p}(\Omega)} + \sum_{1 \leqslant i,j \leqslant n} \|\overline{b}_{ij} - b_{ij}\|_{L_p(\Omega)}.$$

The quantity Δ_p characterizes the closeness of the fundamental tensors of these surfaces. Let $A_i(x)$ and $\overline{A}_i(x)$, $i = 1, 2, \ldots, n$, be matrix-valued functions defined for these surfaces by the equalities (6.1). The following estimate holds:

$$\delta_p(A, \overline{A}) \leqslant C\Delta_p(\mathbf{r}, \overline{\mathbf{r}}), \tag{6.4}$$

where C is a constant.

From the result of Sect. 5 we obtain the following assertion.

Theorem 6.1. *Let $\Omega \subset \mathbb{R}^n$ be a star-shaped domain in $\mathbb{R}^n$ relative to the ball $B(a, r)$, and let the functions $\mathbf{r} : \Omega \to \mathbb{R}^{n+1}$ and $\overline{\mathbf{r}} : \Omega \to \mathbb{R}^{n+1}$ be hypersurfaces of class W^2_p, where $p > n$. Let $ds^2 = g_{ij}$ and $ds^2 = \overline{g}_{ij}$ be the first quadratic forms of these surfaces. Denote by $g(x)$ and $\overline{g}(x)$ the discriminants of these quadratic forms. Assume that there exists a constant $\gamma > 0$ such that $g(x) > \gamma$ and $\overline{g}(x) > \gamma$ for all $x \in \Omega$. Then there are constants $\varepsilon_0 > 0$ and $C < \infty$ such that, in the case $\Delta_p(\mathbf{r}, \overline{\mathbf{r}}) < \varepsilon$, it is possible to find a motion Φ of the space $\mathbb{R}^{n+1}$ such that*

$$\|\Phi \circ \overline{\mathbf{r}} - \mathbf{r}\|_{W^2_p(\Omega)} < C\Delta_p(\mathbf{r}, \overline{\mathbf{r}}).$$

The proof of this theorem can be found in [5, Theorem 3].

Surfaces of class $\mathcal{C}^3$ were considered in [2, 4, 3], where the following assertion was proved: If the fundamental quadratic forms of surfaces are close in the topology of the space $\mathcal{C}^l$, $l \geqslant 2$, then the surfaces are also close in the topology of the space $\mathcal{C}^{l+2}$. By Theorem 6.1, a similar assertion remains valid under so weak assumptions as it is allowed by the fundamentals of the theory of surfaces.

References

1. Borovskii, Yu. E.: Completely integrable Pfaffian systems (Russian). Izv. VUZ. Ser. Mat. no. 2, c. 29-40 (1959); no. 2, 35-38 (1960)
2. Ciarlet, Ph.G.: A surface is a continuous function of its fundamental forms. C. R. Acad. Sci. Paris, Ser. I **335**, 609-614 (2002)
3. Ciarlet, Ph.G.: The continuity of a surface as a function of its two fundamental forms. J. Math. Pures Appl. **82**, 253-274 (2003)
4. Ciarlet, Ph.G., Larsonneur F.: On the recovery of a surface with prescribed first and second fundamental forms. J. Math. Pures Appl. **81**, 167-185 (2002)
5. Reshetnyak, Yu.G.: On the stability in Bonnet's theorem of surface theory. Georgian Math. J. **14**, no. 3, 543-564 (2007)
6. Sobolev, S.L.: Some Applications of Functional Analysis in Mathematical Physics (Russian). 1st ed. Leningrad State Univ., Leningrad (1950); 3rd ed. Nauka, Moscow (1988); English transl. of the 1st Ed.: Am. Math. Soc., Providence, RI (1963); English transl. of the 3rd Ed. with comments by V. P. Palamodov: Am. Math. Soc., Providence, RI (1991)

Counting Schrödinger Boundstates: Semiclassics and Beyond

Grigori Rozenblum and Michael Solomyak

Abstract This is a survey of the basic results on the behavior of the number of the eigenvalues of a Schrödinger operator, lying below its essential spectrum. We discuss both fast decaying potentials, for which this behavior is semiclassical, and slowly decaying potentials, for which the semiclassical rules are violated.

1 Introduction

The outstanding personality of Sergey L'vovich Sobolev determined the development of Analysis in the XXth century in many aspects. One of his most influential contributions to Mathematics is the invention of the function spaces now named after him and the creation of the machinery of embedding theorems for these spaces. The ideology and the techniques based upon these theorems enabled S.L. Sobolev and his followers to find comprehensive and exact solutions to many key problems in Mathematical Physics. The paper to follow is devoted to a survey of results in one of such problems. This problem concerns the behavior of the discrete part of the spectrum of a Schrödinger operator with negative potential.

The classical Weyl lemma states that the essential spectrum of a self-adjoint operator **H** in a Hilbert space is stable under perturbations by a compact operator. This lemma has many important generalizations. In particular, if

Grigori Rozenblum
Chalmers University of Technology and The University of Gothenburg S-412 96, Gothenburg, Sweden, e-mail: grigori@math.chalmers.se

Michael Solomyak
The Weizmann Institute of Science Rehovot, 76100, Israel, e-mail: michail.solomyak@weizmann.ac.il

V. Maz'ya (ed.), *Sobolev Spaces in Mathematics II*, *International Mathematical Series*.
doi: 10.1007/978-0-387-85650-6,

$\mathbf{H}$ is nonnegative, the result survives if the perturbation is only relatively compact with respect to $\mathbf{H}$, in the sense of quadratic forms.

The leading and most inspiring example in spectral theory, where the Weyl lemma plays the key role, concerns the discrete spectrum of a Schrödinger operator

$$\mathbf{H}_V = -\Delta - V$$

on $\mathbb{R}^d$. Here, $V = V(x)$ is a real-valued measurable function on $\mathbb{R}^d$ (the potential), which we assume to decay at infinity, in a certain appropriate sense. Then the operator can be defined via the corresponding quadratic form considered on the Sobolev space $H^1(\mathbb{R}^d)$. We assume for simplicity that $V \geqslant 0$. Results for general real-valued potentials can be then derived by using the variational principle. In this paper, we do not touch upon the results which take into account the interplay between the positive and the negative parts of the potential.

For the description of the spectrum of the operators involved we use the following notation. Let $\sigma(\mathbf{H})$ and $E_{\mathbf{H}}(\cdot)$ stand for the spectrum and the spectral measure of a self-adjoint operator $\mathbf{H}$. We call the number

$$\mathrm{bott}(\mathbf{H}) := \inf\{\lambda : \lambda \in \sigma(\mathbf{H})\}$$

the *bottom* of the operator $\mathbf{H}$. We put

$$N_-(\beta; \mathbf{H}) = \dim E_{\mathbf{H}}(-\infty, \beta), \qquad \beta \in \mathbb{R}.$$

The relation

$$N_-(\beta; \mathbf{H}) < \infty$$

means that the spectrum of $\mathbf{H}$ on the half-line $(-\infty, \beta)$ is discrete, moreover, finite and $N_-(\beta; \mathbf{H})$ gives the number of the eigenvalues of $\mathbf{H}$ counted according to their multiplicities and lying on this half-line.

The spectrum of the free Laplacian $\mathbf{H}_0 = -\Delta$ in $L^2(\mathbb{R}^d)$ is the half-line $[0, \infty)$, and, by the Weyl lemma, the negative spectrum of $\mathbf{H}_V$ is discrete. However, this lemma gives no quantitative information about the negative spectrum: it does not allow one to find out, whether the quantity $N_-(0; \mathbf{H}_V)$ is infinite, or finite, and, in the latter case, it gives no control of its size. It is often important to answer these questions. In order to make the problem more transparent, it is useful to insert a real positive parameter (the *coupling constant*) and to study the above questions for the family

$$\mathbf{H}_{\alpha V} = -\Delta - \alpha V, \qquad \alpha > 0. \tag{1.1}$$

The function $N_-(0; \mathbf{H}_{\alpha V})$ grows together with α, and this growth of the number of negative eigenvalues can be interpreted as birth of new bound states from the edge of the continuous spectrum as the exterior field grows. At the same time, $N(0; \mathbf{H}_{\alpha V}) = N(0; -\alpha^{-1}\Delta - V)$, so the behavior of this quantity as $\alpha \to \infty$ describes simultaneously the semiclassical behavior of the eigenvalues as the "Planck constant" $\alpha^{-\frac{1}{2}}$ tends to 0.

Along with $N_-(0; \mathbf{H}_{\alpha V})$, one often studies the function $N_-(-\gamma; \mathbf{H}_{\alpha V})$, where $\gamma > 0$. If the assumptions about V guarantee discreteness of the negative spectrum of $\mathbf{H}_{\alpha V}$, then the latter number is always finite. If, in addition, $N_-(0; \mathbf{H}_{\alpha V}) = \infty$, the behavior of $N_-(-\gamma; \mathbf{H}_{\alpha V})$ as $\gamma \to 0+$, for α fixed, is an important characteristics of the operator.

The main contents of the present paper is a survey and a certain detailing of the known results on the behavior of the function $N_-(-\gamma; \mathbf{H}_{\alpha V})$, $\gamma \geqslant 0$, for the Schrödinger operator (1.1) and its generalizations – such as Schrödinger operators on manifolds, or in domains $\Omega \subset \mathbb{R}^d$. Note that, in the latter cases, the bottom of the Laplacian is not necessarily equal to zero. Then we discuss the behavior of $N_-(\beta; \mathbf{H}_{\alpha V})$ for a fixed value of $\beta \leqslant \operatorname{bott}(-\Delta)$ (we refrain from using the notation $N_-(-\gamma, \mathbf{H}_{\alpha V})$ except for the cases where $\operatorname{bott}(\mathbf{H}_0) = 0$).

Our starting point is the Weyl asymptotic law, which allows one to realize what sort of results is desirable.

If the potential V is nice (say, C_0^∞), then for any $\gamma \geqslant 0$ the function $N_-(-\gamma; \mathbf{H}_{\alpha V})$ exhibits the semiclassical, or Weyl, asymptotic behavior, i.e.,

$$N_-(-\gamma; \mathbf{H}_{\alpha V}) \sim w_d \alpha^{\frac{d}{2}} \int\limits_{\mathbb{R}^d} V^{\frac{d}{2}} dx, \qquad \alpha \to \infty. \tag{1.2}$$

Here, $w_d = v_d (2\pi)^{-d}$, where v_d stands for the volume of the unit ball in $\mathbb{R}^d$. (The word "semiclassical" is used in order to indicate that the expression on the right-hand side in (1.2) is proportional to the volume of the region in the classical phase space $\mathbb{R}^{2d}$, where the classical Hamiltonian $p^2 - \alpha V(x)$ is negative.) In particular, the asymptotic formula (1.2) hints that for any potential $V \in L^1_{\text{loc}}(\mathbb{R}^d)$ the function $N_-(-\gamma; \mathbf{H}_{\alpha V})$ cannot grow (in α) slower that $O(\alpha^{\frac{d}{2}})$. But can it grow faster?

In this connection, the following questions arise in a natural way.

A. *To describe the classes of potentials that guarantee the estimate*

$$N_-(-\gamma; \mathbf{H}_{\alpha V}) = O(\alpha^{\frac{d}{2}}), \qquad \alpha \to \infty. \tag{1.3}$$

Another important question is this:

B. *Suppose that for a given potential V we have* (1.3). *Does this imply the asymptotic formula* (1.2)?

One more natural question:

C. *What can be said about the eigenvalues for such potentials that the negative spectrum of $\mathbf{H}_{\alpha V}$ is still discrete, but* (1.3) *is violated?*

In the paper, we discuss the present situation with answers to these questions. The answers heavily depend on the dimension. In particular, the answer to the question **B** is YES if $d \geqslant 3$, and it is NO if $d = 1, 2$.

We also discuss the analogues of these problems for the Laplacian on a manifold and, more briefly, on domains $\Omega \subset \mathbb{R}^d$ and on the lattice $\mathbb{Z}^d$. Note that in all these cases the situation is understood up to a much lesser extent, than for $\mathbb{R}^d$.

The number $N_-(0; \mathbf{H}_V)$ can be interpreted as the borderline value, for $r = 0$, of the quantity

$$S_r(V) = \sum_{\lambda_j(\mathbf{H}_V)<0} |\lambda_j(\mathbf{H}_V)|^r, \qquad r > 0.$$

Estimating such sums is important for Physics, and this is the main subject in the so-called Lieb–Thirring inequalities. In this paper, we do not touch upon this popular topic; see [15] for a survey and [11] for newer results.

2 Operators on $\mathbb{R}^d$, $d \geqslant 3$

2.1 The Rozenblum–Lieb–Cwikel estimate

In the case considered, the answer to the questions **A, B** is given by the so-called Rozenblum–Lieb–Cwikel (RLC) estimate, named after the mathematicians who gave the first independent proofs of the result. In the form given below, the result is due to Rozenblum [22, 23]. Other authors [18] and [6] did not discuss the necessity of the condition on V.

Theorem 2.1. *Let $d \geqslant 3$. Then there exists a constant $C = C(d)$ such that for any $V \in L^{\frac{d}{2}}(\mathbb{R}^d)$, $V \geqslant 0$, and any $\gamma \geqslant 0$*

$$N_-(-\gamma; \mathbf{H}_{\alpha V}) \leqslant C(d)\alpha^{\frac{d}{2}} \int_{\mathbb{R}^d} V^{\frac{d}{2}} dx; \tag{2.1}$$

moreover, the asymptotic formula (1.2) holds.

Conversely, suppose that $d \geqslant 3$, for a certain $V \geqslant 0$ the operator $\mathbf{H}_{\alpha V}$ is well defined for all $\alpha > 0$, and for some $\gamma \geqslant 0$ the function $N_-(-\gamma; \mathbf{H}_{\alpha V})$ is $O(\alpha^{\frac{d}{2}})$ as $\alpha \to \infty$. Then $V \in L^{\frac{d}{2}}(\mathbb{R}^d)$, and, therefore, the estimate (2.1) and asymptotic formula (1.2) are fulfilled for an arbitrary $\gamma \geqslant 0$.

Evidently, the estimate (2.1) for any $\gamma > 0$ and $\alpha > 0$ is a consequence of its particular case for $\gamma = 0$ and $\alpha = 1$. The asymptotic formula (1.2) is proved first by elementary methods (Dirichlet–Neumann bracketing) for potentials $V \in C_0^\infty(\mathbb{R}^d)$. It extends to the general case by a machinery, known as "completion of spectral asymptotics" and presented in detail in the book [3], see especially Lemma 1.19 there.

The proofs given by Rozenblum, by Lieb, and by Cwikel used different techniques. Rozenblum's approach was based upon the Sobolev embedding theorem in combination with Besicovitch type covering theorem; Cwikel applied harmonic analysis and theory of interpolation of linear operators. Both these proofs apply to much more general classes of operators than just to the Laplacian, but only in the $\mathbb{R}^d$-setting. The first proof which admits generalization to other situations, say to operators on manifolds, is due to Lieb, who used the semigroup theory, in the form of path integrals.

Later several other proofs were suggested, including the ones given by Fefferman [10] and by Li and Yau [17]. For us, the latter is especially remarkable since it shows in an extremely transparent form the deep connection between the "global Sobolev inequality" and the RLC estimate. The techniques in [17] uses semigroup theory in a somewhat more direct way than in [18]. Like Lieb's proof, it admits far-reaching generalizations.

3 The General Rozenblum–Lieb–Cwikel Inequality

3.1 The approach by Li and Yau

What we present below is an abstract version of the Li-Yau result. It was established in the paper [16] whose authors aimed at finding the most general setting in which the approach of [17] applies. The classical notion of sub-Markov semigroup is used in the formulation.

Let (Ω,σ) be a measure space with sigma-finite measure. We denote $L^q(\Omega) = L^q(\Omega,\sigma)$ and $\|\cdot\|_q = \|\cdot\|_{L^q(\Omega)}$. Suppose that a nonnegative quadratic form $Q[u]$ is defined on a dense in $L^2(\Omega)$ linear subset Dom $[Q]$. We assume that Q is closed and the corresponding self-adjoint operator $\mathbf{A} = \mathbf{A}_Q$ generates a symmetric, positivity preserving semigroup. In this situation, we say that the operator $\mathbf{A}$ is a *sub-Markov generator*. We also suppose that there exists an exponent $q > 2$ and a positive constant K such that

$$\|u\|_q^{\frac{2}{q}} \leqslant KQ[u] \qquad \forall u \in \text{Dom } Q. \tag{3.1}$$

Theorem 3.1. *Let $Q[u]$ be the quadratic form of a sub-Markov generator in $L^2(\Omega)$. Suppose that the estimate* (3.1) *is satisfied with some $q > 2$. Let*

$$0 \leqslant V \in L_p(\Omega), \qquad p = (1 - \frac{2}{q})^{-1}. \tag{3.2}$$

Then the quadratic form

$$Q_V[u] := Q[u] - \int_\Omega V|u|^2 d\sigma, \qquad u \in \mathrm{Dom}\,[Q],$$

is bounded from below in $L^2(\Omega)$ and is closed. The negative spectrum of the corresponding self-adjoint operator $\mathbf{A} - V$ in $L^2(\Omega)$ is finite, and

$$N_-(0; \mathbf{A} - V) \leqslant C(p)K^p \int_\Omega V^p d\sigma, \qquad C(p) = e^2(\frac{p}{2})^p. \tag{3.3}$$

We will call (3.3) the *general RLC inequality.*

It is well known that for any d the (minus) Laplacian on $\mathbb{R}^d$ is a sub-Markov generator. The inequality (3.1) is satisfied if $d \geqslant 3$ with $q = \frac{2d}{d-2}$, so that $p = \frac{d}{2}$ in (3.2). This is the so-called "global Sobolev inequality," and the sharp value of the constant K is known (see, for example, [19, Sect. 2.3.3, the inequality (3)]). So, Theorem 3.1 implies the RLC estimate (2.1) with an explicitly given constant. For the case $d = 3$, which is the most interesting for Physics, this constant is slightly greater than the best value $C(3) = .116$ in (2.1), known up to now. It should be compared with the constant $w_3 = .078$ in the asymptotic formula (1.2). This best value is given by Lieb's approach (see Subsect. 3.2 below). Note that the sharp value of the constant $C(d)$ in (2.1), even for $d = 3$, is unknown up to now.

3.2 The approach by Lieb

Below, we present the main result of the paper [24], where an abstract version of Lieb's approach was elaborated.

Any nonnegative self-adjoint operator $\mathbf{A}$ in $L^2(\Omega)$ generates a contractive semigroup $e^{-t\mathbf{A}}$. We suppose that this semigroup is $(2,\infty)$-bounded, which means that for any $t > 0$ the operator $e^{-t\mathbf{A}}$ is bounded as acting from $L^2(\Omega)$ to $L^\infty(\Omega)$. We write

$$\mathbf{A} \in \mathcal{P}$$

if the semigroup $e^{-t\mathbf{A}}$ is $(2,\infty)$-bounded and positivity preserving.

Let $K(t; x, y)$ be the integral (Schwartz) kernel of $e^{-t\mathbf{A}}$. Then $K(t; x, x)$ is well defined on $\mathbb{R}_+ \times \Omega$, and it belongs to $L^\infty(\Omega)$ for each $t > 0$. We put

$$M_{\mathbf{A}}(t) = \|K(t; \cdot)\|_\infty.$$

The main result is a *parametric estimate* (see (3.5) below): it involves an arbitrary function $G(z)$ of a certain class, as a parameter. The class $\mathcal{G}$ of admissible functions G is defined as follows.

The function G is continuous, convex, nonnegative, grows at infinity no faster than a polynomial, and is such that $z^{-1}G(z)$ is integrable at zero. With each $G \in \mathcal{G}$ we associate another function,

$$g(\lambda) = \int_{\mathbb{R}_+} z^{-1} G(z) e^{-\frac{z}{\lambda}} dz, \qquad \lambda > 0. \tag{3.4}$$

Theorem 3.2. *Suppose that* $\mathbf{A} \in \mathcal{P}$ *is such that the function* $M_{\mathbf{A}}(t)$ *is integrable at infinity and is* $o(t^{-a})$ *at zero with some* $a > 0$. *Fix a function* $G \in \mathcal{G}$, *and define* $g(\lambda)$ *as in* (3.4). *Then*

$$N_-(0; \mathbf{A} - V) \leqslant \frac{1}{g(1)} \int_{\mathbb{R}_+} \frac{dt}{t} \int_{\Omega} M_{\mathbf{A}}(t) G(tV(x)) d\sigma, \tag{3.5}$$

whenever the integral on the right is finite.

Note that the finiteness of the integral in (3.5) guarantees that the relative bound of V with respect to the quadratic form of the operator $\mathbf{A}$ is smaller than 1, so that the operator $\mathbf{A} - V$ is well defined via its quadratic form.

If (Ω, σ) is $\mathbb{R}^d$ with the Lebesgue measure, and $\mathbf{A} = -\Delta$, then the semigroup $e^{-t\mathbf{A}}$ is positivity preserving and $(2,\infty)$-bounded, and $M_{-\Delta}(t) = (2\pi)^{-\frac{d}{2}} t^{-\frac{d}{2}}$. Since $M(t)$ is a pure power, the choice of $G \in \mathcal{G}$ is indifferent, within the value of the constant factor in the estimate. Indeed, by a change of variables, the estimate (3.5) reduces to the form

$$N_-(0; \mathbf{A} - V) \leqslant C(G) \int_{\mathbb{R}^d} V^{\frac{d}{2}} dx, \tag{3.6}$$

where

$$C(G) = \frac{1}{g(1)(2\pi)^{\frac{d}{2}}} \int_0^{\infty} z^{-(\frac{d}{2}+1)} G(z) dz.$$

The assumptions about G and the finiteness of $C(G)$ dictate the restriction $d \geqslant 3$. The optimal choice of $G \in \mathcal{G}$ was pointed out by Lieb [18].

The relation between Theorems 3.2 and 3.1 is based upon the deep connection between the Sobolev type inequality (3.1) and the estimate

$$M_{\mathbf{A}}(t) \leqslant C t^{-\frac{d}{2}}; \qquad t \in (0, \infty) \tag{3.7}$$

for the heat kernel corresponding to the operator $\mathbf{A} = \mathbf{A}_Q$. This connection was established by Varopoulos (see [26, Sect. II.2] or [7, Theorem 2.4.2]).

Theorem 3.3. *If the quadratic form* $Q[u]$ *generates a symmetric sub-Markov semigroup on the measure space* (Ω, σ), *then the estimate* (3.7) *with* $d > 2$ *is equivalent to the inequality* (3.1) *with* $q = \frac{2d}{d-2}$.

So, the result of Theorem 3.2 yields the general RLC inequality (3.3) and thus, is stronger than Theorem 3.1. Indeed, in the general setting, the behavior of the function $M_{\mathbf{A}}(t)$ is not necessarily expressed by the inequality (3.7) with the same exponent d both as $t \to 0$ and $t \to \infty$. In many cases,

$$M_{\mathbf{A}}(t) \leqslant C_0 t^{-\frac{\delta}{2}}, \qquad t < 1; \tag{3.8}$$

$$M_{\mathbf{A}}(t) \leqslant C_\infty t^{-\frac{D}{2}}, \qquad t > 1, \tag{3.9}$$

with $D \neq \delta$. In [26], such estimates were studied for the sub-Laplacian on nilpotent groups, and δ and D were called there dimensions at zero and at infinity respectively. We also use these terms. One encounters a similar situation when studying the Laplacian on manifolds or domains in $\mathbb{R}^d$.

If the estimates (3.8), (3.9) are known with $\delta, D > 2$, the eigenvalue estimates obtained from (3.5) vary essentially, depending on which dimension, δ or D is larger.

We formulate the corresponding results, not trying to find best possible constants, however we include the coupling parameter α.

Theorem 3.4. *Under the conditions of Theorem* 3.2, *suppose that the inequalities* (3.8), (3.9) *are satisfied with some* $\delta, D > 2$. *Then the following eigenvalue estimates hold*:

$$N_-(0; \mathbf{H}_{\alpha V}) \leqslant C_0' \alpha^{\frac{\delta}{2}} \int_\Omega V^{\frac{\delta}{2}} d\sigma + C_\infty' \alpha^{\frac{D}{2}} \int_\Omega V^{\frac{D}{2}} d\sigma \tag{3.10}$$

if $\delta \geqslant D$ *and*

$$N_-(0; \mathbf{H}_{\alpha V}) \leqslant C_0'' \alpha^{\frac{\delta}{2}} \int_{\alpha V \geqslant 1} V^{\frac{\delta}{2}} d\sigma + C_\infty'' \alpha^{\frac{D}{2}} \int_{\alpha V < 1} V^{\frac{D}{2}} d\sigma \tag{3.11}$$

if $\delta \leqslant D$.

Remark 3.1. If $\delta \leqslant D$, the inequalities (3.8) and (3.9) imply (3.7) with $d = D$, and hence

$$N_-(0; \mathbf{H}_{\alpha V}) \leqslant \widetilde{C} \alpha^{\frac{D}{2}} \|V\|_{\frac{D}{2}}^{\frac{D}{2}}, \qquad D > 2. \tag{3.12}$$

It is often important that the assumption $\delta > 2$, appearing in Theorem 3.4, here is unnecessary.

We discuss applications of the estimates (3.10) and (3.11) in Sects. 7–9.

4 Operators on $\mathbb{R}^d$, $d \geqslant 3$: Non-Semiclassical Behavior of $N_-(0; \mathbf{H}_{\alpha V})$.

Now, suppose that $d \geqslant 3$ but $V \notin L^{\frac{d}{2}}(\mathbb{R}^d)$, though $V(x)$ vanishes as $|x| \to \infty$, again in some appropriate sense. Then the negative spectrum of $-\Delta - \alpha V$

is still discrete, but the RLC inequality becomes useless. In this situation, some estimates for the quantity $N_-(0; \mathbf{A} - \alpha V)$ can be obtained by using interpolation between the RLC inequality (2.1) and a remarkable result due to Maz'ya [19, Sect. 2.3.3]. This result gives the necessary and sufficient conditions on a weight function $V \geqslant 0$ for the Hardy type inequality

$$\int_{\mathbb{R}^d} V|u|^2 dx \leqslant C(V) \int_{\mathbb{R}^d} |\nabla u|^2 dx \qquad \forall u \in C_0^\infty(\mathbb{R}^d)$$

to be satisfied.

Here, we present only a particular case of the general class of estimates obtained by this approach (see [4] for details).

Theorem 4.1. *Let $d \geqslant 3$. Suppose that for some $q > \frac{d}{2}$ the potential V satisfies the condition*

$$[[V]]_q^q := \sup_{t>0} \Big(t^q \int_{|x|^2 V(x) > t} \frac{dx}{|x|^d} \Big) < \infty. \tag{4.1}$$

Then for any $\alpha > 0$ the operator $-\Delta - \alpha V$ on $\mathbb{R}^d$ is bounded from below, its negative spectrum is finite, and the estimate

$$N_-(0; \mathbf{H}_{\alpha V}) \leqslant C(d,q)\alpha^q [[V]]_q^q \tag{4.2}$$

is satisfied.

The condition (4.1) means that the function $|x|^2 V(x)$ belongs to the so-called *weak L^q-space*, usually denoted by L^q_w, with respect to the measure $|x|^{-d}dx$ on $\mathbb{R}^d$. The functional $[[V]]_q$ is equivalent to the norm in this space, but it does not meet the triangle inequality itself. The space L^q_w is nonseparable, and it contains the usual space L_q with respect to the same measure. Replacing in (4.2) the functional $[[V]]_q$ by the norm in L^q coarsens the estimate, and we come to the inequality

$$N_-(0; \mathbf{H}_{\alpha V}) \leqslant C'(d,q)\alpha^q \int_{\mathbb{R}^d} V^q |x|^{2q-d} dx, \qquad 2q > d, \tag{4.3}$$

which looks simpler than (4.2). The estimate (4.3) was established in [9] by a direct approach, generalizing the one in [23]. However, (4.3) is knowingly not exact: it is easy to see that the finiteness of the integral in (4.3) implies

$$N_-(0; \mathbf{H}_{\alpha V}) = o(\alpha^q), \qquad \alpha \to \infty. \tag{4.4}$$

Indeed, this is certainly the case for the potentials $V \in C_0^\infty(\mathbb{R}^d)$. Such potentials are dense in L^q with weight $|x|^{2q-d}$, and the procedure of completion of spectral asymptotics, mentioned in the paragraph next to Theorem 2.1, shows that (4.4) extends to all V from this space. This nice reasoning is due to Birman (private communication). It easily extends to the general situation, and it shows that *any order-sharp estimate of order* $q > \frac{d}{2}$ *for the quantity* $N_-(0; \mathbf{H}_{\alpha V})$ *must involve some nonseparable class of potentials.*

In contrast to (4.3), the estimate (4.2) is order-sharp: say, for the potential V which for large $|x|$ is equal to

$$V(x) = |x|^{-2}(\log|x|)^{-\frac{1}{q}}, \qquad 2q > d, \tag{4.5}$$

the condition (4.2) is satisfied, and for such potentials the asymptotics

$$N_-(0; \mathbf{H}_{\alpha V}) \sim c_q \alpha^q, \ c_q > 0, \qquad \alpha \to \infty$$

is known (see [4]).

The condition (4.1) allows local singularities of V at the point $x = 0$, which are stronger than those allowed by the inclusion $V \in L^{\frac{d}{2}}(\mathbb{R}^d)$. The weight function $|x|^2$ and the measure $|x|^{-d}dx$ in (4.1) can be replaced by functions and measures in a rather wide class (see [4]). In particular, this allows one to control effects coming from singularities of V distributed on submanifolds in $\mathbb{R}^d$. For example, suppose that we are interested in the potentials with singularities at the sphere $|x| = 1$. Then, instead of (4.2), one can use the estimate

$$N_-(0; \mathbf{H}_{\alpha V}) \leqslant C\alpha^q \sup_{t>0} \int\limits_{V(x)||x|-1|^{\frac{2}{d}} > t} \frac{dx}{||x|-1|}, \qquad 2q > d \geqslant 3.$$

Both this estimate and (4.2) are particular cases of Theorem 4.1 in [4].

We do not think that a unified condition on the potential, which is necessary and sufficient for $N_-(0; \mathbf{H}_{\alpha V}) = O(\alpha^q)$ with a prescribed value of $q > \frac{d}{2}$, does exist.

5 Operators on the Semi-Axis

5.1 Semiclassical behavior

In the case $d = 1$, it is natural to deal with the operators on the semi-axis $\mathbb{R}_+$ defined as

$$\mathbf{H}_{\alpha V} u(x) = -u''(x) - \alpha V(x) u(x), \qquad u(0) = 0. \tag{5.1}$$

An accurate definition can be given via the corresponding quadratic form. The case of operators on the whole axis is easily reduced to this one by imposing the additional Dirichlet condition at $x = 0$ and adding up the two similar estimates for the operators acting on the positive and the negative semi-axis. The term +1 must be included in the right-hand side of the resulting estimate since imposing this boundary condition means the passage to a subspace of codimension 1 in $H^1(\mathbb{R})$. Appearing of the term +1 reflects the fact that $\lambda = 0$ is a resonance point for the operator $-\frac{d^2}{dx^2}$ in $L^2(\mathbb{R})$. This means that for an arbitrary nontrivial potential $V \geqslant 0$ at least one eigenvalue exists for any $\alpha > 0$. Hence no estimate homogeneous in α is possible.

The character of estimates for the operator (5.1) is quite different from the RLC inequality which governs the case $d \geqslant 3$. The necessary and sufficient condition for the semiclassical order

$$N_-(0; \mathbf{H}_{\alpha V}) = O(\alpha^{\frac{1}{2}}) \tag{5.2}$$

is given by Theorem 5.1 below. However, this condition hardly can be reformulated in purely function-theoretic terms.

With any function $0 \leqslant V \in L^1_{\text{loc}}(\mathbb{R}_+)$ we associate the sequence $\boldsymbol{\eta}(V) = \{\eta_j(V)\}$, $j \in \mathbb{Z}$, where

$$\eta_j(V) = 2^j \int_{I_j} V(x)dx, \qquad I_j = (2^j, 2^{j+1}), \ j \in \mathbb{Z}. \tag{5.3}$$

It is not difficult to show that

$$N_-(0; \mathbf{H}_{\alpha V}) \leqslant C\alpha^{\frac{1}{2}} \sum_{j \in \mathbb{Z}} \eta_j^{\frac{1}{2}}(V),$$

so that the condition

$$\boldsymbol{\eta}(V) \in \ell^{\frac{1}{2}} \tag{5.4}$$

is sufficient for the estimate (5.2). It also guarantees the validity of the Weyl asymptotics, which in this case takes the form

$$N_-(0; \mathbf{H}_{\alpha V}) \sim \pi^{-1}\alpha^{\frac{1}{2}} \int_{\mathbb{R}_+} V^{\frac{1}{2}} dx, \qquad \alpha \to \infty. \tag{5.5}$$

However, the condition (5.4) is not necessary either for (5.2), or for (5.5).

In order to write the necessary and sufficient condition, let us consider the family of eigenvalue problems on the intervals I_j, $j \in \mathbb{Z}$:

$$-\lambda u''(x) = V(x)u(x) \text{ on } I_j, \qquad u(2^j) = u(2^{j+1}) = 0. \tag{5.6}$$

Here, it is convenient for us to put the spectral parameter in the left-hand side, then for each j the eigenvalues $\lambda_{j,k}$, $k = 1, 2, \ldots,$ of the problem (5.6) correspond to a compact operator. Let $n_j(\lambda)$ stand for their counting function:

$$n_j(\lambda) = \#\{k : \lambda_{j,k} > \lambda\}, \qquad \lambda > 0.$$

Each function $n_j(\lambda)$ satisfies the estimate

$$\lambda^{\frac{1}{2}} n_j(\lambda) \leqslant C\left(2^j \eta_j(V)\right)^{\frac{1}{2}} = C2^j\Big(\int\limits_{I_j} V dx\Big)^{\frac{1}{2}} \tag{5.7}$$

and exhibits the Weyl asymptotic behavior:

$$\lambda^{\frac{1}{2}} n_j(\lambda) \to \pi^{-1} \int\limits_{I_j} V^{\frac{1}{2}} dx. \tag{5.8}$$

The estimate (5.7) is uniform in j (i.e., the constant C does not depend on j), but the asymptotics (5.8) is not. This is reflected in the fact that the potential V is involved in (5.7) and in (5.8) in two different ways.

The following result was obtained in [20].

Theorem 5.1. *Let $0 \leqslant V \in L^1_{loc}(\mathbb{R}_+)$, and let $\mathbf{H}_{\alpha V}$, $\alpha > 0$, be the family of operators* (5.1). *The two conditions*

$$\#\{j \in \mathbb{Z} : \eta_j(V) > \lambda\} = O(\lambda^{-\frac{1}{2}}), \qquad \lambda + \lambda^{-1} \to \infty, \tag{5.9}$$

and

$$\sup_{\lambda > 0} \sum_j \lambda^{\frac{1}{2}} n_j(\lambda) < \infty \tag{5.10}$$

are necessary and sufficient for the semiclassical order (5.2) *of the quantity $N_-(0; \mathbf{H}_{\alpha V})$.*

The condition (5.9) means, by definition, that the sequence $\boldsymbol{\eta}(V)$ belongs to the *weak $\ell^{\frac{1}{2}}$-space* (notation $\ell^{\frac{1}{2}}_w$). This condition is much weaker than (5.4).

The conditions (5.9) and (5.10) do not guarantee the Weyl asymptotics (5.5). The necessary and sufficient condition on V for the validity of this asymptotics was also established in [20]; we do not duplicate it here. Note only that, in [20], a series of examples was constructed of potentials V for which the estimate (5.2) holds but the asymptotic formula is valid with the coefficient different from the one in (5.5). This is impossible in dimension $d \geqslant 3$.

5.2 Non-semiclassical behavior of $N_-(0; \mathbf{H}_{\alpha V})$

The situation is rather simple. The criterion for $N_-(0; \mathbf{H}_{\alpha V}) = O(\alpha^q)$ with a given $q > \frac{1}{2}$ can be expressed in terms of the same sequence (5.3).

Theorem 5.2. *Let $0 \leqslant V \in L^1_{loc}(\mathbb{R}_+)$, and let $2q > 1$. The condition*

$$\#\{j \in \mathbb{Z} : \eta_j(V) > \lambda\} = O(\lambda^{-q}), \qquad \lambda + \lambda^{-1} \to \infty, \tag{5.11}$$

is necessary and sufficient for $N_-(0; \mathbf{H}_{\alpha V}) = O(\alpha^q)$, *and the inequality*

$$N_-(0; \mathbf{H}_{\alpha V}) \leqslant C_q \alpha^q \sup_{\lambda > 0} \lambda^q \#\{j \in \mathbb{Z} : \eta_j(V) > \lambda\} \tag{5.12}$$

is satisfied.

The condition similar to (5.11) *with* $o(\lambda^{-q})$ *on the right-hand side, is necessary and sufficient for* $N_-(0; \mathbf{H}_{\alpha V}) = o(\alpha^q)$.

In particular, the condition (5.11) with $q = 1$ is fulfilled provided that

$$\int_{\mathbb{R}_+} xV(x)dx < \infty.$$

The inequality

$$N_-(0; \mathbf{H}_{\alpha V}) \leqslant \alpha \int_{\mathbb{R}_+} xV(x)dx$$

is the classical Bargmann estimate (see, for example, [21]). So, the inequality (5.12) covers this result, within the value of the constant factor. Note that, under the Bargmann condition, one always has $N_-(0; \mathbf{H}_{\alpha V}) = o(\alpha)$. The argument is the same as in Sect. 4.

The proof of Theorem 5.2 can be found in [3], where actually more general multi-dimensional problems were analyzed, and in [2] (see also [1], where the result is presented without proof).

Theorem 5.2 turns out to be quite useful for the estimation of $N_-(\gamma; \mathbf{H}_{\alpha V})$ for such multi-dimensional problems, where an additional "channel" can be singled out, that contributes independently to the behavior of this function for large values of α. This happens, for instance, in many problems on manifolds (see Sect. 8). Another, may be the most striking example, is connected with the Laplacian on $\mathbb{R}^2$. We discuss this case in the following section.

6 Operators on $\mathbb{R}^2$

6.1 Semiclassical behavior

In the borderline case $d = 2$, the exhaustive description of the class of potentials such that $N_-(0; \mathbf{H}_{\alpha V}) = O(\alpha)$, or at least

$$N_-(-\gamma; \mathbf{H}_{\alpha V}) = O(\alpha) \qquad \forall \gamma > 0, \tag{6.1}$$

is not known up to present. On the technical level, this is a consequence of the fact that the embedding theorem $H^1(\mathbb{R}^d) \subset L^q(\mathbb{R}^d)$, $q = \frac{2d}{d-2}$, fails for $d = 2$ (when $q = \infty$), or of the equivalent fact that $M_\Delta(t) = ct^{-1}$ (see Sect. 3.2), so the integral in (3.5) diverges. There are various sufficient conditions on the potential which ensure the order-sharp in α estimate for the function (6.1), but all of them are not sharp in the function classes for V. Even the most general sufficient condition of this type, known up to now (formulated in terms of Orlicz spaces), see [25], is not necessary. What is more, there are problems of a rather close nature, for which the RLC-like condition $V \in L^1(\mathbb{R}^2)$ turns out to be sufficient (see [13, 14]). So, for $d = 2$ the situation is not well understood up to now.

Below, we present a comparatively simple sufficient condition, which was found in [1]. Fix a number $\varkappa > 1$, and with any potential $0 \leqslant V \in L^{\varkappa}_{\text{loc}}(\mathbb{R}^2)$ let us associate the sequence $\boldsymbol{\theta}(V, \varkappa) = \{\theta_j(V, \varkappa)\}$, $j = 0, 1, \dots$, where

$$\theta_0(V, \varkappa)^{\varkappa} = \int\limits_{|x|<1} V^{\varkappa} dx,$$

$$\theta_j(V, \varkappa)^{\varkappa} = \int\limits_{2^{j-1}<|x|<2^j} |x|^{2(\varkappa-1)} V^{\varkappa} dx, \qquad j \in \mathbb{N}.$$

Theorem 6.1. *For any fixed numbers $\varkappa > 1$ and $\gamma > 0$ there exists a constant $C(\gamma, \varkappa) > 0$ such that as soon as*

$$\boldsymbol{\theta}(V, \varkappa) \in \ell^1, \tag{6.2}$$

the operator $\mathbf{H}_{\alpha V}$ is bounded from below for any $\alpha > 0$, its negative spectrum is discrete, and

$$N_-(-\gamma; \mathbf{H}_{\alpha V}) \leqslant C(\gamma, \varkappa)\alpha \|\boldsymbol{\theta}(V, \varkappa)\|_1.$$

The constant $C(\gamma, \varkappa)$ may blow up as $\gamma \to 0$, and the assumption (6.2) does not guarantee the semiclassical behavior of $N_-(0; \mathbf{H}_{\alpha V})$. It turns out that for the analysis of this behavior one has to consider separately the subspace $\mathcal{F}$ of radial functions, $u(x) = f(|x|)$. On $\mathcal{F}$, the quadratic form of $\mathbf{H}_{\alpha V}$ generates a second order ordinary differential operator whose spectrum is not controlled by the sequence (4.1). In order to control it and to have the semiclassical order $N_-(0; \mathbf{H}_{\alpha V}) = O(\alpha)$ for the original operator, one uses Theorem 5.2 with the exponent $q = 1$. In the following theorem, we present the final result which can be obtained by means of this approach; in formula (6.3) below, we express the potential V in the polar coordinates.

Theorem 6.2. *Let $V \geqslant 0$ satisfy the conditions of* Theorem 6.1. *Consider an auxiliary "effective potential" on $\mathbb{R}_+$,*

$$F_V(t) = e^{2t} \int_{-\pi}^{\pi} V(e^t, \varphi) d\varphi, \qquad t > 0. \tag{6.3}$$

Let $\{\eta_j(V)\}$ *be the sequence* (5.3) *for the potential* F_V. *Then* (6.1) *holds for* $\gamma = 0$ *if and only if the additional condition* (5.11) *with* $q = 1$ *is fulfilled.*

Note that by changing (5.11) to a stronger condition, with $o(\lambda^{-1})$ on the right, we come to a condition ensuring the Weyl asymptotics (1.2) for $d = 2$ (see [25] and, especially, [1] for more details and discussion).

This effect (appearance of an additional differential operator in a lower dimension, which contributes to the behavior of $N_-(0; \mathbf{H}_{\alpha V})$ in an independent way) we meet in several other problems discussed in Sect. 8. This can be interpreted as opening of an additional channel which affects the behavior of the system studied.

6.2 Non-semiclassical behavior

It is easy to see that the condition $\boldsymbol{\theta}(V, \varkappa) \in \ell^\infty$ is sufficient for form-boundedness in $H^1(\mathbb{R}^2)$ of the multiplication by V. The next result follows from this property and Theorems 6.1, 6.2 by interpolation.

Theorem 6.3. 1. *Suppose that for some* $q > 1$

$$\#\{j \in \mathbb{N} : \theta_j(V, \varkappa) > \lambda\} = O(\lambda^{-q}).$$

Then for any $\gamma > 0$ *and* $\alpha > 0$

$$N_-(-\gamma; \mathbf{H}_{\alpha V}) \leqslant 1 + C_{\gamma,q} \alpha^q \sup_{\lambda > 0} \lambda^q \#\{j \in \mathbb{N} : \theta_j(V, \varkappa) > \lambda\}. \tag{6.4}$$

2. *Moreover, suppose that the sequence* $\eta_j(V)$ *introduced in* Theorem 6.2 *satisfies the condition* (5.11) *with the same value of* q. *Then* $N_-(0; \mathbf{H}_{\alpha V}) = O(\alpha^q)$ *and the function* $N_-(0; \mathbf{H}_{\alpha V})$ *is controlled by the expression as on the left-hand side of* (6.4) *plus the additional term*

$$\alpha^q \sup_{\lambda > 0} \lambda^q \#\{j \in \mathbb{Z} : \eta_j(V, \varkappa) > \lambda\}.$$

7 Schrödinger Operator on Manifolds

7.1 Preliminary remarks

Let $\mathcal{M} = \mathcal{M}^d$ be a smooth Riemannian manifold of dimension $d \geqslant 3$, and let dx stand for the volume element on $\mathcal{M}$. In this section, we discuss the behavior of the function $N_-(\beta; -\Delta_{\mathcal{M}} - \alpha V)$, where $\Delta_{\mathcal{M}}$ is the Laplacian on $\mathcal{M}$ (i.e., the Laplace–Beltrami operator). In order to avoid any ambiguity, here we do not use the shortened notation $\mathbf{H}_{\alpha V}$. As a rule, we suppose that $\mathcal{M}$ is not compact. Otherwise, the spectrum of $-\Delta_{\mathcal{M}}$ is discrete, and it makes no sense to speak about birth of eigenvalues of $-\Delta_{\mathcal{M}} - \alpha V$ from the essential spectrum of $-\Delta_{\mathcal{M}}$.

For a complete Riemannian manifold $\mathcal{M}$ the operator $-\Delta_{\mathcal{M}}$, defined initially on $C_0^\infty(\mathcal{M})$, is essentially self-adjoint and generates a sub-Markov semigroup. Thus, the results of Theorems 3.1 and 3.2 can be applied as soon as one has sufficient information about the embedding theorem on $\mathcal{M}$, or about estimates of the heat kernel. The global Sobolev inequality (3.1) with the correct order $q = 2d(d-2)^{-1}$ holds only in some special cases, and for general manifolds, probably, the only existing approach is based upon heat kernel estimates of the type (3.8), (3.9). Usually (though, not always) (3.8) is satisfied with $\delta = d$. For example, this is the case for the manifolds of bounded geometry (see, for example, [12]). In the discussion below, we will assume that

$$M_{-\Delta_{\mathcal{M}}}(t) \leqslant C_0 t^{-\frac{d}{2}}, \qquad t < 1. \tag{7.1}$$

On the other hand, D in (3.9) reflects the global geometry of the manifolds, however, rather roughly, and any relation $d > D$, $d = D$, or $d < D$ is possible. The results that follow from such estimates are given by Theorem 3.4, where one should take $\delta = d$.

An important difference from the case $\mathcal{M} = \mathbb{R}^d$ is that now the possibility of a positive $\beta_{\mathcal{M}} := \operatorname{bott}(-\Delta_{\mathcal{M}})$ is not excluded. May be, the only general result which holds true for any manifold subject to (7.1) is the following elementary, but useful statement.

Theorem 7.1. *Let $\mathcal{M} = \mathcal{M}^d$ be a smooth complete Riemannian manifold, $d \geqslant 3$. Suppose that the inequality (7.1) is satisfied. Then for any $0 \leqslant V \in L^{\frac{d}{2}}(\mathcal{M})$ and $\beta < \beta_{\mathcal{M}}$ the following inequality holds:*

$$N_-(\beta; -\Delta_{\mathcal{M}} - \alpha V) \leqslant C(\mathcal{M}, \beta)\alpha^{\frac{d}{2}} \int_{\mathcal{M}} V^{\frac{d}{2}} dx \qquad \forall \alpha > 0. \tag{7.2}$$

Along with the estimate (7.2), the Weyl asymptotic formula

$$N_-(\beta; -\Delta_{\mathcal{M}} - \alpha V) \sim w_d \alpha^{\frac{d}{2}} \int_{\mathcal{M}} V^{\frac{d}{2}} dx, \qquad \alpha \to \infty$$

is satisfied.

We only outline the proof of the inequality (7.2). For any $\beta < 0$ the semigroup $e^{-t(-\Delta_{\mathcal{M}} - \beta)}$ is sub-Markov (together with $e^{t\Delta_{\mathcal{M}}}$) and the function $M_{-\Delta_{\mathcal{M}} - \beta}(t) = e^{\beta t} M_{\Delta_{\mathcal{M}}}(t)$ satisfies the same estimate (7.2). Moreover, this function exponentially decays as $t \to \infty$, and hence (3.9) is fulfilled with any D. So, applying (3.12) with $D = d$ to the semigroup generated by the operator $-\Delta_{\mathcal{M}} - \beta$, we justify (7.2) for any $\beta < 0$. It extends to any values $\beta < \beta_{\mathcal{M}}$ by the standard variational argument. One should only take into account that for all $\beta < \beta_{\mathcal{M}}$ the quadratic forms

$$\int_{\mathcal{M}} (|\nabla u|^2 - \beta |u|^2) dx$$

generate mutually equivalent metrics on the Sobolev space $H^1(\mathcal{M})$.

The main issue in this type of problems is whether the estimate (7.2) remains valid for $\beta = \beta_{\mathcal{M}}$. Just such an estimate, rather than (7.2) for $\beta = 0$, should be considered as the genuine generalization of the RLC inequality (2.1) to the operators on manifolds. The answer to this question is positive only in some special cases. The Hyperbolic Laplacian is one of these cases.

7.2 Hyperbolic Laplacian

Let us consider the d-dimensional Hyperbolic space $\mathbb{H}^d$ for $d \geqslant 3$ in the upper half-space model. This means that $\mathbb{H}^d$ is realized as $\mathbb{R}^d_+ := \mathbb{R}^{d-1} \times \mathbb{R}_+$ with the metric

$$ds^2 = z^{-2}(|dy|^2 + dz^2), \qquad y \in \mathbb{R}^{d-1},\ z \in \mathbb{R}_+.$$

The corresponding volume element is $dv_{\mathrm{hyp}} = z^{-d} dy dz$. Recall that the Hyperbolic Laplacian is given by

$$\Delta_{\mathrm{hyp}} = z^2(\Delta_y + \partial_z^2) - (d-2) z \partial_z,$$

where Δ_y stands for the Euclidean Laplacian on R^{d-1}. The bottom of $-\Delta_{\mathrm{hyp}}$ is the point $\beta_0(d) = \frac{(d-1)^2}{4}$.

The following result, which can be called the RLC estimate for the Hyperbolic Laplacian, was obtained in [16].

Theorem 7.2. *Let $d \geqslant 3$, and let $0 \leqslant V \in L^{\frac{d}{2}}(\mathbb{H}^d)$. Then*

$$N_-(\beta_0(d); -\Delta_{\mathrm{hyp}} - \alpha V) \leqslant C(d) \alpha^{\frac{d}{2}} \int_{\mathbb{H}^d} V^{\frac{d}{2}} dv_{\mathrm{hyp}}.$$

For the proof one considers the quadratic form of $-\Delta_{\mathrm{hyp}}$ which is

$$Q[u] = \int_{\mathbb{H}^d} (|\nabla_{\mathrm{hyp}} u|^2 - \beta_0(d)|u|^2) dv_{\mathrm{hyp}} = \int_{\mathbb{R}^d_+} (|\nabla u|^2 - \beta_0(d)|u|^2) z^{2-d} dy dz.$$

The function $\varphi(y,z) = z^{\frac{d-1}{2}}$ satisfies the equation $-\Delta_{\text{hyp}}\varphi = \beta_0(d)\varphi$. The standard substitution $u = w\varphi$ reduces $Q[u]$ to the form

$$Q[u] = \int_{\mathbb{R}^d_+} |\nabla w|^2 z dy dz.$$

For this quadratic form the lower bound is already $\beta = 0$. Now, the global Sobolev inequality, which allows to apply Theorem 3.1 and leads to the estimate in Theorem 7.2, follows from [19, Corollary 2.1.6/3].

8 Operators on Manifolds: Beyond Theorem 3.4

A theory, allowing one to describe the potentials V on a general manifold, which ensure the semiclassical behavior $N_-(\beta_{\mathcal{M}}; -\Delta_{\mathcal{M}} - \alpha V) = O(\alpha^{\frac{d}{2}})$, does not exist. The situation simplifies if one has a more detailed information about the manifold than that given by the values of the exponents δ and D in the inequalities (3.8), (3.9). We illustrate this by several examples. We start with the simple case of a compact manifold.

Example 8.1. Let $\mathcal{M}$ be a compact and connected Riemannian manifold of dimension $d \geqslant 3$. Then the spectrum of $\mathbf{A} = -\Delta_{\mathcal{M}}$ is discrete. The number $\lambda_0 = 0$ is a simple eigenvalue of $-\Delta_{\mathcal{M}}$, the corresponding eigenspace $\mathcal{F}$ is formed by constant functions on $\mathcal{M}$. So, we have $\beta_{-\Delta_{\mathcal{M}}} = 0$. The estimate (7.2) for $\beta = 0$ certainly fails, which immediately follows from the analytic perturbation theory. Indeed, it shows that for any nontrivial $V \geqslant 0$ and $\alpha > 0$ the operator $-\Delta_{\mathcal{M}} - \alpha V$ has at least one negative eigenvalue. On the contrary, (7.2) with $\beta = 0$ would give $N_-(0; -\Delta_{\mathcal{M}} - \alpha V) = 0$ for α sufficiently small.

It is easy to show that, instead of (7.2), we have in this example

$$N_-(0; -\Delta_{\mathcal{M}} - \alpha V) \leqslant 1 + C(\mathcal{M})\alpha^{\frac{d}{2}} \int_{\mathcal{M}} V^{\frac{d}{2}} dx. \tag{8.1}$$

The estimate (8.1) has the same properties as the RLC estimate for $\mathbb{R}^d$: it gives the correct order in $\alpha \to \infty$ and it involves the sharp class of potentials for which this order is correct.

In general, for noncompact manifolds, one or both of these properties can be lost and some additional reasoning must be used.

In the case $d > D > 2$, the estimate (3.10) implies (3.3) with $2p = d$ for any compactly supported V, and, similarly to the case of a compact manifold, this result is sharp. Next, if the support of V has infinite measure and $V \in L^{\frac{d}{2}} \cap L^{\frac{D}{2}}$, neither of the terms in (3.10) majorizes the other one for a fixed α, however as $\alpha \to \infty$, the first term in (3.10) dominates. This indicates that it is possible to relax the condition of finiteness of the expression in

(3.10) and still have the semiclassical order in large coupling parameter. This difference in the dimensions d and D may generate an additional channel which can contribute to the behavior of $N_-(\beta_{\mathcal{M}}; -\Delta_{\mathcal{M}} - \alpha V)$ in a nontrivial way.

In the following example, $\mathcal{M}$ is a product manifold.

Example 8.2. Let $\mathcal{M} = \mathcal{M}_0 \times \mathbb{R}^m$, where $\mathcal{M}_0$ is a compact, connected smooth manifold of dimension $d-m$. We suppose that $d \geqslant 3$. Denote by (x, y) points on $\mathcal{M}$, where $x \in \mathcal{M}_0$ and $y \in \mathbb{R}^m$. Further, dx and dy stand for the volume element on $\mathbb{R}^m$ and on $\mathcal{M}_0$ respectively, then the volume element on $\mathcal{M}$ is $d\sigma = dxdy$. The heat kernel on $\mathcal{M}$ is the product of heat kernels on $\mathcal{M}_0$ and $\mathbb{R}^m$. Easy calculations show that $\delta = d$ and $D = m$. If $m > 2$, the estimate (3.10) applies, as soon as $V \in L^{\frac{d}{2}} \cap L^{\frac{m}{2}}$. However, this condition on V is not sharp since the first term in (3.10) majorizes the second one as $\alpha \to \infty$. For $m \leqslant 2$ we simply cannot apply (3.10). The reasoning below demonstrates a typical way to handle such situations.

The Laplacian on $\mathcal{M}_0$ has the lowest eigenvalue $\lambda_0 = 0$, simple, with the corresponding eigenspace consisting of constants. Let λ_1 be the first nonzero eigenvalue on $\mathcal{M}_0$. Consider the orthogonal decomposition of the space $L^2(\mathcal{M})$,

$$L^2(\mathcal{M}) = \mathfrak{F} \oplus \widetilde{L}^2(\mathcal{M}), \tag{8.2}$$

where $\mathfrak{F}$ consists of functions depending only on y, i.e., $u(x, y) = v(y)$, $v \in L^2(\mathbb{R}^m)$. Given a function $u \in L^2(\mathcal{M})$, its orthogonal projection onto $\mathfrak{F}$ is

$$v(y) = \frac{1}{\operatorname{vol} \mathcal{M}_0} \int\limits_{\mathcal{M}_0} u(x, y) dx,$$

which implies that $\widetilde{L}^2(\mathcal{M})$ consists of functions $\widetilde{u}(x, y)$ with zero integral over $\mathcal{M}_0$ for almost all $y \in \mathbb{R}^m$. The decomposition (8.2) is orthogonal also in the metric of the Dirichlet integral,

$$\int\limits_{\mathcal{M}} |\nabla u|^2 dxdy = \int\limits_{\mathcal{M}} |\nabla \widetilde{u}|^2 dxdy + \int\limits_{\mathbb{R}^m} |\nabla v(y)|^2 dy.$$

Denote by $\widetilde{H}^1(\mathcal{M})$ the space of those $\widetilde{u} \in \widetilde{L}^2(\mathcal{M})$ that belong to $H^1(\mathcal{M})$. On $\widetilde{H}^1(\mathcal{M})$, the metric generated by the Dirichlet integral is equivalent to the standard metric in $H^1(\mathcal{M})$:

$$\int\limits_{\mathcal{M}} |\nabla \widetilde{u}|^2 dxdy \geqslant \frac{1}{2}\Big(\int\limits_{\mathcal{M}} |\nabla \widetilde{u}|^2 dxdy + \lambda_1 \int\limits_{\mathcal{M}} |\widetilde{u}|^2 dxdy\Big) \qquad \forall \widetilde{u} \in \widetilde{H}^1(\mathcal{M}). \tag{8.3}$$

We also have

$$\int_{\mathcal{M}} V|u|^2 dx \leqslant 2\Big(\int_{\mathcal{M}} V|\widetilde{u}|^2 dx + \int_{\mathbb{R}^m} W(y)|v(y)|^2 dy \Big), \tag{8.4}$$

where the "effective potential" $W(y)$ is given by

$$W(y) = \int_{\mathcal{M}_0} V(x,y)dy. \tag{8.5}$$

The inequalities (8.3), (8.4), being combined with the variational principle, show that

$$N_-(0; \varDelta_{\mathcal{M}} - \alpha V) \leqslant N_-(-\lambda_1; \varDelta_{\mathcal{M}} - c\alpha V) + N_-(0; -\varDelta_{\mathbb{R}^m} - c\alpha W), \tag{8.6}$$

where $c > 0$ is some constant depending only on the value of λ_1, and the second term corresponds to the Schrödinger operator on $\mathbb{R}^m$ with the potential $-c\alpha W(y)$. For the first term in (8.6) we can use the estimate (7.2). The appearing of the second term in (8.6) can be interpreted as opening of a new channel in the system under consideration. For estimating this term, we can use Theorem 4.1, 6.2, or 5.2, depending on the dimension m. We would like to emphasize that here we need just the estimates of order $O(\alpha^{\frac{d}{2}})$. For the Laplacian on $\mathbb{R}^m$ such estimates are non-semiclassical.

Moreover, suppose that $V \in L^{\frac{d}{2}}$, but the effective potential W given by (8.5) satisfies the conditions of one of these theorems with some $q > \frac{d}{2}$. Then it may happen that the second term in (8.6) is stronger than the first one. In particular, if $m \geqslant 3$ and the potential W is like in (4.5), this second term, in fact, gives the correct asymptotic behavior of the function $N_-(0, \mathbf{H}_{\alpha V})$.

Recall that for the operators on the half-line Theorem 5.2 gives the necessary and sufficient condition for the behavior $N_-(0; \mathbf{H}_{\alpha V}) = O(\alpha^q)$ with a prescribed value of $q > \frac{1}{2}$; this condition extends to the operators on the whole line in an obvious way. So, for $m = 1$ the above construction gives more than for $m \geqslant 2$. Namely, it leads to the following result.

Theorem 8.1. *The two conditions:* $V \in L^{\frac{d}{2}}(\mathcal{M})$ *and*

$$\#\{j \in \mathbb{Z} : 2^j \int_{2^j \leqslant |y| \leqslant 2^{j+1}} W(y)dy\} = O(\lambda^{-\frac{d}{2}}), \ \lambda + \lambda^{-1} \to \infty$$

are necessary and sufficient for the semiclassical behavior of the function $N_-(0; -\varDelta_{\mathcal{M}} - \alpha V)$, *where* $\mathcal{M} = \mathcal{M}_0 \times \mathbb{R}$ *is a* d*-dimensional cylinder,* $d \geqslant 3$.

Note that the inclusion $V \in L^{\frac{d}{2}}(\mathcal{M})$ does not imply any restrictions on the behavior of W. Actually, under some additional assumptions about W, the function $N_-(0; \mathbf{H}_{\alpha V})$ may have regular asymptotic behavior of order $\alpha^{\frac{d}{2}}$, but with the asymptotic coefficient different from classical Weyl formula.

This kind of results can be easily extended to manifolds with cylindric ends.

In order to better understand the mechanism lying behind such two-term estimates, let us consider the free Laplacian $-\Delta_{\mathcal{M}}$ in Example 8.2. The separation of variables shows that $-\Delta_{\mathcal{M}}$ is unitary equivalent to the orthogonal sum of the operators $-\Delta_{\mathbb{R}^m} + \lambda_k$, $k = 0, 1, \ldots$, where λ_k are the eigenvalues of $-\Delta_{\mathcal{M}_0}$; recall that $\lambda_0 = 0$. So, the structure of the spectrum of $-\Delta_{\mathcal{M}}$ on $[0, \lambda_1)$ is determined by the m-dimensional Laplacian. This makes it clear, why the behavior of the function $N_-(0; -\Delta_{\mathbb{R}^m} - \alpha W)$ in dimension $m < d$ may affect the behavior of $N_-(\beta_{\mathcal{M}}; -\Delta_{\mathcal{M}} - \alpha V)$ for the Laplacian on a manifold of dimension d. The result of Theorem 7.1 shows that this effect does not appear for the function $N_-(\beta; -\Delta_{\mathcal{M}} - \alpha V)$ with $\beta < \beta_{\mathcal{M}}$.

This can be considered as manifestation of the "threshold effect" in this type of problems.

This effect exhibits in many other problems. One of them concerns the behavior of $N_-(0; -\Delta - \alpha V)$ on $\mathbb{R}^2$, discussed in Sect. 6. Note that this is the problem where the effect of appearance of an additional channel was observed and explained for the first time (see [25] and [1]). It is worth noting also, that in the latter problem the mechanism behind this effect is rather latent. Indeed, unlike Example 8.2, here removing the "bad" subspace of radial functions does not lead to the shift of the spectrum of the unperturbed operator.

Another class of problems where the threshold effect has to be taken into account, concerns various periodic operators, perturbed by a decaying potential. In this connection see [2, 5].

One meets similar effects when studying the behavior of $N_-(\beta; -\Delta_\Omega - \alpha V)$, where Δ_Ω is the Dirichlet Laplacian in an unbounded domain $\Omega \subset \mathbb{R}^d$. For the corresponding heat kernel the estimate (3.8) with $\delta = d$ always holds. Again, it may happen that the bottom of $-\Delta_\Omega$ is a point $\beta_0 > 0$. Suppose that $d \geqslant 3$. Then for any $\beta < \beta_0$ Theorem 7.1 applies. So, the problem consists in finding the estimates and the asymptotics of $N_-(\beta_0; -\Delta_\Omega - \alpha V)$. The general strategy here is the same as for manifolds, and examples like Example 8.2 can be easily constructed.

9 Schrödinger Operator on a Lattice

The techniques based upon Theorems 3.1 and 3.2 applies also to the discrete Laplacian. Below, we present some results for the simplest case where the underlying measure space (Ω, σ) is $\mathbb{Z}^d$ with the standard counting measure, so that $\sigma(E) = \#E$ for any subset $E \subset \mathbb{Z}^d$. For the sake of definiteness, we discuss only the case $d \geqslant 3$. The discrete Laplacian is

$$(\mathbf{A}_d u)(x) = \sum_j (u(x+1_j) + u(x-1_j) - 2u(x)), \qquad x \in \mathbb{Z}^d,$$

where 1_j is the multiindex with all zero entries except 1 in the position j. This is a bounded operator, and its spectrum is absolutely continuous and coincides with the segment $[0, 2d]$. The corresponding heat kernel can be found explicitly, it is bounded as $t \to 0$ and is $O(t^{-\frac{d}{2}})$ as $t \to \infty$, thus $\delta = 0$ and $D = d$. The inequality (3.12) applies, and we obtain the discrete RLC estimate

$$N_-(0; \mathbf{A}_d - \alpha V) \leqslant C\alpha^{\frac{d}{2}} \int_{\mathbb{Z}^d} V^{\frac{d}{2}} d\sigma \quad \forall \alpha > 0; \qquad d \geqslant 3.$$

On the contrary to the continuous case, this estimate cannot be order-sharp since the assumption $V \in L^{\frac{d}{2}}(\mathbb{Z}^d)$ immediately yields

$$N_-(0; \mathbf{A}_d - \alpha V) = o(\alpha^{\frac{d}{2}}), \qquad \alpha \to \infty.$$

Indeed, this is certainly true for any V with bounded support, since for such V the number $N_-(0; \mathbf{A}_d - \alpha V)$ is no greater than the number $\#\{x \in \mathbb{Z}^d : V(x) \neq 0\}$. The set of all such V is dense in $L^{\frac{d}{2}}(\mathbb{Z}^d)$. Therefore, the result extends to all $V \in L^{\frac{d}{2}}$.

We do not know even a single example of a potential V on $\mathbb{Z}^d$ such that $N_-(0; \mathbf{A}_d - \alpha V) = O(\alpha^{\frac{d}{2}})$, but $\neq o(\alpha^{\frac{d}{2}})$.

One more important difference with the continuous case is that for the discrete operators the behavior $N_-(0; \mathbf{A}_d - \alpha V) = O(\alpha^q)$ with $2q < d$ is possible; in the continuous case, it never occurs in dimensions $d \geqslant 3$ and, probably, also in $d = 2$. In the case $d = 1$, the order $O(\alpha^q)$ with $2q < 1$ is possible if one allows potentials which are distributions supported by a subset of zero Lebesgue measure.

For a given potential $V \geqslant 0$ on $\mathbb{Z}^d$ one cannot formally use (3.11) with $\delta = 0$ since the value $\delta = 0$ lies outside the set admissible by Theorem 3.4. However, by using the variational principle and (3.11) written for the potential V restricted to the set $\{x : \alpha V(x) < 1\}$, it is not difficult to show that, in this particular case, (3.11) holds for any $\alpha > 0$ even with $\delta = 0$. If we introduce the distribution function of V

$$m(\tau) = \#\{x \in \mathbb{Z}^d : V(x) > \tau\}, \qquad \tau > 0,$$

this line of reasoning leads to the inequality

$$N_-(0; \mathbf{A}_d - \alpha V) \leqslant C\Big(m(2\alpha^{-1}) + \alpha^{\frac{d}{2}} \int_{\alpha V(x) < 1} V^{\frac{d}{2}} d\sigma\Big). \tag{9.1}$$

By estimating the integral in (9.1), we come to the following result which has no continuous analogue.

Theorem 9.1. *Suppose that $d \geqslant 3$. Then for any $\nu > 2$*

$$N_-(0; \mathbf{A}_d - \alpha V) \leqslant C(d,\nu)\alpha^{\frac{d}{\nu}} \sup_{\tau>0}\left(\tau^{\frac{d}{\nu}} \#\{x \in \mathbb{Z}^d : V(x) > \tau\}\right). \qquad (9.2)$$

The class of discrete potentials $V(x)$, for which the functional on the right-hand side of (9.2) is finite, is nothing, but the cone of all positive elements in the "weak" space $\ell_w^{\frac{d}{\nu}}(\mathbb{Z}^d)$. The assumption $\nu > 2$ yields

$$\ell_w^{\frac{d}{\nu}}(\mathbb{Z}^d) \subset L^{\frac{d}{2}}(\mathbb{Z}^d),$$

so that the estimate (9.1) applies. The inequality (9.2) gives a better estimate than (9.1), and it is possible to show that, unlike (9.1), it is order-sharp.

Theorem 9.1 applies to the potentials decaying no slower than $|x|^{-\nu}$, $\nu > 2$, and gives the order $O(\alpha^{\frac{d}{\nu}})$. For potentials decaying more slowly (but still faster that $|x|^{-2}$), so that the integral in (9.1) diverges, the following result applies. It is the direct analogue of Theorem 4.1; its proof is also based upon the interpolation theory.

Theorem 9.2. *Let $d \geqslant 3$ and $2q > d$. Suppose that the potential $V(n) \geqslant 0$ is such that*

$$|V|_q^q := \sup_{\tau>0}\left(\tau^q \int\limits_{(|x|^2+1)V(x)>\tau} \frac{d\sigma}{|x|^2}\right) < \infty.$$

Then the estimate (4.2) holds for the operator $\mathbf{H}_{\alpha V} = \mathbf{A}_d - \alpha V$.

In connection with Theorems 9.1 and 9.2, we note that for any *beta* > 0 the potential $V(x) = (|x|+1)^{-\beta}$ belongs to the class $\ell_w^{\frac{d}{\beta}}(\mathbb{Z}^d)$, and the potential V that for large $|x|$ behaves as

$$V(x) = |x|^{-2}(\log|x|) - 1/q$$

meets the property $|V|_q < \infty$. So, these theorems embrace the cases of estimates of orders, respectively, smaller and larger than $\alpha^{\frac{d}{2}}$. It is unclear at the moment whether a sharp estimate of the order $\alpha^{\frac{d}{2}}$ is possible. This indicates that the notion of "semiclassical" order is not applicable here.

The above results can be extended to combinatorial Schrödinger operators on arbitrary infinite graphs, as soon at the heat kernel estimates (3.8), (3.9) are known with $\delta = 0$, $D > 2$.

10 Some Unsolved Problems

In this concluding section, we list some problems in this field, which remain unsolved up to present. In our opinion, their solution would be important for the further progress in the field.

In the first place, this is the study of the Schrödinger operator on $\mathbb{R}^2$. Here, we mean an exhaustive description of potentials V ensuring the semiclassical behavior $N_-(0;\mathbf{H}_{\alpha V}) = O(\alpha)$. As was mentioned in Sect. 6, the situation here is unclear, and many natural conjectures fail to be true.

The next class of problems concerns manifolds. In particular, we believe that the class of d-dimensional manifolds, $d \geqslant 3$, for which the structure of the potentials V, guaranteeing the semiclassical estimate $N_-(\beta_{\mathcal{M}}; -\Delta_{\mathcal{M}} - \alpha V) = O(\alpha^{\frac{d}{2}})$, can be exhaustively described, can be considerably widened compared with Theorem 8.1.

Theorem 9.1 indicates that the problems for the continuous and the discrete Schrödinger operators have rather different nature, and the expected results for these two parallel classes of operators should essentially differ. It would be useful to understand the discrete case up to a greater extent.

Finally, we mention the problems of the type discussed for the metric graphs (quantum graphs in other terminology), in particular, for the metric trees. The few existing results (see, for example, [8]) still do not give the adequate understanding of the effects which appear when studying the Schrödinger operator on graphs.

References

1. Birman, M.Sh., Laptev, A.: The negative discrete spectrum of a two-dimensional Schrödinger operator. Commun. Pure Appl. Math. **49**, no. 9, 967–997 (1996).
2. Birman, M.Sh., Laptev, A., Suslina, T.: The discrete spectrum of a two-dimensional second order periodic elliptic operator perturbed by a decreasing potential. I. A semi-infinite gap (Russian). Algebra Anal. **12**, no. 4, 36–78 (2000); English transl.: St. Petersbg. Math. J. **12**, no. 4, 535–567 (2001).
3. Birman, M.Sh., Solomyak, M.: Quantitative Analysis in Sobolev Imbedding Theorems and Applications to Spectral Theory (Russian). Tenth Mathem. School, Izd. Inst. Mat. Akad. Nauk Ukrain. SSR, Kiev, 5–189 (1974); English transl.: Am. Math. Soc., Providence, RI (1980).
4. Birman, M.Sh., Solomyak, M.: Estimates for the number of negative eigenvalues of the Schrödinger operator and its generalizations. Adv. Sov. Math. **7**, 1–55 (1991)
5. Birman, M.Sh., Solomyak, M.: On the negative discrete spectrum of a periodic elliptic operator in a waveguide-type domain, perturbed by a decaying potential. J. Anal. Math. **83**, 337–391 (2001)
6. Cwikel, M.: Weak type estimates for singular values and the number of bound states of Schrödinger operators. Ann. Math. (2) **106**, no. 1, 93–100 (1977)
7. Davies, E.B.: Heat Kernels and Spectral Theory. Cambridge Univ. Press, Cambridge (1989)

8. Ekholm, T., Frank, R., Kovarik, H.: Eigenvalue Estimates for Schrödinger Operators on Metric Trees. arXive:0710.5500.
9. Egorov, Yu.V., Kondrat'ev, V.A.: On the estimation of the number of points of the negative spectrum of the Schrödinger operator (Russian). Mat. Sb. **134**, no. 4, 556–570 (1987); English transl.: Math. USSR–Sb. **62**, no. 2, 551–566 (1989)
10. Fefferman, C.: The uncertainty principle. Bull. Am. Math. Soc. **9**, no. 2, 129–206 (1983)
11. Frank, R., Lieb, E., Seiringer, R.: Hardy–Lieb–Thirring Inequalities for Fractional Schrödinger Operators. arXive:0610.5593.
12. Grigor'yan, A.: Heat kernels and function theory on metric measure spaces. In: Heat Kernels and Analysis on Manifolds, Graphs, and Metric Spaces (Paris, 2002), pp. 143–172. Am. Math. Soc., Providence, RI (2003)
13. Laptev, A.: The negative spectrum of the class of two-dimensional Schrödinger operators with potentials that depend on the radius (Russian). Funkt. Anal. Prilozh. **34**, no. 4, 85–87 (2000); English transl.: Funct. Anal. Appl. **34**, no. 4, 305–307 (2000)
14. Laptev, A., Netrusov, Yu.: On the negative eigenvalues of a class of Schrödinger operators. In: Differential Operators and Spectral Theory, pp. 173–186. Am. Math. Soc., Providence, RI (1999)
15. Laptev, A., Weidl, T.: Recent Results on Lieb–Thirring Inequalities. J. "Équat. Deriv. Partielles" (La Chapelle sur Erdre, 2000), Exp. no. 20, Univ. Nantes, Nantes (2000)
16. Levin, D., Solomyak, M.: The Rozenblum–Lieb–Cwikel inequality for Markov generators. J. Anal. Math. **71**, 173–193 (1997)
17. Li, P., Yau, S.-T.: On the Schrödinger equation and the eigenvalue problem. Commun. Math. Phys. **88**, no. 3, 309–318 (1983)
18. Lieb, E.: The number of bound states of one-body Schrödinger operators and the Weyl problem. Bull. Am. Math. Soc. **82**, 751–753 (1976)
19. Maz'ya, V.G.: Sobolev Spaces. Springer-Verlag, Berlin–Tokyo (1985)
20. Naimark, K., Solomyak, M.: Regular and pathological eigenvalue behavior for the equation $-\lambda u'' = Vu$ on the semiaxis. J. Funct. Anal. **151**, no. 2, 504–530 (1997)
21. Reed, M., Simon, B.: Methods of Modern Mathematical Physics. IV. Analysis of Operators. Academic Press, New York-London (1978)
22. Rozenblum, G.: Distribution of the discrete spectrum of singular differential operators (Russian). Dokl. Akad. Nauk SSSR **202**, 1012–1015 (1972); English transl.: Sov. Math., Dokl. **13**, 245–249 (1972)
23. Rozenblum, G.: Distribution of the discrete spectrum of singular differential operators (Russian). Izv. VUZ, Ser. Mat. no. 1(164), 75–86 (1976); English transl.: Sov. Math., Izv. VUZ **20** no 1, 63–71 (1976)
24. Rozenblum, G., Solomyak, M.: The Cwikel–Lieb–Rozenblum estimates for generators of positive semigroups and semigroups dominated by positive semigroups (Russian). Algebra Anal. **9** , no. 6, 214–236 (1997); English transl.: St. Petersbg. Math. J. **9**, no. 6, 1195–1211 (1998)
25. Solomyak, M.: Piecewise-polynomial approximation of functions from $H^l((0,1)^d)$, $2l = d$, and applications to the spectral theory of the Schrödinger operator. Israel J. Math. **86**, no. 1-3, 253–275 (1994)
26. Varopoulos, N.Th., Saloff-Coste, L., Coulhon, T.: Analysis and Geometry on Groups. Cambridge Univ. Press, Cambridge (1992)

Function Spaces on Cellular Domains

Hans Triebel

Abstract A Lipschitz domain in $\mathbb{R}^n$ is called *cellular* if it is the finite union of diffeomorphic images of cubes. Bounded C^∞ domains and cubes are prototypes. The paper deals with spaces of type B^s_{pq} and F^s_{pq} (including Sobolev spaces and Besov spaces) on these domains. Special attention is paid to traces on (maybe nonsmooth) boundaries and wavelet bases.

1 Introduction and Preliminaries

1.1 Introduction

Let $A^s_{pq}(\mathbb{R}^n)$ with $A \in \{B, F\}$ and $s \in \mathbb{R}$, $0 < p, q \leqslant \infty$ ($p < \infty$ for the F-spaces) be the nowadays well-known function spaces on the Euclidean n-space $\mathbb{R}^n$. Recall that

$$H^s_p(\mathbb{R}^n) = F^s_{p,2}(\mathbb{R}^n), \qquad 1 < p < \infty, \quad s \in \mathbb{R}, \tag{1.1}$$

are the (fractional) *Sobolev spaces* with the *classical Sobolev spaces*

$$W^k_p(\mathbb{R}^n) = H^k_p(\mathbb{R}^n), \qquad k \in \mathbb{N}_0, \quad 1 < p < \infty, \tag{1.2}$$

as special cases. Furthermore,

$$B^s_{pq}(\mathbb{R}^n), \qquad s > 0, \quad 1 < p < \infty, \quad 1 \leqslant q \leqslant \infty, \tag{1.3}$$

Hans Triebel
Mathematisches Institut, Friedrich-Schiller-Universität Jena, D-07737 Jena, Germany,
e-mail: triebel@minet.uni-jena.de

V. Maz'ya (ed.), *Sobolev Spaces in Mathematics II*, *International Mathematical Series*.
doi: 10.1007/978-0-387-85650-6,

are the *classical Besov spaces*. Elements of these spaces can be expanded by (sufficiently) smooth orthonormal wavelet bases creating isomorphic maps onto distinguished sequence spaces (unconditional bases if $p < \infty$, $q < \infty$). Let $A^s_{pq}(\Omega)$ be the restriction of $A^s_{pq}(\mathbb{R}^n)$ to the domain (= open set) Ω in $\mathbb{R}^n$. The search for wavelet bases in these spaces is a rather tricky business. It attracted some attention, especially for $L_p(\Omega)$ with $1 < p < \infty$ and some spaces of type (1.2), (1.3) with Ω in place of $\mathbb{R}^n$. In the present paper, we wish to contribute to this subject. We are especially interested in the interplay between admitted domains Ω and the parameters s, p, q in $A^s_{pq}(\Omega)$. However, this paper is not a survey. Just on the contrary. It might be considered as the direct continuation of the survey [17] and the two books [15, 16]. There we collected (with some care as we hope) relevant references. This will not be repeated here. We rely mainly on [16]. On the other hand, we tried to make this paper independently readable as far as definitions and assertions are concerned and to present a reliable account about the topics indicated. For this purpose, we take over a few results from [16], but new ones will be proved. Our main new results are Theorems 4.1, 4.4 and Corollaries 4.3, 4.5 about wavelet bases and Proposition 3.1, Theorem 3.3 which deal with traces and wavelet-friendly extension operators for function spaces in cubes which is the crucial new instrument (and might be of self-contained interest).

1.2 Definitions

We use standard notation. Let $\mathbb{N}$ be the collection of all natural numbers, and let $\mathbb{N}_0 = \mathbb{N} \cup \{0\}$. Let $\mathbb{R}^n$ be Euclidean n-space, where $n \in \mathbb{N}$. Put $\mathbb{R} = \mathbb{R}^1$, whereas $\mathbb{C}$ is the complex plane. Let $S(\mathbb{R}^n)$ be the usual Schwartz space, and let $S'(\mathbb{R}^n)$ be the space of all tempered distributions on $\mathbb{R}^n$. Furthermore, $L_p(\mathbb{R}^n)$ with $0 < p \leqslant \infty$ is the standard quasi-Banach space with respect to the Lebesgue measure in $\mathbb{R}^n$, quasinormed by

$$\|f \,|L_p(\mathbb{R}^n)\| = \left(\int_{\mathbb{R}^n} |f(x)|^p \, dx \right)^{1/p}$$

with the usual modification if $p = \infty$. As usual, $\mathbb{Z}$ is the collection of all integers and $\mathbb{Z}^n$, where $n \in \mathbb{N}$, denotes the lattice of all points $m = (m_1, \ldots, m_n) \in \mathbb{R}^n$ with $m_j \in \mathbb{Z}$. Let $\mathbb{N}^n_0$, where $n \in \mathbb{N}$, be the set of all multiindices,

$$\alpha = (\alpha_1, \ldots, \alpha_n) \quad \text{with} \quad \alpha_j \in \mathbb{N}_0 \quad \text{and} \quad |\alpha| = \sum_{j=1}^n \alpha_j.$$

If $x = (x_1, \ldots, x_n) \in \mathbb{R}^n$ and $\beta = (\beta_1, \ldots, \beta_n) \in \mathbb{N}^n_0$, then we put

$$x^\beta = x_1^{\beta_1} \cdots x_n^{\beta_n} \qquad \text{(monomials)}.$$

If $\varphi \in S(\mathbb{R}^n)$, then

$$\widehat{\varphi}(\xi) = (F\varphi)(\xi) = (2\pi)^{-n/2} \int_{\mathbb{R}^n} e^{-ix\xi}\, \varphi(x)\, dx, \qquad \xi \in \mathbb{R}^n, \tag{1.4}$$

denotes the Fourier transform of φ. As usual, $F^{-1}\varphi$ and $\varphi^\vee$ stand for the inverse Fourier transform, given by the right-hand side of (1.4) with i in place of $-i$. Here $x\xi$ denotes the scalar product in $\mathbb{R}^n$. Both F and F^{-1} are extended to $S'(\mathbb{R}^n)$ in the standard way. Let $\varphi_0 \in S(\mathbb{R}^n)$ with

$$\varphi_0(x) = 1 \text{ if } |x| \leqslant 1 \quad \text{and} \quad \varphi_0(y) = 0 \text{ if } |y| \geqslant 3/2,$$

and let

$$\varphi_k(x) = \varphi_0\left(2^{-k}x\right) - \varphi_0\left(2^{-k+1}x\right), \qquad x \in \mathbb{R}^n, \quad k \in \mathbb{N}.$$

Then $\sum\limits_{j=0}^{\infty} \varphi_j(x) = 1$ in $\mathbb{R}^n$ is a dyadic resolution of unity. The entire analytic functions $(\varphi_j \widehat{f})^\vee(x)$ make sense pointwise.

Definition 1.1. Let $\varphi = \{\varphi_j\}_{j=0}^\infty$ be the above dyadic resolution of unity.

(i) Let $0 < p \leqslant \infty$, $0 < q \leqslant \infty$, $s \in \mathbb{R}$. Then $B^s_{pq}(\mathbb{R}^n)$ is the collection of all $f \in S'(\mathbb{R}^n)$ such that

$$\|f\,|B^s_{pq}(\mathbb{R}^n)\|_\varphi = \left(\sum_{j=0}^\infty 2^{jsq} \|(\varphi_j \widehat{f})^\vee\,|L_p(\mathbb{R}^n)\|^q \right)^{1/q} < \infty \tag{1.5}$$

(with the usual modification if $q = \infty$).

(ii) Let $0 < p < \infty$, $0 < q \leqslant \infty$, $s \in \mathbb{R}$. Then $F^s_{pq}(\mathbb{R}^n)$ is the collection of all $f \in S'(\mathbb{R}^n)$ such that

$$\|f\,|F^s_{pq}(\mathbb{R}^n)\|_\varphi = \left\| \left(\sum_{j=0}^\infty 2^{jsq}\, |(\varphi_j \widehat{f})^\vee(\cdot)|^q \right)^{1/q} |L_p(\mathbb{R}^n) \right\| < \infty \tag{1.6}$$

(with the usual modification if $q = \infty$).

Remark 1.2. The theory of these spaces may be found in [11, 12, 15]. In particular, these spaces are independent of admitted resolutions of unity φ (equivalent quasinorms). This justifies our omission of the subscript φ in (1.5), (1.6) in the sequel. As usual nowadays, we write

$$A^s_{pq}(\mathbb{R}^n) \qquad \text{with} \quad A \in \{B, F\}$$

if the assertions considered apply both to $B^s_{pq}(\mathbb{R}^n)$ and $F^s_{pq}(\mathbb{R}^n)$. Recall that the Sobolev and Besov spaces mentioned in (1.1)–(1.3) are special cases. One may consult [15, Sect. 1.2] for well-known classical norms in these distinguished spaces.

Open sets Ω in $\mathbb{R}^n$ are called *domains*. As usual, $D(\Omega) = C_0^\infty(\Omega)$ stands for the collection of all complex-valued infinitely differentiable functions in $\mathbb{R}^n$ with compact support in Ω. Let $D'(\Omega)$ be the dual space of all distributions in Ω. Let $g \in S'(\mathbb{R}^n)$. Then we denote by $g|\Omega$ its restriction to Ω,

$$g|\Omega \in D'(\Omega): \qquad (g|\Omega)(\varphi) = g(\varphi) \quad \text{for} \quad \varphi \in D(\Omega).$$

Definition 1.3. Let Ω be a domain in $\mathbb{R}^n$, and let

$$0 < p \leqslant \infty, \quad 0 < q \leqslant \infty, \quad s \in \mathbb{R},$$

with $p < \infty$ for the F-spaces.

(i) Then

$$A^s_{pq}(\Omega) = \left\{ f \in D'(\Omega) : f = g|\Omega \text{ for some } g \in A^s_{pq}(\mathbb{R}^n) \right\},$$

$$\|f \,|A^s_{pq}(\Omega)\| = \inf \|g \,|A^s_{pq}(\mathbb{R}^n)\|, \tag{1.7}$$

where the infimum is taken over all $g \in A^s_{pq}(\mathbb{R}^n)$ with $g|\Omega = f$.

(ii) Let $\overset{\circ}{A}{}^s_{pq}(\Omega)$ be the completion of $D(\Omega)$ in $A^s_{pq}(\Omega)$.

(iii) Let

$$\widetilde{A}^s_{pq}(\overline{\Omega}) = \left\{ f \in A^s_{pq}(\mathbb{R}^n) : \operatorname{supp} f \subset \overline{\Omega} \right\}.$$

Then

$$\widetilde{A}^s_{pq}(\Omega) = \left\{ f \in D'(\Omega) : f = g|\Omega \text{ for some } g \in \widetilde{A}^s_{pq}(\overline{\Omega}) \right\},$$

$$\|f \,|\widetilde{A}^s_{pq}(\Omega)\| = \inf \|g \,|A^s_{pq}(\mathbb{R}^n)\|,$$

where the infimum is taken over all $g \in \widetilde{A}^s_{pq}(\overline{\Omega})$ with $g|\Omega = f$.

Remark 1.4. Part (i) is the usual definition of $A^s_{pq}(\Omega)$ by restriction. The spaces $\widetilde{A}^s_{pq}(\overline{\Omega})$ are closed subspaces of $A^s_{pq}(\mathbb{R}^n)$. By the usual abuse of notation, they can be identified with $\widetilde{A}^s_{pq}(\Omega)$ if

$$\left\{ h \in A^s_{pq}(\mathbb{R}^n) : \operatorname{supp} h \subset \partial\Omega \right\} = \{0\},$$

where $\partial\Omega$ is the boundary of Ω. In general, this is not the case. But we return to this point later on.

1.3 Wavelet systems and sequence spaces

We introduced in [16, Sect. 5.1.5] sequence spaces and wavelet systems. We need now the following modifications.

Definition 1.5. Let Ω be a bounded domain in $\mathbb{R}^n$, and let

$$\mathbb{Z}^\Omega = \left\{ x_l^j \in \Omega : j \in \mathbb{N}_0;\ l = 1, \ldots, N_j \right\}, \tag{1.8}$$

typically with $N_j \sim 2^{jn}$, such that for some $c_1 > 0$,

$$|x_l^j - x_{l'}^j| \geqslant c_1\, 2^{-j}, \qquad j \in \mathbb{N}_0, \quad l \neq l'. \tag{1.9}$$

Let χ_{jl} be the characteristic function of the ball $B(x_l^j, c_2 2^{-j}) \subset \mathbb{R}^n$ (centered at x_l^j and of radius $c_2 2^{-j}$) for some $c_2 > 0$. Let $s \in \mathbb{R}$, $0 < p, q \leqslant \infty$. Then $b^s_{pq}(\mathbb{Z}^\Omega)$ is the collection of all sequences

$$\lambda = \left\{ \lambda_l^j \in \mathbb{C} : j \in \mathbb{N}_0;\ l = 1, \ldots, N_j \right\} \tag{1.10}$$

such that

$$\|\lambda \,| b^s_{pq}(\mathbb{Z}^\Omega)\| = \left(\sum_{j=0}^{\infty} 2^{j(s-\frac{n}{p})q} \left(\sum_{l=1}^{N_j} |\lambda_l^j|^p \right)^{q/p} \right)^{1/q} < \infty$$

and $f^s_{pq}(\mathbb{Z}^\Omega)$ is the collection of all sequences (1.10) such that

$$\|\lambda \,| f^s_{pq}(\mathbb{Z}^\Omega)\| = \left\| \left(\sum_{j=0}^{\infty} \sum_{l=1}^{N_j} 2^{jsq} |\lambda_l^j \chi_{jl}(\cdot)|^q \right)^{1/q} |L_p(\Omega) \right\| < \infty \tag{1.11}$$

(obviously modified if $p = \infty$ and/or $q = \infty$).

Remark 1.6. Here $L_p(\Omega)$ with $0 < p \leqslant \infty$ is the standard Lebesgue space quasinormed by

$$\|f \,| L_p(\Omega)\| = \left(\int_\Omega |f(x)|^p \, dx \right)^{1/p}$$

(obviously modified if $p = \infty$). In what follows, we are interested in C^∞ domains, Lipschitz domains and cellular domains. Then one can replace χ_{jl} in (1.11) by the characteristic function of $B(x_l^j, c_2 2^{-j}) \cap \Omega$ (equivalent quasi-norms). This is well known and covered by [15, Sect. 1.5.3, pp. 18-19] and the references given there.

In [16], we dealt first with wavelet bases for function spaces $A^s_{pq}(\mathbb{R}^n)$ on $\mathbb{R}^n$, where we used the usual tensor products of the one-dimensional compactly supported Daubechies wavelets related to the dyadic lattices $2^{-j}\mathbb{Z}^n$. The standard references are [2, 6, 18]. Switching from $\mathbb{R}^n$ to a domain Ω in $\mathbb{R}^n$ the set $\mathbb{Z}^\Omega$ in (1.8) is an adapted substitute of the lattices $\{2^{-j}\mathbb{Z}^n; j \in \mathbb{N}_0\}$. Similarly one has to modify the original $\mathbb{R}^n$-wavelets which results in the following more qualitative version. This was done in [16] in several steps. We rely here on the final outcome. Let Ω be an arbitrary (not necessarily bounded) domain in $\mathbb{R}^n$ and $u \in \mathbb{N}_0$. Then $C^u(\Omega)$ is the collection of all complex-valued functions f having classical derivatives up to order u inclusively in Ω such that any function $D^\alpha f$ with $|\alpha| \leqslant u$ can be extended continuously to $\overline{\Omega}$ and

$$\|f\,|C^u(\Omega)\| = \sum_{|\alpha|\leqslant u} \sup_{x\in\Omega} |f(x)| < \infty. \tag{1.12}$$

Let

$$C^\infty(\Omega) = \bigcap_{u=0}^{\infty} C^u(\Omega). \tag{1.13}$$

We are mainly interested in bounded domains Ω, where (1.12) follows from what had been said before. But the above version will be of some use for us later on.

Definition 1.7. Let Ω be a bounded domain in $\mathbb{R}^n$, and let $\mathbb{Z}^\Omega$ be as in (1.8), (1.9). Let $u \in \mathbb{N}$.

(i) Then

$$\Phi = \left\{\Phi^j_l : j \in \mathbb{N}_0;\ l = 1, \ldots, N_j\right\} \subset C^u(\Omega)$$

is called a *u-wavelet system* in $\overline{\Omega}$ if for some $c_3 > 0$ and $c_4 > 0$

$$\operatorname{supp} \Phi^j_l \subset B\left(x^j_l, c_3 2^{-j}\right) \cap \overline{\Omega}, \qquad j \in \mathbb{N}_0;\ l = 1, \ldots, N_j,$$

and

$$|D^\alpha \Phi^j_l(x)| \leqslant c_4\, 2^{j\frac{n}{2}+j|\alpha|}, \qquad j \in \mathbb{N}_0;\ l = 1, \ldots, N_j, \tag{1.14}$$

for all $\alpha \in \mathbb{N}^n_0$ with $0 \leqslant |\alpha| \leqslant u$.

(ii) The above u-wavelet system Φ is called an *interior u-wavelet system* in Ω if, in addition, all wavelets Φ^j_l have compact support in Ω.

Remark 1.8. As said, the wavelets Φ^j_l originate from L_2-normalized wavelets in $\mathbb{R}^n$ of Daubechies type. This explains the normalizing factor $2^{j\frac{n}{2}}$ in (1.14).

1.4 Domains

This paper is a continuation of [16, Chapt. 5], where we dealt mainly with wavelet frames and wavelet bases in function spaces of positive smoothness in smooth domains. First we fix the types of domains considered. For $2 \leqslant n \in \mathbb{N}$

$$\mathbb{R}^{n-1} \ni x \mapsto h(x') \in \mathbb{R} \tag{1.15}$$

is called a *Lipschitz function* (on $\mathbb{R}^{n-1}$) if there is a number $c > 0$ such that

$$|h(x') - h(y')| \leqslant c\,|x' - y'| \quad \text{for all} \quad x' \in \mathbb{R}^{n-1}, \quad y' \in \mathbb{R}^{n-1}. \tag{1.16}$$

Furthermore, h is called a *bounded C^∞ function* if it is a C^∞ function and all the derivatives are bounded. Recall that a one-to-one map ψ from a bounded domain Ω in $\mathbb{R}^n$ onto a bounded domain ω in $\mathbb{R}^n$,

$$\psi : \quad \Omega \ni x \mapsto y = \psi(x) \in \omega,$$

is called *diffeomorphic* if

$$\psi_l \in C^\infty(\Omega) \quad \text{and} \quad (\psi^{-1})_l \in C^\infty(\omega), \qquad l = 1, \ldots, n,$$

for the components of ψ and of its inverse ψ^{-1} (where $C^\infty(\Omega)$ and $C^\infty(\omega)$ are given by (1.13)) and if the Jacobian $\left(\frac{\partial \psi}{\partial x}\right)$ is positive in $\overline{\Omega}$. Let

$$Q = \{x \in \mathbb{R}^n : 0 < x_l < 1\}$$

be the unit cube in $\mathbb{R}^n$. If Γ is a set in $\mathbb{R}^n$, then Γ^0 is the largest open set in $\mathbb{R}^n$ with $\Gamma^0 \subset \Gamma$ (the interior of Γ).

Definition 1.9. Let $2 \leqslant n \in \mathbb{N}$.

(i) A *special Lipschitz domain* (*C^∞ domain*) in $\mathbb{R}^n$ is the collection of all points $x = (x', x_n)$ with $x' \in \mathbb{R}^{n-1}$ such that

$$h(x') < x_n < \infty,$$

where $h(x')$ is a Lipschitz function according to (1.15), (1.16) (a bounded C^∞ function).

(ii) A *bounded Lipschitz domain* (*C^∞ domain*) in $\mathbb{R}^n$ is a bounded domain Ω in $\mathbb{R}^n$, where the boundary $\Gamma = \partial\Omega$ can be covered by finitely many open balls B_j in $\mathbb{R}^n$ with $j = 1, \ldots, J$, centered at Γ such that

$$B_j \cap \Omega = B_j \cap \Omega_j \qquad \text{for} \quad j = 1, \ldots, J,$$

where Ω_j are rotations of suitable special Lipschitz domains (C^∞ domains) in $\mathbb{R}^n$.

(iii) A *bounded cellular domain* in $\mathbb{R}^n$ is a bounded Lipschitz domain Ω which can be represented for some $L \in \mathbb{N}$ as

$$\Omega = \Big(\bigcup_{l=1}^{L} \overline{\Omega_l} \Big)^{\circ}, \qquad \Omega_l \cap \Omega_{l'} = \varnothing \quad \text{if} \quad l \neq l',$$

such that for each Ω_l there is a diffeomorphic map ψ^l of a neighborhood of $\overline{\Omega_l}$ onto a neighborhood of $\overline{Q}$ with $\psi^l(\Omega_l) = Q$.

Remark 1.10. In [16, Chapt. 5], we dealt preferably with bounded C^∞ domains in $\mathbb{R}^n$. But we introduced in [16, Definition 5.40] also cellular domains and cellular manifolds. According to [16, Remark 5.41], one has the following assertion.

Proposition 1.11. *Let* $2 \leqslant n \in \mathbb{N}$. *A bounded* C^∞ *domain in* $\mathbb{R}^n$ *is cellular.*

Remark 1.12. For bounded Lipschitz domains one can speak about traces of functions belonging to some function spaces in the context of trace spaces. But it is not so clear how to extend the corresponding assertions to derivatives. If a bounded Lipschitz domain is even cellular, then it makes sense to ask for traces of derivatives of the corresponding functions at the smooth faces and edges of the domain. This is just our point of view in this paper. But first we collect a few more general assertions.

1.5 Some properties

A function $m \in L_\infty(\mathbb{R}^n)$ is said to be a *pointwise multiplier* in a space $A^s_{pq}(\mathbb{R}^n)$ according to Definition 1.1 and Remark 1.2 if

$$f \mapsto mf \quad \text{generates a bounded map in } A^s_{pq}(\mathbb{R}^n).$$

A careful discussion how to understand pointwise multipliers may be found in [8, Sect. 4.2]. We are only interested in characteristic functions χ_Ω of domains Ω as pointwise multipliers. This distinguished case attracted a lot of attention. One may consult [8], [15, Sect. 2.3] and the literature mentioned there.

Proposition 1.13. *Let* Ω *be a bounded cellular domain in* $\mathbb{R}^n$ *according to Definition* 1.9(iii), *and let* $A^s_{pq}(\mathbb{R}^n)$ *and* $A^s_{pq}(\Omega)$ *be the spaces introduced in Definitions* 1.1 *and* 1.3(i).

(i) *Then* χ_Ω *is a pointwise multiplier in* $A^s_{pq}(\mathbb{R}^n)$ *if and only if*

$$0 < p \leqslant \infty, \quad 0 < q \leqslant \infty, \quad \max\left(n\left(\frac{1}{p}-1\right), \frac{1}{p}-1\right) < s < \frac{1}{p} \tag{1.17}$$

($p < \infty$ for the F-spaces).

(ii) *Let $\sigma \in \mathbb{R}$, $k \in \mathbb{N}$, and $s = \sigma + k$. Then*

$$A^s_{pq}(\Omega) = \left\{f \in A^\sigma_{pq}(\Omega) : D^\alpha f \in A^\sigma_{pq}(\Omega),\ |\alpha| \leqslant k\right\}$$

and

$$\|f\,|A^s_{pq}(\Omega)\| \sim \sum_{|\alpha|\leqslant k} \|D^\alpha f\,|A^\sigma_{pq}(\Omega)\|$$

(equivalent quasinorms).

Remark 1.14. Part (i) is well known for half-spaces and, hence, also for cubes. A proof and historical comments may be found in [8, Sect. 4.6, pp. 208, 258]. This carries over to bounded cellular domains. Part (ii) may be found in [16, Proposition 4.21] even for the more general class of bounded Lipschitz domains.

Proposition 1.15. *Let Ω be a bounded cellular domain in $\mathbb{R}^n$ according to Definition* 1.9(iii).

(i) *Let p, q, s be as in* (1.17) *with $p < \infty$ for the F-spaces. Then*

$$A^s_{pq}(\Omega) = \widetilde{A}^s_{pq}(\Omega) = \widetilde{A}^s_{pq}(\overline{\Omega}) \tag{1.18}$$

for the corresponding spaces from Definition 1.3 *and the interpretation given in Remark* 1.4.

(ii) *Let*

$$0 < p < \infty, \quad 0 < q < \infty, \quad \max\left(n\left(\frac{1}{p}-1\right), \frac{1}{p}-1\right) < s < \frac{1}{p}.$$

Then

$$A^s_{pq}(\Omega) = \widetilde{A}^s_{pq}(\Omega) = \mathring{A}^s_{pq}(\Omega) \tag{1.19}$$

for the corresponding spaces from Definition 1.3.

Proof. Step 1. According to Remark 1.4, the second equality in (1.18) is justified if

$$\left\{g \in A^s_{pq}(\mathbb{R}^n) : \operatorname{supp} g \subset \partial\Omega\right\} = \{0\}. \tag{1.20}$$

Recall that

$$A^s_{pq}(\Omega) \hookrightarrow L_1(\Omega) \qquad \text{if} \quad s > \sigma_p = n\left(\frac{1}{p}-1\right)_+. \tag{1.21}$$

Since $|\partial\Omega| = 0$, one gets (1.20) for these cases. It remains to prove (1.20) for the admitted spaces with $s \leqslant 0$. It is sufficient to deal with

$$B^s_{pp}(\mathbb{R}^n), \qquad 1 < p < \infty, \quad \frac{1}{p} - 1 < s < 0.$$

We use the duality [11, Sect. 2.11.2]

$$B^s_{pp}(\mathbb{R}^n)' = B^{-s}_{p'p'}(\mathbb{R}^n), \qquad \frac{1}{p} + \frac{1}{p'} = 1, \quad 0 < -s < \frac{1}{p'}.$$

Let $\varphi \in S(\mathbb{R}^n)$. Then from Proposition 1.13(i) it follows that both $\varphi\chi_\Omega$ and $\varphi(1-\chi_\Omega)$ belong to $B^{-s}_{p'p'}(\mathbb{R}^n)$. On the other hand, the translation

$$g(\cdot) \mapsto g_h(\cdot) = g(\cdot + h) \qquad \text{with } h \in \mathbb{R}^n$$

is continuous in $B^{-s}_{p'p'}(\mathbb{R}^n)$. To justify this claim, we recall that $D(\mathbb{R}^n)$ is dense in $B^{-s}_{p'p'}(\mathbb{R}^n)$. If $\psi \in D(\mathbb{R}^n)$ approximates $g \in B^{-s}_{p'p'}(\mathbb{R}^n)$, then ψ_h approximates g_h (uniformly in h) and the claimed continuity of the translation operator follows from

$$\begin{aligned} &\|f - f_h \,|B^{-s}_{p'p'}(\mathbb{R}^n)\| \\ &\leqslant c\|f - \psi \,|B^{-s}_{p'p'}(\mathbb{R}^n)\| \\ &\quad + c\|\psi - \psi_h \,|B^{-s}_{p'p'}(\mathbb{R}^n)\| + c\|(f-\psi)_h \,|B^{-s}_{p'p'}(\mathbb{R}^n)\| \\ &\leqslant c'\|f - \psi \,|B^{-s}_{p'p'}(\mathbb{R}^n)\| + c'\|\psi - \psi_h \,|B^{-s}_{p'p'}(\mathbb{R}^n)\|. \end{aligned} \tag{1.22}$$

Then one gets that $D(\mathbb{R}^n \setminus \partial\Omega)$ is dense in $B^{-s}_{p'p'}(\mathbb{R}^n)$. Let now $g \in B^s_{pp}(\mathbb{R}^n)$ with supp $g \subset \partial\Omega$, and let $\varphi \in S(\mathbb{R}^n)$ be approximated in $B^{-s}_{p'p'}(\mathbb{R}^n)$ by $\varphi^k \in D(\mathbb{R}^n \setminus \partial\Omega)$. Then one gets

$$g(\varphi) = \lim_{k\to\infty} g(\varphi^k) = 0 \qquad \text{for any} \quad \varphi \in S(\mathbb{R}^n).$$

Hence $g = 0$. This justifies the second equality in (1.18). The first equality in (1.18) follows now from Proposition 1.13.

Step 2. If $p < \infty$ and $q < \infty$, then one gets from the standard localization (multiplication with a smooth resolution of unity) and the continuity of the translation operator as indicated above that $D(\Omega)$ is dense in $\widetilde{A}^s_{pq}(\Omega)$. This proves (1.19). □

Remark 1.16. Assertions of type (1.18), (1.19) under the indicated conditions for p, q, s have some history. As far as bounded C^∞ domains are concerned, one may consult [13, Sects. 5.3–5.6, pp. 44–50], where also the above duality argument comes from. In the case of bounded Lipschitz domains, we quoted what is known in [16, Proposition 5.19] referring to [15, Sect. 1.11.6] and (in greater detail) to [14, Proposition 3.1, Sect. 5.2]. But there is no full generalizations of Proposition 1.15 to bounded Lipschitz domains. In particular, it seems to be unknown whether there is a full counterpart of the pointwise multiplier assertions in Proposition 1.13(i) for bounded Lipschitz domains. What is known so far can be found in [3, Corollary 13.6] and in [14]. This may support even at this level (before discussing boundary values of derivatives) to specify bounded Lipschitz domains to bounded cellular domains.

Proposition 1.17. *Let Ω be a bounded cellular domain in $\mathbb{R}^n$ according to Definition* 1.9(iii). *Let $k \in \mathbb{N}$, and let*

$$0 < p < \infty, \quad 0 < q < \infty, \quad \max\Big(n\big(\frac{1}{p} - 1\big), \frac{1}{p} - 1\Big) < \sigma < \frac{1}{p}.$$

Then

$$\widetilde{A}^s_{pq}(\Omega) = \mathring{A}^s_{pq}(\Omega) \qquad \text{with} \quad s = \sigma + k \tag{1.23}$$

(equivalent quasinorms) for the corresponding spaces according to Definition 1.3.

Proof. If $f \in \widetilde{A}^s_{pq}(\Omega)$, then from Proposition 1.13(ii) it follows that

$$D^\alpha f \in A^\sigma_{pq}(\Omega) = \widetilde{A}^\sigma_{pq}(\Omega), \qquad |\alpha| \leqslant k$$

(equivalent quasinorms). Now one gets (1.23) from (1.19) and the translation argument in connection with (1.22). □

Remark 1.18. Assertions of this type for the Sobolev spaces and the classical Besov spaces in (1.1)–(1.3) in bounded C^∞ domains are known since a long time and may be found in [10, Sect. 4.3.2]. The extension of these assertions to more general spaces A^s_{pq}, especially with $p < 1$ is a somewhat tricky game. One may consult [15, Sect. 1.11.6, pp. 66-67] and the references given there.

1.6 Frames

It is our main aim to study wavelet bases for function spaces $A^s_{pq}(\Omega)$ as introduced in Definition 1.3 in bounded cellular domains. But, to provide a better understanding, it seems to be reasonable to recall some assertions about wavelet frames for spaces in bounded Lipschitz domains. We follow closely [16, Sect. 5.4.1]. Let

$$\sigma_p = n\Big(\frac{1}{p} - 1\Big)_+ \quad \text{and} \quad \sigma_{pq} = \max(\sigma_p, \sigma_q), \quad 0 < p, q \leqslant \infty, \tag{1.24}$$

where $b_+ = \max(b, 0)$ for $b \in \mathbb{R}$. Let $a^s_{pq}(\mathbb{Z}^\Omega)$ with $a \in \{b, f\}$ be the sequence spaces as introduced in Definition 1.5, and let $\left\{\Phi^j_l\right\}$ be the u-wavelet systems according to Definition 1.7(i). If $\{a_i\}$ and $\{b_i\}$ are positive numbers for an index set I, then $a_i \sim b_i$ (equivalence) means that there are two positive numbers c_1 and c_2 such that

$$c_1 a_i \leqslant b_i \leqslant c_2 a_i \qquad \text{for all} \quad i \in I.$$

Theorem 1.19. *Let Ω be a bounded Lipschitz domain in $\mathbb{R}^n$ according to Definition* 1.9(ii). *For any $u \in \mathbb{N}$ there is a u-wavelet system $\Phi = \left\{\Phi^j_l\right\}$ with the following property. Let*

$$0 < p < \infty, \quad 0 < q < \infty, \quad s \in \mathbb{R},$$

and let

$$u > \max(s, \sigma_p - s) \ B\text{-spaces}, \qquad u > \max(s, \sigma_{pq} - s) \ F\text{-spaces}.$$

Then $f \in D'(\Omega)$ is an element of $A^s_{pq}(\Omega)$ if and only if it can be represented as

$$f = \sum_{j=0}^{\infty} \sum_{l=1}^{N_j} \lambda^j_l \, 2^{-jn/2} \, \Phi^j_l, \qquad \lambda \in a^s_{pq}(\mathbb{Z}^\Omega), \tag{1.25}$$

unconditional convergence being in $A^s_{pq}(\Omega)$. Furthermore,

$$\|f \,|A^s_{pq}(\Omega)\| \sim \inf \|\lambda \,|a^s_{pq}(\mathbb{Z}^\Omega)\|,$$

where the infimum is taken over all admissible representations (1.25) *(equivalent norms). Any $f \in A^s_{pq}(\Omega)$ can be represented as*

$$f = \sum_{j=0}^{\infty} \sum_{l=1}^{N_j} \lambda^j_l(f) \, 2^{-jn/2} \, \Phi^j_l, \tag{1.26}$$

where $\lambda^j_l(\cdot) \in A^s_{pq}(\Omega)'$ are linear and continuous functionals in $A^s_{pq}(\Omega)$ and

$$\|f \,|A^s_{pq}(\Omega)\| \sim \|\lambda(f) \,|a^s_{pq}(\mathbb{Z}^\Omega)\|$$

(u-wavelet frame).

Remark 1.20. Details, references and further results may be found in [16, Sect. 5.4.1]. In particular, the above assertions can be extended to $p = \infty$ and/or $q = \infty$ for the B-spaces and $q = \infty$ for the F-spaces.

1.7 Bases

A set $\{b_j\}_{j=1}^{\infty} \subset B$ in a separable (infinite-dimensional) complex quasi-Banach space B is called a *basis* if any $b \in B$ can be uniquely represented as

$$b = \sum_{j=1}^{\infty} \lambda_j \, b_j, \qquad \lambda_j \in \mathbb{C} \quad \text{(convergence in } B\text{)}. \tag{1.27}$$

A basis $\{b_j\}_{j=1}^{\infty}$ is called *unconditional* if for any rearrangement σ of $\mathbb{N}$ (one-to-one map of $\mathbb{N}$ onto itself) $\{b_{\sigma(j)}\}_{j=1}^{\infty}$ is again a basis and

$$b = \sum_{j=1}^{\infty} \lambda_{\sigma(j)} \, b_{\sigma(j)} \qquad \text{(convergence in } B\text{)}$$

for any $b \in B$ with (1.27). Since all wavelet bases in this paper are isomorphic to standard bases in the sequence spaces $a^s_{pq}(\mathbb{Z}^\Omega)$, they are unconditional.

Proposition 1.21. *Let Ω be an arbitrary bounded domain in $\mathbb{R}^n$. For any $u \in \mathbb{N}$ there are interior u-wavelet systems $\left\{\Phi^j_l\right\}$ according to Definition* 1.7(ii) *which are orthonormal bases in $L_2(\Omega)$ (orthonormal u-wavelet bases).*

Remark 1.22. This is the main result of [16, Sect. 2.4.1]. It can be extended to arbitrary (not necessarily bounded) domains Ω and to $L_p(\Omega)$ with $1 < p < \infty$ [16, Theorem 2.36]. For our purpose the following observation is of interest.

Theorem 1.23. *Let Ω be a bounded cellular domain in $\mathbb{R}^n$ according to Definition* 1.9(iii). *Let for $u \in \mathbb{N}$*

$$\Phi = \left\{\Phi^j_l : j \in \mathbb{N}_0;\ l = 1, \ldots, N_j\right\}$$

be an interior orthonormal u-wavelet basis in $L_2(\Omega)$ according to Proposition 1.21 *and Definition* 1.7(ii). *Let $A^s_{pq}(\Omega)$ and $a^s_{pq}(\mathbb{Z}^\Omega)$ with $a \in \{b, f\}$ be as introduced in Definitions* 1.3 *and* 1.5. *Let $0 < p, q < \infty$, and let*

$$-\infty < s < \min\left(\frac{1}{p}, \frac{n}{n-1}\right). \tag{1.28}$$

Let $u > u(A^s_{pq})$, where $u(B^s_{pq}) = u(s,p)$ and $u(F^s_{pq}) = u(s,p,q)$ are sufficiently large natural numbers. Then $f \in D'(\Omega)$ is an element of $A^s_{pq}(\Omega)$ if and only if it can be represented as

$$f = \sum_{j=0}^{\infty} \sum_{l=1}^{N_j} \lambda^j_l \, 2^{-jn/2} \, \Phi^j_l, \qquad \lambda \in a^s_{pq}(\mathbb{Z}^\Omega), \tag{1.29}$$

unconditional convergence being in $A^s_{pq}(\Omega)$. *Furthermore, if* $f \in A^s_{pq}(\Omega)$, *then the representation* (1.29) *is unique,*

$$f = \sum_{j=0}^{\infty} \sum_{l=1}^{N_j} \lambda^j_l(f)\, 2^{-jn/2}\, \Phi^j_l,$$

where $\lambda^j_l(\cdot) \in A^s_{pq}(\Omega)'$ *are linear and continuous functionals on* $A^s_{pq}(\Omega)$ *and*

$$f \mapsto \left\{\lambda^j_l(f)\right\} \quad \text{is an isomorphic map of } A^s_{pq}(\Omega) \text{ onto } a^s_{pq}(\mathbb{Z}^\Omega)$$

(*interior u-wavelet basis*).

Remark 1.24. This coincides essentially with [16, Theorem 5.4.3], where one finds proofs, explanations and also some generalizations. We inserted Theorems 1.19 and 1.23 to provide a better understanding of what follows in the next sections. Since $p < \infty$, $q < \infty$, one gets from the above theorem that $D(\Omega)$ is dense in all spaces $A^s_{pq}(\Omega)$ with (1.28). In particular, these spaces have no traces at $\partial\Omega$. If $s > 1/p$, then the situation changes drastically. The corresponding spaces have traces and representations of type (1.25), (1.26) in terms of interior wavelet systems cannot be expected. The boundary $\partial\Omega$ must be taken into account. One could try to start with frames from Theorem 1.19. But there is no hope to convert these frames into wavelet bases as considered in this paper. We follow here a method first developed in [16, Chapt. 5] and applied there to very specific domains such as balls in $\mathbb{R}^n$. Now we deal with bounded cellular domains what causes some additional problems in connection with traces and extensions, which might be of self-contained interest and which will be considered first.

2 A Model Case

2.1 Traces and extensions

Let $n \in \mathbb{N}$ and $l \in \mathbb{N}$ with $l < n$. Let $\mathbb{R}^n = \mathbb{R}^l \times \mathbb{R}^{n-l}$ with $x = (y, z) \in \mathbb{R}^n$,

$$y = (y_1, \ldots, y_l) \in \mathbb{R}^l, \qquad z = (z_1, \ldots, z_{n-l}) \in \mathbb{R}^{n-l}, \tag{2.1}$$

where $\mathbb{R}^l$ is identified with the hyperplane $x = (y, 0)$ in $\mathbb{R}^n$. Let Q_l be the unit cube

$$Q_l = \{x = (y, z) \in \mathbb{R}^n : z = 0,\ 0 < y_m < 1;\ m = 1, \ldots, l\} \tag{2.2}$$

in this hyperplane. Let

$$Q_l^n = \{x = (y,z) \in \mathbb{R}^n : y \in Q_l,\ z \in \mathbb{R}^{n-l}\} \tag{2.3}$$

be the related cylindrical domain in $\mathbb{R}^n$. Let $\mathrm{tr}_{\,l}$ be the trace operator

$$\mathrm{tr}_{\,l}: \quad f(x) \mapsto f(y,0), \qquad f \in A^s_{pq}(\mathbb{R}^n), \tag{2.4}$$

on $\mathbb{R}^l$ or Q_l (if exists) defined in the usual way by completion of pointwise traces of smooth functions with respect to spaces specified below. Let

$$\mathbb{N}_l^n = \{\alpha = (\alpha_1, \ldots, \alpha_n) \in \mathbb{N}_0^n : \alpha_1 = \cdots = \alpha_l = 0\}. \tag{2.5}$$

Then

$$D^\alpha f = \frac{\partial^{|\alpha|} f}{\partial z_1^{\alpha_{l+1}} \cdots \partial z_{n-l}^{\alpha_n}}, \qquad \alpha \in \mathbb{N}_l^n, \tag{2.6}$$

are the derivatives perpendicular to Q_l. Let

$$\sigma_p^l = l\Big(\frac{1}{p} - 1\Big)_+, \qquad 0 < p \leqslant \infty, \quad l \in \mathbb{N},$$

be as in (1.24) indicating now l. Let $A^s_{pq}(\mathbb{R}^n)$ be the spaces as introduced in Definition 1.1, and let $B^\sigma_{pq}(Q_l)$ be the spaces from Definition 1.3 in the context of $\mathbb{R}^l$.

Proposition 2.1. *Let $l \in \mathbb{N}$, $n \in \mathbb{N}$ with $l < n$, and $r \in \mathbb{N}_0$. Let $0 < p \leqslant \infty$ ($p < \infty$ for the F-spaces), $0 < q \leqslant \infty$, and*

$$s > r + \frac{n-l}{p} + \sigma_p^l. \tag{2.7}$$

Let

$$\mathrm{tr}^{\,r}_{\,l}: \quad f \mapsto \{\mathrm{tr}_{\,l} D^\alpha f : \alpha \in \mathbb{N}_l^n,\ |\alpha| \leqslant r\}. \tag{2.8}$$

Then

$$\mathrm{tr}^{\,r}_{\,l}: \quad B^s_{pq}(\mathbb{R}^n) \hookrightarrow \prod_{\substack{\alpha \in \mathbb{N}_l^n, \\ |\alpha| \leqslant r}} B_{pq}^{s - \frac{n-l}{p} - |\alpha|}(Q_l) \tag{2.9}$$

and

$$\mathrm{tr}^{\,r}_{\,l}: \quad F^s_{pq}(\mathbb{R}^n) \hookrightarrow \prod_{\substack{\alpha \in \mathbb{N}_l^n, \\ |\alpha| \leqslant r}} B_{pp}^{s - \frac{n-l}{p} - |\alpha|}(Q_l) \tag{2.10}$$

(continuous embedding).

Remark 2.2. Assertions of this type have a long history. Although S.L. Sobolev dealt in his famous book [9] with boundary value problems for elliptic equations, he did not care for the exact traces of Sobolev spaces. This came in the 1950s and 1960s and may be found in [7, 1]. A detailed account of the history of assertions of type (2.9), (2.10) was given in [10, Sect. 2.9.4] and [11, Sect. 3.3.3] including the remarkable observation that the trace of the F-space is independent of q. Nowadays, one can prove (2.9), (2.10) rather quickly. One expands $f \in A^s_{pq}(\mathbb{R}^n)$ into atoms or wavelets and takes the required derivatives. Restrictions to $\mathbb{R}^l$ give again atomic expansions, where (2.7) ensures that one does not need moment conditions for the resulting atoms. It is also (more or less) known that (2.9), (2.10) are maps *onto* the corresponding target spaces and that there are linear and bounded extension operators. But for our purpose one needs rather specific *wavelet-friendly extension operators*. We dealt with related constructions in [16, Sect. 5.1.3] for bounded C^∞ domains in $\mathbb{R}^n$. Then one has $l = n-1$ in the above context. But one can modify the technique developed there such that it fits first to the above model case and finally to bounded cellular domains.

We describe the corresponding modifications of the constructions in [16, Sect. 5.1.3] adapted to our later needs. First we apply [16, Theorem 3.13] to the spaces

$$\widetilde{B}^\sigma_{pq}(Q_l), \qquad 0 < p \leqslant \infty, \quad 0 < q \leqslant \infty, \quad \sigma > \sigma^l_p,$$

according to Definition 1.3 (with l in place of n). Let

$$\left\{\Phi^j_m(y) : j \in \mathbb{N}_0; m = 1, \ldots, N_j\right\} \subset C^u(Q_l) \tag{2.11}$$

with $N_j \sim 2^{jl}$ be an interior orthonormal u-wavelet basis for $L_2(Q_l)$ according to Proposition 1.21 (with l in place of n) and Definition 1.7(ii). Let $u > \sigma$. Then one has for $g \in \widetilde{B}^\sigma_{pq}(Q_l)$ the wavelet expansion

$$g = \sum_{j=0}^{\infty} \sum_{m=1}^{N_j} \lambda^j_m(g)\, 2^{-jl/2}\, \Phi^j_m \tag{2.12}$$

with

$$\lambda^j_m(g) = 2^{jl/2} \int_{Q_l} g(y)\, \Phi^j_m(y)\, dy \tag{2.13}$$

and the isomorphic map

$$g \mapsto \left\{\lambda^j_m(g)\right\} \quad \text{of } \widetilde{B}^\sigma_{pq}(Q_l) \text{ onto } b^\sigma_{pq}(\mathbb{Z}^{Q_l}).$$

In particular, if $p < \infty$ and $q < \infty$, then $\left\{\Phi^j_m(y)\right\}$ in (2.11) is an interior unconditional u-wavelet basis in $\widetilde{B}^\sigma_{pq}(Q_l)$. We incorporate temporarily $p = \infty$ and $q = \infty$. Then (2.12) converges in any space $\widetilde{B}^{\sigma-\varepsilon}_{pq}(Q_l)$ with $\varepsilon > 0$. We refer for technical details to [16, Sect. 3.2.2]. Next we modify the construction of the extension operators as used in the proof of [16, Theorem 5.14] adapted to the right-hand sides of (2.9), (2.10). Let

$$\chi \in D(\mathbb{R}^{n-l}), \quad \operatorname{supp} \chi \subset \{|z| \leqslant 2\}, \quad \chi(z) = 1 \text{ if } |z| \leqslant 1, \tag{2.14}$$

with $z \in \mathbb{R}^{n-l}$ as in (2.1). One may assume that χ satisfies sufficiently many moment conditions,

$$\int_{\mathbb{R}^{n-l}} \chi(z)\, z^\beta \, dz = 0 \quad \text{if} \quad |\beta| \leqslant L. \tag{2.15}$$

Let Φ^j_m be the wavelets according to (2.11), and let

$$\Phi^{j,\alpha}_m(x) = 2^{j|\alpha|}\, z^\alpha\, \chi(2^j z)\, 2^{(n-l)j/2}\, \Phi^j_m(y) \tag{2.16}$$

with $\alpha \in \mathbb{N}^n_l$ according to (2.5). Let $r \in \mathbb{N}_0$. Then

$$\begin{aligned} &\operatorname{Ext}^{r,u}_l \{g_\alpha : \alpha \in \mathbb{N}^n_l, \ |\alpha| \leqslant r\}\,(x) \\ &= \sum_{|\alpha| \leqslant r} \sum_{j=0}^{\infty} \sum_{m=1}^{N_j} \frac{1}{\alpha!}\, \lambda^j_m(g_\alpha)\, 2^{-j|\alpha|}\, 2^{-jn/2}\, \Phi^{j,\alpha}_m(x) \end{aligned} \tag{2.17}$$

is the counterpart of the wavelet-friendly extension operator used in the proof of [16, Theorem 5.27], where $\{g_\alpha\} \subset L_1(Q_l)$ and $\lambda^j_m(g_\alpha)$ as in (2.13) with g_α in place of g. Let $\widetilde{A}^s_{pq}(Q^n_l)$ be the spaces as introduced in Definition 1.3, where Q^n_l is the cylindrical domain in (2.3).

Theorem 2.3. *Let $\{\Phi^j_m\}$ be the interior orthonormal u-wavelet basis in $L_2(Q_l)$ according to (2.11). Let χ be as in (2.14), (2.15) (with $L \in \mathbb{N}_0$ sufficiently large in dependence on q for the F-spaces). Let $r \in \mathbb{N}_0$, and let $\operatorname{Ext}^{r,u}_l$ be given by (2.17) with (2.13). Then*

$$\operatorname{Ext}^{r,u}_l : \quad \{g_\alpha : \alpha \in \mathbb{N}^n_l, \ |\alpha| \leqslant r\} \mapsto g$$

is an extension operator

$$\begin{aligned} &\operatorname{Ext}^{r,u}_l : \prod_{\substack{\alpha \in \mathbb{N}^n_l, \\ |\alpha| \leqslant r}} \widetilde{B}^{s-\frac{n-l}{p}-|\alpha|}_{pq}(Q_l) \hookrightarrow \widetilde{B}^s_{pq}(Q^n_l), \\ &0 < p \leqslant \infty, \ 0 < q \leqslant \infty, \ u > s > r + \frac{n-l}{p} + \sigma^l_p, \end{aligned} \tag{2.18}$$

and

$$\begin{aligned} &\operatorname{Ext}^{r,u}_l : \prod_{\substack{\alpha \in \mathbb{N}^n_l, \\ |\alpha| \leqslant r}} \widetilde{B}^{s-\frac{n-l}{p}-|\alpha|}_{pp}(Q_l) \hookrightarrow \widetilde{F}^s_{pq}(Q^n_l), \\ &0 < p < \infty, \ 0 < q \leqslant \infty, \ u > s > r + \frac{n-l}{p} + \sigma^l_p, \end{aligned} \tag{2.19}$$

with

$$\mathrm{tr}\,{}_l^r \circ \mathrm{Ext}\,{}_l^{r,u} = \mathrm{id} \quad \textit{identity in} \quad \prod_{\substack{\alpha\in\mathbb{N}_l^n,\\ |\alpha|\leqslant r}} \widetilde{B}_{pq}^{s-\frac{n-l}{p}-|\alpha|}(Q_l). \tag{2.20}$$

Proof. Since $s > \sigma_p^l$ and for $p < 1$

$$s > \frac{n-l}{p} + \sigma_p^l = \frac{n}{p} - l > \sigma_p^n, \tag{2.21}$$

from Definition 1.3 it follows that one can identify $\widetilde{B}_{pq}^\sigma(Q_l)$ in (2.18), (2.19) with $\widetilde{B}_{pq}^\sigma(\overline{Q_l})$, considered as subspaces of $B_{pq}^\sigma(\mathbb{R}^l)$, and $\widetilde{A}_{pq}^s(Q_l^n)$ with $\widetilde{A}_{pq}^s(\overline{Q_l^n})$ as subspaces of $A_{pq}^s(\mathbb{R}^n)$. The case $l = n-1$ is essentially covered by Theorem 5.14 in [16] and its proof. There we dealt with a bounded C^∞ domain Ω in place of Q_l^n and its boundary $\partial\Omega$ in place of Q_{n-1}. Since we replaced $B_{pq}^\sigma(\partial\Omega)$ now by $\widetilde{B}_{pq}^\sigma(Q_{n-1})$, we avoided any difficulty which may be caused by the boundary ∂Q_{n-1} of Q_{n-1}. But otherwise, one can follow the arguments in [16, Theorem 5.14] for all $l \in \mathbb{N}$ with $l < n$. In particular, the functions $\Phi_m^{j,\alpha}$ in (2.16) are the adapted atoms. By (2.21), one does not need moment conditions in case of the B-spaces. If $\varepsilon \leqslant q < \infty$ for some ε with $0 < \varepsilon < 1$, then one may choose $L \in \mathbb{N}$ with $L \geqslant n\big(\frac{1}{\varepsilon} - 1\big)$ in (2.15) in the case of F-spaces. □

Remark 2.4. It comes out that $\mathrm{Ext}\,{}_l^{r,u}$ is a common extension operator for given r, u and $0 < \varepsilon \leqslant 1$ with $\varepsilon \leqslant q$ in (2.19) in place of $0 < q$.

2.2 Approximation, density, decomposition

Let Q_l and Q_l^n be as in (2.2), (2.3). By Definition 1.3, the completion of $D(Q_l^n)$ in $A_{pq}^s(Q_l^n)$ is denoted by $\mathring{A}_{pq}^s(Q_l^n)$. Proposition 1.17 extended to Q_l^n shows that

$$\widetilde{A}_{pq}^s(Q_l^n) = \mathring{A}_{pq}^s(Q_l^n) \tag{2.22}$$

(equivalent quasinorms) if

$$1 \leqslant p < \infty, \quad 0 < q < \infty, \quad 0 < s - \frac{1}{p} \notin \mathbb{N}. \tag{2.23}$$

If $p < 1$, then one has to exclude some (p,s)-regions in order to get (2.22). This is a first hint that it might be reasonable to restrict the further considerations to $p \geqslant 1$. But we add a second more substantial argument. For

$0 < p < \infty$, $0 < q < \infty$ and the largest admitted $r \in \mathbb{N}_0$ in (2.7) one may ask under which circumstances

$$D(Q_l^n \setminus Q_l) \quad \text{is dense in} \quad \left\{ f \in \widetilde{A}^s_{pq}(Q_l^n) : \operatorname{tr}{}_l^r f = 0 \right\}. \tag{2.24}$$

Of course, $Q_l^n \setminus Q_l$ is considered as a set in $\mathbb{R}^n$. As for the classical cases, we refer to [10, Sect. 2.9.4, p. 223] complemented by [11, Sect. 3.4.3, p. 210]. We begin with a preparation. We use the same notation as in (2.1)–(2.3) and Proposition 2.1.

Proposition 2.5. *Let $l \in \mathbb{N}$, $n \in \mathbb{N}$ with $l < n$. Let $r \in \mathbb{N}_0$, and let*

$$0 < p, q < \infty, \qquad s > r + \frac{n-l}{p} + \sigma_p^l. \tag{2.25}$$

Let $u \in \mathbb{N}$ with $u > s$. Then

$$\{g \in C^u(\mathbb{R}^n) : \operatorname{supp} g \subset \{y \in Q_l\} \times \{|z| < 1\},\ \operatorname{tr}{}_l^r g = 0\} \tag{2.26}$$

is dense in

$$\left\{ f \in \widetilde{A}^s_{pq}(Q_l^n) : \operatorname{tr}{}_l^r f = 0 \right\}.$$

Proof. By the same decomposition and translation arguments as in connection with (1.22) now for $h = (h_l, 0)$ (parallel to $\mathbb{R}^l$), it follows that it is sufficient to approximate

$$f \in A^s_{pq}(\mathbb{R}^n), \quad \operatorname{supp} f \subset Q_l^n, \quad \operatorname{tr}{}_l^r f = 0$$

(hence vanishing near ∂Q_l^n). For given $\varepsilon > 0$ one can approximate f in $A^s_{pq}(\mathbb{R}^n)$ by $f_\varepsilon \in S(\mathbb{R}^n)$ with

$$\operatorname{supp} f_\varepsilon \subset Q_l^n \quad \text{and} \quad \| \operatorname{tr}{}_l^r f_\varepsilon \mid \operatorname{tr}{}_l^r A^s_{pq}(\mathbb{R}^n) \| \leqslant \varepsilon,$$

where $\operatorname{tr}{}_l^r A^s_{pq}(\mathbb{R}^n)$ stands for the trace spaces according to (2.9), (2.10). Let

$$f^\varepsilon = \operatorname{Ext}{}_l^{r,u} \{ \operatorname{tr}{}_l^r f_\varepsilon \}$$

with $\operatorname{Ext}{}_l^{r,u}$ as in (2.17) and Theorem 2.3. By the properties of $\operatorname{Ext}{}_l^{r,u}$, one gets

$$f^\varepsilon \in C^u(\mathbb{R}^n), \quad \operatorname{supp} f^\varepsilon \subset Q_l^n, \quad \|f^\varepsilon \,|A^s_{pq}(\mathbb{R}^n)\| \leqslant c\varepsilon$$

for some $c > 0$ which is independent of ε. Then $f_\varepsilon - f^\varepsilon$ is the desired approximation. □

Remark 2.6. We return to the question (2.24). By Proposition 2.5, it is sufficient to approximate the functions g from (2.26). Let

$$S_j = \left\{ x = (y, z) : y \in Q_l,\ 2^{-j-1} \leqslant |z| \leqslant 2^{-j+1} \right\}, \qquad j \in \mathbb{N}_0,$$

and let

$$1=\sum_{j=0}^{\infty}\sum_{m=1}^{N_j}\varphi_{jm}(x),\quad N_j\sim 2^{jl},\quad y\in Q_l,\quad 0<|z|<1,$$

be a resolution of unity by suitable C^∞ functions φ_{jm} with

$$\operatorname{supp}\varphi_{jm}\subset B_{jm}\subset S_j,\qquad j\in\mathbb{N}_0,\quad 1\leqslant m\leqslant N_j,$$

where B_{jm} is a ball of radius 2^{-j}. Let g be as in (2.26) (in particular, g vanishes near the boundary ∂Q_l). Then

$$g=\sum_{j=0}^{\infty}\sum_{m=1}^{N_j}\lambda_{jm}\,2^{j(r+1)}\,2^{-j(s-\frac{n}{p})}\,\varphi_{jm}\,g \tag{2.27}$$

is an atomic decomposition of g in $B^s_{pq}(\mathbb{R}^n)$ with p,q,s as in (2.25) (not to speak about immaterial constants). No moment conditions are needed, but we used $\operatorname{tr}{}^r_l g=0$ according to (2.8) for $g\in C^u(\mathbb{R}^n)$ and related Taylor expansions in the z-directions, where the factor $2^{j(r+1)}$ comes from. We have

$$\lambda_{jm}\sim 2^{-j(r+1-s+\frac{n}{p})}.$$

Let g_J with $J\in\mathbb{N}$ be given by (2.27) with $j\geqslant J$ in place of $j\in\mathbb{N}_0$. Then it follows that

$$\begin{aligned}\|g_J\,|B^s_{pq}(\mathbb{R}^n)\|&\leqslant c\Big(\sum_{j=J}^{\infty}\Big(\sum_{m=1}^{N_j}|\lambda_{jm}|^p\Big)^{q/p}\Big)^{1/q}\\&\leqslant c'\Big(\sum_{j=J}^{\infty}2^{-jp(r+1-s+\frac{n}{p}-\frac{l}{p})q/p}\Big)^{1/q}\\&\leqslant c''\,2^{-J\delta}\quad\text{if}\quad\delta=r+1-s+\frac{n-l}{p}>0.\end{aligned} \tag{2.28}$$

Then $g-g_J$ approximates g in $B^s_{pq}(\mathbb{R}^n)$. These functions vanish near Q_l. An additional mollification gives (2.24) for $\widetilde{B}^s_{pq}(Q^n_l)$. By elementary embedding, one gets this assertion also for $\widetilde{F}^s_{pq}(Q^n_l)$. The largest admitted $r\in\mathbb{N}_0$ in (2.25) is given by

$$r=s-\frac{n-l}{p}-\sigma^l_p-\varepsilon\quad\text{for some }\varepsilon\text{ with }0<\varepsilon\leqslant 1.$$

If $p\geqslant 1$, then

$$\delta=1-\varepsilon>0\quad\text{if and only if}\quad s-\frac{n-l}{p}\notin\mathbb{N}.$$

If $p < 1$, then (2.28) requires

$$\delta = 1 - \varepsilon - \sigma_p^l = 1 - \varepsilon - l\Big(\frac{1}{p} - 1\Big) > 0. \tag{2.29}$$

Although there are some $p < 1$ and $0 < \varepsilon < 1$ with (2.29) the situation is not very satisfactory. Together with (2.22) and (2.23), it is a second good reason to restrict what follows to $p \geqslant 1$.

After these discussions we are now in a similar situation as in [16, Corollary 5.16, Theorem 5.21]. There we dealt with bounded C^∞ domains Ω (instead of Q_l^n) and its boundary $\partial\Omega$ (instead of Q_l). We describe briefly the (more or less technical) changes. First we remark that one gets by Proposition 2.1 for the same parameters as there that

$$\operatorname{tr}{}_l^r : \quad \widetilde{B}^s_{pq}(Q_l^n) \hookrightarrow \prod_{\substack{\alpha\in\mathbb{N}_l^n,\\ |\alpha|\leqslant r}} \widetilde{B}_{pq}^{s-\frac{n-l}{p}-|\alpha|}(Q_l) \tag{2.30}$$

and

$$\operatorname{tr}{}_l^r : \quad \widetilde{F}^s_{pq}(Q_l^n) \hookrightarrow \prod_{\substack{\alpha\in\mathbb{N}_l^n,\\ |\alpha|\leqslant r}} \widetilde{B}_{pp}^{s-\frac{n-l}{p}-|\alpha|}(Q_l). \tag{2.31}$$

Combining this assertion with Theorem 2.3, one gets for the same parameters p, q, s (and r, u) as there that

$$P_r^l = \operatorname{Ext}{}_l^{r,u} \circ \operatorname{tr}{}_l^r : \quad \widetilde{A}^s_{pq}(Q_l^n) \hookrightarrow \widetilde{A}^s_{pq}(Q_l^n).$$

By (2.20), from

$$(P_r^l)^2 = \operatorname{Ext}{}_l^{r,u} \circ \operatorname{tr}{}_l^r \circ \operatorname{Ext}{}_l^{r,u} \circ \operatorname{tr}{}_l^r = P_r^l$$

it follows that P_r^l is a projection onto its range, denoted by $P_r^l\widetilde{A}^s_{pq}(Q_l^n)$. Then $T_r^l = \mathrm{id} - P_r^l$ is also a projection and

$$T_r^l\widetilde{A}^s_{pq}(Q_l^n) = \Big\{ f \in \widetilde{A}^s_{pq}(Q_l^n) : \operatorname{tr}{}_l^r f = 0 \Big\}. \tag{2.32}$$

Furthermore, $\operatorname{Ext}{}_l^{r,u}$ is an isomorphic map of the right-hand sides of (2.30), (2.31) onto $P_r^l\widetilde{A}^s_{pq}(Q_l^n)$, hence

$$\operatorname{Ext}{}_l^{r,u} \prod_{\substack{\alpha\in\mathbb{N}_l^n,\\ |\alpha|\leqslant r}} \widetilde{B}_{pq}^{s-\frac{n-l}{p}-|\alpha|}(Q_l) = P_r^l\widetilde{B}^s_{pq}(Q_l^n) \tag{2.33}$$

and a similar assertion for the F-spaces. Now we clip together the above considerations. Recall that Q_l and related spaces refer to $\mathbb{R}^l$, whereas Q_l^n and $Q_l^n \setminus \overline{Q_l}$ and related spaces are considered in $\mathbb{R}^n$.

Theorem 2.7. *Let $l \in \mathbb{N}$ and $n \in \mathbb{N}$ with $l < n$. Let*

$$1 \leqslant p < \infty, \quad 0 < q < \infty, \quad 0 < s - \frac{n-l}{p} \notin \mathbb{N}, \quad s - \frac{1}{p} \notin \mathbb{N}. \tag{2.34}$$

Let Q_l and Q_l^n be as in (2.2), (2.3). Then

$$\widetilde{A}^s_{pq}(Q_l^n) = \mathring{A}^s_{pq}(Q_l^n). \tag{2.35}$$

Let $r = \left[s - \frac{n-l}{p}\right] \in \mathbb{N}_0$ be the largest integer smaller than $s - \frac{n-l}{p}$. For $s < u \in \mathbb{N}$ let $\mathrm{Ext}\,_l^{r,u}$ be the extension operator according to (2.17), (2.13), and Theorem 2.3. Then

$$\mathring{B}^s_{pq}(Q_l^n) = \mathring{B}^s_{pq}(Q_l^n \setminus \overline{Q_l}) \times \mathrm{Ext}\,_l^{r,u} \prod_{\substack{\alpha \in \mathbb{N}_l^n, \\ |\alpha| \leqslant r}} \widetilde{B}_{pq}^{s - \frac{n-l}{p} - |\alpha|}(Q_l) \tag{2.36}$$

and

$$\mathring{F}^s_{pq}(Q_l^n) = \mathring{F}^s_{pq}(Q_l^n \setminus \overline{Q_l}) \times \mathrm{Ext}\,_l^{r,u} \prod_{\substack{\alpha \in \mathbb{N}_l^n, \\ |\alpha| \leqslant r}} \widetilde{B}_{pp}^{s - \frac{n-l}{p} - |\alpha|}(Q_l) \tag{2.37}$$

(*complemented subspaces*).

Proof. First we remark that (2.35) follows from (2.22), (2.23). By (2.32) and (2.33), one can decompose $\widetilde{A}^s_{pq}(Q_l^n)$ into the second factors on the right-hand sides of (2.36), (2.37) and the space on the right-hand side of (2.32). The assumptions (2.34) ensure that one can apply the considerations in Remark 2.6 resulting in (2.24). This proves the theorem. □

3 Spaces on Cubes and Polyhedrons

Let

$$Q = \{x \in \mathbb{R}^n : x = (x_1, \ldots, x_n),\ 0 < x_m < 1;\ m = 1, \ldots, n\} \tag{3.1}$$

be the unit cube in $\mathbb{R}^n$, where $2 \leqslant n \in \mathbb{N}$. We wish to extend Theorem 2.7 from Q_l^n to Q. This is a technical matter which needs some care. It is based on the crucial observation that the corresponding extension operators of type (2.16), (2.17) from an l-dimensional face of Q of type Q_l into Q do not interfere with the boundary data at all other faces (edges, vertices) of Q. This can be extended without essential changes to bounded polyhedral domains in $\mathbb{R}^n$. But this will not be done here in detail. One may consider Q as a prototype of a polyhedron in $\mathbb{R}^n$. The boundary $\Gamma = \partial Q$ of Q can be represented as

$$\Gamma = \bigcup_{l=0}^{n-1} \Gamma_l, \qquad \Gamma_l \cap \Gamma_{l'} = \emptyset \quad \text{if} \quad l \neq l', \tag{3.2}$$

where Γ_l collects all l-dimensional faces (edges, vertices). Then Γ_l consists of finitely many l-dimensional cubes of type Q_l as in (2.2) (open as a subset of $\mathbb{R}^l$). We wish to apply Theorem 2.7, where we have now faces of dimensions $l = 0, \ldots, n-1$. This suggests the restriction

$$1 \leqslant p < \infty, \quad 0 < q < \infty, \quad s > \frac{1}{p}, \quad s - \frac{k}{p} \notin \mathbb{N}_0 \text{ for } k = 1, \ldots, n.$$

The trace (2.4) is now given by the restriction

$$\mathrm{tr}_{\,l} : \quad f(x) \mapsto f|\Gamma_l, \qquad f \in A^s_{pq}(\mathbb{R}^n),$$

of f to Γ_l. If $\alpha \in \mathbb{N}_0^n$ and $\gamma \in \Gamma_l$, then $D^\alpha_\gamma f$ denotes the obvious counterpart of (2.6), where only derivatives referring to directions perpendicular in $\mathbb{R}^n$ to Γ_l in γ are admitted. One replaces (2.8) by

$$\mathrm{tr}^{\,r}_{\,l} : \quad f \mapsto \left\{ \mathrm{tr}_{\,l} D^\alpha_\gamma f : |\alpha| \leqslant r \right\}, \qquad l = 0, \ldots, n-1, \tag{3.3}$$

for traces on Γ_l. Let $1 \leqslant p < \infty$, and let be either

$$0 < s - \frac{n}{p} \notin \mathbb{N} \quad \text{or} \quad 0 < s - \frac{n - l_0}{p} < \frac{1}{p} \text{ for some } l_0 = 1, \ldots, n-1.$$

If $s > n/p$, then we put $l_0 = 0$. Let

$$\overline{r} = \left(r^{l_0}, \ldots, r^{n-1} \right) \quad \text{with} \quad r^l = \left[s - \frac{n-l}{p} \right], \quad l_0 \leqslant l \leqslant n-1. \tag{3.4}$$

Then

$$\mathrm{tr}^{\,\overline{r}}_{\,\Gamma} : \quad f \mapsto \left\{ \mathrm{tr}_{\,l} D^\alpha_\gamma f : \gamma \in \Gamma_l, \ |\alpha| \leqslant r^l; \ l_0 \leqslant l \leqslant n-1 \right\} \tag{3.5}$$

is the appropriate modification of (2.8) for (maximal) traces on Γ.

Proposition 3.1. *Let Q be the cube in* (3.1) *with the boundary $\Gamma = \partial Q$ according to* (3.2). *Let*

$$1 \leqslant p < \infty, \quad 0 < q < \infty, \quad s > \frac{1}{p}, \quad s - \frac{k}{p} \notin \mathbb{N}_0 \textit{ for } k = 1, \ldots, n. \tag{3.6}$$

Let $l_0 = 0$ *if* $s > n/p$ *and otherwise* $l_0 = 1, \ldots, n-1$ *such that*

$$0 < s - \frac{n - l_0}{p} < \frac{1}{p}.$$

Let $\mathrm{tr}\,^{\overline{r}}_{\Gamma}$ *be as in* (3.4), (3.5). *Then*

$$\mathrm{tr}\,^{\overline{r}}_{\Gamma}: \quad B^s_{pq}(\mathbb{R}^n) \hookrightarrow \prod_{l=l_0}^{n-1} \prod_{|\alpha| \leqslant r^l} B^{s - \frac{n-l}{p} - |\alpha|}_{pq}(\Gamma_l) \tag{3.7}$$

and

$$\mathrm{tr}\,^{\overline{r}}_{\Gamma}: \quad F^s_{pq}(\mathbb{R}^n) \hookrightarrow \prod_{l=l_0}^{n-1} \prod_{|\alpha| \leqslant r^l} B^{s - \frac{n-l}{p} - |\alpha|}_{pp}(\Gamma_l). \tag{3.8}$$

One can replace $\mathbb{R}^n$ *on the left-hand sides of* (3.7), (3.8) *by* Q.

Proof. This follows immediately from Proposition 2.1 and the above considerations. □

Remark 3.2. One can extend the above assertion to $q = \infty$ in (3.6) and also to $p < 1$ with appropriately adapted restrictions for s as in Proposition 2.1. But our main concern is the restriction of Theorem 2.7 to the above cube Q. For this purpose one has to clip together the extension operators $\mathrm{Ext}\,^{r,u}_{l}$ as used in Theorems 2.3 and 2.7 with a reference to (2.17). Recall that Γ_l consists of l-dimensional (open as subsets of $\mathbb{R}^l$) disjoint cubes of type Q_l in (2.2),

$$\Gamma_l = \overline{\Gamma_l} \setminus \bigcup_{k<l} \Gamma_k, \qquad l = 1, \ldots, n-1$$

(corner points if $l = 0$). In particular, $D(\Gamma_l)$ is the disjoint union of sets of type $D(Q_l)$. In the same way, one has to interpret $\mathring{B}^{\sigma}_{pq}(\Gamma_l)$ and $\widetilde{B}^{\sigma}_{pq}(\Gamma_l)$. With p, q, s as in (3.6), one gets, by Proposition 1.17, that

$$\mathring{B}^{s - \frac{n-l}{p} - |\alpha|}_{pq}(\Gamma_l) = \widetilde{B}^{s - \frac{n-l}{p} - |\alpha|}_{pq}(\Gamma_l), \qquad l = 1, \ldots, n-1, \quad |\alpha| \leqslant r^l.$$

Let $\mathrm{Ext}\,^{r,u}_{\Gamma_l}$ be the obvious generalization of (2.17) from Q_l to Γ_l, and let

$$\mathrm{Ext}\,^{\overline{r},u}_{\Gamma} = \left\{ \mathrm{Ext}\,^{r^l,u}_{\Gamma_l} : l_0 \leqslant l \leqslant n-1 \right\} \tag{3.9}$$

with $\overline{r}$ and l_0 as in (3.4). If $s > n/p$, then $l = l_0 = 0$ is admitted. These are the corner points of Q. Then one has to modify (2.16), (2.17) (one may consult [16, Sect. 5.2.3], where we dealt with the one-dimensional case and traces in points).

Theorem 3.3. *Let Q be the cube in (3.1) with the boundary $\Gamma = \partial Q$ according to (3.2). Let*

$$1 \leqslant p < \infty, \quad 0 < q < \infty, \quad s > \frac{1}{p}, \quad s - \frac{k}{p} \notin \mathbb{N}_0 \text{ for } k = 1, \ldots, n. \tag{3.10}$$

Let $s < u \in \mathbb{N}$. Let l_0 and $\overline{r}$ be as in (3.4) and Proposition 3.1. Let $\operatorname{Ext}_\Gamma^{\overline{r},u}$ be given by (3.9). Then

$$\operatorname{Ext}_\Gamma^{\overline{r},u} : \quad \{g^l_\alpha : l_0 \leqslant l \leqslant n-1, \ |\alpha| \leqslant r^l\} \mapsto g$$

is an extension operator for

$$\operatorname{Ext}_\Gamma^{\overline{r},u} : \quad \prod_{l=l_0}^{n-1} \prod_{|\alpha| \leqslant r^l} \widetilde{B}_{pq}^{s - \frac{n-l}{p} - |\alpha|}(\Gamma_l) \hookrightarrow B^s_{pq}(Q) \tag{3.11}$$

and

$$\operatorname{Ext}_\Gamma^{\overline{r},u} : \quad \prod_{l=l_0}^{n-1} \prod_{|\alpha| \leqslant r^l} \widetilde{B}_{pp}^{s - \frac{n-l}{p} - |\alpha|}(\Gamma_l) \hookrightarrow F^s_{pq}(Q). \tag{3.12}$$

Let $\operatorname{tr}_\Gamma^{\overline{r}}$ be the trace operator according to (3.5). Then

$$\operatorname{tr}_\Gamma^{\overline{r}} \circ \operatorname{Ext}_\Gamma^{\overline{r},u} = \mathrm{id} \quad \textit{identity in} \quad \prod_{l=l_0}^{n-1} \prod_{|\alpha| \leqslant r^l} \widetilde{B}_{pq}^{s - \frac{n-l}{p} - |\alpha|}(\Gamma_l). \tag{3.13}$$

Furthermore,

$$B^s_{pq}(Q) = \widetilde{B}^s_{pq}(Q) \times \operatorname{Ext}_\Gamma^{\overline{r},u} \prod_{l=l_0}^{n-1} \prod_{|\alpha| \leqslant r^l} \widetilde{B}_{pq}^{s - \frac{n-l}{p} - |\alpha|}(\Gamma_l) \tag{3.14}$$

and

$$F^s_{pq}(Q) = \widetilde{F}^s_{pq}(Q) \times \operatorname{Ext}_\Gamma^{\overline{r},u} \prod_{l=l_0}^{n-1} \prod_{|\alpha| \leqslant r^l} \widetilde{B}_{pp}^{s - \frac{n-l}{p} - |\alpha|}(\Gamma_l) \tag{3.15}$$

(complemented subspaces).

Proof. We use Theorem 2.7 and an induction by dimension $l \geqslant l_0$. Let $l_0 \geqslant 1$. Then, by Proposition 1.15,

$$B_{pq}^{s - \frac{n-l_0}{p}}(\Gamma_{l_0}) = \widetilde{B}_{pq}^{s - \frac{n-l_0}{p}}(\Gamma_{l_0}).$$

Now, by the above considerations with $r^{l_0} = 0$, one gets that

$$\operatorname{Ext}_{\Gamma_{l_0}}^{r^{l_0},u} : \quad B_{pq}^{s - \frac{n-l_0}{p}}(\Gamma_{l_0}) \hookrightarrow B^s_{pq}(Q).$$

If $l_0 = 0$, then one has to modify according to [16, Sect. 5.2.3], where we dealt with the one-dimensional case. This decomposes $B^s_{pq}(Q)$ as in (2.32), (2.33),

$$B^s_{pq}(Q) = \left\{ f \in B^s_{pq}(Q) : \mathrm{tr}^{r^{l_0}}_{\Gamma_{l_0}} f = 0 \right\} \times \mathrm{Ext}^{\,r^{l_0},u}_{\Gamma_{l_0}} B^{s-\frac{n-l_0}{p}}_{pq}(\Gamma_{l_0}) \qquad (3.16)$$

into complemented subspaces. Let $l = l_0 + 1$. The trace space of the first factor on the right-hand side of (3.16) is now $\widetilde{B}^{s-\frac{n-l}{p}}_{pq}(\Gamma_l)$. But this observation ensures that one can apply the splitting technique in connection with Theorem 2.7. Then one gets the factor with $l = l_0 + 1$ on the right-hand side of (3.14). It remains a complemented subspace now with $\mathrm{tr}^{r^l}_{\Gamma_l} f = 0$ for $l = l_0$ and $l = l_0 + 1$. We remark again that, according to the construction (2.17), the extensions at different values of l do not interfere. Iteration ends up finally with (3.14), (3.15), where one has to use the counterpart of (2.35) or Proposition 1.17 with $\Omega = Q$ both for the B-spaces and F-spaces. This proves also (3.11)-(3.13). □

Remark 3.4. Beyond the technical details the proof relies on the following two remarkable facts:

1. With p, q, s as in (3.10), the boundary data of $f \in A^s_{pq}(Q)$ on different faces of Γ_l, but also for different Γ_l and $\Gamma_{l'}$ with $l \neq l'$ are totally decoupled.
2. The wavelet-friendly extension operator for a face of Γ_l respects this observation and does not interfere with the boundary data of other faces of Γ_l and with the boundary data of faces of $\Gamma_{l'}$ with $l \neq l'$.

4 Wavelet Bases

4.1 Cubes and polyhedrons

Theorem 3.3 brings us in the same position as in [16, Chapt. 5], where we asked for wavelet bases in spaces $A^s_{pq}(\Omega)$ in the case of bounded C^∞ domains Ω. Let $a^s_{pq}(\mathbb{Z}^\Omega)$ with $a \in \{b, f\}$ be the sequence spaces as introduced in Definition 1.5, and let $\left\{\Phi^j_l\right\}$ be the u-wavelet systems according to Definition 1.7(i). We recalled at the beginning of Sect. 1.7 what is meant by a (unconditional) basis. If a basis in $A^s_{pq}(\Omega)$ generates an isomorphic map onto the standard basis of some sequence space $a^s_{pq}(\mathbb{Z}^\Omega)$, then it is unconditional. We will not stress this point in what follows.

Theorem 4.1. *Let $\Omega = Q$ be the cube (3.1) in $\mathbb{R}^n$ with $n \geqslant 2$. Let*

$$1 \leqslant p < \infty, \qquad s > \frac{1}{p}, \qquad s - \frac{k}{p} \notin \mathbb{N}_0 \text{ for } k = 1, \dots, n, \qquad (4.1)$$

$0 < q < \infty$ for the B-spaces, $1 \leqslant q < \infty$ for the F-spaces. Let $u \in \mathbb{N}$ with $s < u$. Then there is a u-wavelet system

$$\Phi = \left\{ \Phi^j_l : j \in \mathbb{N}_0;\ l = 1, \ldots, N_j \right\} \subset C^u(\Omega), \qquad N_j \in \mathbb{N}$$

(where $N_j \sim 2^{jn}$) in $\overline{\Omega}$ according to Definition 1.7(i) such that $f \in D'(\Omega)$ belongs to $A^s_{pq}(\Omega)$ if and only if it can be represented as

$$f = \sum_{j=0}^{\infty} \sum_{l=1}^{N_j} \lambda^j_l \, 2^{-jn/2} \, \Phi^j_l, \qquad \lambda \in a^s_{pq}(\mathbb{Z}^\Omega), \tag{4.2}$$

unconditional convergence being in $A^s_{pq}(\Omega)$.

The representation (4.2) is unique,

$$f = \sum_{j=0}^{\infty} \sum_{l=1}^{N_j} \lambda^j_l(f) \, 2^{-jn/2} \, \Phi^j_l,$$

where $\lambda^j_l(\cdot) \in A^s_{pq}(\Omega)'$ are linear and continuous functionals on $A^s_{pq}(\Omega)$ and

$$f \mapsto \left\{ \lambda^j_l(f) \right\} \quad \text{is an isomorphic map of } A^s_{pq}(\Omega) \text{ onto } a^s_{pq}(\mathbb{Z}^\Omega)$$

(u-wavelet basis).

Proof. By [16, Theorem 3.13], all the spaces on the right-hand sides of (3.14) and (3.15) have interior u-wavelet bases according to Definition 1.7(ii) (where the spaces on Γ_l are considered in $\mathbb{R}^l$). This is the point where one has to replace $0 < q < \infty$ in the case of F-spaces by $1 \leqslant q < \infty$. In [16, Proposition 5.34, Theorem 5.27], we described in the case of bounded C^∞ domains how wavelet bases for $\widetilde{A}^s_{pq}(\Omega)$ can be complemented by transferred wavelet bases for trace spaces such that one gets wavelet bases for $A^s_{pq}(\Omega)$. This applies also to the above situation now based on (3.14), (3.15), resulting in the theorem. □

Remark 4.2. In [16, Chapt. 5], we used the above decomposition technique to prove the existence of u-wavelet bases in planar C^∞ domains (in $\mathbb{R}^2$) and in those bounded C^∞ domains Ω in $\mathbb{R}^n$ with $n \geqslant 3$, where all connected boundary components are diffeomorphic to the $(n-1)$-dimensional sphere $\mathbb{S}^{n-1}$.

Corollary 4.3. *Theorem 4.1 remains valid for bounded polyhedrons Ω in $\mathbb{R}^n$.*

Proof. The proof of Theorem 3.3 is based on the extension operators (2.16), (2.17), clipped together by (3.9). But one does not need that the respective faces intersecting each other are orthogonal. It is sufficient to know that they intersect with nonzero angles. Similarly, one can replace the perpendicular derivatives D^α_γ in (3.3) by oblique derivatives. This does not change the arguments and one gets the above assertion. □

4.2 Cellular domains

What follows might be considered as the main result of the paper.

Theorem 4.4. *Let $n \geqslant 2$. Theorem* 4.1 *remains valid for any bounded cellular domain Ω in $\mathbb{R}^n$ according to Definition* 1.9(iii).

Proof. Diffeomorphic maps of cubes onto the corresponding domains generate isomorphic maps both for the related spaces and also for wavelet bases. The extension operators used in Theorem 3.3 extend boundary spaces (and wavelets) from Γ_l to $\mathbb{R}^n$, in particular to all adjacent cells. Hence the constructions resulting in Theorem 3.3, but also the indicated constructions taken over from [16] apply simultaneously to all cells (diffeomorphic images of cubes) resulting in the above theorem. □

Corollary 4.5. *Let $n \geqslant 2$. Theorem* 4.1 *remains valid for any bounded C^∞ domain Ω in $\mathbb{R}^n$ according to Definition* 1.9(ii).

Proof. This follows from Theorem 4.4 and Proposition 1.11. □

4.3 Comments

As described in Theorem 1.19, the spaces $A^s_{pq}(\Omega)$ in bounded Lipschitz domains Ω have common wavelet frames. If $D(\Omega)$ is dense in $A^s_{pq}(\Omega)$, then it makes sense to ask for interior wavelet systems according to Definition 1.7(ii) which are frames or bases. We described related assertions in Theorem 1.23. If $D(\Omega)$ is no longer dense in $A^s_{pq}(\Omega)$ (with $q < \infty$, $p < \infty$), then one may ask whether these spaces have not only wavelet frames, but even wavelet bases. This was the subject of [16, Chapt. 5] complemented now by the assertions in the above Sects. 4.1, 4.2. The question arises to which extent the diverse restrictions in Theorems 4.1, 4.4 and Corollaries 4.3, 4.5 are natural. We add a few comments.

Comment 4.6. Theorem 4.1 relies on the total decoupling of the boundary data of $A^s_{pq}(Q)$ on the faces Γ_l of Q as described in Theorem 3.3. This decoupling is no longer valid if one admits the exceptional values $s - \frac{k}{p} \in \mathbb{N}_0$. We discussed this point in [16, Remark 5.50] with a reference to [4, 5]. It comes out that this disturbing effect even happens for the classical Sobolev spaces $W_2^1(Q)$, where Q is a square in the plane $\mathbb{R}^2$. Then $s - \frac{n}{p} = 1 - \frac{2}{2} = 0$ is an exceptional value. In other words, our method cannot be extended to the critical values $s - \frac{1}{p} \in \mathbb{N}_0$ in (4.1).

Comment 4.7. In continuation of the discussion in the preceding comment one may ask whether the exclusion of $s - \frac{k}{p} \in \mathbb{N}_0$ in (4.1) is an artifact of our method which can be removed by other arguments. But it comes out that at least $s - \frac{1}{p} \notin \mathbb{N}_0$ lies in the nature of the problem. The second main point in Theorem 3.3 (besides the decoupling) is the total reduction of the spaces $A^s_{pq}(Q)$ to $\widetilde{A}^s_{pq}(Q)$ and $\widetilde{B}^\sigma_{pq}(\Gamma_l)$. This reduces the question for wavelet bases in $A^s_{pq}(Q)$ to wavelet bases in spaces of type $\widetilde{A}^\sigma_{pq}(\Omega)$. But for these spaces one has a rather satisfactory theory, subject to [16, Chapt. 3]. We relied especially in Theorem 2.7 on Proposition 1.17 which again excludes the above exceptional values. We discussed in [16, Chapt. 6] in a larger context what can be said about spaces $A^s_{pq}(\Omega)$ if s is an exceptional value. We quote a few assertions which are related to the subject of this paper. Let Ω be a bounded C^∞ domain in $\mathbb{R}^n$, and let

$$A^s_{pq}(\Omega), \qquad 1 < p, q < \infty, \qquad s - \frac{1}{p} \in \mathbb{N}_0,$$

be exceptional spaces. Let $\mathring{A}^s_{pq}(\Omega)$ and $\widetilde{A}^s_{pq}(\Omega)$ be the corresponding spaces according to Definition 1.3. Then $D(\Omega)$ is dense in $\widetilde{A}^s_{pq}(\Omega)$ and (by definition) dense in $\mathring{A}^s_{pq}(\Omega)$. But, in contrast to Proposition 1.17, one has now in these exceptional cases

$$\mathring{A}^s_{pq}(\Omega) \neq \widetilde{A}^s_{pq}(\Omega), \qquad 1 < p, q < \infty, \qquad s - \frac{1}{p} \in \mathbb{N}_0. \tag{4.3}$$

Even worse, whereas the spaces $\widetilde{A}^s_{pq}(\Omega)$ have interior wavelet bases, it comes out that the spaces

$$\mathring{A}^s_{pq}(\Omega) \text{ in (4.3) do not have interior wavelet bases}$$

according to Definition 1.7(ii). In other words, the situation is totally different compared with the corresponding assertions based on Proposition 1.17 and [16, Theorem 3.13].

Comment 4.8. By the above consideration, it is clear that $s - \frac{1}{p} \notin \mathbb{N}_0$ is a natural restriction in the context of the above theory. But what about the additional restrictions $s - \frac{k}{p} \notin \mathbb{N}_0$ with $k = 2, \dots, n$ in Theorems 4.1, 4.4 and Corollaries 4.3, 4.5? The approach in [16, Chapt. 5] is restricted to special domains such as balls and tori. This improves the situation as far as wavelet bases on the boundary, for example spheres, are concerned. We describe a special, but remarkable example. Let

$$H^s_p(\Omega) = F^s_{p,2}(\Omega), \qquad 1 < p < \infty, \quad s \in \mathbb{R},$$

be the distinguished (fractional) Sobolev spaces briefly mentioned in (1.1), (1.2) in the case of $\mathbb{R}^n$. Let $\mathbb{B}^3$ be the unit ball in $\mathbb{R}^3$ and $\mathbb{M}^3$ be the torus of revolution in $\mathbb{R}^3$. Then from [16, Theorems 5.35, 5.38, Remark 5.36] and Corollary 4.5 it follows that one has wavelet bases in

$$H^s_p(\mathbb{M}^3) \quad \text{if} \quad 1 < p < \infty, \quad 0 < s - \frac{1}{p} \notin \mathbb{N},$$

in

$$H^s_p(\mathbb{B}^3) \quad \text{if} \quad 1 < p < \infty, \quad 0 < s - \frac{1}{p} \notin \mathbb{N}, \quad s - \frac{2}{p} \notin \mathbb{N}_0, \tag{4.4}$$

and in

$$H^s_p(\Omega) \quad \text{if} \quad 1 < p < \infty,\ 0 < s - \frac{1}{p} \notin \mathbb{N},\ s - \frac{2}{p} \notin \mathbb{N}_0,\ s - \frac{3}{p} \notin \mathbb{N}_0, \tag{4.5}$$

for bounded C^∞ domains Ω in $\mathbb{R}^3$. But it is unlikely that the additional exceptional values in (4.4) (or (4.5)) are naturally related to the different topologies of $\partial\mathbb{B}^3$ and $\partial\mathbb{M}^3$. The situation is especially curious if $p = 2$. By (4.4), it is unclear whether the very classical Sobolev spaces

$$W^k_2(\mathbb{B}^3), \qquad k \in \mathbb{N},$$

have wavelet bases.

References

1. Besov, O.V., Il'in, V.P., Nikol'skij, S.M.: Integral Representations of Functions and Embedding Theorems (Russian). Nauka, Moskva (1975); Second ed. (1996); English transl.: Wiley, New York (1978/79)
2. Daubechies, I.: Ten Lectures on Wavelets. CBMS-NSF Regional Conf. Series Appl. Math. SIAM, Philadelphia (1992)
3. Frazier, M., Jawerth, B.: A discrete transform and decompositions of distribution spaces. J. Funct. Anal. **93**, 34-170 (1990)
4. Grisvard, P.: Elliptic Problems in Nonsmooth Domains. Pitman, Boston (1985)
5. Grisvard, P.: Singularities in Boundary Value Problems. Springer, Berlin (1992)
6. Meyer, Y.: Wavelets and Operators. Cambridge Univ. Press, Cambridge (1992)
7. Nikol'skij, S.M.: Approximation of Functions of Several Variables and Embedding Theorems (Russian). Nauka, Moskva (1969); Second ed. (1977); English transl.: Springer, Berlin (1975)
8. Runst, T., Sickel, W.: Sobolev Spaces of Fractional Order, Nemytskij Operators, and Nonlinear Differential Equations. W. de Gruyter, Berlin (1996)
9. Sobolev, S.L.: Some Applications of Functional Analysis in Mathematical Physics (Russian). 1st ed. Leningrad State Univ., Leningrad (1950); 3rd ed. Nauka, Moscow

(1988); German transl.: Einige Anwendungen der Funktionalanalysis auf Gleichungen der mathematischen Physik. Akademie-Verlag, Berlin (1964); English transl. of the 1st Ed.: Am. Math. Soc., Providence, RI (1963); English transl. of the 3rd Ed. with comments by V. P. Palamodov: Am. Math. Soc., Providence, RI (1991)

10. Triebel, H.: Interpolation Theory, Function Spaces, Differential Operators. North-Holland, Amsterdam (1978); Second ed. Barth, Heidelberg (1995)
11. Triebel, H.: Theory of Function Spaces. Birkhäuser, Basel (1983)
12. Triebel, H.: Theory of Function Spaces II. Birkhäuser, Basel (1992)
13. Triebel, H.: The Structure of Functions. Birkhäuser, Basel (2001)
14. Triebel, H.: Function spaces in Lipschitz domains and on Lipschitz manifolds. Characteristic functions as pointwise multipliers. Rev. Mat. Complutense **15**, 475-524 (2002)
15. Triebel, H.: Theory of Function Spaces III. Birkhäuser, Basel (2006)
16. Triebel, H.: Function Spaces and Wavelets on Domains. European Math. Soc. Publishing House, Zürich (2008) [To appear]
17. Triebel, H.: Wavelets in Function Spaces. [Submitted]
18. Wojtaszczyk, P.: A mathematical Introduction to Wavelets. Cambridge Univ. Press, Cambridge (1997)

Index

Printed in the United States
144911LV00001B/9/P